An Introduction to Separation Science

AN INTRODUCTION TO SEPARATION SCIENCE

BARRY L. KARGER

Department of Chemistry
Northeastern University
Boston, Massachusetts

LLOYD R. SNYDER

Research Department
Technicon Instruments Corporation
Tarrytown, New York

CSABA HORVATH

Yale University
New Haven, Conneticut

A WILEY-INTERSCIENCE PUBLICATION

JOHN WILEY & SONS
New York · **London** · **Sydney** · **Toronto**

Library of Congress Cataloging in Publication Data

Karger, Barry L 1939–
 An introduction to separation science.

 "A Wiley-Interscience publication."
 Includes bibliographies.
 1. Separation (Technology) I. Snyder, Lloyd R.,
joint author. II. Horvath, Csaba, 1930–   joint
author. III. Title.

QD63.S4K37   544   73–4016
ISBN 0–471–45860–0

10 9 8 7 6 5 4 3 2

21.50

CONTRIBUTORS

MILAN BIER
Veterans Administration Hospital
Tuscon, Arizona

JUA-YI CHUANG
Department of Chemistry and
    Institute of Materials Science
University of Connecticut
Storrs, Connecticut

R. A. CROSS
Romicon, Inc.
Woburn,
Massachusetts

HENRY FREISER
Department of Chemistry
The University of Arizona
Tuscon, Arizona

JULIAN F. JOHNSON
Department of Chemistry and
    Institute of Materials Science
University of Connecticut
Storrs, Connecticut

ROBERT H. McCORMICK
College of Engineering
The Pennsylvania State University
University Park, Pennsylvania

HERBERT L. ROTHBART
Eastern Regional Research
    Laboratory
U.S. Department of Agriculture
Philadelphia, Pennsylvania

HEINER STRATHMANN
Physikalische Technisches
    Laboratorium
Berghof GMBH
Tubingen, Germany

WILLIAM R. WILCOX
Chemical Engineering and
    Materials Science Departments
University of Southern California
Los Angeles, California

PREFACE

Over the last 20 years the field of analytical separations has undergone an explosive growth. Although the most prominent developments have come from the numerous chromatographic procedures, other methods such as electrophoresis, ultracentrifugation, and extraction have also achieved impressive advances. Unfortunately, little or no interaction has occurred between the specialists in these areas. Consequently, the fields of separations seems to consist of a series of unrelated methods.

We believe, however, that common underlying principles of separations exist and that the understanding of these fundamentals can result in a fuller appreciation of the advantages and disadvantages of the specific methods. We also believe that these principles lead to the logical establishment of a field of separation science. This book provides an introduction to separation science, with special emphasis on the analytical scale aspects of the methods.

Chapter 1 outlines the organization of the book. The fundamentals of the field are then presented (Chapters 2 to 6), followed by their application in the specific separation methods (Chapters 7 to 19). This approach minimizes the problem of the book becoming rapidly outdated.

To our knowledge this book is the only up-to-date treatment of separations that approaches the subject in the fashion described above. The text should be useful in an upper-level undergraduate or graduate course in separation science. We hope that its appearance will encourage the establishment of courses in this subject, which has such great practical value for many areas of chemistry and related fields. We also hope that the book will be useful to the practitioners of separation methods in research and development laboratories.

As might be expected when one brings together areas that have not over-lapped in the past, a nomenclature problem results. In many instances the same symbol has been used in the literature to represent different parameters in several separation methods. An ad hoc committee on nomenclature of separation methods, set up by the Analytical Chemistry Division of the IUPAC, has addressed itself to this problem and has put forth certain recommendations.

For the most part we have tried to follow the recommendations of the IUPAC committee. For example, the symbol K is used when the distribution is independent of solute concentration over a given range (the distribution constant). A different symbol $\tilde{K}$ is used when the distribution can be, or is, sample concentration dependent (the distribution coefficient). We use k' for the capacity factor, since the symbol k has not been recommended by the IUPAC committee.

To overcome the problems of different meanings for a given symbol and the introduction of new symbols for a particular method the following procedure is adopted in this book. A master nomenclature list is provided at the beginning of the text defining all the symbols used in Chapters 2 to 6. In general the symbols for the second part of the book are consistent with the first part. However, in several instances a symbol was in such common use for a particular method that it did not seem wise to define a new symbol to agree with the first part of the book. To prevent confusion a nomenclature list is provided at the end of each chapter dealing with the specific methods (Chapters 7 to 18). It defines these symbols and presents other symbols not included in the first part of the book.

Finally, we thank the contributing authors for their participation in this project. Their efforts in presenting the salient features of a particular method and in relating that method to the underlying principles of separation science have greatly strengthened the book.

BARRY L. KARGER
LLOYD R. SNYDER
CSABA HORVATH

Boston, Massachusetts
May 1973

CONTENTS

MASTER SYMBOL LIST

This list applies fully to Chapters 2 to 6 and generally to Chapters 7 to 19. For the latter group additions and alterations are assembled as separate lists at the end of each chapter. The references in parenthesis indicate the first appearance of the symbol.

English Symbols

a	Activity of chemical species (Eq. 2.3)
a	Volume ratio of two phases in distribution equilibria (p. 115)
A	Constant in repulsive term of equation for the net energy of interaction of two atoms (Eq. 2.25)
A	Term in van Deemter equation to represent eddy diffusion (Eq. 5.37)
A	Peak area (Eq. 3.31)
A_S	Surface area of adsorbent
B	Constant in attractive term of equation for the net energy of interaction of two atoms (Eq. 2.25)
B^0	Specific permeability (Eq. 3.70)
B/v	Term in van Deemter equation to represent longitudinal diffusion (Eq. 5.37)
c	Constant (Eq. 3.76)
c_i	Concentration of solute in a given equilibrium stage (Eq. 4.20)
$C, C',$ C'', C'''	Constants
C	Molar concentration
C_0	Initial concentration
C	Number of components in Gibbs phase rule (Eq. 2.11)

C^* Equilibrium concentration (Eq. 3.54)

C_A^*, C_B^* Measures of soft acid and base strengths, respectively, of acid-base interactions (Eq. 2.33)

C_S, C_M Concentration of solute in the stationary and mobile phases, respectively (Section 5.22)

Cv Term in van Deemter equation to represent stationary and mobile phase mass transfer terms (Eq. 5.37)

d Characteristic diffusion distance (Eq. 5.26)

d_R Retention in distance units in open-bed chromatography (Eq. 5.13 and Fig. 5.6)

d_p Support particle diameter (Sections 3.62 and 5.34)

D Distribution ratio (Section 2.13)

D Diffusion coefficient (Eq. 3.1)

D_{eff} Effective diffusion coefficient (Eq. 3.47)

D_0 Temperature-independent portion of the diffusion coefficient (Eq. 3.42)

D_M, D_S Diffusion coefficients in the mobile and stationary phases, respectively (Sections 5.33 and 5.34)

D_s Surface diffusion coefficient (Eq. 3.50)

e Number of interactions between a molecule i and its nearest neighbors (Section 2.22)

E Activation energy for diffusion (Eq. 3.42)

E_{ij} Total interaction energy of molecule i with surrounding phase (Section 2.21)

E_{ab} Acid-base interaction energy (Section 2.21, Eq. 2.33)

E_d Dispersion interaction energy (Section 2.21, Eq. 2.29)

E_i Induction interaction energy (Section 2.21, Eq. 2.32)

E_0 Dipole orientation interaction energy (Section 2.21, Eq. 2.31)

E_A^*, E_B^* Measures of hard acid and base strengths, respectively, of acid-base interactions (Eq. 2.33)

E_{x-j} Interaction energy of group X with solvent j (Eq. 2.47)

ΔE^S Solution energy (Eq. 2.40a)

ΔE^V Vaporization energy (Section 2.21)

f Fugacity (Section 2.11)

f^0 Fugacity in the standard state (Section 2.11)

F Number of degrees of freedom in Gibbs phase rule (Eq. 2.11)

F Volumetric flow rate

$\mathscr{F}$ Faraday's constant (Eq. 3.46)

F_K Volumetric flow rate of a gas with Knudsen flow (Eq. 3.75)

F_A Force of molecule A on B (Eq. 3.38)

g Constant (Eq. 3.5)

G Free energy (Section 2.11)

h	Partial molar enthalpy (Eq. 2.10)
h	Peak height of a band (Eq. 3.34)
h	Reduced plate height (Eq. 5.16)
h^0	Partial molar enthalpy in the standard state (Eq. 2.10a)
Δh^e	Partial molar excess enthalpy (Eq. 2.39)
H	Height equivalent to a theoretical plate (Eq. 5.14) ($HETP$)
H_D	$HETP$ due to mobile phase diffusion effects (Eq. 5.31)
H_F	$HETP$ due to mobile phase flow effects (Eq. 5.30)
H_L	$HETP$ due to longitudinal diffusion (Eq. 5.18)
H_M	$HETP$ due to mobile phase mass transfer (Section 5.34)
H_{SM}	$HETP$ due to stagnant mobile phase diffusion (Eq. 5.34)
H_{min}	Minimum $HETP$ (Eq. 5.38)
ΔH	Molar enthalpy change (Eq. 2.12)
ΔH^F	Molar heat of fusion (Section 2.22)
ΔH^M	Molar heat of mixing (Eq. 2.42b)
ΔH^V	Molar heat of vaporization (Section 2.21)
i^x	Solute i in phase x (Section 2.11)
I	Ionizational potential (Eq. 2.26)
J_x	Flux in x direction (Eq. 3.1)
k	Boltzman constant (Eq. 2.31)
k'	Capacity factor (Eq. 2.20)
k_A, k_B	Solute mass transfer coefficients for phases A and B, respectively, (Eq. 3.53)
k_a, k_d	Rate constants for adsorption and desorption, respectively, (Section 5.33)
K	Distribution constant (Section 2.11)
K^0	Thermodynamic distribution constant (Section 2.11)
K_a	Acid ionization constant (Eq. 2.17d)
$\tilde{K}_g$	Distribution coefficient involving gas-liquid or gas-solid as two phases (Eqs. 2.9a and 2.9b)
$\tilde{K}_c$	Distribution coefficient with molarity as concentration unit (Eq. 2.1b)
$\tilde{K}_x$	Distribution coefficient with mole fraction as concentration unit (Eq. 2.1a)
l	Length of a step in the random walk model (Eq. 5.19)
L	Column length (Section 5.21)
M	Total quantity of diffusing material (Eq. 3.6)
M_A, M_B	Molecular weights of components A and B (Eq. 3.36)
n	Number of moles (Eq. 2.2a)
n	Refractive index (Eq. 2.30)
n	Tube number of a cascade of $n + 1$ tubes (Section 4.81)
n	Number of steps in random walk model (Eq. 5.19)

n_S, n_M	Number of moles of solute in the stationary and mobile phases, respectively, (Section 5.21)
N	Avogadro's number (Eq. 2.30)
N	Number of theoretical plates for column chromatography (Section 5.31)
N'	Number of theoretical plates for open-bed chromatography (Section 5.31)
N_{eff}	Effective number of theoretical plates (Eq. 5.45)
p	Fraction of solute in upper phase on equilibration (Section 4.81)
P	Number of phases in Gibbs phase rule (Eq. 2.11)
P	Pressure (Section 2.11)
P^0	Vapor pressure of pure component (Section 2.11)
P_c	Critical pressure (Section 2.12)
P_i	Partial pressure of component i (Section 2.11)
P_i	Inlet pressure (Section 3.61)
$P_{n,r}$	Fraction of solute in tube n after transfer r times (Eq. 4.10)
P_0	Outlet pressure (Section 3.61)
q	Fraction of solute in lower phase upon equilibration (Section 4.81)
q	Configuration factor to take account of shape of dispersed liquid on solid support (Eq. 5.28)
r	Internuclear distance (Eq. 2.25)
r	Radius of tube (Section 3.61), for open tubular gas chromatographic columns (Eq. 5.29)
r	Number of upper phase transfers in a multistage separator (Section 4.81)
r_A	Radius of solute molecule (Eq. 3.39)
r_e	Equilibrium internuclear distance (Section 2.21)
r_h	Hydraulic radius of solute molecule (Eq. 3.69)
R	Gas constant (Section 2.11)
R	Fraction of solute in mobile phase (Eq. 5.1)
Re	Reynold's number (Eq. 3.60)
R_f	Retention parameter in open-bed chromatography (Eq. 5.9)
R_M	Retention parameter in open-bed chromatography (Eq. 5.10)
R_s	Resolution (Eq. 4.7 and Section 5.41)
R_T	Thermal diffusivity ratio (Eq. 3.52)
s	Partial molar entropy (Eq. 2.10)
s^0	Partial molar entropy in the standard state (Eq. 2.10a)
Sc	Schmidt number (Eq. 3.37)
ΔS^F	Molar entropy of fusion (Section 2.22)
ΔS^M	Molar entropy of mixing (Section 2.22)

t	Time
t_a, t_d	Mean adsorption and desorption times, respectively (Section 5.33)
t_D	Mean diffusion time (Eq. 5.26)
t_0	Retention time of an unretained component (Section 5.21)
t_R	Retention time (Section 5.21)
T	Temperature
T_b	Boiling point temperature (Section 2.21)
T_c	Critical temperature (Section 2.12)
T_f	Melting point temperature (Section 2.22)
U_k, U_a	Mobility of cation and anion, respectively (Eq. 3.46)
v	Mobile-phase velocity (Section 5.21)
v_i	Empirical atomic diffusion volumes (Eq. 3.36)
v_{max}	Maximum velocity (Eq. 3.62)
v_0	Superficial velocity (Eq. 3.67)
v_{opt}	Mobile-phase velocity at H_{min} (Eq. 5.39)
v_s	Band velocity (Eq. 5.2)
v_M, v_S	Volume per unit column length of mobile and stationary phases, respectively (Eqs. 5A.2 and 5A.3)
v_U	Volume of upper phase per stage (Eq. 4.15)
$\bar{V}$	Molar volume (Section 2.11)
V^x, V^y	Volume of phases x and y, respectively (Eq. 2.19)
V_L, V_U	Volume of lower and upper phases in equilibrium stage separation system, respectively (Section 4.81)
V_M, V_S	Volume of mobile and stationary phases, respectively (Eq. 5.6)
V_N	Net retention volume (Eq. 5.7)
V_R	Retention volume (Eqs. 4.15 and 5.6)
ΔV^M	Volume change on mixing (Section 2.22)
W	Total moles of a binary sample for distillation (Section 4.3)
W_1, W_2	Initial and final moles of undistilled liquid (Section 4.3)
W_b	Peak width at the baseline (Table 5.2)
W_i	Peak width at the inflection points (Table 5.2)
$W_{0.5}$	Peak width at half height (Table 5.2)
W_S	Weight of stationary phase (Section 5.21)
x	Phase notation
x	Direction, one dimension (Eq. 3.1)
x'	Position after time t (Eq. 3.27)
X	Mole fraction (Section 2.11)
y	Phase notation (Section 2.11)
z	Exponent for attractive term of E_{ij} (Eq. 2.25)
z	Distance from column inlet (Appendix, Chapter 5)

z_f	Capillary length for flow in open-bed chromatography (Eq. 3.73)
Z	Valence of ion (Eq. 3.46)

Greek Symbols

α	Separation factor (Eq. 2.22)
α	Polarizability (Eq. 2.26)
α	Thermal diffusion coefficient (Eq. 3.51)
$\alpha_e{}^v$	Polarizability per unit volume (Eq. 2.30)
γ	Activity coefficient (Section 2.11)
γ	Tortuosity factor to account for nonlinear diffusion (Eq. 3.47)
γ	Surface tension (Eq. 3.73)
δ	Solubility parameter (Section 2.22)
δ_m	Effective film thickness (Eq. 3.59)
ϵ	Interparticle porosity (Eq. 3.68)
η	Arbitrary function (Eq. 3.20)
η	Viscosity (Eq. 3.37)
θ	Fractional coverage of adsorbed sites by solute (Eq. 2.16)
κ	Rate constant for diffusion (Eq. 3.43)
Λ	Packing structure factor (Eq. 5.30)
μ	Chemical potential (Section 2.11)
μ^0	Standard chemical potential (Section 2.11)
μ_s	Stage number of maximum concentration (Eq. 4.11)
$\bar{\mu}$	Dipole moment of molecule (Eq. 2.31)
ν	Kinematic viscosity (Eq. 3.65)
ν	Reduced velocity (Eq. 5.17)
ν_A	Steady-state velocity of molecule A (Eq. 3.38)
ξ	Arbitrary function (Eq. 3.7)
ρ	Density (Eq. 3.37)
$\sigma_{AB}{}^2$	Collisional cross-sectional area of molecules A and B (Eq. 3.35)
σ_s	Standard deviation of band after r operations in terms of number of stages (Eq. 4.12)
σ_V	Standard deviation of band in volume units (Eq. 4.17)
σ_d	Standard deviation in distance units (Eq. 5.13)
σ_t	Standard deviation in time units (Eq. 5.11)
$\sigma_L{}^2$	Band variance in (distance)^2 units for longitudinal diffusion (Eq. 5.18)
$\sigma_S{}^2$	Band variance in (distance)^2 units for slow adsorption-desorption kinetics (Eq. 5.22)
τ	Number to define band width for unit resolution (Eq. 4.7)
ϕ	Fraction of space available to the mobile species (Eq. 3.47)
ϕ	Proportionality constant (Eq. 3.74)

ϕ	Fraction of total mobile phase in porous structure (intraparticle space) (Eq. 5.34)
ϕ	Peak capacity (Section 5.52)
ϕ^x	Fraction of solute in phase x (Eq. 2.19)
ϕ'	Volume fraction (Eq. 2.42b)
φ	Arbitrary function (Eq. 5A.6)
ψ	Electrical potential (Eq. 3.48)
ψ_B	Association factor of solvent B (Eq. 3.45)
Ω	Packing structure factor (Eq. 5.31)
Ω_D	Dimensionless function (Eq. 3.35)

An Introduction to Separation Science

One

INTRODUCTION

What is separation? The meaning of the word may be obvious to all, yet defining "separation" precisely and comprehensively is quite difficult. We carry out separations in so many ways, with so many types of mixtures, for so many purposes, and on such vastly different scales, that the common features of these procedures are difficult to uncover and specify. In general, separation is an operation by which a mixture is divided into at least two fractions having different compositions. Most commonly the goal of this operation is to increase the mole fraction(s) of one component (or more) of the original mixture in relation to the other components.* The actual separation is usually achieved by physical means, although chemical reactions may be involved in the process.

The important role of separation in various branches of science and technology demands—and the common principles underlying the various separation processes have made possible—a comprehensive and integrated treatment of the subject. Hence a science of separation is now emerging that deals with the fundamental physical and chemical phenomena involved in the attainment of separations, as well as with the development and application of various separation processes. These systems are often, beyond their practical usefulness, excellent examples to illustrate the underlying physicochemical principles. An introduction to separation science, therefore, is an introduction to thermodynamics and transport phenomena as well.

---

* In some separation processes, notably chromatography, the concentration of all the components decreases from dilution by the mobile phase; however, the concentration of one or more components *relative* to the other components in the mixture increases.

This book attempts to present a comprehensive treatment of both the fundamental aspects and the practical applications of separation with major emphasis on laboratory and analytical separations. Since these have been developed, employed, and treated within many often unrelated disciplines, there is a great need for a unified treatment. Large-scale commercial applications of separation, along with the technological and economical problems involved, have been and will remain the province of chemical engineering. Our emphasis on laboratory-scale and analytical applications is a result of the desire to provide a coherent and comprehensive treatment of separation techniques scattered over many disciplines, and to serve the reader who wants to become familiar also with practical aspects of separation methods outside his own field. A discussion of these small-scale separations that are used by so many and are employed so frequently in various areas of science and technology makes it possible to demonstrate the diversity and the unity of separation science better than a treatment of the few separation processes that have found large-scale industrial application.

1.1　Historical Perspective

Almost from the beginning of history, man's efforts to understand his environment and to change it to his benefit have been closely linked to the separation of chemical substances. Chemistry, which has played a major role in this human endeavor, is a discipline based on a variety of separation processes. Indeed, *scheikunde*, the Dutch word for chemistry, means literally "the art of separation." Separation is the opposite of mixing, a process favored by the second law of thermodynamics. Like other attempts to cope with the consequences of this natural law, separation often demands the utmost of human ingenuity.

The historian can point to numerous examples of separation, from Biblical through medieval times. Moses' conversion of bitter water into sweet by the immersion of tree branches may have been the first recorded application of ion exchange. Metallurgy, since its beginning, has been based on separation. The distillation of alcohol, the extraction of natural drugs, and the isolation of dyes are also separation processes that date from very early times.

The industrial and scientific revolutions brought about the need for a variety of separation methods and led to their development. Since then most major advances in chemistry, chemical technology, and the life sciences have been related to advances in separation. The essential role of the petroleum industry in our lives today would have been impossible without separation. The atomic age was ushered in only after the difficult technological problem of separating ^{235}U from ^{238}U had been achieved. Unraveling the details of biochemical processes became possible only after the development of such separation methods as chromatography, electrophoresis, and ultracentrifugation. Alternatively the study of separation processes in the laboratory is of great significance in understanding phenomena taking place *in vivo*, such as the selective permeation of cell membranes. Modern medicine also takes advantage of separation methods, as is manifested, for example, by the widespread use of the artificial kidney.

On the dark side of modern technology, it is now apparent that "progress" has led to a number of major problems. Among these is the pollution of our environment. Solutions to the problems of air and water fouling will depend to a considerable extent on finding economical ways to solve the corresponding separation problems. Removal of contaminants from the biosphere will necessitate the application of separation processes on a vast scale. Similarly the detection and control of pollution require new developments in analytical chemistry, which in turn will be based on corresponding advances in separation and isolation procedures.

1.2 The Profusion of Separation Methods

Tables 1.1, 1.2, and 1.3 list more than thirty different separation methods that have found practical application to the present time. Each of these individual methods can be subdivided further into different, clearly distinguishable techniques. This great variety of individual separation procedures can be attributed to three factors: (1) the many distinct separation goals, (2) the diversity of the mixtures to be separated, and (3) the variety of physical and chemical phenomena that can be involved in separation.

Table 1.1. Separation Methods Based on Phase Equilibria
The components of a mixture are distributed between the two phases indicated.

Gas-Liquid	Gas-Solid	Liquid-Liquid	Liquid-Solid
Distillation	Adsorption	Extraction	Precipitation
Gas-liquid	Sublimation	Liquid-liquid	Zone melting
chromatography	Molecular	chromatography	Fractional
Foam	sieves	Exclusion	crystallization
fractionation			Ion exchange
			Adsorption
			Exclusion
			Molecular
			sieves
			Clathration

Table 1.2. Separation Methods based on Rate Processes[a]

Barrier Separation	Field Separations	Other
Membrane filtration	Electrophoresis	Molecular distillation
Dialysis	Ultracentrifugation	Enzyme degradation[b]
Ultrafiltration	Thermal diffusion	Destructive distillation[b]
Electrodialysis	Electrodeposition[b]	
Electro-osmosis	Mass spectrometry	
Reverse osmosis		
Gaseous diffusion		

[a] Under certain circumstances some of the physicochemical phenomena underlying these processes can be utilized to attain separation under equilibrium conditions, as in, for example, ultracentrifugation.
[b] These processes involve chemical reaction.

All of the various separation goals have as a common objective the adequate segregation of constituents of interest with maximum speed and minimum effort, and with as large a capacity as possible. A major goal is the isolation or purification of certain substances or materials by production-scale processes or on a smaller scale in the laboratory. Separation also plays a vital role in the analysis of mixtures; examples are the removal of interfering constituents before the determination of one or more known compounds, the isolation of unknown compounds for their further characterization, and the complete analysis of complex, unknown mixtures by separation of the entire sample into individual constituents.

Mixtures to be separated can vary greatly with respect to molecular weight, volatility, or other properties. They can range from mixtures of atomic species through organic molecules and macromolecules to molecular aggregates (particles). In some cases the properties of the individual components are so different that very simple separation techniques can be used (e.g., mixtures of salt and water). In other cases, the species to be separated may be scarcely distinguishable (e.g., pairs of optical isomers). The relative complexity of the mixture—with respect to either the number of constituents or their physical and chemical properties—markedly affects the difficulty of

Table 1.3. Particle Separation Methods

Filtration	Particle electrophoresis
Sedimentation	Electrostatic precipitation
Elutriation	Flotation
Centrifugation	Screening

separation and the type of technique that is applicable. Finally, the separation technique of choice depends also on the amount of mixture to be separated, which may range from tons in industrial-scale processes to a few molecules in a sensitive analytical method.

To attain separation a large number of physical or physicochemical phenomena have been utilized. These can be divided into two major categories: equilibrium and rate processes. Equilibrium processes are based on differences in the equilibrium properties of the individual components to be separated. The most common equilibrium phenomena involve the distribution of substances between two phases, and indeed equilibrium separation processes are often based on phase equilibria. In a few cases, however, an equilibrium distribution in a single phase resulting from the action of a force field is also utilized for separation, as in equilibrium ultracentrifugation. Rate processes are based on differences in the kinetic properties of the components of the mixture. These include diffusion rates through barriers and migration velocities in an electric, gravitational, or thermal field.

So far we have considered only the separation of chemical species. However, the separation of particles is also of great technical significance. Some of the methods used for this purpose are listed in Table 1.3. Since there is no sharp discontinuity between very large molecules, colloid particles, and macroscopic particles, and no fundamental difference in the underlying principles applied, the recognition of distinct particle separation methods takes into account practical aspects. These methods are often referred to as mechanical separation processes; examples are filtration and classification when particles are separated according to their size or shape, as in screening or elutriation.

1.3 About the Book

In the following chapters we will examine the various separation methods both individually and collectively, review the basic theory and practice essential to their application, and evaluate and compare individual methods in terms of practical objectives. Our approach is to start with the common underlying principles of separation, and then apply these principles to the understanding of individual methods. This plan provides the reader with the physical basis of separation and an appreciation of the uniqueness of each method. Moreover, a given principle need be learned but once, rather than many times in connection with each individual separation method.

We begin in Chapter 2, with a look at the thermodynamics of separation The fundamentals of thermodynamics that are pertinent to separation systems are reviewed, and specific relationships of broad applicability are derived. The dynamic aspects of separation, which involve diffusion and mass transport, are examined in Chapter 3, where basic diffusion and flow processes are reviewed for later application to both equilibrium and nonequilibrium separation systems. The operational aspects of separation receive attention in Chapter 4, where the various separation modes and their practical significance are discussed. In Chapter 5 we conclude our discussion of general principles with a detailed review of the most important laboratory separation process: chromatography. Chapter 6 provides a preview of subsequent chapters by outlining the separation characteristics of individual methods.

The specific separation methods (Chapters 7 to 18) are subdivided into equilibrium and rate processes. The equilibrium processes covered in Chapters 7 through 15 involve the distribution of different sample constituents between two phases. The rate processes, such as those based on restricted diffusion or molecular migration, as well as a few miscellaneous procedures, are discussed in Chapters 16 through 18. The classification of separation techniques and their organization into different chapters have been based on the particular physical and/or chemical phenomena that are responsible for separation (e.g., vapor-liquid equilibrium). Differences in technique or operational mode have been recognized within individual chapters.

Our treatment begins with distillation and gas-liquid chromatography (Chapters 7 and 8). Both procedures are based on the equilibrium distribution of sample components between a liquid and a gas phase. Similarly, extraction (Chapter 9) and liquid-liquid chromatography (Chapter 10) rely on equilibrium distribution between two liquid phases.

Chapters 11 through 13 discuss separations based on various liquid-solid equilibria. In the methods described in Chapter 11, such as crystallization, the solid phase is formed by a component of the mixture to be separated. In turn, methods involving ion exchange and adsorption (treated in Chapters 12 and 13) make use of pre-existing solid phases that interact differently with the individual components.

Chapter 14 is devoted to several other equilibrium methods based on adsorption: gas-solid adsorption and adsorptive bubble processes such as foam fractionation, flotation, and solvent sublation. Chapter 15, which deals with exclusion processes, concludes our treatment of equilibrium separation methods. Here we examine separation by molecular sieve adsorption, gel chromatography, and clathration. These methods, which simultaneously exhibit equilibrium behavior and restricted access to the solid phase, form a bridge between the preceding equilibrium methods and the restricted-migration methods of Chapter 16.

The various barrier separation methods, for example, dialysis, membrane diffusion, and ultrafiltration, are discussed in Chapter 16. Electrophoresis is treated in Chapter 17. Chapter 18 examines briefly several less easily classifiable processes: ultracentrifugation, thermal diffusion, mass spectrometry, enzyme degradation, and particle separations.

Finally, in Chapter 19, we examine the problem of combining individual methods into an integrated, multistep separation scheme for the analysis of very complex mixtures. We consider how the separation characteristics outlined in Chapter 6 relate to the choice of methods for a separation scheme, and the order of methods within the scheme. In conclusion, several examples of successful, multistep separation schemes are presented and discussed.

Part One

Fundamentals

Two

SEPARATION EQUILIBRIA

Separation represents a change in the *relative* concentrations of two or more sample components within a defined region, as a result of the transfer of chemical species (components) from one region to another. Either rate or equilibrium processes can be involved in this transfer. In this chapter we give a general account of equilibrium as it relates to separation. For a similar treatment of rate processes in separation, see Chapter 3.

Equilibrium separation systems are necessarily heterogeneous, involving a partitioning or transfer of components between two or more distinct— usually immiscible*—phases. These phases may be formed entirely from the components that we wish to separate, as in the distillation of a two-component mixture. We will refer to such separations as based on *phase equilibria*. Alternatively the phases may be constituted largely of chemical species that are not part of the mixture to be separated. Such separations are said to be based on *distribution equilibria*. Separations based on equilibrium can be achieved only if there is a difference in the distribution of components between the two immiscible phases. These differences in distribution (and separation) can be further enhanced by a variety of operational procedures described in Chapter 4. Separations based on phase or distribution equilibria can be subclassified according to the types of phases involved (assuming only two phases): (1) gas-liquid (e.g., distillation); (2) liquid-liquid (e.g., extraction); (3) gas-solid (e.g., sublimation); and (4) liquid-solid (e.g., precipitation); see Table 1.1. We can also distinguish between bulk phase and surface (e.g., adsorption) equilibria.

* The main exceptions are described in Chapter 15, for molecular exclusion processes.

The interpretation and prediction of separation equilibria can be approached in two different ways: via (1) thermodynamics or (2) simple models based on the molecular structure of a separation system. Thermodynamics is characterized by rigor and generality. Models based on molecular structure permit more detailed predictions than are possible with thermodynamics, but they are less accurate and often apply to only a limited range of separation systems. Separation thermodynamics is discussed in Section 2.1. The molecular basis of equilibrium separations is treated in Section 2.2.

2.1   Thermodynamics of Separation

2.11   GENERAL

We will first provide a brief review of certain fundamental thermodynamic principles of importance in separation. The reader is assumed to have had the usual undergraduate course in thermodynamics (for a review, see Ref. 1 or 2 or any basic physical chemistry text). Initially we will emphasize the separation of species whose concentrations in the equilibrium system are small (i.e., distribution equilibria).

Consider the distribution of a solute i between two immiscible phases, x and y, at constant temperature and pressure:

$$i^y \rightleftharpoons i^x$$

Separation is determined by the ratio of solute concentrations in the two phases, which we can define by an equilibrium *concentration distribution coefficient* $\tilde{K}_x$ or $\tilde{K}_c$*

$$\tilde{K}_x = \frac{X^x}{X^y} \tag{2.1a}$$

$$\tilde{K}_c = \frac{C^x}{C^y} \tag{2.1b}$$

---

* Over a given solute concentration range, the ratios in Eqs. 2.1a and 2.1b may or may not be dependent on the initial solute concentration in one of the two phases. An IUPAC ad hoc committee on nomenclature of separation methods (D. M. Hume, private communication) has recommended that the term *distribution constant* be used only in those cases where K is independent of concentration and that the term *distribution coefficient* be used in all other cases. We follow this practice in this book. In the completely general case where the concentration ratio may or may not be constant, we have chosen the term distribution coefficient. In addition, we attempt to follow the recommendations of this committee and use the symbol K for the distribution constant. The symbol $\tilde{K}$ will apply

Here X^x and X^y refer to mole fractions of i in phases x and y, respectively, and C^x and C^y are corresponding molar concentrations.* For the separation of two components to occur as a result of this distribution process, their distribution coefficients ($\tilde{K}_c$ or $\tilde{K}_x$) must differ. For dilute solutions, the above distribution coefficients are related as †

$$\tilde{K}_c = \tilde{K}_x \left(\frac{\bar{V}^y}{\bar{V}^x} \right)$$

where $\bar{V}^y$ and $\bar{V}^x$ refer to the molar volumes of the compounds that constitute phases y and x, respectively. In the rest of this section we will concentrate on $\tilde{K}_x$; however, parallel relationships can be derived for $\tilde{K}_c$ (see Ref. 3).

We consider next the variation of $\tilde{K}_x$ with the compositions of phases x and y, including the possible effect of solute concentration on $\tilde{K}_x$. The *chemical potential* μ (partial molar free energy) of the solute i in a given phase can be defined:

$$\mu = \left(\frac{\partial G}{\partial n_i} \right)_{T,P} \tag{2.2a}$$

Here G is the free energy of the total phase, and n_i refers to the number of moles of i present in the phase. At equilibrium the net change in free energy for transfer of solute i between phases x and y must be zero, so that the chemical potentials for i in each phase must be equal:

$$\mu^x = \mu^y \tag{2.2b}$$

The value of μ in any phase can be expressed as

$$\mu = \mu^0 + RT \ln a \tag{2.3}$$

where a is defined as the *activity* of the solute in that phase, and μ^0 refers to the chemical potential of the solute in some chosen *standard state*. Since the standard state is uniquely described by its composition, temperature, and pressure, the quantity μ^0 is a constant, and $a = 1$ for the solute in the standard state (i.e., $\mu \equiv \mu^0$).

Combination of Eqs. 2.2b and 2.3 gives (at equilibrium)

$$\mu^{0,x} + RT \ln a^x = \mu^{0,y} + RT \ln a^y$$

---

to the distribution coefficient. The symbol K^0 (Eq. 2.5) applies to the thermodynamic distribution constant, that is, ratio of activities. The symbol D(e.g., Eq. 2.18a) represents the distribution when chemical equilibria take place in one or both phases and is called the distribution ratio.

* The assumption is made that the phases are either gas or liquid. The concentration units may change if the distribution involves adsorption or penetration of a gel matrix.

† This relationship follows from the fact that $C = X/\bar{V}$, for dilute solutions of i in x or y.

which rearranges to

$$RT \ln \left(\frac{a^x}{a^y} \right) = \mu^{0,y} - \mu^{0,x}$$

$$= -\Delta\mu^0 \qquad (2.4)$$

Here $\mu^{0,x}$ and $\mu^{0,y}$ are μ^0 values for x and y, respectively, as standard states, and a^x and a^y refer to solute activities in phases x and y. Since $\mu^{0,y}$ and $\mu^{0,x}$ are each constant, $\Delta\mu^0$ is likewise constant. This in turn requires that the activity ratio a^x/a^y be a constant, K^0, which we will call the *thermodynamic distribution constant*. Consequently,

$$RT \ln K^0 = -\Delta\mu^0 \qquad (2.5)$$

In separation we are interested in the distribution coefficient $\tilde{K}_x$, rather than K^0. The fact that K^0 is a thermodynamic constant, however, makes it easier to derive than $\tilde{K}_x$. For this reason, we will be concerned with the relationship between $\tilde{K}_x$ and K^0.

According to Eq. 2.5, the choice of standard states for the solute in each phase determines K^0 and the solute activities a^x and a^y. The significance of standard states and solute activities is a frequent source of confusion in the application of thermodynamics to separation systems. We are free to define standard states (real or hypothetical) in any way we choose, and this flexibility often makes possible a particularly convenient representation of the thermodynamics of a given separation system. However, this same flexibility can lead us into error if we are not careful. The following discussion is intended to reduce the possibility of error in handling standard states and solute activities.

Consider first the case of a liquid solution of solute i. We will assume that the solute activity a varies with solute mole fraction X in some manner, as illustrated by the solid curve in Fig. 2.1a. Three reasonably distinct regions of this curve can be distinguished: (1) X small, where a varies linearly with X; (2) $X \approx 1$, where a is also proportional to X, but the proportionality constant differs from that at small values of X; and (3) intermediate values of X, where a is not proportional to X. At this point it is convenient to introduce the solute *activity coefficient* γ:

$$a = \gamma X$$

The activity coefficient can be regarded as a correction factor, which when applied to concentration X gives activity a. Since the slope of the a versus X plot of Fig. 2.1a does not remain constant, it follows that γ changes with X. In Section 2.2 we will interpret this change in γ with X in terms of corresponding changes in the interactions between a molecule of i and surrounding molecules (i or the other component—x or y—of the phase).

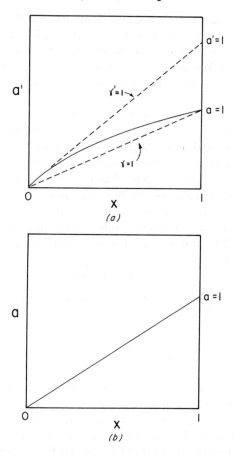

Figure 2.1. Activity versus mole fraction plots to illustrate two standard-state definitions (*a*) and an ideal solution (*b*). In (*a*), the values of *a* and *a′* are not equal on the ordinate axis.

For solutes in a liquid phase, two standard states are in common use: one based on pure solute, and the other based on dilute solutions of the solute. We will consider first the pure solute standard state. In this case μ^0 in Eq. 2.3 represents the chemical potential of the pure solute ($X = 1$). Since $a = 1$ for the standard state, it follows that $\gamma = 1$. In the region $X \approx 1$, where a is proportional to X, γ remains constant ($\gamma = 1$). In Fig. 2.1*a* this linear region of the a versus X plot is extrapolated (lower dashed line) to $X = 0$. From the definition of γ, this dashed line corresponds to an a versus X plot for $\gamma = 1$ over the entire range of X ($0 \leqslant X \leqslant 1$). For $X < 1$, the actual solute activity a is greater than predicted by this line, so that $\gamma > 1$ for the actual plot. For values of X close to zero, where a is again proportional to X,

γ is constant but has a higher value than that for the region $X \approx 1$. In the intermediate region $0 < X < 1$, γ changes with X.

Consider now a standard state based on dilute i, that is, the region $X \approx 0$ in Fig. 2.1a. We define the solute activity coefficient $\gamma' = 1$ in this region.* The linear a' versus X plot for $X \approx 0$ is n w extrapolated to $X = 1$ (upper dashed curve of Fig. 2.1a). The standard state (for $a' = 1$) again occurs at the concentration $X = 1$, but in this case we have a *hypothetical* standard state: pure liquid solute with the same thermodynamic properties (i.e., $\gamma' = 1$) as an infinitely dilute solution of the solute in a particular solvent or phase. This hypothetical standard state can be called an extrapolated infinite dilution standard state. Since $\gamma' = 1$ for the upper dashed line of Fig. 2.1a, the actual value of γ' is less than unity for $X > 0$.

Let us next compare these two standard states. For a given value of X, different activity coefficient values result; for example, $\gamma > 1$, $\gamma' < 1$, in the example of Fig. 2.1a. Since the chemical potential μ must be the same regardless of standard state (it is a *real* property of the system), from Eq. 2.3 and the definition of γ or γ' ($=a/X$) we have

$$\mu = \mu^0 + RT \ln \gamma + RT \ln X = \mu^{0'} + RT \ln \gamma' + RT \ln X$$

or

$$RT \ln \frac{\gamma}{\gamma'} = \mu^{0'} - \mu^0 \tag{2.6}$$

That is, the ratio of these two activity coefficients (for different standard states) is a constant, equal to $e^{(\mu^{0'} - \mu^0)/RT}$.

The *reference state* refers to a solution composition in which the activity coefficient equals unity. For the pure liquid as standard state, the standard and reference states are the same (i.e., $X = 1$). For the extrapolated infinite dilution standard state, however, the reference state ($X = 0$) and the standard state ($X = 1$) differ. If the reference state and the concentration units† are specified, the standard state is thereby defined.

Solutions in which activity is equal to mole fraction for all component concentrations are called *ideal* or are said to follow *Raoult's law*. Figure 2.1b provides an illustration for one component (solute) of an ideal solution. We

---

*Here γ' and a' are used to denote the solute activity coefficient and solute activity in this second standard state to avoid confusion with the first standard state. In later sections of the book γ and a will be used interchangeably for both standard states, but the standard states will be specified.

† So far our discussion has been based on mole fractions as concentration units, but other concentration units can be used. Note that the standard state in the case that the concentration unit is molarity and the reference state is dilute concentration ($C = 0$) is the hypothetical state of $C = 1M$ extrapolated from infinite dilution.

see that γ is constant and equal to 1 for all values of X. The two previously defined standard states (Fig. 2.1a) are equivalent for ideal solutions; that is, the same value of the activity results for either standard state. Ideal solutions and Raoult's law are discussed further in Section 2.22.

Next we consider solutes in the gas phase, that is, gaseous mixtures that include i. The usual standard state is taken as the *ideal gas* at 1 atmosphere and 25° (alternatively, the temperature of the system can be used). This corresponds to an extrapolated infinite dilution standard state, since the ideal gas law ($PV = nRT$) holds only at low pressures (i.e., very dilute gas). The activity of a gaseous solute i is equal to the *fugacity* (f) of i.* If the fugacity of i is plotted versus the partial pressure (P_i) of i in units of atmospheres, a plot similar to that of Fig. 2.1a results. The present standard state is defined by the upper dashed line, representing the extrapolation of the linear f versus P_i (or a versus X) plot, at small P_i, to $P_i = 1$. The fugacity of i in the standard state f^0 is equal to 1, and in the example of Fig. 2.1a the value of f for $P_i = 1$ is seen to be less than 1. At low values of P_i, $f = P_i$. We can define a gas activity coefficient γ'' for the solute i (comparable to the liquid activity coefficient γ'):

$$f = \gamma'' P_i$$

In the example of Fig. 2.1a, $\gamma'' = 1$ for $P_i \approx 0$, and $\gamma'' < 1$ for $P_i > 0$.

In the case of the solid phase, the pure crystalline solute i is generally chosen as the standard state for i.

Having described the various standard states used for gas, liquid, and solid mixtures, we can now define the relationship of the concentration distribution coefficient $\tilde{K}_x$ to the thermodynamic distribution constant K^0. We should note that K^0 is a function of the chosen standard states, whereas $\tilde{K}_x$ is not. Equations 2.4 and 2.5 can be combined to give

$$K^0 = \frac{a^x}{a^y} \tag{2.7}$$

For the case of two immiscible liquid phases x and y, we can substitute γX for a in Eq. 2.7 to give

$$K^0 = \frac{\gamma^x X^x}{\gamma^y X^y} \tag{2.7a}$$

and from Eq. 2.1a

$$\tilde{K}_x = \left(\frac{\gamma^y}{\gamma^x}\right) K^0 \tag{2.8}$$

---

* The relationship of activity and fugacity is $a = f/f^0$, where $f^0 = $ fugacity in the standard state. Since $f^0 = 1$, $a = f$. Note, however, that a and F differ dimensionally.

For the extrapolated infinite dilution reference state (each phase), K^0 in Eq. 2.8 is given by Eq. 2.5, with $\Delta\mu^0$ calculated for the transfer of a mole of solute from dilute solution in x to dilute solution in y. For dilute solutions of solute in each phase, $\gamma^x = \gamma^y = 1$, and[*]

$$K_x = K^0 \tag{2.8a}$$

For the pure solute as standard state (each phase), K^0 in Eq. 2.8 is equal to 1,[†] so Eq. 2.8 simplifies to

$$K_x = \frac{\gamma^y}{\gamma^x} \tag{2.8b}$$

Much of the confusion concerning standard states and related activity coefficients has arisen from the use of Eqs. 2.8a and 2.8b (and related expressions) without specifying standard states. The common assumption that activity coefficients equal unity in dilute solution is valid only for the infinitely dilute reference state (however, γ^y/γ^x is always constant in dilute solution).

For the distribution of solute between a liquid phase y and a gas phase x we can define a distribution coefficient $\tilde{K}_g$ equal to P_i/X^y (P_i is the partial pressure of solute in the gas phase). For the infinitely dilute reference state (each phase) a^x is simply P_i, and a^y is $\gamma^y X^y$. Substituting these terms into Eq. 2.4, followed by combination with the above definition of $\tilde{K}_g$, gives

$$\tilde{K}_g = K^0 \gamma^y \tag{2.9a}$$

which for dilute solutions simplifies to Eq. 2.8a. Again K^0 is given by Eq. 2.3, with $\Delta\mu^0$ corresponding to the transfer of a mole of solute from dilute solution in x to the pure solute as ideal gas at 1 atm. If we use the pure solute as the reference state for each phase, a^y is again given as $\gamma^y X^y$. Now a^x is P_i/P^0, where P^0 is the vapor pressure of the solute over pure solute. As was true previously for the case of two liquid phases, $K^0 = 1$, so that

$$\tilde{K}_g = \gamma^y P^0 \tag{2.9b}$$

Essentially similar expressions can be derived readily for gas-solid and liquid-solid equilibria. Here the pure solute is usually taken as the standard state for the solid phase. For solids that do not cocrystallize the activity of a solid component is unity, which considerably simplifies expressions for $\tilde{K}_c$ or $\tilde{K}_x$.

---

[*] In Eqs. 2.18a and 2.18b, we have changed terms from $\tilde{K}_x$ to K_x because the concentration ratio is constant and the distribution constant must therefore be used. (See footnote on p. 12.)

[†] This follows from Eq. 2.3. Whenever the same standard state is used for the solute in each phase, $\Delta\mu^0$ must equal zero since $\mu^{0,x} = \mu^{0,y}$.

The temperature dependence of K^0, γ, $\tilde{K}$, etc., can be derived from the standard thermodynamic relationships:

$$\mu = h - Ts \qquad (2.10)$$

and

$$\Delta\mu^0 = \Delta h^0 - T\,\Delta s^0 \qquad (2.10a)$$

where h and s are the partial molar heat and the entropy, respectively, of the solute in a given phase; h^0 and s^0 are the corresponding quantities in the standard states; and Δh^0 and Δs^0 are the differences between standard states. Thus the temperature dependence of K^0 is obtained from Eqs. 2.3 and 2.10a

$$\frac{d \ln K^0}{d(1/T)} = \frac{-\Delta h^0}{R} \qquad (2.10b)$$

The significance of these heat and entropy terms is examined in detail in Section 2.2.

2.12   PHASE EQUILIBRIA

Phase equilibria separations—in contrast to distribution equilibria—involve phases that are composed of the mixture to be separated. Such separations are favored in preparative and production-scale applications since no species that require subsequent removal are added. Phase equilibria separations are also best suited for relatively simple mixtures containing only a few components of interest. In this section we will review certain elementary aspects of one-, two-, and three-component phase equilibrium systems (see Ref. 4 for more details).

The study of phase equilibria begins with the *Gibbs phase rule*:

$$P + F = C + 2 \qquad (2.11)$$

Given the number of components C in a system, we can predict the number of phases P that can coexist for a specified number of degrees of freedom F. The degrees of freedom correspond to the independent intensive variables of the system (temperature, pressure, or the concentrations of the components in the given phases). Alternatively, given C and P, we can specify the number of variables that can vary independently of each other. The phase rule is completely general, being applicable to all types of phase behavior. The significance of this rule for understanding phase equilibria will become apparent as we examine specific examples.

Systems of One Component

Within the restrictions of the phase rule the intensive variables of a system can range over wide limits. In order to adequately describe a phase equilibrium system from the standpoint of using that system for separation, we

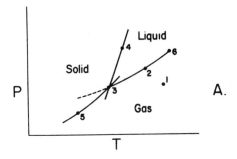

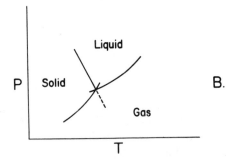

Figure 2.2. Typical phase diagrams for a single component.

need a *phase diagram*. Figures 2.2a and 2.2b show phase diagrams for a single component in terms of two variables: pressure P and temperature T. From the phase rule, a single component and one phase require that two degrees of freedom (temperature and pressure) be specified to define the system.

The interpretation of these phase diagrams is straightforward. In Fig. 2.2a point *1* represents a pressure and temperature at which only the gas phase exists. Similar domains for the liquid and solid phases are also shown. The lines represent pressures and temperatures at which two phases are in equilibrium with each other (with two phases specified, $F = 1$, so that a value of P or T determines the other variable). Point *2*, for example, corresponds to a pressure and temperature at which gas and liquid coexist in equilibrium with one another. Two-phase equilibria are also indicated by points *4* (liquid-solid) and *5* (gas-solid). Point *3* is the triple point, at which all three phases are in equilibrium. Since $F = 0$, the triple point is unique, occurring at only one temperature and pressure. In both Fig. 2.2a and Fig. 2.2b the gas-liquid line terminates at the *critical point* (point *6* in 2.2a), corre-·sponding to the critical temperature T_c and critical pressure P_c of the compound in question. The properties of the gas and liquid become equivalent at the critical point; above the critical temperature only one phase exists (the so-called *supercritical fluid*) regardless of pressure. Supercritical fluids have recently been used in certain separation processes (see, e.g., Sections 10.4 and 13.5).

Figure 2.2 also illustrates the phenomena of supercooling (Fig. 2.2*a*) and superheating (Fig. 2.2*b*) of liquids (dashed curves). Rapid changes in temperature can in some cases maintain a liquid below its freezing point (2.2*a*) or above its boiling point (2.2*b*). In each case the metastable (nonequilibrium) liquid state coexists with the gas phase for a temperature and a pressure that fall on the dashed curves. In principle, phase equilibria separations could be based on the supercooled or superheated liquid, despite the fact such systems are not strictly at equilibrium.

The slopes of the various lines in the phase diagrams of Fig. 2.2 are given by the Clapeyron equation:

$$\frac{dP}{dT} = \frac{\Delta H}{T\,\Delta\bar{V}} \tag{2.12}$$

Here ΔH refers to the enthalpy change per mole as a result of the phase change, and $\Delta\bar{V}$ is the corresponding change in molar volume. If ΔH and $\Delta\bar{V}$ have the same sign, the slope of the two-phase line is positive; otherwise the slope is negative. For the processes of sublimation and vaporization dP/dT is always positive, since ΔH^V (the heat of vaporization) and $\Delta\bar{V}$ are always positive. For most substances the slope of the liquid-solid curve is also positive, as in Fig. 2.2*a*, because ΔH^F (the heat of fusion) and $\Delta\bar{V}$ are generally positive. Only for a few compounds (notably water) is there a decrease in volume on melting, with a resulting negative slope of the liquid-solid line (Fig. 2.2*b*).

The Clapeyron equation can be modified for sublimation and evaporation by noting that the gas molar volume is much larger than the molar volume of the liquid or solid. In Eq. 2.12 $\Delta\bar{V}$ can then be replaced by $\bar{V}$, the gas molar volume, which can be expressed in terms of the ideal gas law (RT/P). The resulting relationship is the *Clausius-Clapeyron equation*:

$$\frac{d(\ln P)}{dT} = \frac{\Delta H}{RT^2} \tag{2.12a}$$

Integration of this equation gives the well-known relationship between the vapor pressure of a pure compound P^0 and the reciprocal absolute temperature:

$$\ln P^0 = \frac{-\Delta H}{RT} + C \tag{2.12b}$$

where C is an integration constant. This equation will be shown to be important in distillation (Chapter 7) and gas chromatography (Chapter 8 and Section 14.1).

Systems of Two Components

The phase diagram becomes more complicated for systems of two components. When only one phase is present, $F = 3$ from the phase rule, and the three independent variables are P, T, and the mole fraction of one component (the mole fraction of the other component is obtained by difference). When two phases are in equilibrium, only two independent variables need be known. In this case it is convenient to take the mole fraction of one component (in one phase) and either temperature or pressure as the independent variable. As examples of such two-component phase diagrams we will consider the important cases of gas-liquid and liquid-solid phase equilibria at constant pressure.

Figure 2.3 shows the three possible types of phase diagram for vapor-liquid equilibrium in a two-component system. Figure 2.3*a* illustrates the usual type of phase equilibrium, for a lower-boiling component j and a higher-boiling component i. The vapor and liquid (single-phase) regions are indicated, with the intermediate region $(L + V)$ being a two-phase region. A liquid of composition $X = a$ will exist as a single phase below temperature T_2. Boiling of the liquid will occur at T_2, so that a gas and a liquid phase will coexist. As the temperature is raised above T_2, an increasing fraction of the total system will exist as gas, until at temperature T_1 all of the original liquid will have been converted to vapor.

Consider next the point d in the two-phase region, corresponding to temperature T_1 and composition (total system) $X_i = c$. At temperature T_1 vapor of composition a is seen to be in equilibrium with liquid of composition

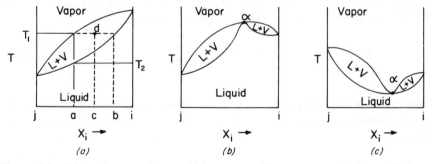

Figure 2.3. Phase diagrams for two components to illustrate vapor-liquid equilibrium. Diagrams (*b*) and (*c*) show azeotropic formation.

b. The horizontal line through point *d* is called the *tie line*. We see that the vapor phase is enriched in the more volatile component *j*, and the liquid phase in the less volatile component *i*. A partial separation thus occurs, and this forms the basis of distillation. The relative amounts of liquid and vapor corresponding to point *d* can be calculated as follows. Let n_v and n_l represent the total moles of *i* plus *j* in the vapor and liquid phases, respectively. For component *i* a material balance gives

$$X_c(n_v + n_l) = X_v n_v + X_l n_l$$

where X_c, X_v, and X_l refer to mole fractions of *i* in the total system, in the vapor phase ($X_i = a$), and in the liquid phase ($X_i = b$), respectively. The above expression rearranges to the so-called *lever rule*, which is general for any point within the two-phase region of a two-component phase diagram:

$$\frac{n_v}{n_l} = \frac{X_l - X_c}{X_c - X_v} \tag{2.13}$$

Figures 2.3*b* and 2.3*c* illustrate the phenomenon of *azeotropic* or constant boiling solutions. At the composition α corresponding to maximum or minimum boiling temperature, the compositions of both vapor and liquid are the same, and the composition of the liquid remains constant as increasing amounts of liquid are converted to vapor. A liquid of composition α thus acts like a pure compound and cannot be separated by distillation at the pressure of the phase diagram. However, in certain cases azeotropic solutions of three components are deliberately created to enhance the separation of a two-component mixture by distillation. The interpretation and use of vapor-liquid phase diagrams such as those of Fig. 2.3 are discussed further in Chapter 7.

Figure 2.4 illustrates a type of phase diagram commonly encountered in liquid-solid phase equilibria for two-component mixtures. Again temperature is plotted versus composition at constant pressure. The two liquids are completely miscible, and no solid solutions are formed.* Given a point *d* in the phase diagram, we can calculate the relative amounts of liquid and solid phase via the lever rule, as previously. The compositions of each phase— *j* as solid (assuming no solid solution), and liquid of composition $X_i = a$— are given by the phase diagram. Note that *j* separates as solid from mixtures to the left of point *e*, and *i* from mixtures to the right of point *e*.

If the mixture of composition *c* is cooled below T_1, more solid *j* forms and the liquid solution composition follows the phase equilibrium line to point *e*,

---

* The case of zero solid solubility probably never occurs. Solid solubility is frequently low and thus may not be measurable; however, it can be very important when purifying by solidification. For simplicity we have chosen to leave this out of the phase diagram; however, see Figure 11.6.

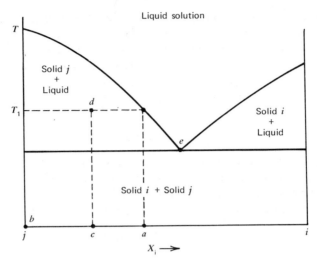

Figure 2.4.   Two-component phase diagram to illustrate liquid-solid equilibrium.

the *eutectic point*. At this point three phases coexist at equilibrium, so that $F = 1$ from the phase rule. This 1 degree of freedom is specified by the pressure assumed in the phase diagram (i.e., the eutectic point is invariant at constant pressure). Further cooling of the eutectic results in the formation of solid i and j in the relative amounts corresponding to e. More complicated phase diagrams result when complex formation occurs or when solid solutions form. For a further discussion of phase equilibria in liquid-solid systems as they apply to separation problems, see Chapter 11.

Systems of Three Components

For a three-component system the phase rule predicts $F = 4$ for one phase and $F = 3$ for two phases. The latter situation is of particular interest in separation applications. The phase diagrams are often based on constant temperature* and pressure (two of our three independent variables, for two phases), with the mole fraction of one component in one of the two phases as the third independent variable. Three-component phase diagrams are commonly encountered in separations by extraction.

Triangular coordinates are used to represent a three-component liquid system, with the mole fractions of all of the components as axes. Figure 2.5*a* illustrates the designation of composition for a hypothetical (completely miscible) liquid system i–j–k. The point A is seen to correspond to 30% i, 50% j, and 20% k. Figure 2.5*b* illustrates a phase diagram for a hypothetical

---

* However, for crystallization, phase diagrams with temperature as a variable are typically used.

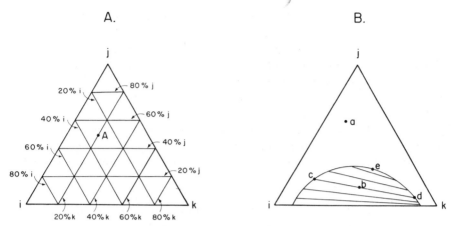

Figure 2.5. (a) Diagram to illustrate the use of triangular coordinates. (b) Ternary-phase diagram to illustrate partial miscibility.

system i–j–k that forms two phases (pressure and temperature constant). Here j is completely miscible with the other two liquids (i and k), but i and k are only partly miscible. An example of such a case is ethanol (j)–benzene–water. The curve shown in Fig. 2.5b is the *binodal curve*; any point falling outside this curve (e.g., a) represents a composition of complete miscibility, while points within the curve (e.g., b) correspond to two-phase systems. The composition and amounts of each phase can be obtained from the tie line through that point (note that these tie lines are not generally parallel to the base of the phase diagram and must be determined experimentally, in contrast to the case of two-component systems as in Figs. 2.3 and 2.4). Thus for point b the compositions of the two phases are given by the intersections of the tie line with the binodal curve (points c and d). The amounts of each phase are given, as previously, by the lever rule. The *plait point e* corresponds to a unique composition, defined by the point on the binodal curve toward which the ends of the tie lines converge.

2.13 DISTRIBUTION EQUILIBRIA

In separations based on distribution equilibria, the two phases are constituted from compounds other than the sample components. These systems are primarily of analytical or preparative interest: first, because the combined volume of the two phases is very much larger than that of the sample and could be unwieldy on a production scale; and, second, because the distribution coefficient is usually constant (independent of sample concentration).

As we shall see in Chapter 5 and following chapters, constant K values are essential in many analytical separations.

Separations based on distribution equilibria have certain advantages over corresponding separations by phase equilibria. In distribution equilibria the composition of one or both phases can be varied over wide limits. This in turn affects the K values of individual sample components and thereby allows otherwise difficult separations to be made. Distribution equilibria (especially when augmented by certain operational modes reviewed in Chapter 4) are much more suitable than phase equilibria for the rapid separation of complex samples containing tens or hundreds of components.

The Distribution Isotherm

Distribution equilibria for a given solute and system at a particular temperature can be described by a plot of solute concentration in one phase x (C^x) versus concentration in the other phase y (C^y). These *distribution isotherms*, as illustrated in Fig. 2.6, are useful because they show how $\tilde{K}$ varies with sample concentration. Since $\tilde{K} = C^x/C^y$, the value of $\tilde{K}$ for a specific solute concentration (e.g., $C^x = a$ in Fig. 2.6a) is given by the slope of the line through the origin and the corresponding point on the isotherm (dashed line in Fig. 2.6a). If $\tilde{K}$ is constant over some range of solute concentration, the corresponding isotherm is linear through the origin. Actual distribution isotherms are seldom linear over a wide range of solute concentration (they can be either concave, as in Fig. 2.6a, or convex, as in Fig. 2.6b). However all isotherms become linear for sufficiently dilute solute concentrations, as can be seen in Fig. 2.6. This linear isotherm region, where the distribution coefficient is converted to the distribution constant, is of major interest in analytical separations.

Linear distribution isotherms are equivalent to *Henry's law* for gas-liquid equilibria (cf. Eq. 2.9a):

$$\frac{P_i}{X_i} = K^0 \gamma^y$$

$$P_i = C'X_i \tag{2.14}$$

and to the *Nernst distribution law* in the case of liquid-liquid equilibria (cf. Eq. 2.8),

$$K_x = K^0 \left(\frac{\gamma^y}{\gamma^x}\right)$$

or

$$K = C'' \tag{2.15}$$

In each case C' and C'' are constants, because the various activity coefficients γ are constant over the solute concentration range where these laws hold

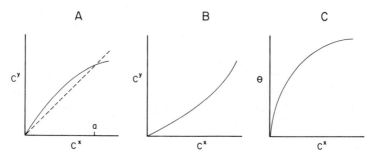

Figure 2.6. Typical shapes of distribution isotherms.

(dilute solutions). A change in solute activity coefficient with concentration is seen to lead to variation in C' or C'', resulting in concave or convex isotherms in the case of liquid or gas phases. These activity coefficients are determined mainly by the intermolecular interactions between solute and solvent molecules (Section 2.2).

In the case of adsorption (usually onto a solid phase), there is often competition among solute molecules for a fixed number of adsorption sites. As these adsorption sites become saturated with adsorbed solute molecules (at high concentrations of solute in the other phase), the concentration of solute in the adsorbed phase approaches a limiting value and the isotherm flattens out as in Fig. 2.6c. Isotherm nonlinearity in this case is not necessarily the result of changing intermolecular interactions, but is essentially statistical in nature. A derivation of the *Langmuir isotherm* will illustrate what we mean.

Adsorption can be represented as the reversible reaction of solute S in the nonadsorbed (gas or liquid) phase with an adsorption site A to form adsorbed S (SA):

$$S + A \rightleftharpoons SA$$

A distribution coefficient $\tilde{K}$ can be defined for this reaction

$$\tilde{K}_x = \frac{X_{SA}}{X_S X_A}$$

where X_{SA}, X_S, and X_A are mole fractions of SA, S, and A. Then $\tilde{K}_x$ is constant if we assume that intermolecular interactions in each phase are constant as X_S and X_{SA} vary. If we define the fractional coverage of sites by adsorbed solute as θ, then $X_{SA} = \theta$ and $X_A = 1 - \theta$. Insertion of these quantities into the equilibrium constant expression yields the Langmuir isotherm (e.g., Fig. 2.6c):

$$\tilde{K}_x = \frac{\theta}{X_S(1 - \theta)}$$

or

$$\theta = \frac{\tilde{K}_x X_S}{1 + \tilde{K}_x X_S} \tag{2.16}$$

For a gas as nonadsorbed phase, P_i replaces X_S. At low solute concentrations, $\theta = K_x X_S$ (K_x = distribution constant) and we have a linear isotherm. For large values of X_S, θ approaches a limiting value of 1, and the distribution coefficient $\tilde{K}_x = \theta/X_S$ approaches zero.

Secondary Chemical Equilibria

The preceding discussion of separation equilibria has emphasized physical mixtures, rather than systems of chemically interacting species. The introduction of a reversible chemical reaction into a separation system (so-called secondary chemical equilibria) provides an additional possibility for the control of the K values of individual sample components. We must keep in mind that separation depends ultimately on differences in the K values of individual sample components. The larger these differences, the easier it will be to achieve separation. We can make use of secondary chemical equilibria to facilitate separation in systems based on either phase or distribution equilibria, but the most important application of secondary chemical equilibria is in distribution systems. Therefore the discussion here will concentrate on this area; the extension of the principles described to phase equilibria should be obvious.

Acid and base extractions provide some of the simplest examples of the use of secondary chemical equilibria to facilitate distribution separations. The conversion of an organic acid or base into a water-soluble salt permits its aqueous extraction from a water-immiscible organic phase. In this way, organic acids or bases can be separated from neutral organic compounds that are preferentially retained in the organic phase. In the same way, control of the pH of the aqueous phase permits the separation of strong acids or bases from weaker acids or bases. Masking agents for the prevention of precipitation constitute another example. Secondary equilibria are often useful also in nonequilibrium separations. Proteins are conveniently separated by electrophoresis, in which species with an overall (net) negative charge migrate toward the positive electrode, and positively charged species migrate toward the negative electrode. The pH of the electrolyte medium controls the net charge on a given protein, which in turn controls its direction and velocity of migration and hence its separation in electrophoresis.

Let us look next at a quantitative description of the effect of secondary chemical equilibria on separation. We will choose as an example the extraction of an organic acid, HA, from water, w, to an organic solvent, o:

$$(HA)_w \rightleftharpoons (HA)_o \tag{2.17a}$$

In the aqueous phase, the acid is capable of dissociating as follows:

$$H_2O + (HA)_w \rightleftharpoons (H_3O^+)_w + (A^-)_w \qquad (2.17b)$$

The distribution constant can be defined from Eq. 2.17a as:

$$K = \frac{[HA]_o}{[HA]_w} \qquad (2.17c)$$

and the acid dissociation constant from Eq. 2.17b as:

$$K_a = \frac{[H_3O^+]_w[A^-]_w}{[HA]_w} \qquad (2.17d)$$

For simplicity, we assume dilute concentrations, so that K is independent of the initial concentration of HA in water.

It is useful to introduce a new term, the *distribution ratio D*, to take account of the influence of chemical equilibria on the overall distribution:

$$D = \frac{\text{stoichiometric concentration of HA in the organic phase}}{\text{stoichiometric concentration of HA in the aqueous phase}}$$

Then:

$$D = \frac{[HA]_o}{[HA]_w + [A^-]_w} \qquad (2.18a)$$

Substitution of Eqs. 2.17c and 2.17d into 2.18a results in

$$D = K\left[\frac{[H_3O^+]}{[H_3O^+] + K_a}\right] \qquad (2.18b)$$

Equation 2.18b can be helpful in understanding the distribution of HA over a whole range of acid and base concentrations in water. First, consider acidic conditions such that $[H_3O^+] \gg K_a$. In this case $D = K$ from Eq. 2.18b. The solution is sufficiently acidic so that no ionization occurs in the aqueous phase, and HA acts as a simple neutral species which is distributed between the two phases.

Next, consider a solution pH such that $[H_3O^+] = K_a$. Here, $D = K/2$ from Eq. 2.18b, and half the previous amount of material is now extracted into the organic phase. The reason for the decrease in the amount extracted is that the concentration of undissociated HA in the aqueous phase is only 50% of the initial HA added to the solution. Since the distribution coefficient (Eq. 2.17c) must be maintained constant, the concentration of HA in the organic phase must decrease by one half of that present when no ionization of HA occurred.

Finally, consider a basic solution such that $[H_3O^+] \ll K_a$. In this case Eq. 2.18b reveals that $D \ll K$, and in fact that little, if any, HA will be extracted. The concentration of HA′ in the aqueous phase is very low because of the ionization of the acid; hence the amount extracted must also be small.

Thus the distribution of HA can be markedly influenced by the concentration of the common ion H_3O^+. Separations of different acids (with different K_a values) are then clearly possible. The larger the difference in K_a for two acids, the easier will be their separation. Furthermore, we see that in the case of secondary chemical equilibria the distribution ratio D gives a much better picture of the distribution process than the distribution coefficients.

Later chapters on individual separation methods will provide numerous examples of the use of secondary chemical equilibria.

Capacity Factor

Separation in a distribution system depends ultimately on the relative amount of solute in each phase, rather than the relative solute concentrations. Thus, in the case of a solute with a K value of 100, equal volumes of each phase mean that 99% of the total solute is in one phase. However, if the volume of the preferred phase is reduced by a factor of 100 (the volume of the other phase being held constant), only 50% of the solute will be contained in this phase. The fraction of total solute ϕ^x in a given phase x can be written as

$$\phi^x = \frac{C^x V^x}{C^x V^x + C^y V^y} \qquad (2.19)$$

where C^x and C^y are the solute concentrations in the two phases, and V^x and V^y are the volumes of phases x and y. The solute distribution constant K is equal to C^x/C^y, so that

$$\phi^x = \frac{K(V^x/V^y)}{1 + K(V^x/V^y)} \qquad (2.19a)$$

Let us now define an important quantity, the capacity factor k':

$$k' = \frac{\text{total amount of solute in phase } x}{\text{total amount of solute in phase } y}$$

$$= \frac{C^x V^x}{C^y V^y}$$

$$= K\left(\frac{V^x}{V^y}\right)^* \qquad (2.20)$$

Combination of Eqs. 2.19a and 2.20 yields a simpler expression for ϕ^x:

$$\phi^x = \frac{k'}{1 + k'} \qquad (2.21)$$

* When secondary equilibria are involved, K is replaced by D, the distribution ratio.

Similarly, the fraction of total i in phase y is given as

$$\phi^y = \frac{1}{1 + k'} \tag{2.21a}$$

In Chapter 5 we will see that the capacity factor k' is a particularly useful parameter in describing chromatographic separation. In chromatography the stationary phase is always selected as phase x.

The interrelationship of ϕ^x, K, and the volume ratio V^x/V^y is illustrated in Fig. 2.7 for two values of V^x/V^y. A change in the volume ratio results in a simple vertical displacement of these curves. The amount of solute in each phase is most sensitive to changes in K or V^x/V^y when k' is close to 0.5, or equal amounts of solute are in each phase. For large or small values of k', θ^x asymptotically approaches one or zero. It thus becomes progressively more difficult to remove the last traces of i from either phase. In other words, when the capacity factor is large or small, large changes in k' have little effect on ϕ^x. Similar relationships, as in Fig. 2.7, hold for adsorption, with the use of surface area instead of volume for the adsorbent phase. In analytical separations it is customary to assume complete removal of solute from one phase when $\phi^x < 0.01$.

Separation Factor

Up to this point we have considered the distribution equilibria of single solute systems. Actual separations that involve two or more solutes can be achieved only when the solute K values differ, that is, there must be some difference in the thermodynamic behaviors of the components. In this connection it is useful to define the separation factor α for a pair of solutes i and j:

$$\alpha = \frac{C_i^x C_j^y}{C_i^y C_j^x}$$

$$= \frac{K_i}{K_j} \tag{2.22}$$

The concentrations C_i^x, C_j^x, C_i^y, and C_j^y refer to solute i or j in phase x or y, and K_i and K_j are the corresponding distribution constants.* We see that large α values favor separation, because they mean larger relative concentrations of solute i in phase x and solute j in phase y. However, separation depends also on the capacity factors for i and j (k_i' and k_j'), as we can see from a simple example.

First note that α is also equal to the capacity factor ratio:

$$\alpha = \frac{k_i'}{k_j'} \tag{2.22a}$$

* When secondary equilibria are involved, K is replaced by D, the distribution ratio.

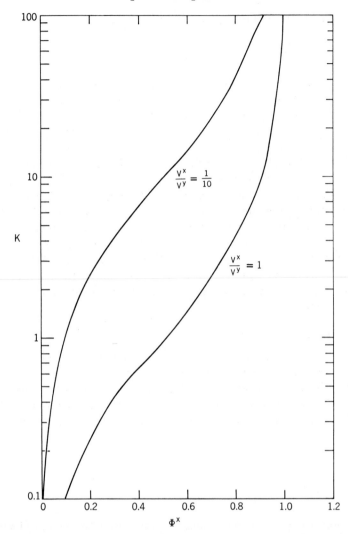

Figure 2.7. Plot of distribution constant K versus fraction extracted into x phase, ϕ^x, as a function of the volume ratio of the two immiscible phases.

Assume next that $k'_i = 100$ and $k'_j = 0.01$. This means that 1% of i is in phase y and 1% of j is in phase x, that is, a good separation. The resulting separation factor α is equal to 10^4. Consider next $k'_i = 1$ and $k'_j = 10^{-4}$. The separation factor is again equal to 10^4, but now 50% of i is distributed into each phase—a poorer separation in terms of the final concentrations. Of course only 0.01% of j is in phase x, but this improved concentration of j into phase x (from 1 to 0.01%) would not normally be significant (unless high purity of i were desired). For a given value of α, optimum separation of two components i and j in a single equilibration occurs when

$$k_i' k_j' = 1 \tag{2.23}$$

The separation factor α can be related to solute chemical potentials μ^0 for dilute solutions:

$$\alpha = \frac{K_{x,i}}{K_{x,j}}$$

where $K_{x,i}$ and $K_{x,j}$ refer to the mole fraction distribution coefficients (Eq. 2.5) for i and j. Using the dilute solution as reference state, and combining the above relationship with Eqs. 2.3 and 2.8a, we have for dilute solutions

$$RT \ln \alpha = -(\Delta\mu_i^0 - \Delta\mu_j^0) \tag{2.24}$$

Here $\Delta\mu_i^0$ and $\Delta\mu_j^0$ are the standard chemical potential differences between phases y and x, for solutes i and j.

Distribution differences can be expressed in terms of $\Delta(\Delta\mu^0)$, in calories per mole. Thus $\alpha = 10^4$ would be -5.5 kcal/mole for $\Delta(\Delta\mu^0)$ at 25°C, whereas $\alpha = 1.01$ is -5.5 cal/mole.

It should be emphasized that α values of 10^4 are often unobtainable for compound pairs that we wish to separate by some distribution process. In some cases this difficulty can be overcome by the use of secondary chemical equilibria. Satisfactory separations based on α values as small as 1.01 can also be achieved, however, using certain operational procedures described in Chapter 4.

The definition of separation factors for phase equilibria systems is essentially similar to the use of α in distribution systems.

2.2 The Molecular Basis of Equilibrium Separations

Section 2.1 has dealt with the thermodynamics of two-phase systems at equilibrium. We have seen that it is possible to calculate the extent of separation in different systems if we are given certain data (vapor pressure, solubility, etc.) for each component as a function of T, P, and phase composition. However, data of this type are often unavailable for a separation system of interest. We must then have some basis for estimating distribution coefficients in order to select the most promising conditions for a given separation. Equilibrium distribution is determined in large measure by the physical and chemical forces between molecules in each phase—the so-called intermolecular interactions. These interactions can in turn be related to the chemical structures of the interacting molecules. Thus it should be possible to estimate phase and distribution equilibria from a knowledge of the structures and concentrations of the compounds comprising an equilibrium system.

In this section we will provide a practical understanding of how molecular structure determines the distribution of a compound between two phases. This knowledge will prove to be a powerful tool in optimizing the separation of known mixtures. Later (e.g., Section 10.11) we will discover that the same general concepts can be applied to the separation of completely unknown mixtures. However, it should be emphasized that the more we know about a particular separation problem—the number of components, their structures, their approximate concentrations, and so forth—the more chance we have of solving it.

2.21 INTERMOLECULAR INTERACTIONS

The partial molar enthalpy h_i of a component i in a given phase is directly related to the forces between i and the surrounding molecules. These h_i values in turn largely determine the corresponding chemical potentials μ_i and the values of K for the distribution of i between different phases. As the inter-molecular interaction energy between i and the surrounding molecules increases, there is a greater tendency for i to concentrate in that phase. If we can estimate the magnitudes of these interaction energies, we can predict the relative distribution of any component between two phases.

The net energy of interaction E_{ij} between two adjacent, nonbonded atoms (or molecules) i' and j' is the result of both attractive and repulsive forces. In general

$$E_{ij} = \frac{A}{r^{12}} - \frac{\sum B}{r^z} \tag{2.25}$$

where r is the distance separating the nuclei of atoms i' and j', and A and B are constants for particular atoms i' and j'. Equation 2.25 recognizes one major repulsive interaction (the universal tendency of two objects to avoid occupying the same space) and several possible attractive interactions (negative E_{ij} represents a net attraction between i' and j'). The coefficient z is generally $\leqslant 6$, so that attractive forces are more important at large interatomic separations, and repulsive forces dominate at small values of r. The r^{-12} dependence of repulsive energy leads to a very steep rise in E_{ij} when r falls below a certain value (r_e), the so-called van der Waals separation* of i' and j' (see Fig. 2.8). Consequently the equilibrium separation r_e is determined mainly by the repulsive term. Similarly the net energy of inter-

---

* That is, the equilibrium separation distance.

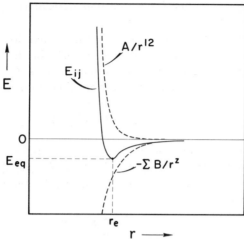

Figure 2.8. Nonbonding interaction energy as a function of separation distance r_e between two adjacent molecules or atoms. E_{ij} = total energy; A/r^{12} = energy of repulsion; $-\sum B/r^2$ = total energy of attraction.

action E_{eq} (at equilibrium) can be regarded as a function mainly of the attractive interactions. This section provides a detailed examination of these all-important attractive interactions.

Dispersion Interactions

Dispersion or London forces exist between any adjacent pair of atoms or molecules. In many cases (e.g., most organic compounds) dispersion forces account for a major part of the total attractive energy, in some cases (e.g., the hydrocarbons) they account for virtually all of E_{ij}, and in a few cases (notably the rare gases) they are the only possible attractive interaction. The origin of dispersion forces is as follows. At any given instant the various electrons of an atom or molecule can assume any of a variety of distinct positions through oscillating movements. Each such instantaneous configuration is characterized by some degree of electrical dissymmetry, in the sense that the atom or molecule possesses an overall dipole (or multipole) moment.* This instantaneous dipole polarizes the electron clouds in adjacent atoms or molecules, inducing a dipole of opposite polarity, which attracts the original dipole. The attractive interaction energy $(E_{ij})_d$ between two atoms i' and j' as a result of these dispersion forces can be expressed as (5)

$$(E_{ij})_d = -\frac{3}{2}\left(\frac{I_i I_j}{I_i + I_j}\right)\left(\frac{\alpha_i \alpha_j}{r^6}\right) \tag{2.26}$$

* We are not concerned here with permanent dipole moments; their interactions with surrounding molecules are treated later in this section.

Here I_i and I_j refer to the first ionization potentials of atoms i' and j', and α_i and α_j are their respective polarizabilities.

Ionization potentials and interatomic distances often do not vary greatly for different pairs of adjacent atoms (particularly in the case of organic compounds). Therefore to a good approximation $(I_i + I_j) \approx 2(I_iI_j)^{1/2}$ and $r = (r_i + r_j) \approx 2(r_ir_j)^{1/2}$. Inserting these expressions into Eq. 2.26, yields

$$(E_{ij})_d = -\frac{3}{8}\left(\frac{I_i\alpha_i^2}{r_i^6}\right)^{1/2}\left(\frac{I_j\alpha_j^2}{r_j^6}\right)^{1/2} \tag{2.27}$$

Corresponding interaction energies $(E_{ii})_d$ and $(E_{jj})_d$ can be defined for interactions between *like* atoms i or j, respectively. Thus, for $(E_{ii})_d$, I_i, α_i, and r_i are substituted into Eq. 2.27 for I_j, α_j, and r_j, respectively. We can then rewrite Eq. 2.27 in the form

$$(E_{ij})_d = (E_{ii})_d^{1/2}(E_{jj})_d^{1/2} \tag{2.27a}$$

which is the so-called *geometric mean rule* of dispersion interactions.

The total dispersion energy E_d for a polyatomic molecule i composed of atoms i' interacting with a surrounding phase j composed of atoms j' is obtained from Eq. 2.27 by summing all possible interactions $(E_{ij})_d$:

$$E_d = \sum^{i'}\sum^{j'} (E_{ij})_d \tag{2.28}$$

If we replace this summation over all interacting atoms with an integration over interacting volume elements (dV_i and dV_j), the electron polarizability per unit volume α_e^v for compounds i and j replaces α_i and α_j. We can then factor $(\alpha_e^v)_i$ and $(\alpha_e^v)_j$ out of $\sum (E_{ij})_d$, while the summation over i' will be proportional to $\bar{V}_i$, the molar volume of compound i; that is,

$$E_d = C'''\bar{V}_i(\alpha_e^v)_i(\alpha_e^v)_j \tag{2.29}$$

where C''' is approximately constant for all molecules i and j*. Equation 2.29 summarizes how dispersion interaction energies vary in different systems as a function of the properties of molecules i and j. The most important factor is the molar volume of i: dispersion interactions per molecule are roughly proportional to the size of the molecule. In addition α_e^v is related to the refractive index n of the compound through the *Lorentz-Lorentz equation*:

$$\alpha_e^v = \tfrac{3}{4}\pi N\frac{n^2 - 1}{n^2 + 2} \tag{2.30}$$

where N is Avogadro's number.

---

* Although C''' is seen to be proportional also to $\sqrt{I_iI_j}$, values of I do not vary sufficiently among common molecules to affect significantly the constancy of C'''. Similarly, variations in r_i and r_j are taken care of in the integration over volume elements.

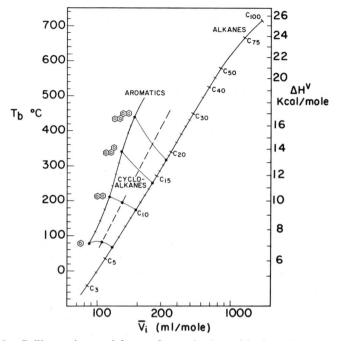

Figure 2.9. Boiling points and heats of vaporization of hydrocarbons versus molar volume.

Dispersion Interactions and Boiling Points

The energy required to remove a molecule i from the pure liquid into the gas phase is a measure of the interactions between molecules i in the liquid phase (gas-phase interactions can normally be ignored). Consequently the boiling points or heats of vaporization of different compounds provide useful information concerning the nature and magnitude of these forces.

Consider first the vaporization of nonpolar molecules, the energy for which is determined almost entirely by dispersion interactions. Figure 2.9 shows a plot of boiling point T_b versus molar volume for various types of hydrocarbons: unsubstituted aromatics, cycloalkanes, and n-alkanes. Heats of vaporization ΔH^v (at the boiling point) are also indicated in Figure 2.9. For nonpolar and slightly polar compounds ΔH^v closely parallels T_b (the *Hildebrand rule*; see Ref. 5). These data illustrate qualitatively the importance of both molar volume $\bar{V}_i$ and volume polarizability α_e^v to ΔH^v and T_b. As measured by the refractive index (Eq. 2.30), α_e^v increases sharply in the sequence n-alkanes, cycloalkanes, and aromatics for constant molar volume.

Table 2.1 shows additional boiling point data for some other nonpolar species. Boiling points for these compounds can be estimated from the curves

Table 2.1. Boiling Points of Some Selected Nonpolar Species

| | | | | T_b | |
| | $\bar{V}_i$ | | | Calc. | Calc. |
Compound	(ml/mole)	n_i	Expt.	(Fig. 2.9)	(Fig. 2.10)
Per-F methylcyclohexane[a]	195	1.285	76	275	41
CCl_4[b]	97	1.460	77	−12	81
Tetrachloroethylene[b]	102	1.505	121	2	108
Bromine[b]	51	1.661	59	−170	50
CS_2[b]	61	1.629	46	−125	65

[a] Cycloalkane curve.
[b] Alkane curve.

of Fig. 2.9 (by taking T_b of the hydrocarbon at the corresponding $\bar{V}_i$), but the resulting values of T_b are relatively unreliable because of differences in α_e^v for these compounds versus the corresponding hydrocarbons. Equations 2.29 and 2.30 suggest that T_b should correlate more closely with the function $\bar{V}_i[(n^2 - 1)/(n^2 + 2)]^2$ than with $\bar{V}_i$, and this relationship is tested in Fig. 2.10. Here the solid curve represents data for the n-alkanes, and we see that typical aromatics and cycloalkanes now fall quite close to this curve. In other words, T_b values for all hydrocarbon types are well correlated by a single function of $\bar{V}_i[(n^2 - 1)/(n^2 + 2)]^2$, as required by Eqs. 2.29 and 2.30 and the Hildebrand rule. We also see in Table 2.1 that T_b values are predicted quite well by the relationship of Fig. 2.10. These examples suggest that the contribution of dispersion forces to T_b and ΔH^v can be estimated reliably from values of $\bar{V}_i$ and n_i (for additional examples, see Refs. 6, 7).

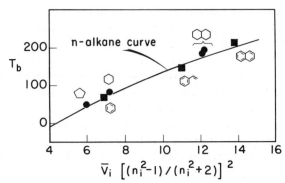

Figure 2.10. Correlation of hydrocarbon boiling points T_b with molar volume $\bar{V}_i$ and refractive index n.

Table 2.2. Boiling Points of Some Polar Compounds—the Role of Dipole Moment and Hydrogen Bonding

Compound	Boiling Point (°C)			Dipole Moment (D[b])
	Expt.	Calc.[a]	Diff.	
Hydrocarbons			0	0
Diethyl ether	35	5	30	1.2
Diethyl sulfide	92	81	11	1.5
Methyl acetate	57	−26	83	1.7
n-Propyl chloride	47	6	41	2.1
Ethyl bromide	38	5	33	2.0
Acetone	56	−31	87	2.8
Nitromethane	101	−46	147	3.5
Acetonitrile	80	−66	146	3.9
o-Dichlorobenzene	179	159	20	2.3
m-Dichlorobenzene	172	159	13	1.5
p-Dichlorobenzene	174	159	15	0.0
cis-1,2-Dichloroethene	60	22	38	1.9
trans-1,2-Dichloroethene	48	22	24	0.0
Methyl alcohol	65	−86	151	1.6
Water	100	−115	215	1.8

[a] From relationship of Fig. 2.10 (dispersion interactions only).
[b] In solution (8).

Table 2.2 extends these correlations to a series of polar compounds. Calculated values of T_b are now seen to be systematically low, implying the importance of intermolecular interactions other than dispersion forces. Indeed the difference between experimental and calculated T_b values is a good measure of the relative importance of these nondispersive interactions, which we now examine.

Specific Interactions

The preceding discussion implies that the separation of compounds having similar values of $\bar{V}_i$ and n_i is difficult, since their boiling points and dispersion interactions per molecule are about the same. However, other attractive interactions exist which can vary markedly among molecules of similar size and polarizability. Because of the selective nature of these nondispersive forces (in contrast to the nonspecific or universal nature of dispersion forces), they are often referred to as *specific* interactions. An understanding of their general nature and potential usefulness in different systems is of the greatest practical importance.

Specific interactions can be subdivided roughly as follows: electrostatic dipole interactions, hydrogen bonding and other electron donor-acceptor interactions, and ionic or covalent bond formation. The physical basis of each of these interactions involves varying contributions from classical electrostatic forces and chemical or covalent bonding, and the corresponding interaction energies can vary from weak physical attractions to strong chemical bonds. We now examine each of these specific interactions.

A molecule with a permanent dipole moment can interact with adjacent molecules in a manner described by classical electrostatic theory. When each of two adjacent molecules possesses a permanent dipole, *dipole orientation* occurs with the positive head of one dipole positioned close to the negative head of the other dipole. For maximum energy of attraction, the two dipole vectors would lie on a straight line, with the positive and negative heads of the dipoles as close together as repulsive forces would allow. However, this optimum alignment of dipoles is opposed by the random thermal motions of individual molecules. A compromise situation results, where at any given time adjacent pairs of dipoles are more often aligned for attraction than for repulsion. This leads on the average to a net attractive energy $(E_{ij})_o$ from dipole orientation (*Keesom forces*; Ref. 5):

$$(E_{ij})_o = -\frac{2\bar{\mu}_i^2 \bar{\mu}_j^2}{3kTr^6} \tag{2.31}$$

Here $\bar{\mu}_i$ and $\bar{\mu}_j$ refer to the permanent dipole moments of adjacent molecules i and j, k is the Boltzmann constant, and T is the absolute temperature.*

A molecule i with a permanent dipole also *induces* temporary dipoles in adjacent molecules j (whether or not j possesses a permanent dipole). A net attractive energy $(E_{ij})_i$ results from this dipole-induced dipole or *Debye interaction* (5), given by

$$(E_{ij})_i = -\frac{\bar{\mu}_i^2 \alpha_j}{r^6} \tag{2.32}$$

When both molecules i and j possess permanent dipoles, the total induction interaction is obtained from Eq. 2.32 by summing terms for each dipole: $-(\bar{\mu}_i^2\alpha_j + \bar{\mu}_j^2\alpha_i)/r^6$.

In the case of pure liquids ($i = j$), Eqs. 2.31 and 2.32 predict increased interaction energies and higher boiling points for compounds with permanent dipole moments (relative to nonpolar compounds). The effect should be more pronounced, the larger the dipole moment. In Table 2.2 the first group of compounds (ending with acetonitrile) illustrates the importance of these

* Equation 2.31 is derived for the case of a dilute gas. In the liquid phase, other observations (7) suggest that $(E_{ij})_o$ is proportional to $\bar{\mu}_i\bar{\mu}_j$, rather than to the square of this quantity.

electrostatic interactions as measured by the differences in experimental and calculated boiling points (based on dispersion forces). As expected, these boiling point differences increase more or less regularly with molecular dipole moment. A perfect correlation is not expected for several reasons. For liquids, Eqs. 2.31 and 2.32 actually apply to individual bond dipoles within the molecule, rather than to the overall molecular dipole moment. Thus an ester group, with a dipole moment of 1.7 D, is composed of a carbonyl bond (2.8 D) and an ether bond (1.2 D), the dipoles of which are so positioned as to cancel partially in the molecular dipole. But adjacent molecules in liquids tend to interact via individual bond dipoles, so that electrostatic interactions in methyl acetate are somewhat larger than we would predict on the basis of its molecular dipole moment.

This bond dipole effect is even more pronounced in the case of molecules with two or more polar substituents, since the substituent dipoles can be far enough apart to interact independently with surrounding molecules. This is illustrated in the second group of compounds in Table 2.2: the isomeric dichlorobenzenes and dichloroethylenes. Despite large differences in the molecular dipole moments among each group of isomers, there is little variation in boiling point or boiling point difference. It is the group dipole moments of these chloro substituents (2.1 D) that control intermolecular interation, not the net dipole moment. Only in the case of the ortho or cis isomers are the two chloro groups sufficiently close for partial addition of the two dipoles, and even here the increase in interaction energy is small. *The molecular dipole moment of a molecule is thus not always a useful measure of its "polarity" or tendency to interact with other "polar" molecules.*

A variety of observations, supported by theoretical calculations, suggest that both induction and dipole orientation play important roles in solution behavior (9). The energy of interaction of a polar molecule with a surrounding nonpolar phase will be generally greater, therefore, than we would expect on the basis of dispersion interactions alone. Consider, for example, the interactions of the solutes propionitrile and propane with the solvents n-hexadecane (nonpolar) and hexadecylnitrile (polar) in dilute solutions. On the basis of molecular size and polarizability, the dispersion interaction energy of propionitrile should be equal in both solvents and similar to that of propane as solute.* In the absence of specific interactions, therefore, the K values for the distribution of propionitrile or propane between either of these solvents and the gas phase should be roughly equal. Gas chromatographic data at 25° (10) show, however, that the K value for propionitrile in hexadecane is about 10-fold greater than that for propane, and about 100-fold greater in hexadecylnitrile. Induction interactions are responsible for the

* The molar volumes and polarizabilities of hexadecane and hexadecylnitrile are about the same, and this is true also for propane and propionitrile.

greater K value (relative to propane) of propionitrile in hexadecane, and dipole orientation interactions account for the further increase in K (relative to propane) for propionitrile in hexadecylnitrile.

The last two compounds of Table 2.2 (methanol and water) show quite large deviations of their experimental boiling points from calculated values, despite rather modest dipole moments (compare these compounds with propyl chloride or ethyl bromide). This behavior is common for compounds with O–H or N–H bonds and has been attributed to *hydrogen bonding* between molecules, for example,

$$CH_3\!-\!O\!-\!H\cdots O\!-\!CH_3 \qquad \overset{\displaystyle H}{\underset{\displaystyle H}{\diagup}}\!N\!-\!H\cdots N\!-\!H\overset{\displaystyle H}{\underset{\displaystyle H}{}} \qquad H\!-\!O\!-\!H\cdots O\!-\!H$$

$$\mathbf{1}\qquad\qquad\qquad\mathbf{2}\qquad\qquad\qquad\mathbf{3}$$

In each case a proton-donor group (e.g., —OH, $\diagup$NH, —SH, H—CCl$_3$) interacts with a proton acceptor, an atom or group with unpaired electrons (e.g., —O—, =N—, —F, —S—, —Cl, $\diagup$C$=$C$\diagdown$, phenyl). The energy of the resulting interaction is usually greatest when the two atoms attached to the bonding proton are nitrogen or oxygen. The basis of hydrogen bonding is largely electrostatic, the normally small O—H or N—H dipole being effectively magnified by the small radius of the hydrogen atom. This allows a close approach of the hydrogen end of the dipole to an interacting proton acceptor, leading to a smaller value of r and a larger interaction energy (Eqs. 2.31 and 2.32).

The various factors that determine the energy of a hydrogen bond are only qualitatively understood (see Refs. 11 and 12). In general, bond strength increases with increasing acidity of the proton-bearing group, and increasing basicity of the proton-acceptor group. The favoured configuration of the hydrogen bond is linear, as in **1** to **3** above. The basicities of the various acceptor groups are roughly proportional to the parameter E_B^* of Table 2.3, decreasing in the following order: amines, neutral oxygen compounds, nitriles, unsaturated hydrocarbons, and sulfides. The acidities of proton donors are similarly proportional to the parameter E_A^* of Table 2.3, decreasing in this order: strong acids, CHCl$_3$, phenol, alcohols, and thiophenol (however, other studies suggest that CHCl$_3$ may be a weaker acid than the alcohols and phenol). For a further discussion of hydrogen bonding, see Section 10.1.

Actually, hydrogen bonding is only one example of a more extensive class of specific interactions: generalized acid-base or electron donor-acceptor reactions of the form

$$A + :B \rightleftharpoons A:B$$

Included in this group are hydrogen bonding, Lewis acid-base reactions (including charge-transfer complexes), and the combination of ions to form salts or complex ions. Pearson (13) has provided a comprehensive qualitative theory of the relative strengths of these acid-base bonds in terms of so-called *hard* and *soft* interactions. Hard interactions are predominantly electrostatic, whereas soft interactions can be regarded as mainly covalent. The acids and bases which can more readily enter into electrostatic interactions are referred to as hard, while soft acids and bases have a greater tendency to form covalent bonds based on orbital overlap. Because of the discrete, fundamentally different nature of these two types of interactions, soft acids prefer to combine with soft bases, and hard acids with hard bases. The characteristics of these various acid and base types are as follows:

Hard acids: small size, high positive oxidation state, no outer electrons that can be excited to higher state.

Soft acids: large size, low or zero positive charge, several easily excited outer electrons.

Hard bases: low polarizability, high electronegativity, hard to reduce, empty orbitals having high energy and hence inaccessible.

Soft bases: high polarizability, low electronegativity, easily oxidized, empty low-lying orbitals.

Table 2.3 provides several examples of hard and soft acids and bases. In general, softness increases regularly as we move down the periodic table within a particular group (e.g., F^- hard, I^- soft).

Hydrogen bonding is a relatively hard acid-base interaction, and strong hydrogen bonding usually requires a hard proton donor and acceptor. Molecular complexation of the picric acid–aromatic hydrocarbon type is a soft interaction, in which such compounds as picric acid, trinitrofluorenone, and tetracyanoethylene play the role of soft acids, and various aromatic compounds act as soft bases. The interactions of silver ion or mercuric ion with organic compounds can be regarded as soft; this explains the relative stability of the Ag^+–olefin, Ag^+–aromatic, and Hg^{2+}–alkyl sulfide complexes.

Selectivity in ion exchange appears to be governed in part by the criterion of hard versus soft interactions. Thus, in cation exchange, the exchange or distribution of ions A^+ and B^+ between a sulfonic acid resin and the surrounding water phase is governed by the relative stability of resin and water complexes of the two ions:

$$\text{Resin-SO}_3^- \ A^+ + B^+ \ (H_2O) \rightleftharpoons \text{resin-SO}_3^- \ B^+ + A^+ \ (H_2O)*$$

The resin-SO_3^- groups can be regarded as bases which compete with water

* Here only a single, exchangeable water molecule is shown of the many that presumably interact with cations in either phase.

Table 2.3. Characteristics of Various Acids and Bases[a]

Acids, in Order of Increasing Softness	E_A^* (hard)	C_A^* (soft)		Bases, in Order of Increasing Softness	E_B^* (hard)	C_B^* (soft)
HF	17.0	0.0		Ketones, R—CO—R	0.7	0.1
CHCl$_3$	5.1	1.0		Alcohols, R—OH	0.8	0.1
Alcohols, R—OH	3.6	0.8		NH$_3$	1.3	0.3
Phenol	4.7	5.7		Amides, R—CO—NH$_2$	1.0	0.3
Thiophenol	1.4	1.7		Sulfoxides, R—SO—R	1.0	0.3
				Nitriles, R—C≡N	0.5	0.2
				Esters, R—CO$_2$—R	0.6	0.2
				Primary amines, R—NH$_2$	1.2	0.6
				Ethers, R—O—R	0.7	0.4
SO$_2$	1.1	7.2		Pyridine	0.9	0.9
				Secondary amines, R$_2$—NH	0.9	0.9
Tetracyanoethylene	1.7	15.0				
Iodine	1.0	10.0				

Acids softness regions: Hard → Intermediate → Soft.
Bases softness regions: Hard → Intermediate → Soft.

Some Other Acids

Hard	Intermediate	Soft
H$^+$, Li$^+$, Na$^+$, K$^+$, Be^{2+}, Mg^{2+}, Al^{3+}, Fe^{3+}, HX	Fe^{2+}, Ni^{2+}, Cu^{2+}, Zn^{2+}, Pb^{2+}	Cu$^+$, Ag$^+$, Hg$^+$, Hg^{2+}, Tl$^+$

Some Other Bases

Hard	Intermediate	Soft
H$_2$O, OH$^-$, F$^-$, Cl$^-$	NO$_2^-$	Br$^-$, I$^-$, SCN$^-$, SO$_3^{2-}$, R—SH, olefins

[a] See Refs. 13 and 14.

44

of hydration in the surrounding aqueous phase for the cationic acids A^+ and B^+. Since water is one of the hardest bases, we expect it to interact preferentially with the hardest acids or cations. As expected, the selectivity or the preferential adsorption of a cation on the resin increases with increasing softness of the cation: $Li^+ < H^+ < Na^+ < K^+ < Rb^+ < Cs^+ < Ag^+ < Tl^+$ (see Table 2.3 and recall that softness increases in going down the periodic table). Similarly, for divalent cations, $Mg^{2+} < Ca^{2+} < Sr^{2+} < Ba^{2+} < Pb^{2+}$. These trends in the stability of ion–ion-exchange resin complexes can be explained in equivalent terms by means of classical electrostatic theory and the charge/radius ratio of the competing ions (see, e.g., Ref. 14 and also Section 12.33).

The interaction energy E_{ab} of these various acid-base interactions can be expressed approximately (12) as

$$E_{ab} = E_A^* E_B^* + C_A^* C_B^* \tag{2.33}$$

Here E_A^* and E_B^* measure the hard (electrostatic) acid and base strengths of a given molecule, respectively, and C_A^* and C_B^* are corresponding measures of soft (covalent) acid and base strengths. Table 2.3 lists values of these parameters for several acids and bases. The separation of a soft from a hard base in an equilibrium distribution is seen to be favored if one of the two phases contains a very hard or soft acid that will interact selectively with one of the two compounds to be separated (i.e., secondary chemical equilibria).

The total energy of interaction of a molecule i with a surrounding phase j is the sum of all possible interactions: dispersion E_d (Eq. 2.29), dipole orientation E_o and induction E_i (from the summation of Eqs. 2.31 and 2.32 for all adjacent molecules j), and acid-base interactions of the type described by Eq. 2.33. We may write this total interaction energy E_{ij} as

$$E_{ij} = E_d + E_o + E_i + E_{ab} \tag{2.34}$$

keeping in mind that the individual interactions are summed over all atoms or groups in the molecule.

For the large, multifunctional molecules that are often encountered in real separation problems, Eq. 2.34 becomes fairly complex. In addition, we generally do not know the numerical values of the various parameters that appear in Eq. 2.34. Furthermore, the equation is approximate at best, and accurate estimates of intermolecular interaction energies are rarely possible. The main value of this equation and the preceding relationships which it summarizes is for qualitative predictions, as well as for correlating and generalizing experimental observations. In addition, Eq. 2.34 serves to pinpoint the main factors that control interaction energy and separation in phase equilibrium and distribution systems: molecular size and polarizability (dispersion interactions), molecular polarity (electrostatic interactions involving permanent dipoles), and the electron donor-acceptor properties of different molecules (Eq. 2.33). When two compounds differ significantly in any of these three properties, their separation is generally straightforward. For such cases the preceding discussion provides a general guide for the selection of the best phases in an equilibrium separation process. Subsequent chapters on individual separation methods will provide specific examples and further details (see, e.g., Section 8.13).

In some cases it may not be obvious that two closely related compounds differ significantly in molecular size, polarity, or electron-sharing properties. However, it is rare for two such compounds to prove inseparable in every equilibrium distribution system. The reason is that virtually every molecular difference necessarily affects the phase distribution properties of the molecule, and modern separation techniques (notably chromatography; see Chapter 5) are capable of taking advantage of extremely small differences in molecular distribution to achieve complete separation of two species. In such cases it is even possible to separate large molecules that differ only in the sign of an optically active center or in an isotopic label, for example, the tritium-labeled compound 4 from its unlabeled derivative (15).

$$
\begin{array}{c}
\text{N} \diagup\diagdown \text{N---H}^3 \\
\text{NH}_2^3 \diagup\diagdown \text{N} \quad \text{N}
\end{array}
$$

4

Equation 2.34 and the preceding discussion emphasize physical interactions, as opposed to strong chemical bonding. A discussion of chemical bonding beyond that given in connection with generalized acid-base reactions is outside the scope of this book. Strong chemical interactions are frequently useful in a separation system (e.g., reaction of acids with bases, Chapter 9). However, we should keep in mind that chemical reaction is often effectively irreversible under the conditions of separation, and irreversibility or slow reversibility can be highly detrimental in an equilibrium distribution system because it seriously limits separation efficiency in multistage processes.*

2.22 QUANTITATIVE MODELS OF THE DISTRIBUTION PROCESS

Having looked at the relationship of intermolecular attraction to molecular structure, we will now re-examine the thermodynamics of certain distribution processes in the light of this and other information. This examination will lead to several useful expressions for predicting the distribution of a given compound between two phases.

---

* In a single stage process, it may often be desirable to invoke strong chemical interactions to achieve separation, for example, selective complex, or precipitate formation.

Ideal Solutions

The ideal solution, like the ideal gas, serves as a starting model for understanding and predicting the behavior of real systems. In a few cases the ideal solution model provides a sufficiently accurate description of the molecular distribution between solutions and other phases. More often, however, it serves as an initial (intentionally rough) estimate of solution behavior, or as a standard for comparing the extent to which different solutions deviate from ideality. We can define an ideal solution as one in which the various system molecules i, j and so forth behave as if they were surrounded by molecules of the same kind; that is, interactions between molecules in an ideal solution are equivalent to corresponding interactions in the pure liquids. Thermodynamically this is equivalent to assuming that there is no heat of mixing ($\Delta H^M = 0$) or change in volume upon mixing ($\Delta V^M = 0$). If all of the various molecules in the solution are of comparable size and shape, it can be shown by certain statistical models (5) that the entropy of mixing of an ideal solution can be written as

$$\Delta S^M = -R \sum^{i} n_i \ln X_i \tag{2.35}$$

where n_i refers to the number of moles of component i, and X_i is the corresponding mole fraction in the final solution. The partial molar entropy of each component i in solution (s_i) relative to the value in pure i (s_i^0) is then obtained from Eq. 2.35 as $(s_i - s_i^0) = -R \ln X_i$. Since ΔH^M is zero for an ideal solution, the latter relationship reduces to $(\mu_i - \mu_i^0) = -T(s_i - s_i^0)$, or

$$\mu_i - \mu_i^0 = RT \ln X_i \tag{2.36}$$

From eq. 2.1 and the definition of γ ($\gamma = a/X$), this is equivalent to $\gamma = 1$ for an ideal solution.

For $\gamma = 1$, Eq. 2.9a for the distribution of a solute i between the solution and gas phases reduces to *Raoult's law*:

$$P_i = X_i P^0 \tag{2.37}$$

Raoult's law is commonly used as a rough approximation for solutions of chemically similar components that do not differ excessively in size. It is often obeyed fairly closely in solutions of nonpolar compounds, where dispersion forces account for most of the intermolecular attraction energies. Separation in vapor-liquid systems that obey Raoult's law is governed strictly by the boiling points of the pure compounds; higher-boiling compounds will be concentrated into the liquid phase at a given temperature. When P_i is experimentally greater than the value predicted by Eq. 2.37, the system is no longer ideal and is said to experience a positive deviation from Raoult's law. Smaller values of P_i are referred to as negative deviations. If we select pure i as the standard state, Eq. 2.37 may be generalized for all solutions (ideal and nonideal) to give

$$P_i = \gamma X_i P_i^0 \qquad (2.37a)$$

For further discussion, see Section 7.2.

The solubility of a solid i in some solvent that forms an ideal solution with i can also be derived from Eq. 2.36. The process of dissolving i can be pictured as occurring in two hypothetical steps; the melting of i to form the supercooled liquid, followed by the mixing of this liquid with the solvent. Assuming that the heat ΔH^F and entropy ΔS^F of fusion of i is temperature independent, we have, at the melting point or fusion temperature T_f, $\Delta H^F = T_f \Delta S^F$ (i.e., $\Delta G^F = 0$). Similarly, at the temperature of interest T, we can write the free energy of fusion as $\Delta G^F = \Delta H^F - T \Delta S^F = \Delta H^F - (T/T_f) \Delta H^F$. The free energy of mixing the supercooled liquid with solvent can be obtained from Eq. 2.35, recalling that ΔH^M is zero (i.e., $\Delta G^M = - T \Delta S^M$). If we assume further that the solubility of i is limited, so that the mole fraction of solvent in the solution is approximately 1, the resulting free energy of mixing simplifies to $\Delta G^M = - RT \ln X_i$. At equilibrium the sum of free energies for these two processes (or the free energy of solution of the solid) is zero; hence $\Delta G^F + \Delta G^M = 0 = \Delta H^F - (T/T_f) \Delta H^F + RT \ln X_i$, and

$$\ln X_i = \frac{\Delta H^F (1/T_f - 1/T)}{R} \qquad (2.38)$$

We see for ideal solutions that solubility always increases with temperature.

Figure 2.11 shows the relationship between ΔH^F and T_f for three general classes of compounds: n-alkane derivatives (hydrocarbons, bromides, and alcohols), alkyl naphthalenes plus disubstituted benzenes (e.g., dinitrobenzene), and alkyl benzenes. Within each group of related compounds the heat of fusion tends to increase with melting point, so that higher-melting compounds are generally less soluble at temperatures below their melting points (Eq. 2.38). The overall scatter of Fig. 2.11 should also be noted, since it implies frequent exceptions to the latter rule. In some cases these exceptions can be exploited in separation processes based on crystallization. For example, the n-alkyl derivatives of Fig. 2.11 tend to have larger ΔH^F values than aromatics of similar melting point, and therefore they will show greater changes in solubility with temperature. This means that the n-alkyl derivatives should tend to crystallize from solution before the aromatic derivatives, other factors being equal (and T_f values comparable). The scatter of Fig. 2.11, in contrast with the close dependence of ΔH^V on T_b (Fig. 2.9), reflects the importance of entropy effects in the crystalline state. These entropy effects are discussed further in Section 2.23.

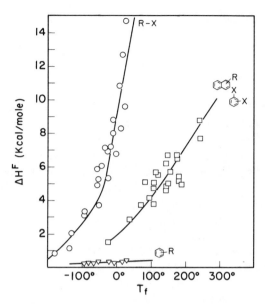

Figure 2.11. Heats of fusion versus melting points for different classes of compounds.

Regular Solutions and Solubility Parameter Theory

In the ideal solution we assume that all intermolecular interactions are equivalent, and the heat of mixing is therefore zero. However, in Section 2.21 abundant evidence was presented that this assumption is often unreliable. In fact, unequal interactions between molecules appear to account for most of the apparent differences in equilibrium distribution systems. The theory of regular solutions (5, 16), focuses on these unequal intermolecular forces and recognizes that the heat of mixing is not necessarily zero. Regular solution theory does assume that the entropy of mixing is ideal (Eq. 2.35) and (in its simplest form) that no volume change occurs on mixing ($\Delta V^M = 0$). It further assumes that only dispersion forces are important; this, as we have seen, is a more serious approximation.

For regular solutions the chemical potential of a component i of the solution is obtained from Eq. 2.36 (ideal solution) by adding the enthalpy term $(h_i - h_i^0)$:

$$\mu_i - \mu_i^0 = (h_i - h_i^0) + RT \ln X_i \tag{2.39}$$

Recalling the definition (Eq. 2.3) of chemical potential in terms of activity a (equal to γX_i), we have

$$\mu_i - \mu_i^0 = RT \ln \gamma X_i$$

which can be combined with Eq. 2.39 to give

$$RT \ln \gamma = h_i - h_i{}^0 = \Delta h^e \tag{2.39}$$

where Δh^e is the partial molar *excess* enthalpy.

We will next relate the enthalpy term $(h_i - h_i{}^0)$ to certain physical properties of the system. We begin by considering the energy of interaction of a molecule i in pure i, versus that of i in a dilute solution of some solvent j. This will lead directly to the heat of mixing of i with j to form a dilute solution. The following derivation (see also Ref. 1) is a simplified version of that generally used in the development of Hildebrand's *solubility parameter theory*, one of the main practical applications of regular solution theory. However, the final expression we obtain will be the same as that provided by the original, more rigorous treatment (see Refs. 5 and 16). It is hoped that the present derivation will facilitate appreciation of the physical basis and limitations of solubility parameter theory.

Solubility parameter theory as initially developed assumes that only dispersion interactions are important in systems which are to be treated. This is a good approximation for solutions of nonpolar or weakly polar compounds, but is less valid for other systems. We will begin by assuming only dispersion interactions; in Chapter 10 we will extend the basic treatment to include the possibility of specific interactions as well.

Consider a molecule i in the pure liquid phase, surrounded by other molecules i, and assume some number (e) of dispersive interactions between i and neighboring molecules. For simplicity assume also that only interactions between nearest neighbors are important, and that i has e nearest neighbors, that is,

Furthermore, assume that each interaction is of constant energy E_{ii}. The energy of vaporization of a molecule i can be visualized as the result of two sequential processes: removal of i from the liquid phase, followed by collapse of the cavity originally occupied by i. This is illustrated in Fig. 2.12 for $e = 6$. The energy of removal is that required for breaking e i-i interactions between i and adjacent molecules: eE_{ii}. Collapse of the resulting cavity requires that each of the molecules originally adjacent to i form a new i-i interaction, leading to the creation of $e/2$ such interactions. The resulting total energy (cavity collapse) is $-(e/2)E_{ii}$, and the net energy of vaporization ΔE_i^V is the sum of the energies for these two processes, times the number of molecules per mole (N):

$$\Delta E_i^V = N\left(\frac{e}{2}\right)E_{ii} \tag{2.40}$$

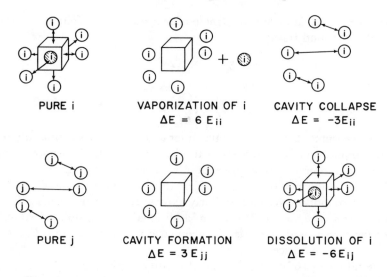

$$\Delta E^M = N\,(3\,E_{ii} - 6\,E_{ij} + 3\,E_{jj}) \cong 3\,N\,(\sqrt{E_{ii}} - \sqrt{E_{jj}})^2$$

Figure 2.12. Hypothetical visualization of the mixing of i with j ($e = 6$); regular solution model.

Now consider the dissolution of the vaporized molecule i in pure j. This process can be viewed as the reverse of vaporization: cavity creation, followed by insertion of i into the cavity. At this point we assume an equal number of intermolecular interactions per unit volume, so that $e/2$ j-j interactions will be broken in creating the cavity (the same number that were formed in the collapse of the previous cavity of equal size). Similarly, e i-j interactions will be formed upon insertion of i into the cavity. If the energies of these individual interactions are defined as E_{jj} and E_{ij}, respectively, then the total energy of dissolution of vaporized i and j (per mole of i) will be

$$\Delta E_{ij}{}^S = N\left[\left(\frac{e}{2}\right)E_{jj} - eE_{ij}\right] \tag{2.40a}$$

Since we assume an equal number of intermolecular interactions per unit volume, we have Ne proportional to $\bar{V}_i$: $Ne = C\bar{V}_i$, where C is a proportionality constant and $\bar{V}_i$ is the molar volume of i. Insertion of this expression into Eq. 2.40 gives

$$\Delta E_i^v = \left(\frac{C}{2}\right)\bar{V}_i E_{ii} \tag{2.40b}$$

Now let us define an important molecular property, the solubility parameter δ_i of compound i:

$$\delta_i^2 = \frac{\Delta E_i^V}{\bar{V}_i}$$

$$= \left(\frac{C}{2}\right) E_{ii} \qquad (2.41)$$

where δ^2 measures the intermolecular interaction energy per unit volume in the pure liquid and is proportional to E_{ii}. Similarly E_{ii} is proportional to E_d (dispersion interactions only), so that from Eqs. 2.29 and 2.41 we see that δ_i should be proportional to the polarizability per unit volume of i (α_e^V) for nonpolar compounds. We can similarly express E_{jj} in terms of the solubility parameter for j: $\delta_j^2 = (C/2)E_{jj}$. An additional assumption of solubility parameter theory is that E_{ij} is the geometric mean of E_{ii} and E_{jj}: $E_{ij} = \sqrt{E_{ii}E_{jj}} = (2/C)\delta_i\delta_j$; see Eq. 2.27a.

The total energy of mixing a mole of i with a large quantity of pure j (to form a dilute solution of i and j) is now given as the sum of ΔE_i^V (Eq. 2.40) and ΔE_{ij}^S (Eq. 2.40a):

$$\Delta E_{ij}^M = N\left(\frac{e}{2}\right)(E_{ii} - 2E_{ij} + E_{jj})$$

$$= \bar{V}_i(\delta_i^2 - 2\delta_i\delta_j + \delta_j^2) \qquad (2.42)$$

Recalling that $\Delta V^M = 0$, so that $\Delta E_{ij}^M = \Delta H_{ij}^M = h_i - h_i^0$, we have

$$h_i - h_i^0 = \bar{V}_i(\delta_i - \delta_j)^2 \qquad (2.42a)$$

Here $h_i - h_i^0$ is the heat of mixing a mole of pure i with a large quantity of j to form a dilute solution. Equation 2.42a is of fundamental importance, not only for its ability to quantitatively describe a considerable range of distribution phenomena, but also for the fundamental insight it provides into the mixing of dissimilar substances. The compound i of Eq. 2.42 can be a solid, liquid, or gas.

When the concentrations of both i and j in solution are significant (i.e. not dilute solutions), Eq. 2.42a is replaced by a similar relationship:

$$\Delta H_{ij}^M = (X_i\bar{V}_i + X_j\bar{V}_j)(\delta_i - \delta_j)^2\phi_i'\phi_j' \qquad (2.42b)$$

where ΔH_{ij}^M is the heat of mixing per mole of final solution, X_i and X_j are the mole fractions of i and j in the final solution, $\bar{V}_i$ and $\bar{V}_j$ are the molar volumes of i and j, ϕ_i' and ϕ_j' are the volume fractions of i and j. Equation 2.42b reduces to Eq. 2.42a when the mole fraction of i becomes small.*

* Solubility parameter theory is less successful when applied to concentrated solutions. For this reason Eq. 2.42b is of more limited value.

According to Eq. 2.42a, deviations from ideality in the mixing of two substances i and j are related to the intermolecular forces in the pure liquids, as measured by the solubility parameter δ. Consequently the solubility parameter of a particular liquid is an important index to its performance in distribution processes. Table 2.4 tabulates δ values for a variety of common solvents (see also Ref. 17). Large values of δ reflect large interaction energies in the pure liquids. In some cases these arise from strong dispersion interactions (e.g., CS_2, CH_2I_2), but usually a large value of δ indicates the existence of strong specific interactions. In these cases it is useful to subdivide the overall solubility parameter into contributions from dispersion, dipole, and hydrogen-bonding interactions (see Sections 8.13 and 10.1).

According to solubility parameter theory, the heat of mixing is always $\geqslant 0$. Therefore ΔH^M is always unfavorable for mixing, becoming more so as the two liquids become less similar in terms of their δ values. This is the physical basis of the well-known rule that "like dissolves like." When the δ values of two liquids become sufficiently different, mixing is largely precluded and two phases can coexist.

The difference in chemical potentials of a compound i in a pure liquid (μ_i^0) and as a dilute, regular solution in j (μ_i) is obtained from Eqs. 2.39 and 2.42:

$$\mu_i - \mu_i^0 = \bar{V}_i(\delta_i - \delta_j)^2 + RT \ln X_i \tag{2.43}$$

Mixing is favored by the entropy term, $RT \ln X_i$, and is opposed by the heat term, $V_i(\delta_i - \delta_j)^2$. The solubility of a liquid i in another liquid j can be calculated from Eq. 2.43, since at equilibrium $\mu_i - \mu_i^0$ is zero. When the solubility of i in j is appreciable, however, the heat term must be replaced by that given in Eq. 2.42a. The activity coefficient γ_i for i dissolved in j can be derived from Eqs. 2.39a and 2.42:

$$\ln \gamma_i = \frac{\bar{V}_i(\delta_i - \delta_j)^2}{RT} \tag{2.44}$$

Table 2.4. Solubility Parameters of Some Common Solvents (at 25°)[a]

Compound	δ	Compound	δ
Perfluorobutane	5.2	Methylene iodide	11.8
n-Pentane	7.1	Acetic acid	12.4
Cyclohexane	8.2	Methanol	12.9
Benzene	9.1	Formamide	17.9
CS_2	10.0	Water	21.0

[a] For additional values, see Table 10.1.

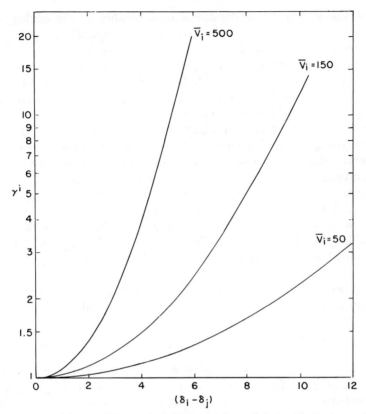

Figure 2.13.   Activity coefficients γ^i of dissolved solutes (infinite dilution) as a function of solute molar volume $\overline{V}_i$ and the solubility parameters of solute (δ_i) and solvent (δ_j).

From the definition of an activity coefficient, $\gamma_i = 1$ in ideal solutions. Thus the definition of an ideal solution in terms of solubility parameter theory is $\delta_i = \delta_j$. Examination of Eq. 2.44 indicates that all deviations from Raoult's law in regular solutions will be positive.

The activity coefficient of a solute is important in describing distribution between two phases. It is useful, therefore, to know how γ varies with solute molar volume and the difference in solute and solvent solubility parameters. Figure 2.13 shows this relationship as calculated from Eq. 2.44.

The solubility of a solid i in a nonideal solvent j can be derived from the ideal solution case (Eq. 2.38) as above. Solubility varies reciprocally with the activity coefficient, so that

$$\ln X_i = \frac{\Delta H^F (1/T_f - 1/T)}{\gamma R} \tag{2.45}$$

According to Eq. 2.44, solubility decreases as the solubility parameters of solute and solvent become increasingly different. This is similar to the case of two liquids.

The distribution of a solute i between two immiscible liquids j and k is defined by the distribution constant $K_x = (X_i)^j/(X_i)^k$. For an ideal system, $K_x = 1$, since the activity coefficients of i in each phase are equal (Eq. 2.8b). For a nonideal system, K_x is equal to γ^k/γ^j and $K = (\bar{V}^k/\bar{V}^j) \times (\gamma^k/\gamma^j)$. Again the activity coefficients can be estimated from Eq. 2.44 or Fig. 2.13 (dilute, regular solutions). When the δ value of the solute is closer to that of solvent j than that of solvent k, K_x will be greater than 1. In other words, i will prefer the phase that is more similar in terms of intermolecular interactions.

The solubility parameter of a solvent mixture (e.g., j plus k) is in general the average of the solubility parameters of the components, weighted according to their volume fractions ϕ':

$$\delta = \sum \phi'\delta \qquad (2.46)$$

for example, $(\phi'_j\delta_j + \phi'_k\delta_k)$ for solvents j and k. It is thus possible to obtain any δ value by mixing two solvents whose δ values bracket the desired value; for example, from Table 2.4 a δ value of 9.5 can be achieved by mixing benzene ($\delta = 9.1$) and CS_2 ($\delta = 10$). Interestingly, Eq. 2.46 predicts that a solvent mixture may exhibit greater solvency for a component i than either of the two components of that mixture. This would be true in the example just given if δ for i were equal to 9.5.

The original derivation of the solubility parameter theory was based on the assumption that only dispersion forces are involved in the various intermolecular interactions. For compounds that also interact by specific forces Eq. 2.42 is generally less accurate. The failure of the geometric mean rule (based on Eq. 2.29 for dispersion forces only) is expected in mixtures of polar and nonpolar molecules, because dipole orientation and acid-base type interactions are precluded between solute and solvent. Failure of the geometric mean rule is particularly pronounced in the interaction of hydrogen-bonding solutes and solvents and other acid-base pairs. Such compound types often mix with the evolution of heat, leading to negative deviations from Raoult's law and activity coefficients of less than 1. This breakdown of the classical solubility parameter treatment can be partially corrected for by an expanded version that recognizes different types of interactions (see Section 10.1 and Refs. 6 and 7).

The Martin Equation: Additivity of Group Interactions

In our discussion of intermolecular interactions in Section 2.21, we saw that the total interaction energy E_{ij} of a molecule i with a surrounding liquid phase j is the sum of interactions for each group $a, b, c, \ldots$ present in the molecule i. For example, consider the case of a dilute solution of propionitrile (i) in n-hexadecane (j). For a propionitrile molecule CH_3—CH_2—$C\equiv N$, E_{ij} is the sum of interaction energies for a methyl group $(E_{CH_3\text{-}j})$, a methylene group $(E_{CH_2\text{-}j})$, and a nitrile group $(E_{CN\text{-}j})$ with surrounding solvent j:

$$E_{ij} = E_{CH_3\text{-}j} + E_{CH_2\text{-}j} + E_{CN\text{-}j}$$

We can generalize this result to any molecule i and any phase j as follows:

$$E_{ij} = \sum^{x} E_{x-j} \tag{2.47}$$

where E_{x-j} refers to the total interaction energy of a group x in the molecule i with surrounding phase j. The summation is taken over all groups x in the molecule i.

For the distribution of i between two phases j and k, the difference in interaction energies of i with each phase $(E_{ik} - E_{ij})$ will be approximately equal to the difference in partial molar enthalpies $(h_{ik} - h_{ij})$, and a similar equality will exist between the *group* interaction energies E_{x-j} and E_{x-k} and the corresponding enthalpy terms h_{x-j} and h_{x-k}. Consequently we can write

$$h_{i-k} - h_{i-j} = \sum^{x} (h_{x-k} - h_{x-j})$$

Combining this relationship with Eqs. 2.8b and 2.39a yields

$$\ln K_x = \frac{\sum^{x} (h_{x-k} - h_{x-j})}{RT}$$

$$\log K_x = \sum^{x} f(x - j, k) \tag{2.48}$$

The group parameter $f(x - j, k)$ equals $(h_{x-k} - h_{x-j})/2.3RT$ and is therefore a function of the group x and the compositions of phases j and k. According to Eq. 2.48, the logarithm of K_x (or K) for distribution of a molecule i between two phases can be expressed as the sum of the group *constants* $f(x - j, k)$ for all groups in the molecule.

For a given set of phases j and k (i.e., a particular separation system), $f(x - j, k)$ is assumed to have the same value for a given group x, whether x is a substituent in different molecules or several groups x are present in the same molecule. This means that the addition of a particular group to any molecule should result in a constant increase in $\log K$, for example, adding —NH_2 to benzene to form aniline, or adding —NH_2 to aniline to form diaminobenzene. Because a large number of different molecules can be formed from a relatively small number of different groups x, Eq. 2.48 is valuable for correlating and predicting K values in distribution systems as a function of the molecular structure of the solute i. Examples of its application in this fashion are given in Chapters 8, 10, and 13. A variation of Eq. 2.48 was first derived by A. J. P. Martin (18) for the correlation of distribution constants in paper chromatography; therefore we will refer to Eq. 2.48 and derived expressions as the *Martin equation*.

The usefulness of the Martin equation in predicting K values is based on the constancy of $f(x - j, k)$ values for a given group x in different molecules or different positions in the same molecule. It should be emphasized, however, that this constancy of the group parameters requires strict comparability of equivalent groups x in regards to their interaction with a surrounding phase. Structural equivalency is not always sufficient in this respect, as illustrated by the two compounds phenol and o-nitrophenol. We might assume that the interaction energies of the —OH group with the surrounding phase are the same in both compounds. However, this interaction energy varies with the dipole moment and acidity of the —OH group. Because of intramolecular hydrogen bonding and resonance interactions between the —OH and —NO_2 groups in o-nitrophenol, the dipole moment and acidity of this —OH group are significantly different from the values in phenol. Consequently $f(x - j, k)$ will not be the same for the —OH groups in these two compounds, leading to an apparent failure of the Martin equation. The Martin equation must be restricted to compounds in which the various groups x are carefully defined, taking into account any adjacent groups that are likely to affect each other's interactions with the surrounding phase. Equation 2.48 is generally reliable for a homologous series CH_3—$(CH_2)_n$—Y, since for $n \leqslant 1$ each additional methylene group is essentially equivalent in regards to its interactions with the surrounding phase. This leads to a general relationship for a homologous series R—Y:

$$\log K = f(CH_3\text{-}j, k) + f(Y\text{-}j, k) + nf(CH_2\text{-}j, k)$$
$$= A + Bn \tag{2.49}$$

Here A and B are constants for a given homologous series and a particular separation system, so that $\log K$ increases linearly with carbon number $(n + 1)$.

Equation 2.48 can be extended to include structural variations in the molecules that constitute the surrounding phase (see, e.g., Ref. 19). The various interactions of phase j with a particular solute group x can be broken down into interactions of x with different groups in a molecule j, in the same way that different groups in i are handled. This approach is much more complicated, however, and therefore less widely used than the Martin equation. So far, it has been applied only to a small number of systems.

2.23   ENTROPY EFFECTS

As we indicated previously, entropy effects are usually less important than enthalpy effects and can often be ignored in separation systems. Even when $T \Delta S$ is comparable in magnitude to ΔH and shows a similar variability with molecular structures of solutes and phases, it frequently happens that these two quantities are linearly related. In these cases the variation of ΔG with molecular structure closely parallels that of ΔH, allowing semiquantitative predictions of K. We will turn now to two major exceptions to the usual unimportance of entropy effects in separation systems: the entropy of fusion, and the entropy of mixing for compounds of greatly dissimilar size.

Entropy of Fusion

The importance of entropy effects in determining the fusion temperature T_f of a compound (and its solubility) is apparent in Fig. 2.11, particularly if we contrast the irregular dependence of T_f on ΔH^F with the close correlation of boiling point with ΔH^V. The entropy of vaporization for many compounds is approximately constant (*Trouton's Rule*); this means that ΔH^V increases regularly with T_b. At the temperature of fusion we have $\Delta G^F = 0 = \Delta H^F - T_f \Delta S^F$, or $\Delta H^F = T_f \Delta S^F$. If, for a particular value of T_f, ΔH^F is larger for one compound than another, then ΔS^F must also be larger. This means that the magnitude of ΔS^F tends to decrease (for a particular value of T_f) in the order n-alkane derivatives > alkyl benzenes. This trend in ΔS^F with molecular structure can be explained in terms of decreasing order in the solid (crystal) phase in going from the n-alkane derivatives to the alkyl benzenes. The more ordered the solid phase, the larger will be the value of ΔS^F, since fusion will then involve a greater increase in disorder and in entropy. For the n-alkane derivatives ΔH^F is maximized by lining up the alkyl chains in a parallel, regular configuration in the solid crystal. However, the achievement of this maximum interaction energy in the crystal requires a high degree of ordering of the individual molecules—the flexible, freely rotatable —CH_2— units must be locked into a discrete configuration. Thus the achievement of a relatively large crystal energy (and value of ΔH^F) for the n-alkane derivatives as a result of this ordering process also requires large values of ΔS^F for a given value of T_f.

In the alkyl naphthalenes and substituted benzenes of Figure 2.11 a large fraction of the molecule is composed of the aromatic ring. In the crystal phase it is advantageous (for maximum interaction energy) for these aromatic rings to overlap, for equivalent interactions of adjacent groups. Because the aromatic ring is a rigid structure in both solid and liquid phases, the degree of ordering required for the overlap matching of —CH= units in the crystal phase is less than in the case of the n-alkanes, and ΔS^F is less for a particular value of T_f. In the case of the alkyl benzenes, the simultaneous ordering of aromatic rings and —CH_2— units is not possible, because the interatomic distances for these two structural groups do not match, and there is no way to segregate aromatic rings from alkyl chains. Consequently the alkyl benzenes (and alkyl naphthalenes for alkyl groups that are large enough) show less ordering in the crystal phase and relatively small ΔS^F values for a given value of T_f.

We should note also the effect of molecular symmetry on the melting point, although this affects ΔH^F rather than ΔS^F.* Molecular symmetry generally favors the optimization of intermolecular interactions in the crystal phase (as in the n-alkanes) and thereby favors higher melting points (all else equal). This accounts for the higher melting points and lower solubilities (Eq. 2.38) of anthracene versus phenanthrene and the para-disubstituted benzenes versus their meta isomers.

The Flory-Huggins Equation—Mixing of Compounds of Different Sizes

Equation 2.35 for the entropy of ideal mixing is based on the assumption of similar sizes and shapes of the involved molecules. This relationship breaks down, therefore, when the molar volumes of the solution components are widely different, as in solutions of polymers in a low-molecular-weight solvent, or in the separation by gas-liquid chromatography of low-molecular-weight solutes with a high-molecular-weight stationary phase (the usual case in gas chromatography). In the limit for large differences in molecular size, the entropy of mixing is often approximated by the Flory-Huggins equation (see Refs. 1 and 5):

$$\Delta S^M = -R \sum^{i} n_i \ln \phi'_i \qquad (2.50)$$

Equation 2.50 differs from Eq. 2.35 for the ideal entropy of mixing only in that mole fraction X_i is replaced by volume fraction ϕ'_i for all solution components i. For solutions of a large solute molecule i in a low-molecular-weight solvent j, or vice versa, Eq. 2.50 reduces to an alternative form:

$$s_i - s_i^0 = -R\left[\ln \phi'_i + \phi'_j\left(1 - \frac{\bar{V}_i}{\bar{V}_j}\right)\right] \qquad (2.50a)$$

where $\bar{V}_i$ and $\bar{V}_j$ are molar volumes of the solute i and the solvent j. The expression in Eq. 2.50a is always positive, indicating that the size-entropy effect is to reduce values of γ_i relative to values calculated in the absence of this effect.

As an illustration of the significance of these size-entropy effects, consider the effect of simply increasing the molecular weight of the stationary phase in gas-liquid chromatography. In the absence of size-entropy effects, assume

* Actually, ΔS^F is also affected, but to a lesser extent.

that K_x remains constant (no enthalpy differences; e.g., two different polymeric stationary phases of the same chemical composition). Since for dilute
solutions of the solute i its concentration $C_i = X_i(\overline{V}_i/\overline{V}_j)$, the distribution
constant K_i should double if the molecular weight of the stationary phase is
doubled. According to Eq. 2.50a, however, size-entropy effects tend to
oppose this increase in K_i. The actual change in K_i as a result of changing the
stationary-phase molecular weight would be less, therefore, than predicted
on the basis of no size-entropy effects.

Temperature Dependence of K

The temperature dependence of K is determined by the partial molar heat
difference Δh^0 for i phases j and k (Eq. 2.10b). When two solutes have the
same K values at one temperature (and hence are inseparable), the possibility always exists that a change in temperature will result in K values that
are no longer the same. For regular solutions, however, and certain other
systems in which Δs^0 is linearly related to Δh^0, compounds having equal
values of $\Delta\mu^0$ (i.e., equal K values) also have approximately equal Δh^0
values. In these cases a change in temperature will not greatly improve
separation. There are two general situations in which equal K values at one
temperature do not correspond to equal K values at all temperatures: solutes
of different shape or size, and solutes whose interactions with one or both
phases are not of the same general type.

We have already seen that Eq. 2.35 for the entropy of ideal mixing is
based on molecules of the same size and shape. Consequently, if two solutes
have K equal at one temperature but are of different shapes, it is likely that
a change in temperature will result in unequal K values and the possibility of
separation. This is particularly true of gas-liquid systems, where this shape-
entropy effect is restricted to one of the two phases (the liquid phase). One
study (20) of temperature effects in gas-liquid chromatography with nonpolar solutes and stationary phase corroborates this generalization nicely. It
was found for compounds of equal K at one temperature that Δh^0 tends to
decrease regularly for more bulky molecules, for example, cycloalkanes versus
alkanes. Thus cycloheptane and n-octane are inseparable at 64° but can be
separated at 58° or 70°. Cycloheptane elutes first at 58°, and n-octane elutes
first at 70°. Similarly alkanes that elute together at one temperature have
higher Δh^0 values in the care of the n-alkanes, and lower Δh^0 values for
highly branched (bulkier) alkanes.

Solutes whose interactions with one or both phases are not of the same
general type usually have one or more functional groups that are not the
same. Thus one group (e.g., —OH) may interact by hydrogen bonding,
another (e.g., —Cl) by dipole orientation, and a third (e.g., —NH_2 +
$H^+ \rightarrow$ —$NH_3{}^+$) by chemical interaction (i.e., secondary chemical equilibria).

Here we often find that compounds with equal K values at one temperature have unequal K values at another temperature, because the temperature dependence of these different interactions is not the same. However, chemically dissimilar compounds are usually easy to separate by a change in the phases of a distribution system, as well as by a change in temperature.

References

1. K. S. Pitzer and L. Brewer, *Thermodynamics*, McGraw-Hill, New York, 1961 (2nd edition of book by G. N. Lewis and M. Randall).
2. K. Denbigh, *The Principles of Chemical Equilibrium*, Cambridge University Press, London, 1966.
3. P. D. Cratin, *Ind. Eng. Chem.*, **60**, 14 (1968).
4. L. O. Case, in *Treatise on Analytical Chemistry*, Part I, Vol. 2, I. M. Kolthoff and P. J. Elving, eds., Interscience, New York, 1961.
5. J. H. Hildebrand and R. I. Scott, *The Solubility of Non-electrolytes*, 3rd ed., Dover Publications, New York, 1964.
6. R. A. Keller, B. L. Karger, and L. R. Snyder, in *Gas Chromatography, 1970*, R. Stock and S. G. Perry, eds., Institute of Petroleum, London, 1971.
7. L. R. Snyder, *Modern Practice in Liquid Chromatography*, J. J. Kirkland, ed., Interscience, New York, 1971.
8. C. P. Smythe, *Dielectric Behavior and Structure*, McGraw-Hill, New York, 1955.
9. E. Meyer and R. A. Ross, *J. Phys. Chem.*, **75**, 831 (1971).
10. E. Kovats and P. B. Weisz, *Ber. Bunsengesell.*, **69**, 812 (1965).
11. R. W. Taft, D. Gurka, L. Joris, P. von R. Schleyer, and J. W. Rakshys, *J. Am. Chem. Soc.*, **91**, 480 (1969).
12. B. B. Wayland and R. S. Drago, *J. Am. Chem. Soc.*, **86**, 5240 (1964); R. S. Drago, G. C. Vogel, and T. E. Needham, *J. Am. Chem. Soc.*, **93**, 6014 (1971).
13. R. G. Pearson, *Science*, **151**, 172 (1966); *J. Am. Chem. Soc.*, **85**, 3533 (1963).
14. F. Helfferich, *Ion Exchange*, McGraw-Hill, New York, 1962, p. 158.
15. P. D. Klein, in *Advances in Chromatography*, Vol. III, J. C. Giddings and R. A. Keller, eds., Marcel Dekker, New York, 1966.
16. J. H. Hildebrand and R. L. Scott, *Regular Solutions*, Prentice-Hall, Englewood Cliffs, N.J., 1962.
17. K. L. Hoy, *J. Paint Tech.*, **42**, 76 (1970).
18. A. J. P. Martin, *Biochem. Soc. Symp.*, **3**, 4 (1949).
19. G. J. Pierotti, C. H. Deal, E. L. Derr, and P. E. Porter, *J. Am. Chem. Soc.*, **78**, 2989 (1956); O. Redlich, E. L. Derr, and G. J. Pierotti, *J. Am. Chem. Soc.*, **81**, 2283 (1959).
20. P. A. Hively and R. E. Hinton, *J. Gas Chromatog.*, **6**, 203 (1968).

Three

DIFFUSION AND MASS TRANSPORT

3.1 Introduction

No separation can take place without transport of mass from one location to another. Mass transport on a molecular scale is called *diffusion* and on the scale of the apparatus is termed *bulk motion*, or more precisely, *convection*. At times, transport on an intermediate scale (e.g., by turbulent eddies) must also be considered. Separation is achieved under conditions in which the molecules of one component transfer to a greater extent than those of another. Such transfers take place by diffusion because other transport processes do not discriminate between molecules of different kinds. Thus diffusion is fundamentally the most important transfer mechanism for the attainment of separation. In most cases, however, bulk flow and convective mass transfer create the necessary concentration gradient for diffusion and are indispensable for carrying out the separation process in practice.

In this chapter we discuss the fundamentals of mass transfer phenomena. Chapter 2 dealt with the *equilibrium aspects* of separation without mentioning how fast a particular phase equilibrium is achieved. This chapter focuses on the *dynamic aspects* which ultimately determine the speed and to a great extent the efficiency of the equilibrium separation process. In addition, several separation techniques are of solely dynamic nature, that is, the separation is achieved by the different transport rates of the components only. Any discussion of these *rate processes*—for example, thermal diffusion, electrophoresis, and centrifugation—must necessarily be based on the underlying mass transport phenomena.

3.2 Diffusion

Mass transport by individual molecular motion is referred to as diffusion. From the point of view of separation, the diffusion of a particular species in a mixture, *interdiffusion*, is of interest, as opposed to *self-diffusion*, in a system containing only identical molecules. The diffusion process is easily demonstrated by the following experiment. Water is poured carefully into a glass cylinder containing a dye solution at the bottom so that no mixing occurs. If the cylinder is left to stand quietly, the dye diffuses into the supernatant water, which gradually becomes colored until after a long period of time the dye is uniformly distributed in the liquid. In this process concentration gradients in the system spontaneously decrease until homogeneity is achieved. Diffusion is thus an irreversible process that demonstrates the second law of thermodynamics, because it leads to maximum entropy by the random distribution of components throughout the solution.

This most common type of diffusion, in which the concentration gradient (more precisely, the chemical potential gradient) is the driving force of molecular motion, is called *molecular* or *ordinary diffusion*. In some separation processes, however, the molecular motion of the species may also be caused by a pressure gradient or a thermal gradient, and these diffusion processes are called *pressure diffusion* and *thermal diffusion*, respectively. If the diffusion is a result of an external force field acting on the component under consideration, it is called *forced diffusion*. The most important processes of this type are *electromigration* and *diffusion in gravitational field*, which are exploited in electrophoresis and in centrifugation, respectively. In flow systems (e.g., chromotography) mass is distributed in the direction of flow because of complex diffusional and flow processes. This phenomenon is called *axial dispersion* in order to reserve the term diffusion for bona fide molecular processes.

3.21 FICK'S FIRST LAW

The diffusion process can be examined in terms of a one-dimensional model as follows. The *diffusional movement of mass* in one direction through a plane is expressed by *Fick's first law*:

$$J_x = -D \frac{dC}{dx} \tag{3.1}$$

where J_x, the flux, is the rate of mass flow per unit area in the x direction (measured in g-moles/cm^2/sec), D is the diffusion coefficient (in cm^2/sec), and C is the concentration of the solute (in g-moles/cm^3). Thus dC/dx is the concentration gradient (in g-moles cm^4) in the direction x, which is normal to the plane. The negative sign indicates that the mass flows in the direction of lower concentration. This equation states that the (mass) flux is equal to the product of the driving force (concentration gradient) and a constant (diffusivity), that is, the flux is directly proportional to the concentration gradient. Similar relationships hold for the flow of momentum (Newton's law) and for the flow of heat (Fourier's law).

Fick's first law rests on an essentially empirical basis and expresses an ideal situation. The diffusion coefficient is not always constant but can depend on the concentration, and the actual driving force is the gradient of chemical potential, which may differ from the concentration gradient. However, in dilute solutions Eq. 3.1 is valid, and we use it as the basis of the following discussion.

Most separation processes can be treated with sufficient accuracy as one-dimensional diffusion processes, because the actual separation occurs as a result of diffusion in a single preferred direction. Forced diffusion by its nature is such a process, because the driving force acts only in one direction and results in the separation of species having different diffusitivities. On the other hand, in equilibrium separation processes mass transfer by molecular diffusion between two phases is important. Then the diffusion in only one direction (normal to the interface) contributes to separation and demands prime consideration.

It should be kept in mind, however, that in a single phase diffusion takes place in every direction in which there is a concentration gradient. Therefore molecular diffusion results in an irreversible isotropic distribution of matter. This three-dimensional nature of molecular diffusion is undesirable from the point of view of separation technology, because it actually leads to mixing instead of separation.

3.22 FICK'S SECOND LAW

Fick's first law is inadequate for the solution of most diffusion problems, because the concentrations are usually unknown. The fundamental equation for the mathematical treatment of diffusion processes is the diffusion equation known as *Fick's second law*, which is written in its one-dimensional form as

$$\frac{dC}{dt} = D \frac{d^2C}{dx^2} \tag{3.2}$$

or generally as

$$\frac{dC}{dt} = D\nabla^2 C \tag{3.3}$$

where t is the time.

Translated into words, this equation states that the time rate of change of concentration is proportional to the spatial rate of change in the direction of the concentration gradient.

When flow and diffusion occur, the time rate of concentration change is given for the one-dimensional case by the following equation:

$$\frac{dC}{dt} = -v\frac{dC}{dx} + D\frac{d^2C}{dx^2} \tag{3.4}$$

where v is the flow velocity in the x direction. This equation is a simple form of the equation of continuity, that is, a mathematical expression of the mass conservation law.

The information conveyed by these equations appears to be meager, because of the quite general nature of partial differential equations. By using appropriate mathematical methods and the specific information available for a given problem, however, we can obtain the solution of that problem. Solutions of the diffusion equation for a large number of problems are given by Crank (2).

3.23   TWO SOLUTIONS OF THE DIFFUSION EQUATION

We next present two solutions of the one-dimensional diffusion equation which are of great practical importance and which demonstrate its application for simple systems. In the first case the diffusion of matter in one direction from a planar source is considered; in the second case, diffusion from a continuous (extended) source. For the sake of simplicity we restrict ourselves to static systems in the mathematical derivation. The results are of interest in many areas because they form a basis for the theoretical treatment of band or front dispersion in elution and frontal chromatography, where, of course, the effect of flow must also be considered.

Diffusion from Planar Source

In the first case, we consider the solute to be present in a thin slab from which it diffuses in both directions away from the slab faces. To solve the diffusion equation for this case we assume that at the start the material is present as a plane of infinitesimal thickness. If D is constant, it is easy to verify that the solution of Eq. 3.2 is given by

$$C = \frac{g}{\sqrt{t}} e^{-x^2/4Dt} \tag{3.5}$$

where C is the concentration expressed in mass per unit length at the position x after time t, and g is an arbitrary constant that can be eliminated as follows. The total quantity of diffusing material M is given by

$$M = \int_{-\infty}^{\infty} C \, dx \tag{3.6}$$

If we now write

$$\xi^2 = \frac{x^2}{4Dt} \tag{3.7}$$

then

$$dx = 2\sqrt{Dt}\, d\xi \tag{3.8}$$

If we rewrite Eq. 3.5 as

$$C = \frac{g}{\sqrt{t}} e^{-\xi^2} \tag{3.9}$$

then Eq. 3.6 becomes

$$M = 2g\sqrt{D} \int_{-\infty}^{\infty} e^{-\xi^2}\, d\xi \tag{3.10}$$

which, on integration, gives

$$M = 2g\sqrt{\pi D} \tag{3.11}$$

Thus

$$g = \frac{M}{2\sqrt{\pi D}} \tag{3.12}$$

and the solution of Eq. 3.2 is

$$C = \frac{M}{2\sqrt{\pi Dt}} e^{-x^2/4Dt} \tag{3.13}$$

which describes the diffusional spreading of M amount of substance present at time $t=0$ in the plane at $x=0$. For unit amount $(M = 1)$ we obtain

$$C = \frac{1}{2\sqrt{\pi Dt}} e^{-x^2/4Dt} \tag{3.14}$$

Equation 3.14 is now compared with the normal (Gaussian) distribution function, which can be written for the concentration distribution when σ is the standard deviation (in length units) in normalized form as

$$C = \frac{1}{\sigma\sqrt{2\pi}} e^{-\frac{1}{2}(x/\sigma)^2} \tag{3.15}$$

The graphical illustration of this function is shown in Fig. 3.1. It is seen that Eqs. 3.14 and 3.15 are equivalent if

$$\sigma^2 = 2Dt \tag{3.16}$$

This equation is commonly known as the *Einstein equation*. In this relationship σ^2, that is, the *variance*, is also called the *mean square displacement* along the x axis.

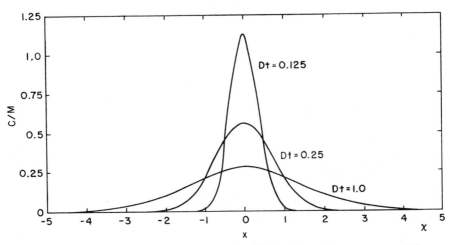

Figure 3.1. Concentration profiles obtained by diffusion from planar source at different Dt values.

The practical significance of the Gaussian distribution and the graphical evaluation of the variance are discussed in Chapter 5.

As a result of the statistical nature of the diffusion process, the spreading of the solute leads to a normal distribution, providing that the molecules are displaced in a random manner very many times in the time period t. Then the variance is the sum of many displacements after time t, the values of which are squared and then averaged.

Figure 3.1 illustrates plots of C/M against x for various values of Dt, constructed from Eq. 3.13. The width of the curves is proportional to the standard deviation σ and is determined entirely by the product Dt. Thus, for example, the same curve is obtained for $D = 10^{-5}$, $t = 10^5$ as for $D = 10^{-1}$, $t = 10$.

The Einstein equation can be used to estimate the "average" time that a molecule needs to diffuse a distance d from its starting point. The root-mean-square average of the diffusion distance in time t_D is the standard deviation σ, and thus d can be set equal to σ. The average time t_D required to achieve such displacement is then obtained as

$$t_D = \frac{d^2}{2D} \tag{3.17}$$

Diffusion from Continuous Source

In the second case we consider the diffusion of a solute from a source that occupies a finite region. For example, in a cylinder two gases are separated by a membrane. We want to calculate the concentration as a function of the

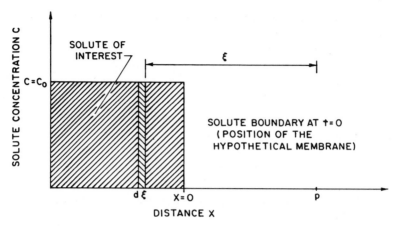

Figure 3.2. Diffusion of solute in direction x from a continuous source.

time and position after the membrane has been carefully removed. Let us consider first an element of the diffusing solute having width $d\xi$ and initial concentration C_0, as shown in Fig. 3.2. This element can be considered as a plane source of amount $C_0 \, d\xi$. From Eq. 3.13 the concentration of this solute at point P (see Fig. 3.2) at time t is given by

$$C = \frac{C_0 \, d\xi}{2\sqrt{\pi Dt}} e^{-\xi^2/4Dt} \tag{3.18}$$

There will be a similar contribution to the concentration at P from every other element $d\xi$ to the left of $x = 0$; therefore the total concentration at P is given by summing over successive elements $d\xi$, that is, by

$$C = \frac{C_0}{2\sqrt{\pi Dt}} \int_x^\infty e^{-\xi^2/4Dt} \, d\xi \tag{3.19}$$

Let

$$\frac{\xi}{2\sqrt{Dt}} = \eta \tag{3.20}$$

so that

$$d\xi = 2\sqrt{Dt} \, d\eta \tag{3.21}$$

Then

$$C = \frac{C_0}{2\sqrt{\pi Dt}} \int_{x/2\sqrt{Dt}}^\infty e^{-\eta^2} \, d\eta \, 2\sqrt{Dt} = \frac{C_0}{\sqrt{\pi}} \int_{x/2\sqrt{Dt}}^\infty e^{-\eta^2} \, d\eta \tag{3.22}$$

Note that, since ξ has been changed to $\xi/2\sqrt{Dt}$, the limit of integration has also been changed to $x/2\sqrt{Dt}$.

Equation 3.22 can be rearranged to

$$C = \frac{C_0}{\sqrt{\pi}} \int_0^\infty e^{-\eta^2} \, d\eta \; - \; \frac{C_0}{\sqrt{\pi}} \int_0^{x/2\sqrt{Dt}} e^{-\eta^2} \, d\eta \qquad (3.23)$$

The terms of this equation can be expressed by the error function, which is given generally as

$$\operatorname{erf} z = \frac{2}{\sqrt{\pi}} \int_0^z e^{-\eta^2} \, d\eta \qquad (3.24)$$

Because this function has the property that $\operatorname{erf} \infty = 1$, we obtain from Eqs 3.23 and 3.24 that

$$C = \frac{C_0}{2} \left(1 - \operatorname{erf} \frac{x}{2\sqrt{Dt}} \right) \qquad (3.25)$$

This equation is usually written as

$$C = \frac{C_0}{2} \operatorname{erfc} \frac{x}{2\sqrt{Dt}} \qquad (3.26)$$

where erfc is referred to as the error-function complement. Values of the error function (the area under a normalized Gaussian curve versus x) are listed in many references. Figure 3.3 shows the change of concentration with distance for various Dt values. The curves are sigmoid in shape, passing from $C = C_0$ at $x = -\infty$ through $C = C_0/2$ at $x = 0$ to $C = 0$ at $x = +\infty$. The

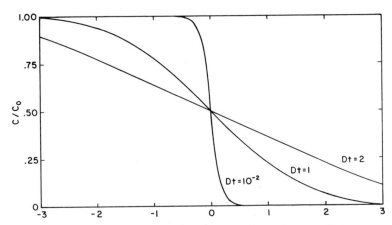

Figure 3.3. Concentration profiles obtained by diffusion from continuous source at different Dt values.

curve rotates counterclockwise with time about the point of inflection at $x = 0$, $C = C_0/2$. Finally, at infinite time, a uniform concentration $C = C_0/2$ is obtained throughout.

Chromatographic Applications

The above solutions of the diffusion equation are applicable also if the source moves in the x direction with a steady velocity. In chromatography, for instance, the spreading of a band (Fig. 3.1) or front (Fig. 3.3) that moves down the column can be treated as a one-dimensional diffusion process. If the center of the band or front moves with a velocity v, we can consider the spreading to take place in a moving coordinate system by making the following transformation:

$$x' = x + vt \tag{3.27}$$

where x' is the position after time t. Then we obtain from Eq. 3.15 that

$$C = \frac{1}{\sigma\sqrt{2\pi}}\, e^{-\frac{1}{2}[(x' - vt)/\sigma]^2} \tag{3.28}$$

and from Eq. 3.26 that

$$C = \frac{C_0}{2}\, \text{erfc}\, \frac{x' - vt}{2\sqrt{Dt}} \tag{3.29}$$

Thus Eqs. 3.28 and 3.29 express the fact that the center of the distribution is not at $x' = 0$ but at $x' = vt$ after time t at flow velocity v. Equation 3.28 is particularly important because it describes the Gaussian concentration peaks which are usually obtained in elution chromatography. Here the solute is injected as a planar source into the stream, moves down the column, and becomes dispersed in a diffusion-like process. At the outlet end of the column, the concentration of the emerging solute is measured and the peak is recorded. The ordinate then represents the concentration C, expressed as amount per volume, and the abscissa is the time t on the recorder chart paper. The Gaussian peak on the chromatogram is essentially the graphical expression of Eq. 3.28. For peak evaluation, however, we have to consider that in Eq. 3.28 x' is equal to the length of the column and the concentration is given in mass per unit length. But on the chart paper we are measuring the time t_R in which the source moving at the peak center passes through the column with the constant velocity v. Therefore $x' - vt$ should be replaced by $t_R - t$ in the exponent, and the standard deviation should be measured in time units, σ_t. The concentration then would be obtained as amount per unit time. In order to convert this concentration into the usual units (amount per unit volume), concentration must be multiplied by the constant flow rate F. For the chromatographic application, therefore, we must modify Eq. 3.28; for M amount injected we then have

$$C = \frac{1}{F}\frac{M}{\sigma_t\sqrt{2\pi}}\,e^{-\frac{1}{2}[(t_R - t)/\sigma_t]^2} \qquad (3.30)$$

The concentration-time relationship can be characterized by the statistical moments of the distribution:

$$\text{Zeroth moment} = \int_0^\infty C\,dt = A \qquad (3.31)$$

$$\text{First moment} = \frac{1}{A}\int_0^\infty Ct\,dt = t_R \qquad (3.32)$$

$$\text{Second moment} = \frac{1}{A}\int_0^\infty Ct^2\,dt - t_R{}^2 = \sigma^2 \qquad (3.33)$$

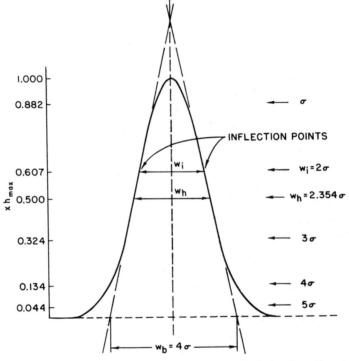

Figure 3.4. The relationship of peak width and standard deviation for a Gaussian peak.

The zeroth moment is the peak area A on the recorder chart paper; A is conveniently obtained by graphical integration of the c-t curve. When the area is known, the injected amount M is given by the product of the area and the (constant) flow rate, according to Eq. 3.30.

The first moment (mean) is identical to the retention time t_R, as can be shown by substituting C from Eq. 3.30 into Eq. 3.32 and integrating. The retention time is obtained as the distance from the starting point of the chromatogram to the peak center. The second moment is the variance and expresses the breadth of the peak. For Gaussian peaks, the square root of the variance or standard deviation is easily obtained graphically. Figure 3.4 shows that the peak width at various fractional heights can be related to the standard deviation. Any one of these relationships can be used for measuring σ. Most often the distance at zero height between the tangents drawn at the inflection points (the so-called baseline band width, $W_b = 4\sigma$) is measured. The maximum concentration, that is, the height of the peak, h, is given by

$$h = \frac{A}{\sigma\sqrt{2\pi}} \tag{3.34}$$

Thus, by measuring the peak area and peak height, the standard deviation can also be evaluated.

The first and second moments are independent of the injected amount, as their expressions (Eqs. 3.39 and 3.40) are normalized to the peak area. Higher statistical moments of Gaussian distributions equal zero.

3.3   Diffusion Rates in Various Media

3.31   DIFFUSION IN GASES

In many separation methods (e.g., gas chromatography and distillation) diffusion in the gas phase is of great interest. Selfdiffusion (interdiffusion of equal molecules) and interdiffusion of similar small gas molecules (e.g., N_2 and CO) are well understood on the basis of the kinetic theory of gases. In practice, however, gas mixtures that contain large molecules and exhibit nonideal behavior are often of importance. The magnitude of the diffusion coefficients is shown in Table 3.1, which is based on experimentally measured diffusivities. It is often difficult, however, to acquire experimental diffusion data. Therefore the reliable estimation of diffusion coefficients in mixtures (and their dependence on temperature and pressure) is significant for the quantitative treatment of many separation processes. A number of formulas have been introduced to approximate the diffusion coefficient D_{AB} for binary gas mixtures. From the kinetic theory, Hirschfelder, Bird, and Spotz arrived at the following equation:

$$D_{AB} = \frac{1.86 \times 10^{-3} T^{3/2}}{P \sigma_{AB}^2 \Omega_D} \left(\frac{1}{M_A} + \frac{1}{M_B}\right)^{1/2} \tag{3.35}$$

where M_A and M_B are molecular weights, T is the temperature in degrees Kelvin, and P is the gas pressure in atmospheres. The value of σ_{AB}, which is measured in angstroms, is interpreted as the minimum distance between the centers of two molecules when they collide with zero initial kinetic energy. The term Ω_D is a dimensionless function of the temperature and of the intermolecular potential field for one molecule of A and one of B. Methods for evaluating σ_{AB} and Ω_D, as well as a detailed discussion of this equation, are given by Bird, Stewart, and Lightfoot (1). This method has a firm theoretical formulation and provides the most reliable estimation of diffusivity data, because it makes possible the prediction of D_{AB} for nonpolar gases within 5%. Because of the difficulties in estimating σ_{AB} and Ω_D, however, its use is cumbersome.

Table 3.1. Diffusivities in Gases at 1 Atm (from Experimental Data)[a]

System	Temperature (°K)	D_{AB}(cm²/sec)
Hydrogen–nitrogen	200	0.401
Hydrogen–nitrogen	273	0.708
Hydrogen–nitrogen	300	0.800
Hydrogen–nitrogen	400	1.270
Hydrogen–nitrogen	573	2.417
Hydrogen–water	307	1.020
Hydrogen–ammonia	293	0.833
Hydrogen–n-hexane	289	0.290
Hydrogen–benzene	311	0.404
Hydrogen–pyridine	319	0.437
Carbon dioxide–water	329	0.211
Carbon dioxide–ethylene oxide	298	0.092
Freon 12–benzene	298	0.038
Freon 12–ethanol	298	0.047
Air–chlorine	293	0.124
Air–water	313	0.277
Air–benzene	298	0.096
Air–butanol	299	0.087
Air–phosgene	273	0.095
Air–chlorobenzene	299	0.079

[a] Condensed from a compilation by E. N. Fuller, P. D. Schettler, and J. C. Giddings, *Ind. Eng. Chem.*, **58** (5), 19 (1966).

A practical method for the calculation of binary gaseous diffusion coefficients was developed by Fuller, Shettler, and Giddings, who obtained the following equation by correlating a large number of diffusivity data:

$$D_{AB} = \frac{1.00 \times 10^{-3} T^{1.75}}{P[(\sum v_i)_A^{1/3} + (\sum v_i)_B^{1/3}]} \left(\frac{1}{M_A} + \frac{1}{M_B}\right)^{1/2} \qquad (3.36)$$

Here T is the temperature in degrees Kelvin, P is the pressure in atmospheres, M_A and M_B are molecular weights. The v_i values are empirical "atomic diffusion volumes" which have to be summed for the molecules of both components to give the pertinent molecular volumes. Table 3.2 contains numerical values for diffusion volumes as obtained from experimental data, which can be used for the calculation of diffusivities.

The above equations are valid at low pressures (a few atmospheres), where the diffusivities are substantially independent of concentration. It is seen that under this condition the diffusivities are inversely proportional to the pressure, that is, the gas density, at constant temperature and increase approximately as $T^{1.7}$ at constant pressure. The viscosity η has a similar temperature coefficient and ideally is independent of the pressure. The ratio of viscosity to

Table 3.2.   Special Atomic Diffusion Volumes[a]

Atomic and Structural Diffusion Volume Increments

C	16.5	(Cl)	19.5
H	1.98	(S)	17.0
O	5.48	Aromatic or heterocyclic	
(N)[b]	5.69	rings	-20.2

Diffusion Volumes of Simple Molecules

H_2	7.07	CO_2	26.9
D_2	6.70	N_2O	35.9
He	2.88	NH_3	14.9
N_2	17.9	H_2O	12.7
O_2	16.6	(CCl_2F_2)	114.8
Air	20.1	(SF_6)	69.7
Ne	5.59		
Ar	16.1	(Cl_2)	37.7
Kr	22.8	(Br_2)	67.2
(Xe)	37.9	(SO_2)	41.1
CO	18.9		

[a] From E. N. Fuller, P. D. Schettler, and J. C. Giddings, *Ind. Eng. Chem.*, **58** (5), 19 (1966).

[b] Parentheses indicate that listed value is based on only a few points.

the product of density and diffusivity is a frequently used dimensionless group termed the *Schmidt number*, Sc:

$$Sc = \frac{\eta}{\rho D} \qquad (3.37)$$

It remains essentially constant under the above conditions. Since viscosity and density data are usually available, the Schmidt number can be used conveniently for crude estimation of the diffusion coefficients at various conditions.

3.32 DIFFUSION IN LIQUIDS

Diffusion processes in liquids play an important role in most separation processes. In contrast to gaseous diffusion, solute diffusivities in liquids are usually strongly dependent on the composition, and their magnitude is determined primarily by solvent properties. Generally, diffusion coefficients in liquids are smaller than those in gases by at least four orders of magnitude (as shown in Table 3.3).

Two crude theories that can be used to approximate diffusivities in liquids relate the diffusivities to the viscosity of the solvent in dilute solutions. The *hydrodynamic theory* is based on the Nernst-Einstein equation:

$$D_{AB} = RT \frac{v_A}{F_A} \qquad (3.38)$$

where D_{AB} is the diffusivity of a solute molecule A through a stationary liquid B, and v_A/F_A is the mobility of the molecule A. The mobility is the steady-state velocity of the particle under the action of unit force. For large, spherical solute molecules, the mobility is obtained from Stokes' law, so that

$$D_{AB} = \frac{RT}{6\pi r_A \eta_B N} \qquad (3.39)$$

where r_A is the radius of the solute molecule, η_B is the viscosity of the solvent, and N is Avogadro's number. This equation is called the *Stokes-Einstein equation*. It can be simplified, assuming spherical molecules with molar volume $\bar{V}_A$, to

$$D_{AB} = \frac{10^{-7} T}{\eta \bar{V}_A^{1/3}} \qquad (3.40)$$

For nonspherical particles a shape factor should be applied, because D_{AB} depends on molecule shape and is, for instance, smaller for cylindrical molecules than for spheres.

The *Eyring rate theory* is based on a simple model of the liquid state which assumes that in a liquid vacancies move among molecules, in contrast to a gas

Table 3.3. Diffusitivities in Liquids for Dilute Solutions at 20°C

Solute (A)	Solvent (B)	D_{AB} (cm²/sec × 10⁵)
O_2	Water	1.80
CO_2	Water	1.77
HCl	Water	2.64
NH_3	Water	1.76
Acetic acid	Water	0.88
Ethanol	Water	1.00
Phenol	Water	0.84
Urea	Water	1.06
Sucrose	Water	0.45
NaCl	Water	1.35
CO_2	Ethanol	3.4
Phenol	Ethanol	0.8
Chloroform	Ethanol	1.23
Phenol	Benzene	1.54
Chloroform	Benzene	2.11
Acetic acid	Benzene	1.92

in which molecules move among vacancies. Therefore liquid molecules are considered to be bound into place by strong intermolecular attractions, and their displacement is considered as an unimolecular reaction that requires activation energy. The diffusion coefficient may then be written as

$$D_{AB} = d^2 \kappa \qquad (3.41)$$

where d is the distance between successive equilibrium positions of the diffusing molecule (cm) and κ is the rate constant (sec⁻¹). If d and κ are approximated from statistical and thermodynamic considerations the Eyring rate theory can be cast into an expression similar to Eq. 3.40.

As an activation process, diffusion has an exponential temperature dependence expressed by the following equation:

$$D_{AB} = D_0 e^{-E/RT} \qquad (3.42)$$

where E is the activation energy of the molecular displacement; D_0 is of the order of 10^{-2} or 10^{-3} cm²/sec and varies only slightly with the temperature. The value of D_0 can be assessed as follows. The reaction rates of unimolecular reactions are invariably of the order of 10^{13} sec⁻¹:

$$\kappa = 10^{13} e^{-E/RT} \qquad (3.43)$$

since the displacement length d can be taken as roughly one molecular diameter, that is 10^{-8} to 10^{-9} cm, the diffusivity is obtained by multiplying by 10^{-16}. Comparing the result with Eq. 3.42, we obtain the right order of magnitude for D_0. Experimental results show that

$$\frac{D_{AB}}{T} \approx \text{constant} \tag{3.44}$$

Hence both diffusion and viscous shear must have very similar exponential temperature dependence, according to Eq. 3.40. Indeed, the activation energy needed to calculate diffusivities at various temperatures by Eq. 3.42 is most conveniently obtained from viscosity data.

Several empirical relations have also been proposed for predicting diffusivities in liquids. The most popular is the *Wilke-Chang equation*:

$$D_{AB} = 7.4 \times 10^{-8} \frac{(\psi_B M_B)^{\frac{1}{2}} T}{\eta \bar{V}_A^{0.6}} \tag{3.45}$$

where $\bar{V}_A$ is the molar volume of the solute A in cubic centimeters per gram-mole (as a liquid at its normal boiling point), M_B is the molecular weight of solvent B, T is the temperature in degrees Kelvin, η is the viscosity of the solution in centipoises, and ψ_B is an "association factor" for the solvent. Recommended values of ψ_B are 1.0 for nonpolar solvents such as benzene, aliphatic hydrocarbons, and ether; 1.5 for ethanol; 1.9 for methanol; and 2.6 for water. This equation gives diffusivities within $\pm 10\%$ for dilute solutions of nondissociating small and medium-size molecules.

Solute diffusivity depends strongly on solvent viscosity, as seen in Eqs. 3.40 and 3.45. Whereas the square roots of the molecular weights of common solvents vary only slightly, the reciprocal viscosities may vary several thousand-fold, thus greatly affecting the diffusivity. The dependence of D_{AB} on solute properties is less pronounced for the usually small solutes, as expressed by the 0.6 power (Wilke-Chang) and 0.33 power (Stokes-Einstein) of the molar volume of solute. The diffusivities of large molecules and macromolecules are of course greatly reduced, as shown in Table 3.4.

It must be stressed that the above considerations are valid only for dilute solutions of small or medium-sized solute and solvent molecules. Large deviations have been observed with high-molecular-weight electrolytic solutes such as proteins, which are hydrated to various degrees depending on experimental conditions, and with solvents consisting of large molecules. Specific solute interactions may also be responsible for irregular behavior. With increasing solute concentration, diffusivity usually decreases because viscosity and nonideality of the mixture increase.

The diffusivity of electrolytes is influenced also by the tendency of cations and anions to migrate with different velocities. A "diffusion potential" is

Table 3.4. Diffusion Constants and Molecular Diameters of Nonelectrolytes[a]

Molecular Weight	Diffusivity (cm²/sec × 10⁵)	Molecular Diameter (Å)
10	2.20	2.9
100	0.70	6.2
1,000	0.25	13.2
10,000	0.11	28.5
100,000	0.05	62.0
1,000,000	0.025	132

[a] From *Chemical Engineers Handbook*, J. H. Perry, ed., 4th ed., McGraw-Hill, New York, 1963, p. 17–44.

generated which retards the motion of the fast-moving ion and accelerates the slow-moving ion until both are moving with equal velocity. In dilute solutions the diffusion coefficient for a binary electrolyte that ionizes into β_+ cations and β_- anions is given by

$$D = \frac{RT}{\mathscr{F}} \frac{\beta_+ + \beta_-}{\beta_+ Z_+} \frac{U_k U_a}{U_k + U_a} \tag{3.46}$$

where U_k and U_a are the mobilities of the cations and anions, respectively, $\mathscr{F}$ is the Faraday number and Z_+ is the electrochemical valence of the cation.

3.33 DIFFUSION IN SOLIDS

Diffusion rates are much smaller in solid materials than in gases or liquids, so that separation processes that require mass transfer in solids are unattractive. According to Eyring's theory, solid diffusion is postulated to be an activated process. Atoms in the solid structure vibrate about their equilibrium positions. A statistical fraction of the atoms have a higher than normal vibration energy and jump to new equilibrium positions or adjacent "holes" in the lattice structure. The value of the diffusion coefficient is proportional to $e^{-\Delta W/RT}$, where W is the activation energy, so that the diffusivity increases rapidly with the temperature. Diffusivities for a few solid systems are given in Table 3.5.

Diffusion in porous solid materials such as ion-exchange resins and membranes is important in many separation processes. The solute diffusion is confined, however, to the pore fluid or to the internal surface in these porous media, as discussed in the next section.

3.34   DIFFUSION IN POROUS MEDIA

Many separation techniques utilize porous media (i.e., gels), and their efficiency is often determined by the diffusion rates in the pores. The diffusion process inside of porous particles or membranes is usually different from that in the bulk phase. Under most conditions the effective solute diffusivity in porous structures is smaller than the bulk diffusivity. A notable exception is the so-called surface diffusion of adsorbable gases in fine pores, which will be discussed separately.

Diffusion within the pores of a solid material is affected in several ways by the geometry of the porous structure and by the interaction of solute molecules with the pore wall. As the cross-sectional area available for diffusion is less than in the same volume of bulk solution, the flux (per unit area of the porous system) is correspondingly reduced for a given concentration gradient. Furthermore, the effective diffusion coefficient is smaller in porous media than in bulk solution because of geometric factors. The most important of these is the tortuosity of the pores, which results in an increased diffusion path length, and the narrowness of the pores, which results in hydrodynamic friction if the diameters of the pores and of the diffusive molecules are similar in magnitude. For examples of different pore structures see Fig. 3.5. When the diffusing molecules are immobilized on the pore wall (e.g., by adsorption or ion exchange), their effective diffusivity is further reduced. In the case of weak sorption, that is, of a linear isotherm, the effective (apparent) diffusivity D_{eff} is given by

$$D_{eff} = \frac{\phi D}{\gamma(K + 1)} \tag{3.47}$$

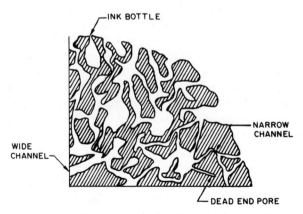

Figure 3.5.   Cross section of a porous structure for some typical pore configuration.

Table 3.5. Diffusion Constants in Solids

Diffusant (A)	Solvent (B)	Temperature (°C)	D_{AB} (cm²/sec)
Phenol	Rubber (weakly cross-linked)	20	1×10^{-8}
Magnesium	Aluminum	395	6×10^{-11}
Silver	Lead	220	1.5×10^{-8}
Gold	Silver	767	3.2×10^{-10}

where ϕ is the fraction of space available to the mobile species, and K is the ratio of solute concentration per unit solid volume to that per unit pore fluid volume, that is, the distribution constant. The diffusivity in the bulk fluid (true diffusivity) is D, and the tortuosity factor is γ, which is usually not larger than $\sqrt{3}$. When the sorption isotherm is nonlinear or irreversible sorption occurs, the measured diffusivity varies with the concentration, and the relationship of the effective and true diffusivities is much more complicated.

When the pore wall carries electric charges, as in ion-exchange resins or membranes, the pore diffusion of the ionic species is also affected by the gradient of the electrical potential. The flux J_{x-e} in the direction of the potential, x is given by

$$J_{x-e} = -U_s ZC \frac{d\psi}{dx} \tag{3.48}$$

where U_s is the electrical mobility of the ionized solute, that is, the rate of solute movement at unit gradient of electrical potential, $d\psi/dx$. The solute concentration in the external solution is C, and the electrochemical valence of the solute is Z. The electrochemical mobility is related to the diffusivity D and can be calculated by the Nernst-Einstein equation:

$$U_s = \frac{D\mathscr{F}}{RT} \tag{3.49}$$

where $\mathscr{F}$ is the Faraday constant, R is the gas constant, and T is the absolute temperature. Electrochemical effects in porous media, including the Donnan potential, which leads to exclusion of certain ionic solutes from the pores, are discussed further in Chapter 12.

Surface Diffusion

Under certain conditions the interaction of diffusing molecules with the pore walls may enhance diffusion rates. At low pressures an adsorbed component of a *gas mixture* transfers significantly faster through porous media having very fine pores than does a nonadsorbed component. This phenomenon is attributed to surface diffusion. If the mean free path length of the diffusing gas molecules is much larger than the channel (pore) diameter, the molecules collide much more frequently with the wall than with each other. As a result, flow or diffusion rates are greatly reduced in this molecular (Knudsen) flow regime, which is discussed in Section 3.64. Under these circumstances, the adsorbed gas molecules can move faster through the channel than the molecules in the gas phase. This phenomenon is considered as diffusion on the surface and is related to a surface diffusion coefficient D_s. Fick's first law is then written as

$$J_{x-s} = -D_s K \frac{dC_s}{dx} \tag{3.50}$$

where J_{x-s} is the flux by surface migration, K is the equilibrium constant for adsorption, and dC_s/dx is the concentration gradient in the adsorption layer. The magnitude of D_s is of the order of 10^{-4} to 10^{-5} cm^2/sec, that is, similar to that of liquid diffusivities but still greater than the diffusivity (or flow) of the nonadsorbed gas molecules.

At low surface concentrations, the surface diffusion process can be considered as a hopping of adsorbed molecules across the surface between adsorption sites. This model leads to an exponential temperature dependence of D_s, similar to that obtained from Eyring's kinetic theory for liquids. Thus, at low surface coverage, surface diffusion is an activation process, because the site-to-site movement of the molecules requires a considerable activation energy, which is roughly half of the heat of adsorption.

At high surface concentrations the adsorbate behaves increasingly like a liquid, it is subject to hydrodynamic laws and if forced through the pores by the pressure gradient. However, the mobility of such liquid films has been found to be lower than that of the bulk liquid. The use of the term surface diffusion is not appropriate for this surface transport phenomenon. Yet its effect is the same: adsorbed molecules move faster through the pore system than unadsorbed molecules. Surface diffusion is particularly important in the separation of gases or vapors by flow through porous membranes. It may also affect mass transfer rates in gas-solid chromatography, in selective adsorption of gases, and in other procedures.

3.4 Thermal Diffusion

The diffusion-like movement of molecules in both gases and liquids under the influence of a temperature gradient is called *thermal diffusion*. Although the effect is quite small, it has been utilized successfully even for production-scale separation processes. Thermal diffusion is well understood on the basis of kinetic theory and irreversible thermodynamics. We will cite only the basic rate law, which relates the flux of a component in a nonisothermal mixture, J_{x-T}, to the temperature gradient, dT/dx:

$$J_{x-T} = \frac{\rho \alpha D}{T} \phi_1' \phi_2' \frac{dT}{dx} \tag{3.51}$$

Here ρ is the density, α is the thermal diffusion constant, D is the molecular diffusivity, T is the temperature, and ϕ_1' and ϕ_2' are fractions of components 1 and 2 in a binary mixture.

The thermal diffusion constant can be approximated by the following expression:

$$\alpha = \frac{105}{118} \frac{M_2 - M_1}{M_2 + M_1} R_T \tag{3.52}$$

where M_1 and M_2 are molecular weights, and R_T is the thermal diffusivity ratio. Experimental data are often reported in terms of R_T, which can be predicted for gases from the temperature coefficient of viscosity.

3.5 Mass Transfer through Interfaces

In many separation processes mass transfer through interfaces is highly important and is often the rate-determining step. As a result, it usually limits the overall efficiency of the separation process. Most commonly the mass transfer through gas-liquid and liquid-liquid interfaces is of interest.

Let us consider the transfer of a solute from phase A into phase B, for example, from a gas into a liquid. The situation is illustrated in Fig. 3.6. It is assumed that solute concentrations in the bulk phases are constant, C_A and C_B, but differ at the interface, C_{As} and C_{Bs}. The concentration differences, $C_A - C_{As}$ and $C_B - C_{Bs}$, provide the necessary driving forces for the solute to diffuse from the bulk phase A to the interface and from the interface to the bulk phase B, respectively.

The flux of the solute through the interface, J_i, can thus be expressed by the following equation:

$$J_i = k_A(C_A - C_{As}) = k_B(C_{Bs} - C_B) \tag{3.53}$$

where k_A and k_B are the so-called mass transfer coefficients for phases A and B, respectively. The mass transfer coefficient, which can be considered as the reciprocal mass transfer resistance, expresses the flux per unit driving force and has the dimension centimeters per second.

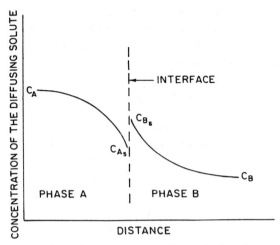

Figure 3.6.   Concentration profiles at the interface.

The measurement of concentrations at the interface is ordinarily impossible, so that k_A and k_B cannot be determined. We assume, however, that complete equilibrium is attained at the interface. Using the equilibrium concentrations C_A^* and C_B^* for the respective interface concentrations, the flux can also be expressed by

$$J_i = k_A^*(C_A - C_A^*) = k_B^*(C_B^* - C_B) \qquad (3.54)$$

where k_A^* and k_B^* are the overall mass transfer coefficients thus defined. As both the flux and the corresponding concentrations are measurable quantities, the overall mass transfer coefficients can be evaluated experimentally. Since the two equilibrium concentrations are related by

$$K = \frac{C_A^*}{C_B} = \frac{C_A}{C_B^*} \qquad (3.55)$$

where K is the equilibrium constant (i.e., the distribution constant), the reciprocal overall mass transfer coefficients, called overall *mass transfer resistances*, are given by the following equations:

$$\frac{1}{k_A^*} = \frac{1}{k_A} + \frac{K}{k_B} \qquad (3.56)$$

$$\frac{1}{k_B^*} = \frac{1}{k_B} + \frac{1}{Kk_A} \qquad (3.57)$$

Thus the overall mass transfer resistance is expressed as the sum of two resistances which are related to the individual mass transfer coefficients.

Although the individual coefficients, k_A and k_B, may be constant throughout the system, Eqs. 3.56 and 3.57 show that the overall mass transfer resistance can vary as a result of variation of K. Therefore the above expressions are valid only if K is indeed constant, that is, at low concentrations and constant temperature.

Although the mass transfer coefficients are of great practical significance, they represent empirical parameters and do not reveal details of the actual interfacial transfer process. A detailed treatment is very difficult because the nature of the interfacial region and the concentration profiles are usually not known. On the other hand, various phenomena in addition to diffusional effects may influence transfer rates. For example, hydrodynamic mixing at the interface may occur and enhance mass transfer, or trace substances may form a barrier at the interface and reduce transfer rates.

In flow systems the transport of a solute from the bulk fluid to the solid surface can be considered as crossing a fluid layer at the surface. The flux is obtained from Fick's first law as

$$J = \frac{D}{\delta_m} (C_A - C_{A_s}) \tag{3.58}$$

where D is the diffusivity of the solute, and δ_m is the fictitious "film" thickness. Then the mass transfer coefficient is given by

$$k_A = \frac{D}{\delta_m} \tag{3.59}$$

The assumption that the mass transfer coefficient is directly proportional to diffusivity has been disproved by experimental data. Nevertheless, even this very simple film theory is useful in practice to estimate k_A by the value of the effective film thickness δ_m, which can be obtained from correlations published in the chemical engineering literature or by experiment.

3.6   Fluid Flow

Most separation processes require mass transport by bulk fluid flow of gases or liquids, for example, from one stage to another in an extraction cascade. Very often the flow is an integral part of the separation process, particularly in differential column processes (e.g., in chromatography). In such cases the flow pattern, that is, the detailed "structure" of the flow, may have a greater influence on the efficiency of the system than any other phenomenon with the exception of diffusion. In essence, the role of flow is not only that of transporting the components through the system, but also that of facilitating diffusional interphase mass transfer by maintaining adequate concentration gradients at the interface.

The quantitative description of fluid flow is the subject of fluid dynamics, a highly developed applied science. In this chapter only some basic concepts that are of particular interest in connection with laboratory-scale separations will be introduced.

3.61   LAMINAR AND TURBULENT FLOW THROUGH TUBES

The following discussion is restricted to Newtonian fluids and to circular tubes. In most separation processes, the gaseous or liquid solutions of interest are Newtonian, that is, their viscosity is constant at a given temperature and is independent of rate of shear and time of application of shear. This does not apply, however, to many polymer solutions and dispersions, which are called non-Newtonian fluids.

The flow through tubes can be laminar or turbulent, depending on the magnitude of the dimensionless Reynolds number Re, which is given by

$$\text{Re} = \frac{\rho v d}{\eta} \tag{3.60}$$

where v is the fluid velocity (cm/sec), d is the tube diameter (cm), ρ is the fluid density (g/cm^3), and η is the fluid viscosity (poise). Physically the Reynolds number reflects the relative strength of the inertial and viscous forces in the moving fluid. Inertial forces increase with density and with the square of velocity (ρv^2). Viscous forces increase with viscosity and velocity and decrease with increasing tube diameter ($\eta v/d$). The Reynolds number is simply the ratio of these two terms.

At low Reynolds numbers (Re < 2100), viscous forces prevail and the flow is called *laminar* or *viscous*. Because inertial forces increase more rapidly with velocity than do viscous forces, inertial forces become dominant, above this critical Reynolds number and turbulence occurs. Whereas in laminar flow the viscous forces fix the fluid elements in well-defined streamlines, turbulent flow is characterized by a random fluctuation of the fluid elements in the form of eddies.

Figure 3.7 illustrates the flow profile in laminar flow through circular tubes, which can be expressed mathematically in a simple form. The velocity of a stream path v_x at a radial position r_x in a tube having a radius r is given by

$$v_x = 2v \left[1 - \left(\frac{r_x}{r} \right)^2 \right] \tag{3.61}$$

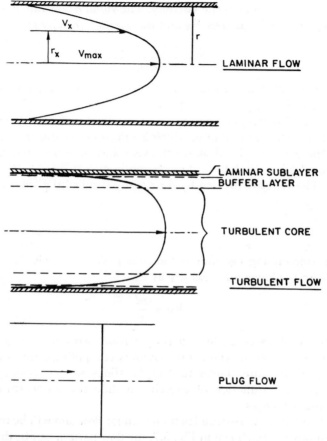

Figure 3.7. Flow profiles in tube flow. The parabolic flow profile is characteristic for laminar flow. The turbulent flow profile is flatter because of radial mixing. Plug flow is unattainable in practice but often represents a convenient model.

where v is the mean velocity. As this is the equation of a parabola, the velocity profile is parabolic. The maximum velocity $v_{\max}$ occurs at $r_x = 0$ (i.e., in the center of the tube) and has the value

$$v_{\max} = 2v \tag{3.62}$$

The mean velocity is given by

$$v = \frac{(P_i - P_o)r^2}{8\eta L} \tag{3.63}$$

where P_o and P_i are the outlet and inlet pressures, respectively; L is the length of the tube; and η is the viscosity. The volume flow rate F is the product of the cross-sectional area $r^2\pi$ and the average velocity v:

$$F = \frac{(P_i - P_o)r^4\pi}{8\eta L} \tag{3.64}$$

This result is called the *Hagen-Poiseuille equation*, and the parabolic flow is often termed *Poiseuille flow*.

As mentioned before, viscosity plays a very significant role in maintaining laminar flow. The cgs unit of η, which is grams per centimeter per second, is called the poise. Table 3.6 shows the viscosities of some common fluids in centipoises (1 cP $= 10^{-2}$ P) for liquids and in micropoises (1 $\mu P = 10^{-6}$ P) for gases. The viscosity divided by the density is called the *kinematic viscosity*, v:

$$v = \frac{\eta}{\rho} \tag{3.65}$$

and its corresponding cgs unit is the stokes. We can write the Reynolds number with the kinematic viscosity as

$$\mathrm{Re} = \frac{vd}{v} \tag{3.66}$$

The viscosity of gases at low density *increases* with temperature and is independent of pressure, whereas the viscosity of liquids usually *decreases* with increasing temperature. Theoretical explanations of these phenomena and formulas for the estimation of viscosity at different temperatures can be found in the references.

Turbulent flow is more complex than laminar flow and will be treated here only qualitatively. As shown in Fig. 3.7, the flow pattern can be divided into three regions. At the wall, in a thin fluid layer called the laminary sublayer, the flow is assumed to be laminar. In the center of the tube a random motion of eddies, which has a radial component and is responsible for the flat flow profile, prevails. This region is called the turbulent core. The transition between the turbulent and laminary flow regions occurs in the buffer layer. The higher the fluid velocity, the more the turbulent core is extended in the radial direction, that is, the flatter is the overall flow profile. Because of dissipation of energy in the eddies and crosscurrents, the energy requirement for maintaining turbulent flow is larger than that for laminar flow. At very high flow velocities the flow profile flattens and is often considered uniform, as shown in Fig. 3.7. This idealized type of flow is termed plug flow.

The mechanism of radial mass transfer is greatly different in laminar and in turbulent flow. In laminar flow the fluid elements are fixed in stream paths which have different velocities. The molecules in the fluid can exchange

Table 3.6. Viscosities of Gases and Liquids at Atmospheric Pressures

Gas	Temperature (°C)	Viscosity (μP)	Liquid	Temperature (°C)	Viscosity (cP)
Nitrogen	0	166	Water	4	1.5674
Nitrogen	50	188	Water	20	1.0050
Nitrogen	100	208	Water	38	0.6814
Nitrogen	200	246	Water	60	0.4688
Nitrogen	400	311	Acetone	20	0.322
Argon	50	242	Benzene	20	0.647
Helium	50	208	Cyclohexane	20	0.97
Carbon dioxide	50	162	Hexane	20	0.320
Oxygen	50	218	Ethanol	20	1.194
Hydrogen	50	94	Carbon tetrachloride	20	0.97
Methane	20	109	Glycerol	20	1069.0
Isobutane	23	76	Sulfuric acid	25	19.15
Mercury	308	654	Mercury	20	1.547

stream paths only by molecular diffusion. In turbulent flow, however, the radial movement of the molecules is facilitated by the eddies and flow, so that in the bulk of the fluid (in the turbulent core and in the buffer layer) mass transfer occurs by this so-called *eddy diffusivity*, and the role of molecular diffusion is actually negligible. Turbulent diffusivity increases roughly in proportion to the mean flow velocity.

The strong radial mixing in turbulent flow is highly important in many large-scale separation processes. In analytical separations, for example, in chromatography, laminar flow occurs most commonly.

3.62 FLOW THROUGH PACKED BEDS

Packed beds (columns) are frequently employed in separation processes such as adsorption, chromatography, and distillation. The structure of the packing is usually quite complex, and the resulting complicated flow pattern in packed beds is not amenable to rigorous theoretical treatment.

The cross section of a packed bed is shown in Fig. 3.8. The arrows illustrate the tortuous paths of the fluid elements through the interstitial space of the bed, which consists of channels (pores) of various diameters. In well-packed columns the diversity of channel diameters and of velocities in the individual channels is small. Then the packed bed can be approximated as a bundle of tortuous capillary tubes. In practice, some wide-diameter channels and gaps in the packing structure may be present where the local flow velocity is relatively low. This results in the undesirable phenomenon of so-called *channeling* of the flow.

The fundamental principle governing the flow of fluids through packed beds is *Darcy's law*, which states that the flow velocity is proportional to the pressure gradient:

$$v_0 = \frac{B^o}{\eta} \frac{P_o - P_i}{L} \tag{3.67}$$

where B^o is the *specific permeability coefficient*, and v_0 is the so-called *superficial velocity*, the average linear velocity the fluid would have in the column if no packing were present. It is calculated by dividing the volume flow rate by the cross-sectional area of the empty tube. Comparison of Eqs. 3.63 and 3.67 shows that the specific permeability coefficient for open tubes is equal to $r^2/8$.

The free cross section of the bed is expressed by the *interparticle porosity* ε. Random packing of equal-size particles usually results in $\varepsilon = 0.4 \pm 0.03$. The *total porosity* of beds packed with porous particles is of course larger

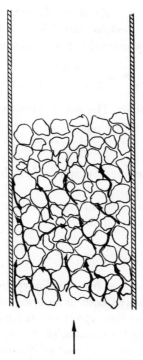

Figure 3.8. Flow through packed bed. The exact flow profile depends on the packing structure and is not known, but in practice this type of flow is often considered plug flow.

because of the *intraparticle space*. The true average fluid velocity is obtained, from Eq. 3.67 as

$$v = \frac{B^o}{\varepsilon\eta} \frac{P_o - P_i}{L} \qquad (3.68)$$

The dimension of the specific permeability B^o is square centimeters but is often given in darcy units (1 darcy = 10^{-8} cm^2).

The hydraulic radius concept is frequently used to calculate flow through channels of different geometry. The hydraulic radius r_h is defined in the following way:

$$r_h = \frac{\text{volume available for flow}}{\text{surface area of particles in contact with fluid}}$$

and the average flow velocity is expressed as

$$v = \frac{(P_o - P_i)r_h^2}{2\eta L} \qquad (3.69)$$

Several equations have been derived to relate the specific permeability to the particle diameter and the bed porosity. The best-known expression is the *Kozeny-Carman equation*, which gives the specific permeability as

$$B^o = \frac{d_p^2 \varepsilon^3}{180(1 - \varepsilon)^2} \tag{3.70}$$

where d_p is the particle diameter. The average fluid velocity is then given by

$$v = \frac{d_p^2 (P_o - P_i) \varepsilon^2}{180 L \eta (1 - \varepsilon)^2} \tag{3.71}$$

This equation is valid for laminar flow and for beds having porosity less than 0.5.

For packed beds the Reynolds number is calculated with particle diameter substituted for tube diameter in Eq. 3.60:

$$\mathrm{Re} = \frac{\rho v d_p}{\eta} \tag{3.72}$$

Turbulence and the transition from laminary to turbulent flow is not nearly as well defined in packed beds as in open tubes. It is assumed that turbulence in packed beds develops gradually as Re increases from 1 to 100. Actually, even at low Reynolds numbers, in packed tubes there is a lateral movement of the fluid elements as a result of stream splitting (see Fig. 3.8). At high flow velocities this amounts to a substantial "convective diffusivity," which is analogous to the eddy diffusivity in turbulent flow. The flow profile can then be approximated as plug flow.

The most uniform flow profile can be obtained when beds are packed carefully with spherical particles of equal size. If the ratio of the tube diameter to the particle diameter is less than 100, this ratio may have a significant effect on the flow profile.

3.63   CAPILLARY FLOW IN PAPER AND THIN-LAYER CHROMATOGRAPHY

The driving force of the eluent flow in paper sheets and adsorbent layers used in paper and thin-layer chromatography is the capillary pressure. Whereas the rise of liquids in capillary tubes is well undestood, the dynamic capillary flow process in porous media can be described only approximately, as the structure of the medium, consisting of interconnected tortuous capillaries of various dimensions, is very complex. The situation is further complicated when a solvent mixture is used, because the properties of the liquid may change because of sorption and evaporation.

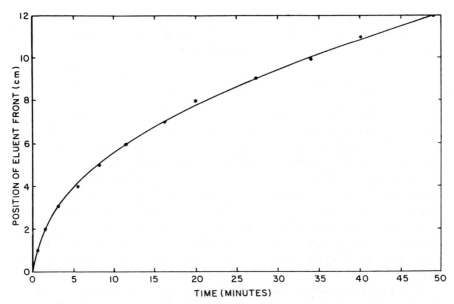

Figure 3.9. Movement of the element front on a thin-layer plate. Adsorbent: 0.3-mm silica gel layer; eluent: chloroform.

Experimental findings indicate that the velocity of the moving eluent front is inversely proportional to the distance advanced by the eluent front. When the position of the front is plotted against time, as shown in Fig. 3.9, a parabola is obtained. Thus the position of the eluent front is proportional to the square root of time.

In order to describe these effects, Giddings (3) and his coworkers used a simple model in which the porous layer is assumed to be a bundle of uniform parallel capillaries. For each capillary the driving force of flow is the capillary pressure, which is proportional to the surface tension γ and is counteracted by the viscosity η. Then the flow velocity (rate of advance of the solvent front located at z_f), dz_f/dt, is proportional to the driving force (surface tension) and inversely proportional to the restraining force (viscosity) multiplied by the length of capillary, z_f, through which liquid is being drawn. Thus

$$\frac{dz_f}{dt} = \text{constant} \times \frac{\gamma}{\eta z_f} \tag{3.73}$$

Integrating from time 0 to t over the flow path from 0 to z_f, we obtain

$$z_f{}^2 = \phi t \tag{3.74}$$

where ϕ is proportional to γ/η and to the radius of the capillaries. Equation 3.73 expresses the parabolic character of the relationship between the advancement of the eluent front and time. In practice, the effect of gravitation forces and of the geometry of the layer has also to be considered, and the situation is more complicated than implied by Eq. 3.73.

3.64   SLIP FLOW AND MOLECULAR FLOW

Table 3.7 shows the mean molecular path lengths of several gases. In many porous materials the diameter of pores are of the same order of magnitude or are smaller than the mean free paths of the gas molecules. Under these conditions the Hagen-Poiseuille equation (Eq. 3.64) does not adequately describe the flow through the pores. If the mean free path lengths of the gas molecules are of the same order of magnitude as the pore diameter, the flow velocity at the pore walls is not zero because the molecules *slip* at the wall, and even at zero pressure drop there is a finite *slip flow*. As a consequence the flow rate through the pores is higher than expected from the Poiseuille equation, and a slip flow correction term has to be added to the viscous flow term.

When the mean free path is very large compared with the tube radius, the flow is called *molecular flow* or *Knudsen flow*. The flow rate of gas, F_K, through a capillary is

$$F_K = \frac{4}{3} \sqrt{\frac{2\pi RT}{M}} \frac{r^3}{h} \frac{\Delta P}{P} \tag{3.75}$$

where R is the gas constant, T is the temperature, M is the molecular weight of the gas, r is the radius, and h is the length of the capillary. The pressure drop is ΔP, and P is the pressure at which the flow rate is measured. According to the kinetic theory, the mean free path λ is given by

$$\lambda = c \frac{\eta}{P} \sqrt{\frac{RT}{M}} \tag{3.76}$$

Table 3.7.   Mean Free Paths of Gas Molecules at 1 Atm and 20°C

Gas	Viscosity (μP)	Mean Free Path (nm)
Carbon dioxide	144	40
Nitrogen	173	60
Methane	108	50
Argon	220	70
Helium	193	190
Hydrogen	87	120

where c is constant approximately equal to z, η is the viscosity, and $\bar{P}$ is the mean pressure. The Knudsen equation can be written with λ measured at the mean pressure as

$$F_K = \frac{4}{3}\sqrt{2\pi}\,\frac{\lambda}{c\eta}\,\frac{r^3}{h}\,\bar{P}\,\frac{\Delta P}{P} \qquad (3.77)$$

References

1. R. B. Bird, W. E. Stewart, and E. N. Lightfoot, *Transport Phenomena*, John Wiley, New York, 1960.
2. J. Crank, *The Mathematics of Diffusion*, Oxford University Press, London, 1956.
3. J. C. Giddings, *Dynamics of Chromatography*, Part I, Marcel Dekker, New York, 1965.
4. J. O. Hirschfelder, C. F. Curtis, and R. B. Bird, *Molecular Theory of Gases and Liquids*, McGraw-Hill, New York, 1958.
5. M. H. Jacobs, *Diffusion Processes*, Springer, New York, 1967.
6. W. Jost, *Diffusion in Solids, Liquids, Gases*, Academic Press, New York, 1965.
7. E. A. Moelwyn-Hughes, *Physical Chemistry*, 2nd ed., Pergamon Press, London, 1961.
8. R. C. Reid and T. K. Sherwood, *The Properties of Gases and Liquids*, McGraw-Hill, New York, 1958.
9. A. E. Scheidegger, *The Physics of Flow through Porous Media*, Macmillan, New York, 1960.
10. R. E. Treybal, *Mass Transfer Operations*, McGraw-Hill, New York, 1968.
11. R. M. Barrer, *Diffusion in and through Solids*, Cambridge University Press, London, 1951.

Four

OPERATIONAL ASPECTS OF SEPARATION

The classification of separation processes in Tables 1.1 and 1.2 is based on the fundamental physicochemical phenomenon by which separation is achieved. The great variety of such phenomena has led to a large number of fundamentally distinct classes of separation methods. In practice, any of the physicochemical phenomena can be exploited for separation under a variety of operational conditions, thus leading to different separation processes. It is useful, therefore, to attempt a classification also from the operational point of view. This chapter will discuss in a general fashion some of the basic schemes and concepts on which separation processes are based. In this way, common features of seemingly unrelated processes can be recognized, and a number of terms that will be used later in this book are introduced and defined.

4.1   Terminology and Classification

Separation processes are carried out on two different scales, depending on the amount of material to be separated or purified: *laboratory* and *production scale*. Laboratory-scale separations can be divided further into *analytical* and *preparative* separations. In analytical work the separation of a small sample is followed by the qualitative or quantitative determination of the separated component or components, and the recovery of the material after analysis is of secondary importance. On the other hand, the aim of preparative-scale separations is the preparation of pure substances and concentrates, including the purification of reagents and solvents for bench-scale work. Separation on the production scale involves the handling of large quantities for commercial purposes. The amount of separated material can be less than a few milligrams in analytical separations and is usually less than a few kilograms in preparative separations. The separation of quantities larger than a few kilograms usually involves production-scale procedures. These quantity limits are somewhat arbitrary, and exceptions occur. For example, the isolation of a few hundred grams of a valuable drug may require a plant-scale process. Conversely, analysis for a trace component may require 1 or more kilograms of sample (Section 19.3).

This book is aimed primarily at laboratory-scale separations, with emphasis on analytical applications. For the sake of completeness, however, frequent reference is made to production-scale operations. Readers interested in the theory and practice of industrial separations should consult the chemical engineering literature given as references at the end of chapters that deal with the individual separation processes.

A major factor that determines how to execute a particular separation is the number of components into which a mixture must be separated. In *binary separations* the mixture is separated into two parts. If the mixture contains only two components (binary mixture), then both are obtained in pure or enriched form. If the original mixture contains several components (multicomponent mixture), then one component is separated from all others in this way. *Purification* is normally a binary separation whereby the *main component* of the mixture is obtained in pure form and the other minor components are removed together. In the laboratory, three of the most common binary separations are crystallization, solvent removal, and continuous water distillation. In industry, distillation and extraction are used most commonly for binary separations. The two products thus obtained are called *light* or *top product* and *heavy* or *bottom product* in distillation, and *raffinate* and *extract* in extraction. These terms are applied sometimes to other processes as well. In *multicomponent separations* the mixture is separated into more than two fractions; such processes are often called *fractionations*. A number of laboratory separation methods, such as chromatography, distillation, and electrophoresis, are particularly suitable for multicomponent separations. In industry, this type of separation requires elaborate processes that are most highly developed in the field of distillation.

Separation processes can be divided into *steady-state* and *unsteady-state* processes. Steady-state processes are characterized by constancy of conditions at a particular point of the system at any time, although conditions are not constant throughout the system. Unsteady-state processes are transient because the process is not invariant in time either. Thus *batch* separations, where the entire quantity of material is charged to the equipment in which the separation is performed, are transient processes. In many unsteady-state separation processes, such as chromatography, fixed-bed adsorption, or ion exchange, there is a continuous flow of one phase through the apparatus. Such processes are called *semibatch* operations. For laboratory work, batch-wise operation is the most economical because it is relatively simple to carry out, and the same equipment can be used for the separation of a variety of substances under appropriate conditions. On the production scale, *continuous* processes are preferred because of their greater productivity. Here the material to be separated is introduced continuously into the equipment, and the product streams are withdrawn continuously. When large quantities are processed, the cost of separating a given amount is generally lower in continuous operation than in batch operation. Since batch processes are generally easier to apply to multicomponent separations than continuous processes, which require a relatively large investment in process design and equipment, batch separations are used quite commonly in industry as well as in the laboratory. Whereas steady-state processes are by definition continuous, continuous processes are not necessarily carried out in steady state. Indeed, recent developments show that the efficiency of a continuous process can be improved by periodically changing flow rate, concentration, or other parameters. This method is called unsteady-state processing, but the same name should not be mistaken for batch processing.

Separation processes can also be distinguished by the design of the equipment, for example, either *stagewise* (discrete equilibrium contact) or *continuous-contact* (differential-contact) operations. Simple separations can be carried out in a *single stage* (e.g., separating funnel, dialyzing bag, or flash evaporator) in order to achieve a desired enrichment. More often, however, the change in composition that can be achieved in a discrete stage is insufficient and several stages are connected in series to form a *cascade*. Distilling columns with trays, the Craig machine (Section 4.8), and gaseous diffusion plants exemplify such *multistage* systems, in which the composition change is amplified from stage to stage. The number of stages necessary to perform a separation depends on the "difficulty" of the separation, expressed usually by a separation factor (Eq. 2.22), and on the efficiency of the individual stages. Continuous-contact operations are carried out in *differential contactors* in which the composition does not change stepwise as in a cascade but continuously, and the degree of separation is gradually augmented down the column. Packed columns used, for instance, in distillation and chromatography, as well as thin layers of paper sheets, used in chromatography and electrophoresis, are examples of differential contactors. In a recently invented separation technique, called parametric pumping, an amplification effect can be achieved in a single stage. Here the mixture is pumped periodically from one end of the stage to the other through a sorbent bed, and this cyclic mass displacement is coupled with a synchronous cycling of the bed temperature or another parametric variable that affects the equilibrium distribution of the components. In the course of this process, the components of the mixture are separated as their concentrations increase gradually at one or the other end of the stage.

The most important classification of separation processes from the operational point of view is based on the *mode of operation*, represented by the fundamental flow sheet arrangement of the process. Fortunately, the various modes of operation can be reduced to a few basic schemes which are usually easy to recognize; (*a*) single-contact (cocurrent) processes; (*b*) differential processes; (*c*) crosscurrent processes; (*d*) countercurrent processes; (*e*) countercurrent processes with reflux; (*f*) differential migration processes; and (*g*) zone refining. This list is probably incomplete, yet these seven operational modes encompass the overwhelming majority of technical applications. Theoretically almost all of the physicochemical phenomena discussed in Chapters 2 and 3 can be exploited for separation, using any of these modes of operation or a combination of them. Indeed, many separation processes such as extraction, crystallization, distillation, and dialysis can be performed in several operational modes. All modes are not applicable, however, to all separation processes. For example, chromatography or zone refining is actually restricted to one mode, and the mode is the dominating characteristic of the technique (although it may be based on a variety of physicochemical phenomena). In this chapter we discuss the principles of the basic operational schemes; practical applications will be described in later chapters dealing with individual separation techniques.

4.2   Single-Contact (Cocurrent) Processes

By simple crystallization a mixture can be separated into crystalline phase and mother liquor, which are in *equilibrium*. The mode of operation is the same when a single-step extraction is carried out in a separating funnel, or a solution is decolorized with added charcoal. These are examples of the single-contact mode of operation, which can be applied to continuous processes as well as the foregoing batch processes. The characteristic feature of this scheme is that *two fractions* are obtained in a *single stage*, and the fractions are (approximately) in *equilibrium* with each other when leaving the system. Hence this mode is restricted to *binary separations* and to batchwise or continuous operation in a *single stage*. The batch process is more common than the continuous process, which is also called a *cocurrent* process. Flash evaporation is a typical example of a continuous process.

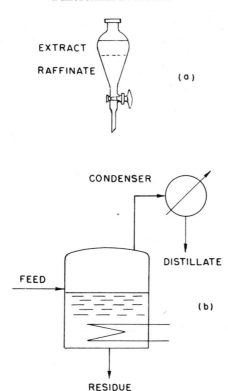

Figure 4.1. Single-stage processes: (*a*) single-step extraction in separating funnel (batch); (*b*) continuous single-stage distillation.

Figure 4.1 schematically illustrates single-contact separation for extraction (Fig. 4.1*a*) and distillation (Fig. 4.1*b*). Continuous distillation as shown is called flash evaporation and is frequently used in industry for removal of volatile components. The efficiency of these single-stage techniques is rather poor; therefore the method is used mainly for "easy" separations, that is, those with large α (separation factor) values.

4.3 Differential Processes

In this operational mode a mixture is also separated in a single stage only. One fraction produced is, however, removed instantaneously, increment by increment. Thus the compositions of both fractions change continuously in the course of the process. This scheme is applied to batch processes and is most commonly used in "pot" distillation in the laboratory. Extraction and processes involving differential crystal melting (e.g., wax "sweating") are other examples. In general, differential operations are single-stage, unsteady-state processes suitable for mutlicomponent separations.

The calculation of the changing composition in the two phases is of practical interest. Assume that W moles of a binary mixture containing a mole fraction X^a of component i undergo differential distillation. After evaporation and removal of a differential amount (dW) of the total liquid, the mole fraction of i in the evaporated amount is X^b. The material balance for component i is

$$X^b\, dW = d(WX^a) = W\, dX^a + X^a\, dW \qquad (4.1)$$

Then

$$\frac{W\, dX^a}{dW} = X^b - X^a \qquad (4.2)$$

and:

$$\int \frac{dW}{W} = \int \frac{dX^a}{X^b - X^a} \qquad (4.3)$$

which is the Rayleigh equation. Integrating from state 1 to state 2, we obtain

$$\ln \frac{W_2}{W_1} = \int_{X^a_1}^{X^a_2} \frac{dX^a}{X^b - X^a} \qquad (4.4)$$

where W_2 and W_1 are the initial and final moles of undistilled liquid, and X_2^a and X_1^a are the initial and final mole fractions of i in these liquid fractions. Under conditions where Henry's law holds, that is, the relative volatility, α, is effectively constant, the right-hand side of Eq. 4.4 can be integrated and becomes

$$\ln \frac{W_2}{W_1} = \frac{1}{\alpha - 1} \ln \frac{X_2^a(1 - X_1^a)}{X_1^a(1 - X_2^a)} + \ln \frac{1 - X_1^a}{1 - X_2^a} \qquad (4.5)$$

This equation, of course, can also be used for differential separation processes other than distillation, if the use of a constant or average α is reasonable.

4.4   Crosscurrent Processes

Multiple extraction in a separating funnel, recrystallization, repeated distillation, and Soxhlet extraction exemplify this mode of operation, which is frequently used in the laboratory. The characteristic feature of these pro-

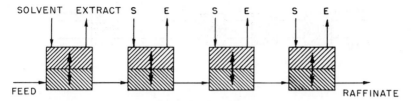

Figure 4.2. Crosscurrent extraction in four stages. This continuous process consists of four consecutive single-step extractions with fresh solvent.

cesses is that the starting solution or solid is repeatedly contacted or washed by successive portions of fresh solvent in a single stage or in a cascade. Therefore this mode of operation can be considered a multistep version of the single-contact mode (Section 4.2) and is usually applied to binary separations. In the limit, however, the crosscurrent process would become a differential process (Section 4.3).

Figure 4.2 shows a four-stage crosscurrent extraction cascade. The feed is introduced into the first stage, where it is equilibrated with fresh solvent. After separation of the phases the extract is removed from the first stage, the raffinate is moved to the next stage. Then the raffinate in the second stage and the feed in the first stage are equilibrated with fresh solvent. After removal of the extract from both stages, the process continues until the raffinate in the fourth stage is removed from the cascade. In the laboratory, the process is conveniently carried out in a single stage. The extraction of a given volume of solution four times with fresh solvent—four-step extraction—represents, then, the single-stage equivalent of the four-stage process described above. In both cases, the final raffinate and the combined extracts can be further processed in order to obtain the separated components.

4.5 Countercurrent Processes

The countercurrent scheme is applied to a large number of separation processes when they are carried out in continuous fashion. The name *countercurrent* indicates that two streams flow in opposite directions. This mode is most frequently used in production-scale multistage processes such as extraction, adsorption, and barrier separations (dialysis). It is also used with differential contactors, as in extraction or gas adsorption. For production-scale separations, countercurrent operation is preferred because of its high efficiency. Even ion exchange is carried out countercurrently with a moving bed. Although true countercurrent processes are very common in industry,

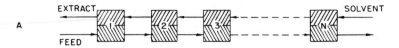

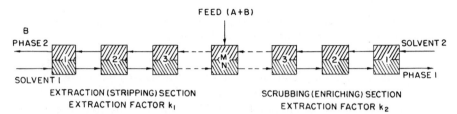

Figure 4.3. Multistage countercurrent extraction. (*a*) Feed at the end. (*b*) Feed at the middle of the cascade.

they are less frequently used in the laboratory. They require elaborate equipment and usually yield only two fractions. It should be mentioned that the Craig distribution, which is frequently called a countercurrent distribution and is discussed in Section 4.8, is not a countercurrent process according to our classification, but is a differential migration process (see Section 4.7).

As an example of countercurrent operations, we shall briefly discuss multistage countercurrent extraction. Two limiting cases are of interest, as shown in Fig. 4.3: first, the feed is introduced at the end of the cascade; second, the feed is introduced in the middle of the cascade. Countercurrent extraction with feed at the end stage is inadequate for the separation of two solutes when α is small. The reason for this is clear: at the feed point both solutes enter the extract phase which is leaving the cascade. This technique is widely used, however, if one or more substances must be removed from one phase to another in a steady-state operation.

If the feed is introduced in the middle of the cascade, a solute pair with a small separation factor can also be separated in a continuous process. This technique is used sometimes in the laboratory (Quick-Fit steady-state distribution machine). It is also called two-solvent extraction, because at each end a different solvent stream enters the cascade. Although only two fractions are obtained here, the separation of multicomponent mixtures can also be achieved by successive runs of the fractions through the cascade.

4.6   Countercurrent Processes with Reflux

In many cases the efficiency of a countercurrent separation process can be increased when a portion of the product leaving the end of the cascade is

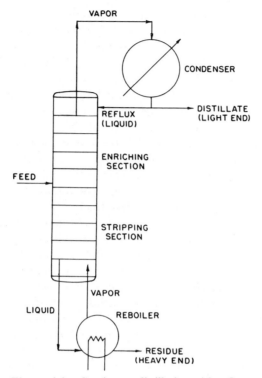

Figure 4.4. Continuous distillation with reflex.

returned to the process. This returned fraction is called *reflux*. The most common (but by no means the sole) application of this mode of operation is found in distillation. It is frequently carried out in the laboratory as a batch process for purification or for fractional distillation. In industry, the continuous process is used most commonly and will be discussed briefly.

Figure 4.4 shows the flow sheet of the process, which is often called *rectification*. Feed is introduced into the middle of a column consisting of discrete *stages* (also called plates or trays). The part of the column above the feed plate is the enrichment section; the part below, the stripping section. The vapor is withdrawn from the top of the column and condensed. Part of it is removed as distillate product, and the remaining liquid (reflux) is returned to the column in order to condense out (scrub) successively the less volatile component from the upflowing vapor on the trays of the enrichment section of the column. In turn, the downflowing liquid is successively enriched in the less volatile component and is finally withdrawn from the bottom of the column. A part of it can be evaporated in the reboiler, however, and the vapor returned into the column in order to boil out (strip) the more volatile component. Although the return of vapor serves in the stripping section a purpose similar to that of the liquid reflux in the enriching section, the term reflux is reserved for the *returned liquid* in the process of distillation. The reflux is usually expressed by the reflux ratio, which is the ratio of the reflux to the withdrawn distillate. In each stage, the liquid and vapor phases are contacted intimately by dispersing the upflowing vapor in the down-flowing liquid. As a result, a fraction of the liquid evaporates, and a fraction of the vapor condenses in the stage, so that the vapor which flows to the upper plate is enriched in the more volatile component, and the liquid which flows to the lower stage is enriched in the less voltatile component.

An *ideal* or *theoretical* plate would allow the establishment of thermo-dynamic equilibrium between phases leaving the stage. In practice, actual plates do not accomplish as large a change in composition as an ideal stage, so that their fractional efficiency is less than 1. As will be discussed in Chapter 7, the minimum number of theoretical plates necessary to obtain a given degree of separation can be easily calculated from equilibrium data for the particular mixture. This minimum number of plates, however, would be adequate to reach the desired composition only if all the top product were returned to the column as reflux, and all the residue were reboiled. Of course, at such *total reflux* no product would be withdrawn from the system. In con-trast, a *minimum reflux* can be defined as the reflux ratio at which an *infinite number of plates* is required for obtaining a desired separation. At reflux ratios lower than this minimum, the desired separation cannot be achieved even with an infinite number of plates. In practice, an intermediate reflux is selected by considering the time of separation, the available equipment, and the required degree of separation.

In the laboratory, rectification is usually carried out in a batch process for the fractionation of a number of components. Here an initial charge is placed in a boiler flask at the bottom of the column and is vaporized at a controlled rate. For difficult separations, the column is first operated at total reflux until the system is in equilibrium. Then product withdrawal begins, and the column can be operated at either *constant* or *variable reflux*. If the reflux ratio remains constant throughout the distillation, equal to a constant fraction of the withdrawn product, the composition of the product will change continuously. Distillate of fixed composition can be obtained from a binary mixture only by continuous increase of the reflux ratio during the course of the process.

4.7 Differential Migration Methods

These techniques are usually semibatch operations used mainly in the laboratory for analytical and preparative-scale separations. A sample mixture is caused to migrate through a porous medium by fluid flow, or across a homogeneous region by application of a force field. Figures 4.5a to 4.5d illustrate the separation of a three-component mixture by a differential migration process. At the beginning of a separation (Fig. 4.5a) the three components are clustered together at the point of sample application. As the sample mixture migrates across the system (4.5b to 4.5d), the three components gradually disengage and eventually are separated. Chromatography is the main example of differential migration with fluid flow, while electrophoresis, ultracentrifugation, and time-of-flight mass spectrometry represent differential migration in a field. Craig distribution, another example of differential migration with fluid flow, is quite similar to chromatography. It differs mainly in being a stagewise rather than a continuous-contact process (see Section 4.8 and Chapter 9).

In the fluid flow methods the migration rate of an individual sample component is determined by the equilibrium distribution of the component between the flowing fluid and the stationary porous medium. Compounds that are distributed mainly into the fluid phase move rapidly; those that are distributed mainly into the porous medium move slowly. Similar migration rate differences arise in the field migration methods as a result of variations in nonequilibrium properties (e.g., diffusion coefficients) among different compounds. In either case, these differences in migration rate among the components of a sample mixture can lead to their separation as in Fig. 4.5.

Differential migration may be stopped while the separated sample components are still in the system (e.g., Fig. 4.5d). Examples of this separation mode include paper and thin-layer chromatography, electrophoresis, and Craig distribution. Alternatively, the individual components may be removed from the system by continued migration (e.g., column chromatography, continuous electrophoresis). In the case of column chromatography, measurement of sample concentration in the column effluent versus time yields a plot (chromatogram) of the form shown in Fig. 4.5e. A similar (but reversed) plot would be obtained for sample concentrations versus migration distance in Fig. 4.5d.

The disengagement of components during migration (Fig. 4.5) is seen to be accompanied by increasing dispersion of the initially sharp zones. This zone dispersion is caused by exchange between phases, ordinary diffusion, mixing, interaction with the medium, and other factors. It leads to a distribution of the species in the direction of travel, and this has a detrimental effect on the separation. Hence, in the absence of zone dispersion, separation would be complete soon after migration is begun (e.g., in Fig. 4.5b). The sample concentration profile that arises from zone dispersion is often Gaussian, and the variance σ^2 of this Gaussian curve is proportional to the migration distance z:

$$\sigma^2 \propto z \qquad (4.6)$$

DIRECTION OF FLUID FLOW

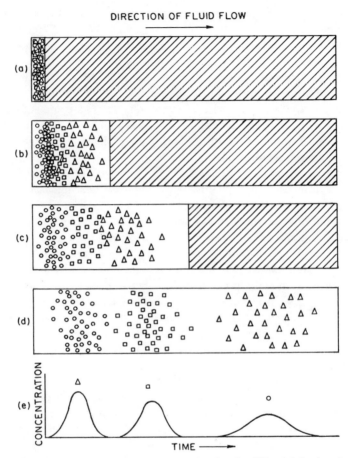

Figure 4.5. Schematic illustration of separation by the differential migration mode.

The proportionality factor σ^2/z is a function of system parameters and operating conditions. It is referred to as plate height in chromatography.

The relative separation of two adjacent zones (sample components) after a given migration distance is seen to depend on two factors: the extent of zone disengagement or distance between the zone centers, and the degree of zone dispersion or width of the zones. Relative separation for two adjacent compounds can be expressed conveniently by a dimensionless number—the *resolution*—defined as

$$R_s = \frac{\Delta z}{\tau \sigma} \tag{4.7}$$

Here Δz is the distance between zone centers, σ is the standard deviation, and

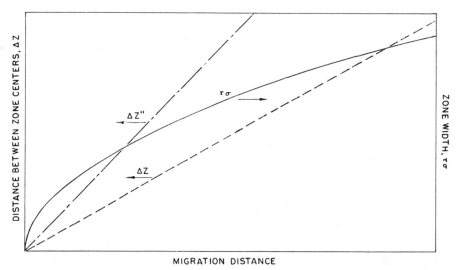

Figure 4.6. Achievement of separation by differential migration. The broken lines $\Delta z'$ and $\Delta z''$ represent the increase of the separation gap with the migration distance for two hypothetical component paris. The curve $\tau\sigma$ illustrates the increase of the width of zones. Unit resolution is obtained at the intersections of the Δz curves with the $\tau\sigma$ curve, which occur at different migration distances, depending on the slopes.

τ is a number (usually 4) which defines the band width for unit resolution. For $\tau = 4$, $R_s = 1$ represents a 98% separation of the two compounds.

The distance between the zone centers increases with the migration distance z. When the components move with steady but different velocities (as is usually the case),

$$\Delta z \propto z \qquad (4.8)$$

Since $\sigma^2 \propto z$ (Eq. 4.6), the standard deviation is proportional to the square root of the migration distance:

$$\sigma \propto \sqrt{z} \qquad (4.9)$$

From Eqs. 4.8 and 4.9 we see that the distance between zone centers increases more rapidly than the zone widths, so that separation improves with greater migration distance. Figure 4.6 illustrates this situation. Separation gaps $\Delta z'$ and $\Delta z''$ for two component pairs are plotted with band width $\tau\sigma$ against the distance migrated. Unit resolution or approximate separation of a component pair is obtained at the intersections of the Δz lines with the $\tau\sigma$ curve. It is seen that component pair $\Delta z''$, which has a larger migration velocity difference, is separated in a shorter migration distance than the pair of components that migrate closer to each other ($\Delta z'$). In principle a pair of components having a very small difference in migration rates can be separated after traveling a sufficiently long distance. This is not always practical, however. Instead a system may be sought in which the difference in migration rates is large. Similarly it is always expedient to select conditions so that relative zone dispersion is small.

4.8 Craig Distribution and Chromatography

Many equilibrium separation techniques can be carried out in either stage or differential column operations. If *discrete stages* (plates, etc.) are used, the process can be modeled as a collection of equilibrium stage processes, although in practice complete equilibrium is not reached in each actual stage. The efficiency of a separating equipment is often expressed by the *number of theoretical stages* (*plates*). This is the number of stages of a hypothetical device that would operate at complete equilibrium on each stage and would produce the same degree of separation as the practical device. For multistage processes the theoretical plate concept does not represent a great departure from physical reality, and it provides a very convenient and simple expression of separating efficiencies. The ratio of the number of theoretical plates to that of actual plates then gives the stage efficiency of the separating equipment.

The treatment of separation processes in *differential column* systems (usually a packed bed) is more complicated because changes in composition are continuous, and hydrodynamic effects and mass transfer in both phases have a great influence on the rate of composition change.

Although a differential column has no stages, the theoretical stage concept is still applicable and leads to a convenient measure of efficiency. From the results obtained with a differential column, the number of theoretical plates that characterizes the efficiency of the device can be calculated by comparison with a hypothetical device as discussed above. Often the length of the column is divided by the number of theoretical plates; this length is termed the *height equivalent of a transfer unit* (HETU) or the *height equivalent of a theoretical plate* (HETP). Obviously the larger this *plate height* is, the lower is the efficiency of the separating device. Conversely, the shorter the column segment needed to achieve the same degree of separation that a theoretical plate would produce, the more efficient the column is.

Although in most cases sophisticated rate theories are available which account for all phenomena taking place in a differential segment of the column, the theoretical plate model remains a useful, convenient approach to describe and design not only stage but also differential column processes. In the following we shall compare a stage and a differential column process. Craig distribution and liquid-liquid chromatography lend themselves as suitable models, since both are batch separation techniques used in the

MOBILE PHASE

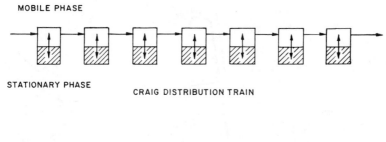

STATIONARY PHASE

CRAIG DISTRIBUTION TRAIN

MOBILE PHASE

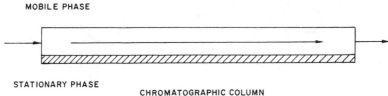

STATIONARY PHASE

CHROMATOGRAPHIC COLUMN

Figure 4.7. Schematic views of Craig countercurrent distribution and liquid-liquid chromatography.

laboratory for the separation of multicomponent mixtures. According to our previous classification, both techniques are of the chromatographic type with respect to the mode of operation. Figure 4.7 illustrates the two processes schematically.

The Craig distribution is carried out in a multistage extraction cascade, in which each stage contains a given volume of the lower phase. After placing the sample to be separated with the upper phase in the first stage, it is shaken to equilibrate the solutes between the upper and lower phases. Then the upper phase is moved to the next stage, fresh upper phase is introduced into the first tube, and both tubes are equilibrated. When the sequence of the discrete equilibrium contacts (transfer and equilibration) continues only until all tubes are occupied by the upper phase, the process is called *fundamental distribution*. When the sequence continues beyond this point and after each equilibration the upper phase is transferred out of the cascade, the process is called *single withdrawal method*.

These operations are carried out in the laboratory with the *Craig machine*. This ingeniously constructed train of glass distribution tubes makes it possible to carry out the individual transfer and equilibration steps with a larger number of tubes (up to 1000) simultaneously, by a very simple movement or automatically in a completely mechanized fashion. Figure 4.8 shows a glass distribution tube of 2-ml capacity. It operates by rocking from position A to B and back for equilibration. The phases separate in position A, and after settling the tube is brought to position B. Then all the upper phase flows into decant tube d through c, since the lower-phase level is at a. When

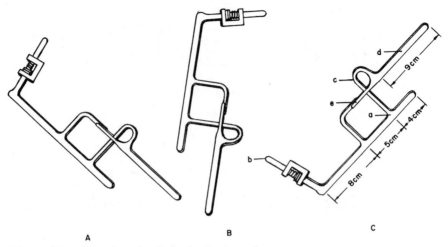

Figure 4.8. Operation of a Craig distribution tube.

the tube is brought to position C, all the upper phases in the decant tube are transferred through e into the next tube, and rocking is repeated for equilibration. Usually ten of these tubes are sealed together through the transfer tube, e, to form a unit. In turn, these units are mounted in series to form a train having the desired number of stages.

In chromatography, differential columns are employed which are usually tubes packed with a support material containing a finely dispersed liquid stationary phase. This stationary phase corresponds to the lower phase in the scheme described above. In liquid-liquid chromatography, the other phase is the eluent liquid, which flows through the column continuously and is also called the mobile phase. The sample is introduced into the stream at one end so that it occupies only a small fraction of the total volume, and it is swept through the column by the eluent. Although the solute molecules are distributed between the two phases, the total sample is not equilibrated as the solute band moves down the column. As a result the band spreads, and the magnitude of spreading is related to the nonequilibrium in this dynamic system. The concentration profile of the solute band emerging from the column at the other end is recorded on the chromatogram as bell-shaped curves.

The efficiency of the chromatographic columns is usually expressed in terms of theoretical plates, which correspond to the equilibrium stages in the Craig distribution. As an intermediate between the discrete equilibrium contacts of the Craig distribution and the differential operation of chromatography, a hypothetical model is considered in which the upper phase would flow continuously through the cascade and complete equilibrium would be reached at each stage.

4.81 DISCRETE EQUILIBRIUM CONTACTS

We start with a cascade of $n + 1$ stages, each containing the same amount of lower phase. The sample is charged to the first stage in a given amount of upper phase. When equilibrium is reached, the fraction q of the solute is in the lower phase, and the fraction p is in the upper phase. Then the upper phase is moved from the first stage to the second stage, so that the solute fractions in the first and second stages are q and p, respectively. Thereafter the same amount of fresh upper phase is introduced into the first stage, and equilibrium is established in both stages. In the next step the upper phases from the first and second stages are transferred into the second and third stages, respectively. At this point the solute in the first stage is q^2, in the second stage $2pq$, and in the third stage p^2. Table 4.1 shows the results of the continuation of this procedure and the distribution of the solute in five stages after four transfers. It can be seen that the solute fractions in the stages are given by the terms resulting from the expansion of $(p + q)^r$, where r is the number of transfers. Thus the discrete equilibrium contact type of operation results in a binomial distribution, which can be explained as follows.

The value p is considered as the probability that a solute molecule will be transferred into the next tube, and q as the probability that it will not be transferred. When these probabilities are constant for all transfers (i.e., k' remains constant), the fraction of the original amount of solute present in the nth tube after r number of transfers, $P_{n\ r}$, is given by the probability that an event will succeed exactly n times out of r number of trials. This is given by the binomial law

$$P_{n,r} = \frac{r!}{n!\,(r-n)!}\,p^n q^{r-n} \tag{4.10}$$

where $P_{n,r}$ is equal to the nth term of the binomial $(p + q)^r = 1$. The values of p and q are calculated from the partition coefficient and from the volume ratio of the two phases in a tube.

The position of the maximum concentration and the number of stages occupied by a solute are calculated as follows. The maximum concentration, that is, the mean of the distribution, is displaced at each transfer p stages. Thus, after r operations, the mean located in stage number μ_s is given by

$$\mu_s = rp \tag{4.11}$$

The standard deviation of the distribution is equal to the root-mean-squared deviation with respect to the mean. After r operations the standard deviation σ_s, expressed as a number of stages, is

$$\sigma_s = \sqrt{rpq} \tag{4.12}$$

Table 4.1. Development of Solute Distribution in the Fundamental Craig Process (Binomial Distribution)

	Stage					Distribution of Solute
	0	1	2	3	4	
Sample introduction and equilibration	p					
	q					
Total amount of solute	1					1
First transfer	0	p				
	q	0				
Total amount of solute in each stage	q	p				$q + p$
Equilibration	pq	p^2				
	q^2	qp				
Second transfer	0	pq	p^2			
	q^2	qp	0			
Total amount of solute in each stage	q^2	$2qp$	p^2			$(q + p)^2$
Equilibration	pq^2	$2qp^2$	p^3			
	q^3	$2q^2p$	qp^2			
Third transfer	0	pq^2	$2qp^2$	p^3		
	q^3	$2q^2p$	qp^2	0		
Total amount of solute in each stage	q^3	$3q^2p$	$3p^2q$	p^3		$(q + p)^3$
Equilibration	pq^3	$3q^2p^2$	$3p^3q$	p^4		
	q^4	$3q^3p$	$3p^2q^2$	p^3q		
Fourth transfer	0	pq^3	$3q^2p^2$	$3p^3q$	p^4	
	q^4	$3q^3p$	$3p^2q^2$	p^3q	0	
Total amount of solute in each stage	q^4	$4q^3p$	$6q^2p^2$	$4p^3q$	p^4	$(q + p)^4$

and the variance is $\sigma_s{}^2 = rpq$. Hence, about 98% of the solute can be found in $4\sigma_s = \sqrt{2rpq}$ number of tubes.

With a large number of tubes the binomial distribution transforms into a Gaussian distribution according to the central limit theorem of statistics. Therefore we can write the distribution as a continuous function of the stage number n as

$$P_{n,r} = \frac{1}{\sqrt{2\pi rpq}} e^{-[(n-rp)^2/2rpq]} \tag{4.13}$$

Equation 4.13 is a very good approximation when the total number of stages, that is, transfer, is larger than 20 or, more precisely, when $rpq \geqslant 3$. Considering that the solute concentration is maximum in the last of the $n + 1$ stages when

$$rp = \mu_s = n + 1 \tag{4.14}$$

we can calculate the total volume of the upper phase (mobile phase) passed through the last stage before the concentration maximum, V_R, as

$$V_R = rv_U = \frac{(n+1)v_U}{p} \tag{4.15}$$

where v_U is the upper-phase volume in one stage, and V_R is called the retention volume. The total volume of upper phase in the cascade, V_U, is $(n + 1)v_U$, and $p = 1/(1 + Ka)$, where K is the distribution constant and a is the phase ratio in the cascade, V_L/V_U. Thus we can write that

$$V_R = V_U \left(1 + K \frac{V_L}{V_U}\right) = V_U + KV_L \tag{4.16}$$

where V_L is the total volume of the lower phase. This equation is a fundamental one that relates the retention volume to the distribution constant and to the volumes of the two phases in the cascade. Of course, it is assumed that K is constant. Equation 4.16 is valid also in chromatography and forms the basis of the equilibrium theory of chromatography.

The standard deviation can also be expressed in volume units of the effluent (upper phase), σ_V. As the upper phase moves $1/p$ times faster through the stages than the solute, we have at the end of the cascade

$$\sigma_V = \frac{\sigma_s v_U}{p} = \frac{v_U \sqrt{rqp}}{p} = \frac{v_U}{p} \sqrt{q(n+1)} \tag{4.17}$$

The ratio

$$\frac{\sigma_V}{V_R} = \sqrt{q/(n+1)} \tag{4.18}$$

thus shows that the breadth of the distribution relative to the mean decreases

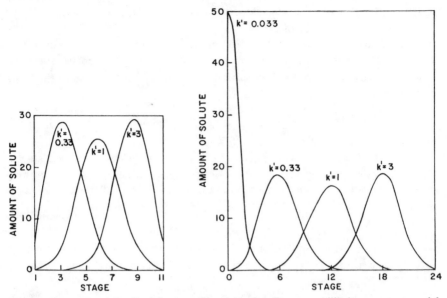

Figure 4.9. Solute distribution as predicted by the discrete equilibrium contact model of chromatography.

as the number of stages increases. It is desirable to have a large number of stages, therefore, in order to obtain relatively narrow solute distribution. Under these conditions the number of stages is given by

$$n + 1 = \frac{V_R^2}{\sigma_V^2} \frac{k'}{1 + k'} \qquad (4.19)$$

It can be shown that the solute distribution in the single withdrawal method also follows the binomial law.

Figure 4.9 illustrates the solute concentration profiles predicted by the above model.

4.82   CONTINUOUS-FLOW OPERATION WITH STAGES

If the cascade is operated as a series of continuation-flow mixers, the solute distribution in each stage follows an exponential law, and the summation for all stages results in a Poisson distribution. Figure 4.10 shows three stages of a cascade that contains a total of $n + 1$ extraction stages. The upper phase flows continuously through the cascade, but in each stage the volumes of the upper phases v_U and the lower phases v_L are the same and remain constant. Complete equilibration is assumed in each stage at any time. The equilibrium is represented by the distribution constant K, so that

$$c_{Li} = K c_{Ui} \qquad (4.20)$$

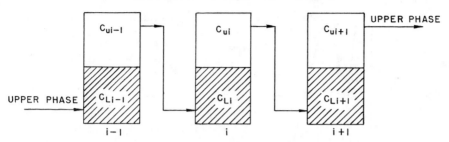

Figure 4.10.   Schematic view of continuous-flow operation with stages.

where c_{Li} and c_{Ui} are the solute concentrations in the lower and upper phases present in stage i, respectively. When a volume dv of the upper phase is moved from stage $i - 1$ to i and from i to $i + 1$, it carries the solute amounts $c_{Ui-1}\, dv$ and $c_{Ui}\, dv$ to and from stage i, respectively. As c_{Ui-1} is not equal to c_{Ui}, the amount of solute in stage i has altered. The change of concentration is given by dc_{Li} in the lower phase and by dc_{Ui} in the upper phase. The mass balance for stage i is given by input minus output, on the one hand, and by the change of solute amount due to changing concentration, on the other hand, that is, by

$$(c_{Ui-1} - c_{Ui})\, dv = v_U\, dc_{Ui} + v_L\, dc_{Li} \qquad (4.21)$$

where v_U and v_L are the volumes of the upper and lower phases, respectively, in one stage.

Assuming that the distribution coefficient is constant, we can eliminate c_{Ui} from Eqs. 4.20 and 4.21, obtaining

$$\frac{dc_{Ui}}{dv} = \frac{c_{Ui-1} - c_{Ui}}{v_U + Kv_L} \qquad (4.22)$$

If solute is present initially in the first stage only and has concentration c_{U0} in the upper phase and c_{L0} in the lower phase, the solution of Eq. 4.22 is given by

$$\frac{c_{Ui}}{c_{U0}} = \frac{e^{-N}N^i}{i!} \qquad (4.23)$$

where

$$N = \frac{V_R}{v_U + Kv_L} \qquad (4.24)$$

as can be verified by differentiation. Here V_R is the total volume of upper phase passed through stage i, and $v_U + Kv_L$ is the so-called effective or equivalent volume of a stage.

Equation 4.23 shows that the continuous-flow model leads to a Poisson distribution of the solute. For small $N - s$ this distribution is asymmetrical, as

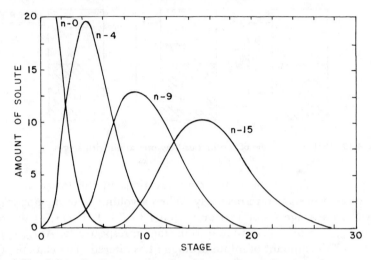

Figure 4.11. Solute distribution as predicted by continuous-flow model.

illustrated in Fig. 4.11, where c_{iU}/c_{oU} is plotted against N for different $i - s$. Maximum concentration of the solute is at $i = N$, as can be seen from Eq. 4.23. If n is large, the Poisson distribution leads also to normal distribution, which can be expressed by an equation similar to 4.13. According to the rules of statistics, however, we obtain from Eq. 4.23 that the mean of the distribution is given by

$$\mu_b = N \tag{4.25}$$

and the variance of the distribution by

$$\sigma_b{}^2 = N \tag{4.26}$$

At the end of the cascade $i = n + 1 = N$ so that

$$\frac{\mu}{\sigma} = \sqrt{N} = \sqrt{n + 1} \tag{4.27}$$

Thus the number of stages is obtained as

$$n + 1 = \frac{\mu^2}{\sigma^2} \tag{4.28}$$

4.9   Zone Melting

This mode of operation was introduced originally for separations utilizing solid-liquid equilibria. The basic concept of the method is illustrated in

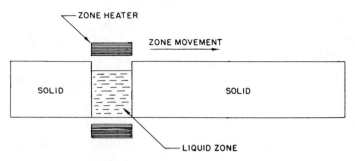

Figure 4.12. Schematic illustration of zone melting.

Fig. 12. A bar of solid material that contains a small amount of impurities is heated at one end by a narrow external source; thus a molten zone is created, while other parts of the bar remain solid. Then the heat source is moved along the bar, causing the molten zone to move with the same velocity. During this process the concentration of impurities that have greater solubility in the melt than in the solid increases gradually in the molten zone. The impurities are, therefore, accumulated in the melt and swept to the other end of the bar, where they are removed. Recently several variations of this original batch process and a number of continuous processes have also been developed. Now zone melting embraces a family of techniques that arise from the concept of moving a *zone of one phase through another*. The process is not restricted to materials that are solid at room temperature, and the moving phase can also be vapor or solid.

Although zone melting was introduced originally as a separation method for ultrapurification, it is now widely used for enrichment of trace components also. For instance, zone melting of frozen dilute solutions of thermally labile volatile materials, such as flavors, yields concentrates without breakdown and loss by evaporation (see Chapter 11 for further details).

References

1. W. L. McCabe and J. C. Smith, *Unit Operations in Chemical Engineering*, McGraw-Hill, New York.
2. H. R. C. Pratt, *Countercurrent Separation Processes*, Elsevier, New York, 1967.
3. L. B. Rogers, in *Treatise on Analytical Chemistry*, Part I, Vol. 2, I. M. Kolthoff and P. J. Elving, eds., Interscience, New York, 1961, pp. 917–955.
4. R. E. Treybal, *Liquid Extraction*, 2nd ed., McGraw-Hill, New York, 1963.
5. R. E. Treybal, *Mass-Transfer Operations*, McGraw-Hill, New York, 1955.

Five

CHROMATOGRAPHY

5.1 Introduction

The great importance of chromatography arises from its speed, resolving power, and ability to handle small amounts of material. A further advantage for many chromatographic applications is marked simplicity of technique, equipment, and practical theory. Figure 5.1 provides an example of the speed of chromatography: the resolution of 10 isomers of heptane in less than 60 sec by gas-liquid chromatography. The resolving power of chromatography is illustrated in Fig. 5.2 for a very complex mixture, and in

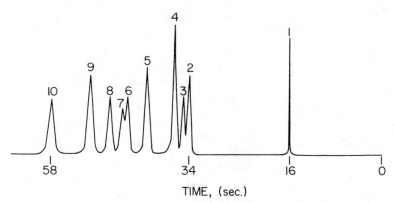

TIME, (sec.)

Figure 5.1. Separation of 10 isomers of heptane by gas-liquid chromatography. Reprinted from Ref. 8, by courtesy of Butterworths Publications, Ltd.

121

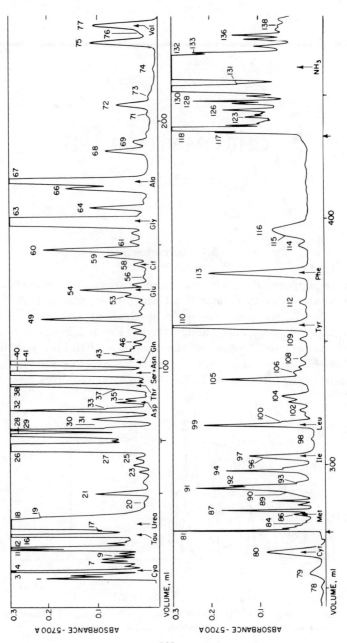

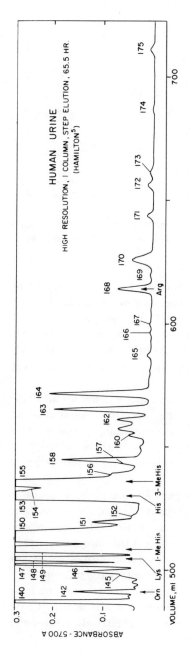

Figure 5.2. Separation of human urine by ion exchange chromatography. Reprinted from Ref. 9, by courtesy of The Chemical Rubber Company.

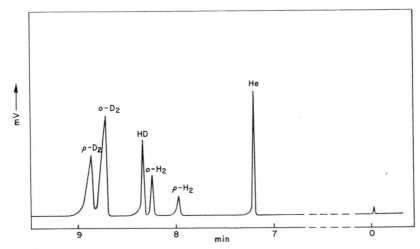

Figure 5.3. Separation of hydrogen isotopes and their nuclear spin isomers by capillary column gas-solid chromatography on silica. Reprinted from Ref. 10, by courtesy of Butterworths Publications, Ltd.

Fig. 5.3 for a mixture of closely similar compounds. Figure 5.2 shows the separation of about 200 amino acids and peptides in a urine extract by ion-exchange chromatography, and Fig. 5.3 shows the separation of two hydrogen isotopes and their nuclear spin isomers by gas-solid chromatography. With regard to sample size, some forms of chromatography are able to achieve routine separation and analysis below the nanogram level (10^{-9} g).

With the techniques of paper and thin-layer chromatography an ultimate in separation simplicity is reached. Here a paper sheet or coated plate is merely spotted with sample and inserted into a developing chamber. The equipment and materials may cost only a few dollars, and the selection of experimental conditions and the interpretation of results can be reduced to a few, easily understood rules.

An initial description of chromatographic separation was given in Sections 4.7 and 4.8 (see Fig. 4.5). Every chromatographic system consists of a moving or *mobile phase* in intimate contact with a *stationary phase*. The stationary phase comprises part or all of the nonmoving portion of the chromatographic system or *bed*. Sample components undergo an equilibrium distribution between these two phases, and this equilibrium determines the velocity with which each component migrates through the system. Band broadening or dispersion of each component in the direction of migration also occurs. These two phenomena, differential migration and band broadening, determine the extent of separation of the starting sample.

Table 5.1. Classification of Chromatographic Methods

Mobile Phase	Stationary Phase	Mechanism	Technique	Sample Development
Gas	Liquid (GLC)	Sorption	Column	Elution
Gas	Solid (GSC)	Adsorption	Open bed	Frontal
Liquid	Liquid (LLC)	Partition		Displacement
Liquid	Solid (LSC)	Exclusion		
Supercritical Fluid	Liquid (SFLC)			
Supercritical Fluid	Solid (SFSC)			

We now examine the general aspects of chromatographic separation in greater detail. Individual chromatographic methods, with their specialized techniques, equipment, and applications, will be considered in following chapters.

5.11 CLASSIFICATION OF CHROMATOGRAPHIC METHODS

Table 5.1 summarizes the classification of chromatographic methods according to (1) type of mobile and stationary phases, (2) mechanism of retention, (3) technique, and (4) method of sample development or movement through the bed. The type of mobile phase used—gas or liquid—leads to the general terms *gas chromatography* and *liquid chromatography*.* Either a liquid or a solid is used as stationary phase. These various possibilities for the stationary and mobile phases lead to three major chromatographic procedures (gas-liquid, liquid-liquid, and liquid-solid chromatography) and three minor procedures (gas-solid, supercritical fluid-liquid, and supercritical fluid-solid chromatography).

In gas-liquid chromatography (GLC) the stationary phase is a nonvolatile liquid (at the operating temperature) coated onto a porous support (e.g., diatomaceous earth) or the walls of a capillary tube. The mobile phase is an inert carrier gas such as helium. Separation results from differences in vapor pressure above the liquid phase for the various sample components—compounds with higher vapor pressures move through the bed more rapidly. Chapter 8 provides a detailed description of GLC. In liquid-liquid chromatography (LLC) the stationary phase is a liquid that is coated on or imbibed within a porous support (e.g., cellulose). The mobile phase is a second liquid that is immiscible with the stationary phase (miscibility would preclude phase distribution and differential migration). Separation arises from differences in the equilibrium distribution of sample components between the two phases. Liquid-liquid chromatography is described in Chapter 10.

* For the (less frequently used) supercritical fluids as mobile phase, we have the term *supercritical fluid chromatography*.

Liquid-solid chromatography (LSC) is based on a solid—usually porous—stationary phase and a liquid mobile phase. Separation of the sample results from adsorption on or exclusion from the porous solid (see below). Liquid-solid chromatography is described in Chapters 12 (ion exchange), 13 (adsorption), and 15 (exclusion).

The three remaining methods of Table 5.1 (gas-solid, dense-gas liquid, and dense gas-solid chromatography) have more limited application at present. Gas-solid chromatography (GSC) uses a solid stationary phase and a gas mobile phase. This technique is currently (but not necessarily) restricted to the separation of quite volatile samples (e.g., fixed gases and lower hydrocarbons) and the characterization of solid surfaces (Chapter 14). The use of a supercritical fluid as mobile phase (SFLC and SFSC) leads to techniques that are intermediate in character with respect to gas and liquid chromatography. The potential importance of these two methods has not yet been established, and they are given only brief mention in Chapters 10 (SFLC) and 13 (SFSC). Practical chromatographic systems sometimes involve more than one active stationary phase (e.g., liquid plus solid), although this is usually unintentional. It complicates the classification scheme of Table 5.1 and can lead to undesirable effects.

The classification of chromatographic methods can also be based on the mechanism of retention in the stationary phase: (1) sorption, which includes both partition and adsorption, and (2) exclusion. Partition, which is the dominant mechanism in both GLC and LLC, is a bulk-phase distribution process. The distributing compound (sample) forms homogeneous solutions in each phase (liquid or gas). Adsorption involves interaction or fixation of sample components at a surface or with fixed sites. The surface or sites are usually defined by a solid stationary phase. In LSC systems adsorption generally involves competition between sample and mobile-phase components for the adsorbed phase. Thus in ion exchange (which is a form of adsorption) the distribution equilibrium takes the following form (for exchange of anions):

$$R^+X^- + Y^- \rightleftharpoons R^+Y^- + X^-$$

Here R^+ represents a positive ion (site) attached to the stationary phase. An anion Y^- in the mobile phase displaces another anion X^- from the stationary phase and is thereby retained at the ion exchange site, R^+. Ligand exchange is another adsorption process, one based on the interaction of different ligands with ions held by an ion exchange matrix or resin.

An exclusion mechanism forms the basis of gel chromatography (gel permeation, gel filtration) and is an important component of adsorption chromatography with molecular sieves as stationary phase. In each case the stationary phase is a porous solid, and the cross section of the pores is similar in size to the sample molecules. Compounds that are small enough to permeate some or all of the pores can distribute into the stationary phase. Relative molecular size then determines the distribution of a compound between mobile and stationary phases. The pores of the molecular sieves (zeolites) are quite small, and hence the zeolites are used for the separation of small molecules (e.g., fixed gases, lower hydrocarbons). Gel chromatography uses wide-pore stationary phases (gels) which are suitable for the separation of the largest molecules, such as polymers and proteins.

We have noted that more than one stationary phase can be involved in a particular chromatographic system (e.g., liquid plus solid). More than one retention mechanism can also operate in a given system (e.g., sorption plus exclusion or partition plus adsorption). Although such effects are often accidental, it may be useful to create deliberately a dual mechanism of retention to enhance differences in migration of the sample components. This is similar to the use of secondary chemical equilibria in distribution systems (Section 2.13).

A third way of classifying chromatographic procedures is by technique. Separation can be carried out in either a column or an open bed. Column chromatography can be further subdivided according to whether a packed column or coated capillary column is used.* Packed columns are prepared by filling an open tube with stationary phase or (in the case of liquid phases) a support coated with stationary phase. Coated capillary columns (also referred to as *open tubular columns*) consist of small-diameter tubes whose inside walls are covered with a thin layer of stationary phase.

Open-bed chromatography (e.g., paper chromatography, thin-layer chromatography) is carried out on a thin, essentially two-dimensional bed. In column separations mobile-phase flow is effected by a pressure drop along the column, but in open-bed chromatography this flow results from capillary wetting. Separation is stopped before the sample components leave the bed, and detection of separated bands is usually accomplished *in situ*. Open-bed chromatography is restricted to the use of liquid mobile phases.

Open-bed chromatography has the advantages of simplicity and flexibility and is most useful for qualitative analysis and the empirical selection of mobile and stationary phases for a given separation. Column chromatography offers greater capacity, easier automation, and potentially higher efficiencies. It is preferred, therefore, for preparative separations, repetitive quantitative analyses, and difficult separations.

A final means of classification is based on the way in which the sample is introduced into the bed and migrates through it. This gives rise to the procedures of elution, frontal development (frontal analysis), and displacement

---

* The further subdivision of column type in gas chromatography is discussed in Chapter 8.

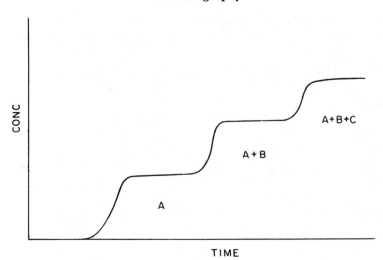

Figure 5.4. The chromatogram in frontal analysis chromatography.

chromatography. In elution the sample is applied to the bed, and mobile-phase flow is begun. Sample components migrate through the bed less rapidly than the mobile phase, so that adjacent bands are usually separated by mobile phase at the conclusion of separation (as in Fig. 4.5). Today elution is used almost exclusively in analytical separations, while frontal development and displacement chromatography are important in preparative- and production-scale operation. Elution separations are subdivided into linear and nonlinear, based on the shape of the isotherm under the conditions of separation (see Fig. 2.6). In linear elution chromatography the distribution constant K is independent of sample concentration, and symmetrical, Gaussian bands are obtained. In nonlinear elution, however, K varies with sample concentration, and the bands are no longer symmetrical. Section 5.22 discusses nonlinear elution in greater detail.

Elution is usually carried out under fixed or constant separation conditions (constant temperature and mobile-phase flow rate, the same mobile and stationary phases throughout separation). However, in some cases it is advantageous to vary the experimental conditions systematically during separation. This gives rise to such techniques as temperature programming, flow programming, and gradient elution (Section 5.5).

In frontal development the sample is introduced into the bed continuously, rather than as a small portion or plug. Thus the sample constitutes the mobile phase. Sample components are selectively retarded as in elution; however, separation results in the formation of *fronts* rather than bands, as in the chromatogram of Fig. 5.4. The least retained component (A) emerges first from the bed, followed by a mixture of A plus the next most strongly retained component (B), and so forth. Frontal development cannot achieve the complete recovery of pure sample components from a mixture, but this technique is useful for concentrating trace impurities and purifying large volumes of gases or liquids (e.g., charcoal treating of air, deionization of water). Frontal analysis is also useful in measuring distribution isotherms.

Displacement chromatography is similar to elution, except that the mobile phase is much more strongly retained by the stationary phase than is the sample. The sample is thus "pushed" through the bed by advancing mobile phase. Displacement provides generally poorer separations of multi-component samples than elution, but greater sample loads can be applied to the bed. It has certain advantages, therefore, in preparative- and production-scale separations.

5.2   Retention and Equilibrium

5.21   LINEAR ELUTION CHROMATOGRAPHY

It is clear from the discussion of Sections 4.7 and 4.8 that differential retardation (or migration) of the various sample components (solutes) has an important effect on their chromatographic separation. When two compounds migrate at the same rate, no separation is possible. We now examine retention in chromatography and relate it to the equilibrium distribution of the solute between the stationary and mobile phases of a given chromatographic system. In this section we assume linear elution chromatography, that is, distribution constants K independent of sample size, which also means Gaussian elution bands.

The fundamental retention parameter in column chromatography is the retention volume V_R, defined as the volume of mobile phase that must flow through the column for elution of a given component. The retention time t_R required for elution of a band (or solute) can be obtained from V_R and the volumetric flow rate F (ml/min) of the mobile phase: $t_R = V_R/F$. The measurement of t_R is illustrated in Fig. 5.5 for a two-component chromatogram (i.e., recorder trace); $t = 0$ corresponds to the introduction of sample to the column, and t_1 and t_2 correspond to t_R values. Note that the time axis in Fig. 5.5 is proportional to the volume of mobile phase V that passes from the column: $t = V/F$. Also shown is t_0, the time for elution of a nonretained component. The corresponding elution volume V_M for a nonretained component is given as t_0F; V_M is also the total volume of mobile phase within the

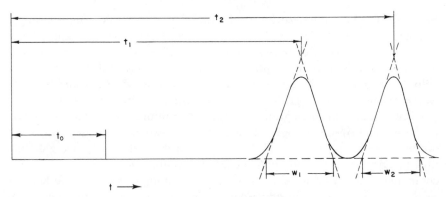

Figure 5.5. A model chromatogram for the separation of two compounds, 1 and 2, by column chromatography.

column at a given time and is often called the column *dead volume*.* Note that the retention time in Fig. 5.5 is measured at the band center or maximum. At the band center the distribution of solute molecules between stationary and mobile phases is approximately at equilibrium (Section 5.3), despite the dynamic nature of the chromatographic process. Consequently we can relate t_R to the equilibrium properties of the system.

A solute molecule migrates down the column only when it is in the mobile phase. Thus a useful measure of retention is R, the fraction of solute in the mobile phase or the probability that a solute molecule will be found in the mobile phase at any given instant. We can express R as a function of the total moles of solute in the stationary and mobile phases: n_S and n_M, respectively:

$$R = \frac{n_M}{n_M + n_S}$$

Since the capacity factor k' has been defined (Section 2.1) as the ratio of total solute distributed between two phases (n_S/n_M), we see that

$$R = \frac{1}{1 + k'} \tag{5.1}$$

The term k' has been related to the distribution constant K through Eq. 2.20: $k' = KV_S/V_M$, where V_S is the volume of the stationary phase.† Since K is

---

* In the present discussion we are assuming that extra-column volumes (Section 5.3) are negligible.

† This assumes a liquid stationary phase. For solid stationary phases K may be defined in terms of the weight of the solid W_S (e.g., moles of solute/g) or its surface area A_S (e.g., moles of solute/m²) (see Section 2.1). Then V_S is replaced by W_S or A_S, respectively.

constant for the systems under discussion (linear elution), we see that R is constant.

Clearly the larger is R, the more rapidly the solute moves through the column. The average migration velocity v_s of the solute is equal to the mobile-phase velocity v times the fraction of solute R in the moving phase at a given time:

$$v_s = vR \tag{5.2}$$

Thus, when R is zero, the solute does not migrate at all $(v_s = 0)$. Similarly, when $R = 1$, the solute moves with the same velocity as the mobile phase $(v_s = v)$. Now $v_s = L/t_R$, where L is column length, and $v = L/t_0$. Inserting these relationships into Eq. 5.2 gives

$$t_R = \frac{t_0}{R} \tag{5.3}$$

or (recalling that $t_R = V_R/F$ and $t_0 = V_M/F$)

$$V_R = \frac{V_M}{R} \tag{5.4}$$

Inserting Eq. 5.1 into Eq. 5.4 now yields a fundamental relationship between retention volume and k' or K:

$$V_R = V_M(1 + k') \tag{5.5}$$

or

$$V_R = V_M + V_S K \tag{5.6}$$

For LSC we note that V_S would be replaced by W_S or A_S, depending on how K is defined. According to Eq. 5.6, V_R is dependent on two factors: V_M and $V_S K$. The column dead volume V_M has no effect on differential migration or separation, since it is the same for all sample components. It is the term $V_S K$ that differs for different compounds and hence determines differential band migration. For this reason the *net retention volume* V_N is often reported:

$$\begin{aligned} V_N &= V_R - V_M \\ &= K V_S \end{aligned} \tag{5.7}$$

If several independent retention mechanisms occur simultaneously (e.g., adsorption plus partition), the observed value of V_N will be the sum of V_N values calculated for all mechanisms.

The fundamental relationship between chromatographic migration and equilibrium within the column given by Eq. 5.6 can be derived also from the principle of conservation of mass (see the Appendix to this chapter). Using either approach, we can now appreciate that the thermodynamic principles

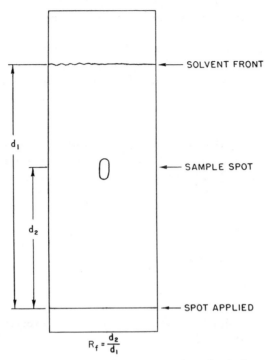

$$R_f = \frac{d_2}{d_1}$$

Figure 5.6. A model chromatogram for the separation of a single compound by open-bed chromatography.

of Chapter 2 can be used to predict separation. Similarly the discussion of Section 2.2 concerning the molecular basis of phase and distribution equilibria can be applied directly to the selection and interpretation of chromatographic systems.

Solute retention or migration in open-bed chromatography is measured in terms of the R_f value: the distance migrated by the solute, divided by the distance migrated by the mobile phase (solvent front). Figure 5.6 illustrates this for a single-component chromatogram. The R_f value is seen to be proportional to the velocity ratio v_s/v, from which we have $R_f = R$, or

$$R_f = \frac{1}{1 + k'} \tag{5.8}$$

$$R_f = \frac{V_m}{V_M + V_S K} \tag{5.9}$$

Equation 5.9 assumes that V_M/V_S and K are constant throughout separation and at all points on the bed (but see discussion of Section 10.13). We can define the useful parameter R_M from this relationship:

$$R_M = \log \left(\frac{1 - R_f}{R_f} \right)$$

$$= \log k'$$

$$= \log K + \log \left(\frac{V_S}{V_M} \right) \tag{5.10}$$

The special value of the R_M function in interpreting chromatographic migration was examined in Section 2.2 and is referred to again in Chapter 10.

5.22 NONLINEAR ELUTION CHROMATOGRAPHY

The preceding discussion of linear elution has assumed that K is independent of sample concentration and that the elution bands are symmetrical and Gaussian. These are consequences of a linear distribution isotherm under the conditions of separation. As discussed in Section 2.1, however, we often encounter nonlinear isotherms in practice, particularly at high sample concentrations. Figure 5.7 illustrates two types of nonlinear isotherm with the resulting chromatographic band shapes. A convex isotherm gives so-called *tailing* of the elution bands, while a concave isotherm gives *fronting* of the bands. Since the average value of $\tilde{K}$ (i.e., C_S/C_M)* decreases with increasing sample concentration in the convex isotherm, V_R also decreases (Eq. 5.6).† Similarly, V_R increases with sample concentration for the concave isotherm. At sufficiently low sample concentrations all isotherms tend toward linearity.

The cause and effect relationship of isotherm type and band shape is illustrated in Fig. 5.8a for a convex isotherm. Solute concentration in the mobile phase is plotted versus position in the column for a moving band. We imagine an initially Gaussian form and see that the band center will tend to migrate more rapidly than the band extremities. This is a consequence of the smaller value of $\tilde{K}$ at larger solute concentrations. Consequently the band becomes unsymmetrical, with the development of a sharp front and an extended tail. This effect becomes more pronounced for larger sample sizes, as illustrated by the series of chromatograms in Fig. 5.8b. We also see V_R (measured from the band peak) decreasing with sample size. The examination of band shape for a concave isotherm proceeds similarly.

* Recall that the symbol $\tilde{K}$ is used to denote a distribution coefficient that is a function of sample size.

† V_R refers the retention volume of the first moment or center of mass of the peak.

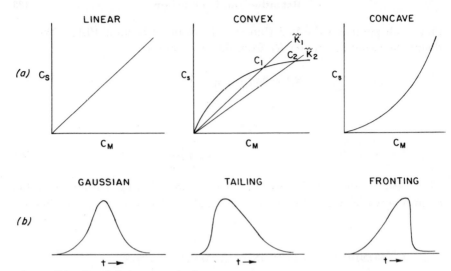

Figure 5.7. Band shape versus isotherm type.

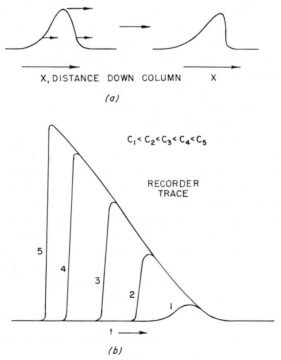

Figure 5.8. Band shape as determined by solute concentration.

When $\tilde{K}$ varies with sample size and elution bands become unsymmetrical (i.e., nonlinear elution), it becomes more difficult to identify eluting compounds from their migration rates (V_R or R_f values), and separation becomes poorer. For these and other reasons, analytical separations are carried out under linear isotherm conditions whenever possible. Some isotherm nonlinearity can usually be tolerated in preparative-scale separations.

Peak asymmetry can arise from other factors besides isotherm nonlinearity. Slow sorption-desorption kinetics can give rise to pronounced tailing. Some tailing of this type is present in almost all elution systems, although it is often below the limits of detection.

5.3 The Origin and Importance of Band Spreading

The broadening of solute bands as they travel through the chromatographic bed is important, because it impedes separation and results in increased dilution of the solute by the mobile phase. The pivotal role of band spreading in chromatography was realized at an early date, and the extent of spreading is an important measure of the performance of a chromatographic system. In this section we examine the origins and control of band spreading. For practical reasons the discussion is restricted to linear elution chromatography.

For the present we are concerned with the broadening of a single solute band as it migrates through the bed. Under ideal conditions the solute will be applied to the bed as a narrow zone; that is, the initial width of the band will be small compared to its final width. Migration of the band begins at a rate v_s, which is equal to the product of mobile-phase velocity v and the fraction of solute in the mobile phase R. However, v_s is not constant in all parts of the bed, and this leads to different overall v_s values for individual solute molecules. These differences constitute the origin of band spreading. There are three major contributions to variation in v_s for different solute molecules. First, the resistance to mass transfer between phases prevents instantaneous equilibration; this is usually the main cause of band spreading. Second, there exists an unevenness of flow (variation in v) with radial position in the column. Third, molecular diffusion in the axial direction (longitudinal diffusion) leads to band broadening, with or without mobile-phase flow. Each of these phenomena will be examined in detail.

5.31 PLATE NUMBER AND PLATE HEIGHT

One of the most important characteristics of a chromatographic system is its *efficiency* or *number of theoretical plates* N. The plate number N can be defined from the chromatogram of a single band, as in Fig. 5.5:

$$N = \left(\frac{t_R}{\sigma_t}\right)^2 \tag{5.11}$$

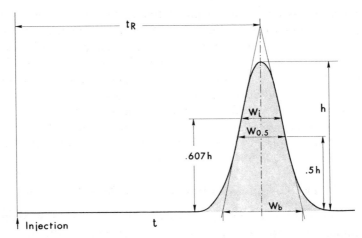

Figure 5.9. Evaluation of a chromatographic peak for the calculation of column efficiency.

where t_R is the time for elution of the band center, and σ_t^2 is the band variance in time units. Note that the plate number N is a dimensionless quantity. The same value of N is obtained from the retention volume V_R and the band variance σ_v^2 in volume units:

$$N = \left(\frac{V_R}{\sigma_v}\right)^2 \tag{5.12}$$

It is also convenient (especially in open-bed chromatography) to define N from distance measurements along the bed: d_R, the distance from the point of sample application to the band center (equal to d_2 in Fig. 5.6); and σ_d^2, the variance of the band in distance units:

$$N' = \left(\frac{d_R}{\sigma_d}\right)^2 \tag{5.13}$$

Now N' refers to the number of plates traversed by the band, not the total plate number of the bed. For elution of the band from a column, N' in Eq. 5.13 would be equal to N for the column. The actual calculation of N from an experimental chromatogram is done more conveniently in terms of band width than of band variance. Table 5.2 in conjunction with Fig. 5.9 summarizes several such relationships, which are equivalent to Eq. 5.11.

The plate number N measures the efficiency of the overall bed or column. Another measure of column efficiency is given by the *plate height H* (also the *height of an equivalent theoretical plate* or *HETP* value):

$$H = \frac{L}{N} \tag{5.14}$$

Table 5.2. Calculation of Plate Number from Chromatogram

Measurements	Conversion to Variance	Plate number
t_R and σ_t	—	$N = (t_R/\sigma_t)^2$
t_R and baseline width W_b	$\sigma_t = W_b/4$	$N = 16(t_R/W_b)^2$
t_R and width at half height $W_{0.5}$[a]	$\sigma_t = W_{0.5}/\sqrt{8 \ln 2}$	$N = 5.54(t_R/W_{0.5})^2$
t_R and width at inflection points (0.607 h) W_i	$\sigma_t = W_i/2$	$N = 4(t_R/W_i)^2$
t_R and band area A and height h	$\sigma_t = A/h\sqrt{2\pi}$	$N = 2\pi(t_R h/A)^2$

[a] This relationship is often used to measure N, because it yields more accurate values of N.

where L is the length of the bed or column, and H has the dimensions of length and is usually reported in millimeters or centimeters. Combination of Eqs. 5.13 and 5.14 for the case of elution from the bed ($N' = N$) gives an equivalent expression for H in terms of the distance variance σ_d:

$$H = \frac{\sigma_d^2}{L} \tag{5.15}$$

The names plate number and plate height originate from the "plate model" of Martin and Synge (1); see also Section 4.8. Here the length of the chromatographic bed is assumed to be divided into segments (theoretical plates) in which a complete equilibration of solute takes place between the two phases. Despite the value of this approach in characterizing the efficiency of distillation columns and extractors, its physical significance in chromatography is questionable. Nevertheless the measured quantities N and H are most useful parameters for characterizing band spreading and the efficiency of a chromatographic system. They can be used independently of the plate model and are then not limited by any deficiencies in that model. The *effective plate number* of a chromatographic system, another useful measure of column performance, is discussed in Section 5.4.

The plate height can also be expressed in dimensionless form as the reduced plate height h; h is the ratio of the plate height H to the particle diameter or tube diameter d, depending on whether we are dealing with packed columns or coated capillary columns:

$$h = \frac{H}{d} \tag{5.16}$$

The *reduced velocity* v, the corresponding dimensionless velocity, is defined by

$$v = \frac{dv}{D_M} \tag{5.17}$$

where ν is the linear flow velocity, and D_M is the solute diffusivity in the mobile phase. The reduced velocity expresses the balance of mass transport by bulk motion (flow) and by molecular motion (diffusion). It can be considered as the ratio of the time required to displace solute molecules a distance equal to one particle or tube diameter to the time needed for the same displacement by molecular diffusion. Although the value of ν in gas chromatography is not much higher than unity, in liquid chromatography columns are operated at reduced velocities of up to several thousands, because D_M is smaller by a factor of 10^4 to 10^5 (compare Tables 3.1 and 3.3).

The main goal of a theory of chromatography is to establish a quantitative relationship between column parameters, solute properties, and operating conditions. Because of the additivity of variances the variance of a chromatographic band is the sum of variances contributed by independent band-broadening processes in the bed. Thus the plate height can be expressed as the sum of incremental plate heights arising from these various processes. We now turn to an examination of the three most important band-broadening phenomena: longitudinal diffusion, slow mass transfer in the mobile and stationary phases, and unevenness of flow. Because of the random, statistical nature of each of these processes, band spreading is similar to a simple axial diffusion process (axial dispersion) and leads to normal or Gaussian distribution of solute molecules along the bed and in the column effluent.

5.32   LONGITUDINAL DIFFUSION

Molecular diffusion of the solute in the direction of flow, or longitudinal diffusion, is usually significant only in the mobile phase. Its contribution σ_L^2 to the total band variance or broadening can be calculated directly from the Einstein equation (Eq. 3.24):

$$\sigma^2 = 2Dt$$

by substituting the appropriate molecular diffusivity and time. On the average the solute molecules spend the time $t = L/v$ in the mobile phase, so that the variance in the mobile phase is given as

$$\sigma_L^2 = \frac{2D_M L}{v}$$

where D_M is the solute diffusion coefficient in the mobile phase. The plate height contribution of longitudinal diffusion H_L is then obtained as

$$H_L = \frac{\sigma_L^2}{L} = \frac{2\gamma D_M}{v} \tag{5.18}$$

where γ is an obstruction factor which recognizes that longitudinal diffusion is hindered by the packing or bed structure. In coated capillary columns γ is unity, and in packed columns it has a value of about 0.6. In normal practice H_L is at most a small fraction of the overall plate height H (see Figs. 5.10 and 5.11). Only when D_M is large and the mobile-phase velocity is small does H_L become significant. In liquid chromatography H_L can normally be assumed to be equal to zero.

5.33 MASS TRANSFER INTO THE STATIONARY PHASE

The rate at which solute molecules transfer into and out of the stationary phase often makes an important contribution to the overall chromatographic efficiency or the value of N. This rate is controlled by diffusion for liquid stationary phases and by adsorption-desorption kinetics in the case of adsorbents. We will calculate the plate height contribution from these mass transfer processes by means of the one-dimensional random walk model of Giddings (2).

As solute molecules move through the bed by flow and diffusion processes, they suffer many random displacements (changes in position) away from the center of this molecular assembly. Displacements in the direction of flow are responsible for band spreading, since these displacements can be regarded as a number of forward and backward steps relative to the band center (which moves continuously through the bed). When the numbers of both steps and solute molecules are sufficiently large, this process leads to a normal distribution of the molecules that started their random walk together at the column inlet. The mean-square displacement of the solute molecules or the band variance σ^2 is then given as

$$\sigma^2 = l^2 n \tag{5.19}$$

where l is the fixed step length, and n is the number of steps.*

Solute retention in the chromatographic bed is the result of a large number of sorption steps in which the solute molecules are captured by the stationary phase. Each sorption step is followed by desorption, and the molecule is released again into the mobile phase and moves further along the bed. In the simplest case we can assume first-order kinetics; that is, the rate of

---

* In order to use this relationship for the calculation of band variance, we must assume that n is large and l is constant. Then l and n must be established from the physical parameters of the system. The number of steps n is equal to the residence time of the solute molecules in the bed, t_R, divided by the time required for a single step. All present theories of band spreading assume that n is large, and this is the case for most chromatographic systems. Large values of n lead to the observed Gaussian bands. Although we have no reason to believe that all steps are of equal length, we can still replace l by the root-mean-square average of the actual step lengths and use this average value.

adsorption and desorption is proportional to the total number of solute molecules involved, and the rate constant is independent of solute concentration. This requirement is usually fulfilled at low solute concentrations. The rate constant is equal to the fraction of the available molecules that is adsorbed or desorbed in unit time, so its dimension is reciprocal time. If k_a and k_d are the rate constants for adsorption and desorption, respectively, we can define $t_a = 1/k_a$ as the mean adsorption time and $t_d = 1/k_d$ as the mean desorption time.

In the random walk model an adsorption step represents a *step backward*, since it leads to immobilization of the adsorbed molecule on the stationary phase while the band center moves on. Desorption steps are *steps forward*, because the molecules are released into the mobile phase and swept along at a faster rate than that of the band center. The total number of steps as the band migrates through the distance L (e.g., the column length) is simply the sum of all forward and backward steps. Since desorption must follow adsorption, and vice versa, the total number of steps is twice the number of adsorption steps.

The number of adsorption steps can be calculated from t_a, the average time a molecule spends in the mobile phase before adsorption. It is equal to the total residence time of a molecule in the mobile phase, L/v, divided by t_a. The total number of steps n is twice this number, or

$$n = \frac{2L}{vt_a} \tag{5.20}$$

The length of each step l can be evaluated from the displacement of solute molecules relative to the band center. The band center migrates with a velocity of $v/(1 + k')$ (Eqs. 5.1 and 5.2), so that during one step of duration t_a the band center will be displaced a distance $vt_a/(1 + k')$. During this same time a molecule in the mobile phase will be migrating at velocity v, so that its displacement before adsorption will be vt_a. The net displacement of the unadsorbed molecule from the band center during this step (i.e., the step length l) is then given as vt_a minus $vt_a/(1 + k')$, or

$$l = \frac{vt_a k'}{1 + k'} \tag{5.21}$$

Substituting n and l (Eqs. 5.20 and 5.21) into Eq. 5.19, we obtain the variance from slow adsorption-desorption as

$$\sigma_s{}^2 = 2 \left(\frac{k'}{1 + k'} \right)^2 vt_a L \tag{5.22}$$

The plate height contribution from the adsorption-desorption process H_S is then $\sigma_S{}^2/L$ (Eq. 5.15) or

$$H_S = 2 \left(\frac{k'}{1 + k'}\right)^2 v t_a \tag{5.23}$$

It is more convenient to express H_S in terms of the mean desorption time t_d rather than t_a. The ratio t_a/t_d is the ratio of times spent in the mobile and stationary phases, respectively. On the average, the fractional time spent by a solute molecule in the mobile phase is $1/(1 + k')$, and in the stationary phase $k'/(1 + k')$. This follows from our discussion of Eq. 5.1. Therefore

$$\frac{t_a}{t_d} = \frac{1}{k'} \tag{5.24}$$

Combining the last two equations gives

$$H_S = \frac{2 v t_d k'}{(1 + k')^2} \tag{5.25}$$

Despite the crudeness of the random walk model, this equation is rigorously correct for adsorption on both uniform and nonuniform surfaces. It shows that H_S increases with mobile-phase velocity and with the time t_d required for desorption of the average solute molecule. Therefore column efficiency will increase with faster desorption or smaller values of t_d. The factor $k'/(1 + k')^2$ has a maximum at $k' = 1$, and approaches zero as k' goes to zero or infinity.

In partition chromatography the stationary phase is a liquid dispersed on the surface of a solid support. Equation 5.25 can be used to approximate the plate height contribution from solute diffusion in the stationary phase, assuming that mass transfer across the interface is not rate limiting (a normally valid assumption). The mean desorption time t_d is replaced by the average diffusion time t_D, which expresses the time required for the solute molecule to diffuse a characteristic distance in the liquid, for example, film thickness or droplet diameter. Denoting this characteristic distance by d and the solute diffusivity in the stationary phase by D_S, we obtain from Eq. 3.16 the following expression for the average diffusion time:

$$t_D = \frac{d^2}{2 D_S} \tag{5.26}$$

When this value is substituted in Eq. 5.25 for t_d, the plate height contribution from diffusion into the stationary phase is given as

$$H_S = \frac{k' d^2 v}{D_S (1 + k')^2} \tag{5.27}$$

This expression is inexact, because the geometrical configuration of the dispersed liquid has not been taken into account. It can be shown by rigorous theoretical derivation that a configuration or shape factor q which depends on the precise shape of the dispersed liquid must be included:

$$H_S = \frac{qk' \, d^2 v}{D_S (1 + k')^2} \tag{5.28}$$

For a uniform liquid film or liquid spheres, q is equal to $\frac{2}{3}$ or $1/30$, respectively. The configuration of the liquid stationary phase is often considered as a thin film on the pores of a support (i.e., as in a coated capillary column), but this is a rough approximation. A spherical configuration is exemplified by a spherical ion-exchange resin, which can be regarded as a liquid sphere supported by the gel matrix.

5.34   MOBILE-PHASE CONTRIBUTIONS TO PLATE HEIGHT

Solute molecules in the mobile phase are displaced not only by diffusion but also by flow. Band spreading is greatly influenced, therefore, by the detailed flow profile and by flow and diffusion rates.

In coated capillary columns the plate height contributions H_M from the flowing mobile phase can be calculated rigorously because the exact flow profile is known (Section 3.6). Then H_M is given as

$$H_M = \frac{1 + 6k' + 11k'^2}{24(1 + k')^2} \frac{r^2 v}{D_M} \tag{5.29}$$

where r is the inner radius of the column, and v is the flow velocity as previously. Despite the complicated mathematics involved, the derivation of Equation 5.29 is facilitated by the fact that in capillary columns radial transport of solute takes place by diffusion only.

In a packed bed the mobile phase flows through the interconnected, tortuous channel system of the bed. Thus lateral mass transport takes place not only by diffusion but also by convection. Furthermore, small- and large-scale flow inequalities are always present as a result of the irregular packing structure of the bed. These flow inequalities also lead to band spreading. The flow pattern in packed beds is too complicated to be capable of rigorous treatment at the present time. Consequently there is considerable uncertainty in formulating H_M for packed beds as v is varied and the relative importance of diffusion and convection changes. It is generally assumed that H_M can be related to plate height contributions from flow profile effects (H_F), often referred to as an *eddy diffusion* contribution, and diffusion effects (H_D). These plate height terms may be written as

$$H_F = \Lambda \, d_p \tag{5.30}$$

and

$$H_D = \frac{\Omega v \, d_p{}^2}{D_M} \tag{5.31}$$

Here d_p is the particle diameter, and Λ and Ω are some functions of the packing structure. In gas chromatography the overall mobile-phase contribution is often given as

$$H_M = H_F + H_D = \Lambda \, d_p + \frac{\Omega v \, d_p{}^2}{D_M} \tag{5.32}$$

This equation predicts a linear increase of H_M with v. Experimental data suggest, however, that at high reduced velocities (such as are often encountered in liquid chromatography) H_M approaches a maximum value because H_F (which is velocity independent) becomes dominating. Thus in liquid chromatography the following relationship may be applicable because of so-called *coupling* of flow and diffusion effects:

$$H_M = \frac{1}{1/H_F + 1/H_D} = \frac{1}{1/\Lambda \, d_p + D_M/\Omega v \, d_p{}^2} \tag{5.33}$$

In general H_M increases with particle diameter and flow velocity and decreases with solute diffusivity. The packing structure, the reduced velocity range (Eq. 5.20), and k' can significantly influence the exact form of the relationship. For a narrow range of conditions, empirical expressions for H_M are sometimes useful (see Chapter 10).

When porous column packings are used, the intraparticle void volume is filled with mobile phase at rest. Solute molecules must diffuse through this *stagnant mobile phase* in order to reach the stationary phase. Thus an additional mobile-phase plate height contribution H_{SM} arises, which is unaffected by flow profile effects and increases linearly with flow velocity. For porous spherical particles it can be calculated as

$$H_{SM} = \frac{(1 - \phi + k')^2 \, d_p{}^2 v}{30(1 - \phi)(1 + k')^2 \gamma D_M} \tag{5.34}$$

where ϕ is the fraction of total mobile phase in the intraparticle space, and γ is the tortuosity factor inside the particle.

In gel chromatography there is (in principle) no sorption. Band spreading arises, therefore, from mobile-phase effects only. In this case the plate height contribution of the immobilized (stagnant) mobile phase in porous spherical particles is found to be

$$H_{SM} = \frac{k'}{30(1 + k')^2} \frac{d_p{}^2 v}{D_M} \tag{5.35}$$

This relationship is formally identical to a stationary-phase plate height contribution, for a spherical configuration of the stationary phase (Eq. 5.28).

5.35 COMPLETE PLATE HEIGHT EQUATION

The magnitude and relative importance of these various plate height contributions can vary greatly from one chromatographic technique to another, and even within a technique as bed geometry and operating conditions are changed. Although the preceding listing of contributions to H is not comprehensive, it encompasses the major band-broadening factors, and the overall plate height can be expressed as their sum:

$$H = H_L + H_S + H_M + H_{SM} \qquad (5.36)$$

To summarize, H is the sum of terms from longitudinal diffusion (H_L), stationary-phase mass transfer (H_S), mobile-phase effects (H_M), and diffusion in or mass transfer across the stagnant mobile phase (H_{SM}). Figure 5.10 illustrates the dependence of H on flow velocity as the sum of individual terms given previously (Eqs. 5.18, 5.28, 5.33, and 5.35). In practice the H versus v relationship is determined experimentally for a given chromatographic system. This provides some insight into the relative importance of H_L, H_S, and the other factors and permits the optimization of experimental conditions for minimum H or maximum N. Such an analysis of experi-

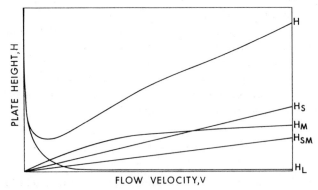

Figure 5.10. Plot illustrating the dependence of the overall plate height H and the plate height increments on the flow velocity in a wide range. Each point on the H curve is given by the sum of the individual plate height contributions at that flow velocity. H_S arises from the stationary-phase effects, and H_M from mobile-phase effects. H_{SM} represents the plate height contribution of the stagnant mobile phase, and H_L that of longitudinal diffusion. The relative magnitudes of the individual plate height contributions vary greatly, depending on the type of chromatography, the column construction, and the capacity ratio as well.

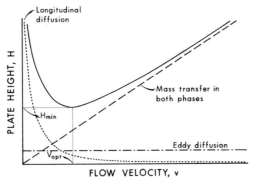

Figure 5.11. Van Deemter plot applied to gas chromatography.

mental H versus v curves has also been a powerful tool in developing a theory of chromatography. In gas chromatography Eq. 5.36 reduces to the well-known *van Deemter equation*, which can be expressed as

$$H = A + \frac{B}{v} + Cv \qquad (5.37)$$

where A, B, and C are constants for a given system. These three terms represent plate height contributions from so-called eddy diffusion (H_F from Eq. 5.30), longitudinal diffusion (Eq. 5.18), and the sum of stationary- and mobile-phase mass transfer terms (Eq. 5.31 plus 5.34). The form of the van Deemter equation is shown in Fig. 5.11. It is observed only at low, reduced velocities, where longitudinal diffusion is significant. A minimum plate height $H_{\min}$ is observed at an intermediate "optimum" value of v (v_{opt}), that is, column efficiency is greatest for a flow velocity equal to v_{opt}. By setting the derivative dH/dv equal to zero for Eq. 5.37, we obtain

$$H_{\min} = A + 2\sqrt{BC} \qquad (5.38)$$

and

$$v_{\mathrm{opt}} = \sqrt{B/C} \qquad (5.39)$$

At v_{opt}, B/v equals Cv. In practice the flow velocity is kept larger than v_{opt} to provide faster separation, so that band broadening is controlled by the mass transfer term Cv. This is particularly true in liquid chromatography, where longitudinal diffusion (i.e., B/v) is insignificant at the large reduced velocities which are normally used. In fact, a minimum in H (at v_{opt}) is seldom observed in liquid chromatography.

5.36 COMPARISON OF GAS AND LIQUID CHROMATOGRAPHY

The fundamental difference between these two techniques, in regard to band spreading, arises from the 10^4 to 10^5 greater solute diffusivities in gases versus liquids. If the speed and efficiency of gas and liquid chromatography are to be comparable in practice, plate heights and flow velocities must also be comparable. This in turn requires drastically reduced diffusion path lengths in liquid chromatography, which are achieved by using columns packed with very fine particles, usually with diameters of 20 to 50 μ, but more recently 5 μ to 10 μ. The small particles in liquid chromatography, combined with much greater mobile-phase viscosities, require large column pressure drops relative to gas chromatography, in order to achieve comparable flow rates (Eq. 3.64). Column pressure drops in high-efficiency liquid chromatography often exceed 3000 psi. It must be emphasized, however, that, despite these differences in the underlying physicochemical phenomena and the resulting differences in equipment and technique, the dynamics of both liquid and gas chromatography are governed by the same fundamental relationships discussed in this section. Recent advances in high-efficiency separations by liquid chromatography are to a considerable extent the result of previous advances in the theory of gas chromatography.

5.4   Resolution

Having examined band migration and band broadening in detail, we turn now to the combined effect of these two processes on separation. We will consider the evaluation of experimental chromatograms in terms of relative band disengagement or resolution, the improvement of separation by systematic variation of experimental conditions, and the control of separation time (i.e., reducing the time of separation to a practical minimum).

It is not unusual to observe over 50 components in a mixture (e.g., Fig. 5.2). In such cases the chromatogram itself provides the most practical description of the adequacy of the separation. It is also highly useful, however, to have a quantitative measure of the degree of separation or *resolution*, particularly for adjacent bands, where the problem of separation is greatest. In this section we develop some general relationships and concepts that describe the resolution of a pair of adjacent bands as a function of the experimental conditions. These results can be easily generalized to the problem of multicomponent separations, as discussed further in Section 5.5.

As previously, we assume linear elution chromatography: Gaussian bands and k' values that are not a function of sample concentration. Band disengagement or resolution is determined by two basic factors: the separation of band centers, and the band widths. This is illustrated in Fig. 5.12. In the top chromatogram there is incomplete separation of the two components. If experimental conditions are adjusted to increase the separation of the band centers, as in the middle chromatogram, band disengagement and improved

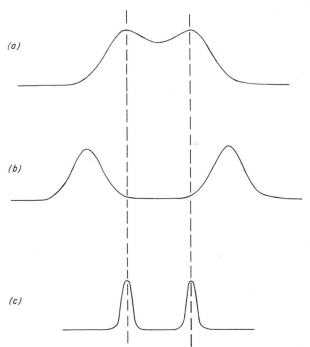

Figure 5.12. Two component chromatograms to illustrate the importance of peak to peak separation and peak widths on separation.

resolution result. The same effect is obtained if we change conditions to reduce band width rather than increase band separation, as shown in the bottom chromatogram of Fig. 5.12. A difficult separation may require a simultaneous increase in the distance between band centers and a decrease in band widths.

In the following discussion we emphasize resolution in column chromatography. Much of what is said will apply equally to open-bed chromatography, and important differences will be pointed out as they arise.

5.41 DEFINITION OF RESOLUTION

The degree of separation or resolution of two adjacent bands is commonly defined as the distance between band centers divided by the average band width. If we measure retention and band width in units of time, as in the chromatogram of Fig. 5.5, resolution R_s is given as

$$R_s = \frac{2(t_2 - t_1)}{W_1 + W_2} \tag{5.40}$$

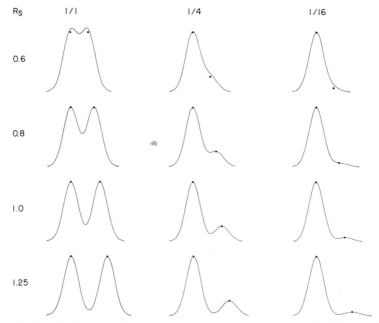

Figure 5.13. Significance of resolution values for two components with different relative concentrations.

where t_1 and t_2 are the t_R values of bands 1 and 2, and W_1 and W_2 are their baseline band widths W_b, defined by the tangents to the inflection points of a given curve. Values of W_b for closely adjacent bands (those where separation is incomplete or barely complete) are approximately constant, that is, $W_1 \approx W_2$.* Since W_b is equal to four standard deviations σ_t for a given band, Eq. 5.40 can also be expressed as

$$R_s = \frac{t_2 - t_1}{4\sigma_t} \tag{5.41}$$

This is equivalent to Eq. 4.6 with $\tau = 4$. The larger is R_s, the greater is the disengagement of the two bands; $R_s = 1$ corresponds to a reasonably good separation. In this case there are four σ_t units between band centers, so that the tail of each band beyond $2\sigma_t$ overlaps the other band. This corresponds to a 2% contamination of each band by the other.

A further illustration of the significance of R_s is provided by Fig. 5.13 and Table 5.3. Figure 5.13 shows the resolution of two bands for different values

* This is implied by the definition of N in terms of W_b (Table 5.2), assuming that N is approximately constant for different solutes. Experimentally, the assumption that W is constant for closely adjacent bands is a good approximation.

Table 5.3.   Variation of Fraction Purity and Recovery with R_s, for Different Relative Concentrations of Two Adjacent Bands (as in Fig. 5.13)

	Relative Concentrations of Two Bands, A/B					
	1/1		3/1		10/1	
R_s	A	B	A	B	A	B
1.5 (6σ separation)						
Cut-point position[a]	3.0σ		3.2σ		3.4σ	
Purity of recovered						
fractions (%)	99.9	99.9	99.9	99.7	100	99.7
Recovery (%)	100	100	100	100	100	99
1.0 (4σ separation)						
Cut-point position[a]	2.0σ		2.3σ		2.8σ	
Purity of recovered						
fraction (%)	97.7	97.7	98.5	96.7	98.9	97.1
Recovery (%)	98	98	99	95	100	88
0.8 (3.2σ separation)						
Cut-point position[a]	1.6σ		2.1σ		—[b]	
Purity of recovered						
fraction (%)	95.5	95.5	95.6	94.1		
Recovery	96	96	98	86		

[a] Position of valley between two bands, relative to band A.
[b] No valley observed.

of R_s and for different relative concentrations of the two bands. When $R_s =$ 1.5, the separation of the two bands is essentially complete (so-called *baseline separation*). The cross contamination of the two bands is normally much less than 1%, and the observed peak center of each band coincides with the true value for that band (heavy dots in Fig. 5.13, for band center in the absence of a second band). Since peak height is proportional to solute concentration in the original sample, and t_R defines the identity of the solute within narrow limits, both quantitative and qualitative analyses are possible when the true band centers coincide with the observed band centers.

For $R_s = 1$, the two bands are readily recognizable over a wide range of relative concentrations (Fig. 5.13). Usually the cross contamination of the two bands is less than 3%, which is suitable for most purposes. The observed values of peak height and t_R differ from the true values (i.e., for a single band) by an insignificant amount over a wide range of relative concentrations for the two bands. Thus qualitative and quantitative analyses are not affected. For $R_s = 0.8$, two distinct bands are just barely recognizable, and only over a limited ratio of relative band concentrations. For lower values of R_s, only a single band is seen in most cases. As long as two distinct band centers are observable (e.g., $R_s = 0.8$, relative band concentrations between $\frac{1}{4}$ and 4), both qualitative and quantitative analyses are relatively unaffected (see Fig. 5.13).

Table 5.3 shows the effect of resolution on the purity and recovery of each solute. As R_s decreases, the recovery of the low-concentration solute (in reasonable purity) decreases significantly. This effect becomes more pronounced as the ratio of concentrations of the two solutes becomes greater. Thus for $R_s = 1$ and a tenfold difference in the concentrations of the two solutes, the recovery of the low-concentration solute is only 88%.

5.42 DERIVATION OF THE FUNDAMENTAL RESOLUTION EQUATION

Equations 5.40 and 5.41 define resolution in a given situation, but they do not show how resolution is related to the conditions of separation. Consequently these relationships cannot be used directly to improve resolution. For this purpose we need a more fundamental form of the resolution equation, which we now derive.

Since $\sigma_1 \approx \sigma_2$, Eq. 5.41 can be expressed as

$$R_s = \frac{t_2 - t_1}{4\sigma_2}$$

where t_2/σ_2 is given in terms of N by Eq. 5.11. Making this substitution in the above equation gives

$$R_s = \left(\frac{\sqrt{N}}{4}\right)\left(1 - \frac{t_1}{t_2}\right)$$

Equations 5.1 and 5.3 give t_1/t_2 as $(1 + k_1')/(1 + k_2')$, and substitution of this quantity into the above equation yields

$$R_s = \left(\frac{\sqrt{N}}{4}\right)\left(1 - \frac{1 + k_1'}{1 + k_2'}\right)$$

$$= \left(\frac{\sqrt{N}}{4}\right)\left(\frac{k_2' - k_1'}{1 + k_2'}\right) \tag{5.42}$$

where k_1' and k_2' refer to the k' values of bands 1 and 2. If we note that the separation factor α is equal to k_2'/k_1' (Eq. 2.22a), then Eq. 5.42 reduces to the important relationship

$$R_s = \left(\frac{\sqrt{N}}{4}\right)\left(\frac{\alpha - 1}{\alpha}\right)\left(\frac{k_2'}{1 + k_2'}\right) \tag{5.43}$$

Resolution is seen to be a function of three separate factors: a column efficiency factor $\sqrt{N}$, a selectivity factor which varies with α, and a capacity factor term which varies with k_2'.

Equation 5.43 assumes that the pair of bands passes completely through the bed (i.e., column chromatography). In open-bed chromatography, where the bands migrate some fractional distance through the bed, an expression essentially similar to Eq. 5.43 can be derived in the same way. The only difference is that N for the entire bed is replaced by the number of plates N' through which the two bands migrate before separation is stopped. Since N is proportional to bed length or migration distance, N' is equal to $N \cdot R_f$; N is the plate number of the entire bed, and it is assumed that separation is stopped when the solvent front reaches the end of the bed. Then R_s for open-bed chromatography is given as

$$
R_s = \left(\frac{\sqrt{N \cdot R_f}}{4}\right)\left(\frac{\alpha - 1}{\alpha}\right)\left(\frac{k_2'}{1 + k_2'}\right)
$$
$$
= \left(\frac{\sqrt{N}}{4}\right)\left(\frac{\alpha - 1}{\alpha}\right)\left[\frac{k_2'}{(1 + k_2')^{3/2}}\right]
$$

$$(5.44)$$

We will now examine Eqs. 5.43 and 5.44 for their practical significance, specifically their ability to relate resolution to the experimental conditions of separation. We should keep in mind that the fundamental parameters N, α, and k_2' can be adjusted more or less independently. Thus N is determined mainly by the dynamics or rates of the various processes that take place during separation, and α and k_2' are thermodynamic or equilibrium properties of the chromatographic system. It is possible to vary N by changing the length of the bed or column, altering the flow rate of the mobile phase, using bed materials of different particle size, and so on (Section 5.3). Changes in α and k_2' are achieved by selecting different stationary and mobile phases or by varying temperature and (less often) pressure. In addition, k_2' can be varied also by changing the relative amounts of mobile and stationary phases within the column. Subsequent chapters on the individual chromatographic methods will review the control of α and k' in detail.

We see from Eqs. 5.43 and 5.44 that R_s approaches zero (resolution is destroyed) as N or k_2' approaches zero, or α approaches 1. In open-bed chromatography R_s also approaches zero as k_2' becomes large. Large k_2' values are also undesirable in column chromatography, since they correspond to long separation times and dilute, hard-to-detect bands. We will show that an optimum value of k_2' (usually between 2 and 5) exists in different chromatographic systems.*

---

* The value of R_s varies only slightly with k' when k' is close to optimum. Therefore we need not specify k' within narrow limits for approximately maximum resolution.

Although each of these three parameters—N, α, and k'_2—is important in controlling resolution, we usually do not attempt their simultaneous optimization in an actual separation. Experimental conditions are selected initially that favor large N values, within the practical limits of convenience plus reasonable separation times. Higher N values always provide improved resolution, other factors being equal, and this is true of analytical or preparative separations, and simple or complex mixtures (Section 5.5). With a reasonable starting value of N, adequate resolution will be attained in most cases if we optimize k'_2 approximately. In gas chromatography an optimum value of k'_2 can be achieved by varying temperature. In liquid chromatography it is more profitable to vary systematically the composition of the mobile phase.

If resolution is still a problem at this point, we can increase either N or α. When α is quite close to 1, it may be impractical to achieve reasonable resolution by further increase in N. This would be indicated by very low values of R_s, that is, severe band overlap after optimization of k'_2. In such cases it is necessary to vary the composition of one or both phases in an effort to increase α,* while simultaneously holding k'_2 approximately constant. This is usually a trial-and-error process which is often time consuming and inconvenient. Moreover, success can rarely be guaranteed in advance. When only a slight increase in resolution is required, it is usually more profitable to increase N. This parameter can always be increased at the expense of longer separation time (see below) in a systematic, predictable manner.

It is often convenient to combine the plate number and capacity factor terms of Eq. 5.43 into a single parameter, column *effective plates* N_{eff}:

$$N_{eff} = N \left(\frac{k'}{1 + k'} \right)^2 \tag{5.45}$$

For α constant, R_s is seen to be proportional to $\sqrt{N_{eff}}$. Hence N_{eff} is a more useful parameter than N for comparing the resolving power of different chromatographic columns. This is particularly true when the same stationary and mobile phases are involved, and the ratio of the two phases within the column varies. Coated capillary columns provide much larger values of N than packed columns in gas chromatography, for example, but corresponding values of N_{eff} are closer in magnitude.† The actual performance of packed versus coated columns correlates well with N_{eff}.

---

* Note that a change in α has a large effect on R_s when α is close to 1.

† This is because V_S/V_M is much lower for capillary columns resulting in smaller k' values relative for packed columns.

5.43 INTERRELATIONSHIP OF TIME AND SEPARATION IN CHROMATOGRAPHY

We turn now to resolution as a function of separation time t, that is, the t_R value of the last eluting band in column chromatography. Our practical goal is maximum resolution per unit time, since this simultaneously favors higher resolution and shorter separation time. As we shall see, resolution can always be improved if more time is allowed for the separation. Alternatively, if we can sacrifice resolution (e.g., if R_s is initially > 1.5), it is always possible to shorten the time of separation.

We have noted that $v = L/t_0$ (see Eq. 5.3) and $H = L/N$. Combining these relationships with Eq. 5.3 gives

$$t = \frac{NH(1 + k')}{v}$$

Eliminating N between the latter equation and Eq. 5.43 now yields

$$t = 16R_s{}^2 \left(\frac{\alpha}{\alpha - 1} \right)^2 \left(\frac{(1 + k')^3}{k'^2} \right) \left(\frac{H}{v} \right) \qquad (5.46)$$

Here we have the time of separation t as a function of the resolution R_s required and the separation conditions (k' refers to the value for the last eluting solute). Equation 5.46 predicts that a doubling of R_s requires a fourfold increase in separation time, if α, k', H, and v are held constant. This could be achieved by a fourfold increase in column length L. Also, Eq. 5.46 emphasizes the importance of the separation factor α on the minimum time of separation. An increase in α from 1.05 to 1.10 permits roughly a fourfold reduction in separation time.

Consider next the capacity factor term of Eq. 5.46: $[(1 + k')^3/k'^2]$ is seen to go to infinity at large and small values of k'. This means that separation time is shortest at an intermediate value of k'. If we assume that α and H are not affected by changes in k', and if, for simplicity, v is held constant (changing v would change H), we can obtain the value of k' for minimum t by differentiating Eq. 5.46 with respect to k' and setting the result equal to zero. The resulting optimum value of k' is found to be equal to 2; that is, $t = 3t_0$. Because of the interrelationship of time and resolution, we see that maximum resolution per unit time also corresponds to a value of $k' = 2$. A similar analysis of resolution in open-bed chromatography, starting with Eq. 5.44, shows that maximum resolution (time is usually held constant) corresponds to $k' = 2$ also. When the possibility of varying v is also considered (column chromatography), and if we take into account the usually minor dependence of H on k', we arrive at optimum values of k' between 1.5 and 5 (see, e.g., Refs. 3 or 4).

According to Eq. 5.46, analysis time is proportional to H/v; this means that in a fixed-time separation R_s will increase as $\sqrt{H/v}$. Thus for an optimized separation (maximum resolution per unit time) we must minimize H/v, not H. We now show that this requires values of v that are greater than v_{opt}, and H values greater than H_{min} (Eqs. 5.38, 5.39).

Consider gas chromatography first. Here H is well represented by the van Deemter equation (5.37), from which H/v is given as $H/v = A/v + B/v^2 + C$. This expression converges on the limiting value C, as v exceeds v_{opt}. Continued increase in v beyond a certain point becomes detrimental, because other factors then operate to increase H beyond the value predicted by the van Deemter equation. In gas chromatography, therefore, it is advantageous to have v greater than v_{opt} (by a factor of about 2), but not too much greater.

In liquid chromatography the mobile-phase diffusion term H_D goes to zero at high flow rates (Eq. 5.33), rather than increasing with v as in gas chromatography. As a result, H/v in liquid chromatography continues to decrease with increase in v. Consequently it is generally preferable in liquid chromatography to operate at relatively high flow rates for a given separation time. Note that this does not mean that resolution is improved simply by increasing v, since H always increases with v in liquid chromatography. Rather, the increase in v must be accompanied by a corresponding increase in column length L to maintain t constant. This is discussed further in Chapter 10. The high mobile-phase velocity required for high-performance liquid chromatography is an additional reason for high-pressure operation with this technique.

In the case of packed columns we should keep in mind that H continues to decrease with decrease in particle size d_p of the packing (equivalent to d in Eq. 5.28). This effect is particularly pronounced and important in liquid chromatography, where d_p is consequently kept as small as practical. However, small values of d_p result in decreased column permeabilities B_o (Eq. 3.70)* and increased column pressure drops for a given mobile-phase velocity v and column length L. Consequently the maximum resolution per unit time in liquid chromatography is limited by the maximum operating pressure of the chromatographic unit. A similar, but less pronounced situation exists in the case of coated capillary columns for gas chromatography. Here H decreases with decreasing column radius r (Eq. 5.29), while B_o is given by $r^2/8$ for ideal open tubular columns. Coated capillary columns have significantly higher permeabilities than corresponding packed columns and therefore provide more rapid analysis.

In view of the interrelationship of separation time and resolution, parameters such as N, N_{eff}, and H do not completely describe the true perform-

---

* To a good approximation we can assume $B_o = d_p^2/1000$.

ance of a chromatographic system. Plates per second N/t and, especially, effective plates per second N_{eff}/t* are more valid criteria in this regard.† In gas chromatographic packed columns N_{eff}/t usually falls between 5 and 30 effective plates per second, while for coated capillary columns N_{eff}/t is usually around 50 to 100. Classical liquid chromatography (i.e., before 1965) was usually characterized by low values, of the order of 0.01 to 0.01, of N_{eff}/t. More recently it has become possible to generate 25 or more effective plates per second in liquid chromatography.

5.5 Multicomponent Separations

Up to this point we have looked only at the separation of two-component mixtures. However, a major advantage of chromatography is the ease with which it can separate samples containing tens—or even hundreds—of individual components. The extension of the preceding treatment to mixtures of three or four components is straightforward, as long as the k' values of these components are not greatly different. Separation can be improved by increasing N, by varying the separation factor α of adjacent bands for further disengagement, and by confining band migration rates to a range that overlaps the optimum value: $k' \approx 2$. The control of separation becomes more complicated, however, for samples that contain many components of widely different k' values. In this section the unique problems of the separation of complex, multicomponent samples receive attention first. Then we turn to several special techniques that have been developed to cope with these problems.

5.51 THE "GENERAL ELUTION PROBLEM"

Consider the separation of a mixture which contains several components of widely different migration rates or k' values. This is illustrated in Fig. 5.14 for elution and in Fig. 5.15 for open-bed chromatography. We assume that sample components *1* and *2* migrate most rapidly, components *3* and *4* less rapidly, and components *5* and *6* least rapidly; that is, their k' values increase in going from *1* to *6*. Now let us further assume some set of experimental conditions A, such that *1* and *2* have k' values close to the optimum value of 2. This means that the k' values for the remaining components are larger than

---

* $N_{eff}/t = N(t_R - t_0)^2/t_R.^3$

† Effective plates per second are a function of a number of variables, so that the use of the values as *true* speed characteristics of, for example, chromatographic packings must be attempted with caution.

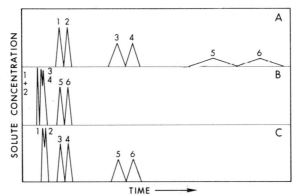

Figure 5.14. The general elution problem in column chromatography.

optimum. In the top portion of Figure 5.14, components *1* and *2* are eluted from the column within a reasonable time as sharp, well-separated bands, *3* and *4* require a longer time for elution, and *5* and *6* are eluted much later. The elution times for the last two bands are excessive, and in some cases strongly retained sample components would not be eluted at all. Bands *5* and *6* have also broadened to the point where they may be difficult to detect in the column effluent. In the case of open-bed chromatography (Fig. 5.15, left) bands *1* and *2* are well separated, because neither N' (equal to NR_f, Eq. 5.13) nor $k'/(1 + k')$ is too small. Bands *3* to *6* migrate only a short distance along the bed, however, and are therefore poorly separated (small N').

To obtain satisfactory separation of components *5* and *6*, we assume some set of experimental conditions *B** such that $k' \approx 2$ for bands *5* and *6* (Figs. 5.14 and 5.15, middle portions). The k' values of bands *1* to *4* are now too small in each case for adequate resolution. If we select an intermediate set of

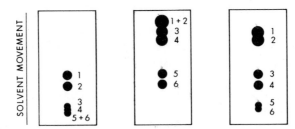

Figure 5.15. The general elution problem in open-bed chromatography.

* Conditions *A*, *B*, and *C* might differ with respect to separation temperature, composition of the stationary or mobile phase, or (less commonly) phase ratio A_S/A_M.

conditions C (Figs. 5.14 lower and 5.15 right) such that the k' values of 3 and 4 are optimum, a compromise situation results as shown.

The situation described above has been referred to as the "general elution problem." It is a widespread phenomenon in separations of multi-component mixtures, and we can summarize its consequences as follows:

- Poor separation of components with relatively small k' values.
- Excessive separation times for components with relatively large k' values in elution from columns; poor separation of such components in open-bed chromatography.
- Broad elution bands that can be difficult or impossible to detect, for components with large k' values.

The general elution problem usually cannot be solved by improvement in bed efficiency N, adjustment of separation selectivity α, or compromise in k'. Actual samples often contain several components whose k' values differ by orders of magnitude for any given experimental conditions. In such cases no single set of operating conditions can yield satisfactory separation of the whole sample. The alternative, as we will see shortly, is to systematically vary conditions *during* the separation. Up to this point we have assumed that separation conditions remain constant.

5.52   THE PROBLEM OF LIMITED "PEAK CAPACITY"

Another general problem that often arises in the separation of multi-component samples is illustrated in Fig. 5.16, for a hypothetical nine-component mixture separated under three different sets of experimental conditions: A, B, and C. In each case we assume that the k' values of the various components are not widely different, so the general elution problem

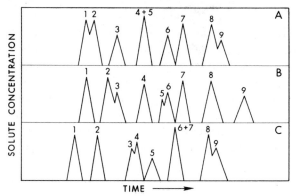

Figure 5.16.   The problem of limited peak capacity.

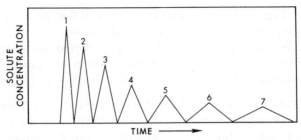

Figure 5.17. The peak capacity of a chromatographic system.

does not arise. In the first separation (conditions A) bands *1* and *2* are incompletely resolved, bands *4* and *5* are unresolved, and bands *8* and *9* are incompletely resolved. Now assume that we know the structure of these components and can predict some set of conditions B such that bands *2, 5,* and *9* are selectively retarded (their k' values are increased relative to those of the remaining components). At first glance this appears to solve our problem, since the resulting α values of band pairs *1–2, 4–5,* and *8–9* now favor their complete resolution. As seen for conditions B, however, the disengagement of the latter three band pairs is accompanied by the merging together of other band pairs: *2* and *3* and *5* with *6*. Similarly, a random change in separation conditions (C) will often result in the separation of bands that were partially resolved in the first attempt at separation, but with simultaneous overlapping of band pairs that were previously well resolved.

The problem in separations of the type shown in Fig. 5.16 is that there are too many components for the space defined by the chromatogram. We have a "molecular traffic jam," in which different bands are constantly getting in each other's way. In this connection it is useful to define a quantity we will call the *peak capacity* of the chromatogram, as shown in Fig. 5.17. If we have a chromatogram that extends from an initial peak *1* to some final peak z (peak *7* in Fig. 5.17), the bed efficiency N will determine the band width of a component at any position in the chromatogram (Eq. 5.11).* Consequently we can calculate the widths of a series of adjacent bands that are just resolved ($R_s = 1$), and thus arrive at the maximum number of separated bands which can be accommodated between the first and last bands in the chromatogram. This number (equal to 7 in the example of Fig. 5.17) is the peak capacity ϕ of the chromatogram. Since band width is proportional to $\sqrt{N}$, ϕ is proportional to $\sqrt{N}$. Similarly ϕ increases with the distance between the first- and the last-eluted bands in the chromatogram, and this distance increases with an increase in the ratio of k' values for the first- and the last-eluted bands (k'_z/k'_1).

---

* Assuming equal N values for all bands (i.e., N independent of k').

When the peak capacity ϕ is large, the crowding of bands (for a given sample) in the resulting chromatogram will be reduced, and problems of the type illustrated in Fig. 5.16 will arise less often. One solution to such problems, therefore, is an increase in peak capacity. However, the required increase in N and/or k'_2/k'_1 may not always be feasible, particularly when ϕ must be increased by a large factor. An alternative solution in cases such as Fig. 5.16, top portion, is to isolate unresolved band pairs after separation as in A, and then resolve these unseparated bands in a second separation under a different set of conditions (e.g., B of Fig. 5.16). Various techniques for achieving such multiple separations without the inconvenience of intermediate handling steps are discussed below under the heading "Multiple Separation."

5.53 SPECIAL TECHNIQUES

Having reviewed the special problems that can arise in the separation of complex, multicomponent samples, we now examine some special experimental techniques for the solution of these problems. For a detailed discussion of the theory of these various techniques, along with an evaluation of their relative merits, see Refs. (5) (gas chromatography), 6 (liquid column chromatography), and 7 (thin-layer chromatography). Their application is further discussed in Chapters 8, 10, 12, and 13.

Change in k' During Separation

The general elution problem (Figs. 5.14 and 5.15) arises because of the large differences in k' for different sample components. For a single set of operating conditions, k' can never be optimum for more than a few components. The obvious solution to this problem is to vary the experimental conditions during separation, so that k' for each sample component is close to optimum during its migration along the bed. To see how this might be achieved, consider the example of Figure 5.14. Suppose that we begin separation with operating conditions A, so that components 1 and 2 are eluted rapidly as narrow, well-resolved bands. As soon as band 2 clears the column, assume that separation conditions change from A to C. The k' values of components 3 and 4 are now reduced into the optimum range (~ 2), and these components are rapidly eluted from the column as narrow, well-resolved bands. Finally, after elution of band 4, experimental conditions are changed again from C to B for optimum separation of bands 5 and 6. The resulting separation is illustrated in Figure 5.18a and should be contrasted with the various separations of Fig. 5.14. Note that bands 1 to 4 do not move far from the column inlet during the initial stage of the sample (conditions A), and that during conditions C bands 5 and 6 undergo little migration. Only under optimum conditions ($k' \approx 2$)

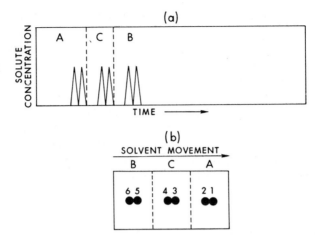

Figure 5.18. Solution of the general elution problem by change of conditions during separation: (*a*) column, (*b*) open-bed.

does significant migration occur for each band, with the result that the overall separation is improved, separation time is reduced, and all band widths are narrow for easy detection.

The same general technique can be applied in open beds, as illustrated in Fig. 5.18*b*. Here we assume that different separation conditions (A, B, and C) prevail over different regions of the chromatographic bed. As a result bands *5* and *6* are significantly retarded in the first bed section ($k' \neq 0$) and are separated there. Bands *3* and *4* pass rapidly through the first section of the bed ($k' \approx 0$) but are retarded in the middle section and separated there. Similarly bands *1* and *2* pass rapidly through the first two bed sections and are retarded in the third section (B). Each pair of bands is resolved in its appropriate section, because k' is equal to neither 0 nor 1 in that section.

The basic technique described above for solving the general elution problem also provides increased peak capacity. Thus in normal elution (e.g., Fig. 5.17) band widths continually increase during the course of separation, so that fewer bands are accommodated per unit eluate volume during the latter stages of elution. If separation conditions are changed during elution, as in Fig. 5.18*a*, however, band widths tend to remain constant throughout separation. As a result a larger number of resolved bands can be accommodated within a given chromatogram, and peak capacity is increased.

Sample k' values can be conveniently altered during separation in one of four ways: change in temperature, change in mobile-phase composition, change in stationary-phase composition, or change in the ratio of moving to stationary phases. These changes in experimental conditions can be either continuous or discontinuous, and they can be used in conjunction with either columns or open beds. A variety of specific techniques for changing k' during separation thereby arise, each of which permits improved separation of multi-component samples. We briefly discuss now several of these techniques that have found practical application.

Deliberate change of temperature during separation, for the purpose of altering sample k' values, is referred to as *temperature programming*. In most chromatographic systems, a solute with a large k' value also has a large heat of sorption. In other words, the large negative free energy that favors reten-tion of the solute in the stationary phase is determined mainly by a large negative enthalpy (Section 2.2). As a result, an increase in temperature generally leads to a decrease in k' (Eq. 2.9), and in most cases temperature programming will involve an increase in temperature during separation. In gas-liquid chromatography, temperature programming is used most often as a solution to the general elution problem. Temperature programming is used less frequently in liquid chromatography (and mainly in ion exchange), because it does not allow sufficiently large variations in k'.

In liquid column chromatography (elution) a change in composition of the mobile phase during separation is referred to as *stepwise elution* if the change or changes are made discontinuously. Continuous changes in mobile-phase composition during separation (i.e., changing composition of solvent mixtures) are referred to as *gradient elution* (or *solvent programming*). In either case the change in mobile-phase composition is such as to give lower k' values for all sample components as separation proceeds. These techniques are widely used in ion-exchange and liquid-solid (adsorption) chromatography, and to a lesser extent in liquid-liquid chromatography. Stepwise and gradient elution are the only techniques in liquid column chromatography by which mixtures with widely different k' values (e.g., k' varying by more than 10^3) can be satisfactorily separated, and these techniques also provide maximum resolution for simpler mixtures.

A change in stationary-phase composition or ratio of stationary to mobile phases during elution from columns can be achieved with the recently intro-duced technique of *coupled columns* (5). Here a short forecolumn prepares rough fractions which are further separated in secondary columns varying in stationary-phase composition or loading. The stationary-phase compositions and/or loadings of the secondary columns are selected to provide optimum separation ($k' \approx 2$) for the rough fraction that is diverted to a particular column. For the example of Fig. 5.14 three secondary columns would be used, giving separation conditions A, B, and C. Bands 1 and 2 from the forecolumn would be sent to the A column, bands 3 and 4 to the C column, and bands 5 and 6 to the B column.

Corresponding techniques for solving the general elution problem in open-beds exist but are less useful. Thus in stepwise or gradient elution from columns the resolution of a given pair of bands (e.g., *3* and *4* in Fig. 5.18*a*) can be equal to that obtained in normal elution with k' optimized for the bands in question (Fig. 5.14, conditions *B*). In corresponding open-bed separations (i.e., with change in mobile phase during separation), the resolution of two adjacent bands will never be as good as can be obtained in normal open-bed chromatography with k' optimized for the two bands. The reason is that in the former case (e.g., Fig. 5.18*b*) the two bands migrate a shorter distance (and N' is smaller) in the region where k' is favorable for separation. An alternative solution to the general elution problem in open beds is to carry out several independent separations of the same sample, as in the three separations of Fig. 5.15. Better separation is thereby obtained than by changing the conditions during separation (Fig. 5.18*b*); moreover, this procedure is experimentally simpler and requires less time and effort.

Flow Programming

When k' is optimized for initially eluting bands in separations on columns, the general elution problem takes the form of very long elution times for strongly retained sample components (e.g., Fig. 5.14, conditions *A*). At the same time the excessive broadening of later eluted bands leads to low peak heights and a corresponding band detection problem. *Flow programming*, a systematic increase in the flow rate of the mobile phase during separation, can be used to solve both of these problems. After the rapid elution of the first components as well-resolved, narrow bands, an acceleration in the flow rate of the mobile phase provides increasingly more rapid elution of strongly retained bands. In this way later-eluting bands appear to emerge from the column in a fashion similar to that shown in Fig. 5.18*a*, when the chromatogram is plotted versus time rather than eluate volume.* If the detector responds to the total quantity of sample per unit time (e.g., a flame ionization detector; see Chapter 8) rather than to sample concentration, later-eluted components will appear as narrow, easily detected bands rather than the broad bands observed in the absence of flow programming. Flow programming can thus provide a reasonable solution to the twin problems of excessive separation time and difficult detection of strongly retained bands.

Flow programming differs fundamentally from changing k' during separation. Column efficiency (value of N) is lowered in the latter part of a flow program, because H is generally larger at high mobile-phase flow rates (Fig. 5.11). Flow programming also provides far less control over band migration rates than is possible by changing k' and thus offers only a limited solution to

---

* Since k' is little affected by changes in column pressure or the flow rate of the mobile phase, band positions are approximately the same in both normal elution and flow programming, when the chromatogram is plotted versus eluate volume.

the general elution problem. Flow programming has found some application, however, in gas chromatography. The resulting lower column
temperatures (relative to temperature programming) favor resolution in
most cases and protect the sample and column from temperature-induced
alteration. On the other hand, flow programming is of limited value in
liquid chromatography. It offers no advantages with respect to stepwise or
gradient elution and coupled-column operation, it provides poorer separation, and it can be used only with samples that have a small range of k'
values. Detection sensitivity with the liquid chromatography detectors
presently available is not improved by flow programming, since all of these
detectors respond to sample concentration in the eluate.

Multiple Separation

A good approach to the problem of limited peak capacity (as in Fig. 5.17)
is the isolation of incompletely resolved groups of bands from an initial
separation, followed by their reseparation under a new set of conditions that
provides different α values. Thus, in the example of Fig. 5.17, a single band
containing components *4* and *5* can be obtained from system *A*. Reseparation
of this single band in system *B* will then yield well-resolved bands of *4* and *5*.
This procedure becomes cumbersome and time consuming, however, when
several pairs of partially resolved bands occur in the initial separation, as is
often the case in dealing with complex, multicomponent samples. A better
alternative is the use of techniques that circumvent the need for intermediate
sample handling: two-dimensional open beds, multiple columns, and elution
open-bed combinations.

   Two dimensional open-bed chromatography is illustrated in Fig. 5.19,
for the example of Fig. 5.16. The sample is applied at one corner of a square
bed (film or plate). Development proceeds in a normal manner along one
edge of the bed, using experimental conditions *A* of Fig. 5.16. Band pairs

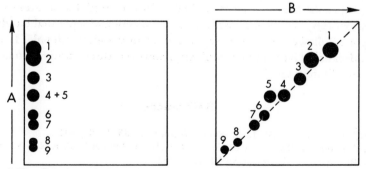

Figure 5.19. Solution of the problem of limited peak capacity by two-dimensional
open-bed chromatography.

1–2, *4–5*, and *8–9* are incompletely resolved in this separation, just as in elution (Fig. 5.16, top). Now a second development is carried out on the same bed, in a direction perpendicular to the first development, with conditions changed from *A* to *B*. Each of the incompletely resolved band pairs from the first development is now resolved, as in Fig. 5.16, middle. The difference between the separations of the latter figure and Fig. 5.19, however, is that no recombination of previously resolved bands can occur in two-dimensional development.

A similar result can be achieved in elution from columns if an initial column is connected through a multiport valve to a series of secondary columns. For the example of Fig. 5.16 the initial column could be operated under conditions *A*, with each of three secondary columns operated under conditions *B*. Unresolved band pairs *1–2*, *4–5*, and *8–9* from the first column (*A*) are then fed in sequence to each of the secondary columns (*B*). Each band pair is thus resolved in a secondary column without the opportunity for remixing with previously unresolved bands. Multicolumn procedures of this kind should not be confused with the previously described technique of coupled columns. The latter aims at a systematic variation in k' during separation, whereas multiple-column separations emphasize variations in α between columns. These two procedures are, however, operationally similar.

A third technique for carrying out automatic, multiple separations is to collect the bands eluting from a column along one edge of a square bed. This results in a distribution of sample along the bed similar to that shown in Fig. 5.19a. The bed is then developed in the same way as in Fig. 5.19b.

Two dimensional open beds have been used widely in paper and thin-layer chromatography (Chapters 10 and 13) because of their simplicity and great separating power (large peak capacity). On the contrary, multi-column procedures as described above have not been employed to any significant extent in either gas or liquid chromatography. This technique is potentially powerful but is unduly complicated to carry out. Combinations of a column plus an open bed have been described. Thus an initial separation by gas chromatography is followed by continuous collection of separated bands on a TLC plate, followed by further separation as in two-dimensional TLC (Fig. 5.19b). Because of its relative complexity, however, this operational mode has found only limited use.

References

1. A. J. P. Martin and R. L. M. Synge, *Biochem. J.*, **35**, 1358 (1941).
2. J. C. Giddings, *Dynamics of Chromatography*, Part 1, Marcel Dekker, New York, 1965.
3. L. R. Snyder, *J. Chromatog. Sci.*, **7**, 352 (1969).
4. B. L. Karger, *Modern Practice of Liquid Chromatography*, J. J. Kirkland, Ed., Interscience, New York, 1971, Chapter 1.

5. W. E. Harris and H. W. Habgood, *Programmed Temperature Gas Chromatography*, John Wiley, New York, 1966.

6. L. R. Snyder, *J. Chromatog. Sci.*, **8**, 692 (1970).

7. L. R. Snyder and D. L. Saunders, *J. Chromatog.*, **44**, 1 (1969).

8. D. H. Desty and A. Goldup, *Gas Chromatography, 1960*, R. P. W. Scott, ed., Butterworths, London, 1960.

9. P. B. Hamilton, *Handbook of Biochemistry*, Selected Data for Molecular Biology, B-47, the Chemical Rubber Co., Cleveland, 1968.

10. M. Mohnke and H. Saffert, *Gas Chromatography, 1962*, Butterworths, London, 1962.

Selected Bibliography

E. Heftmann, *Chromatography*, 2nd ed., Reinhold, New York, 1967.

J. C. Giddings, *Dynamics of Chromatography*, Part 1, Marcel Dekker, New York, 1965.

A. B. Littlewood, *Gas Chromatography*, 2nd ed., Academic Press, New York, 1970.

Appendix   Derivation of Solute Retention Equation in Chromatography

The relationship between the retention volume and column parameters, including the equilibrium constant, can be derived from the principle of mass conservation. Let V = volume of eluent passed through the column (ml), z = distance from inlet end of the column (cm), C_S = solute concentration in the stationary phase (g-mole/ml), C_M = solute concentration in the mobile phase (g-mole/ml), v_S = volume of stationary phase per unit length of column (cm²), and v_M = volume of mobile phase per unit length of column (cm²). Consider an infinitesimal cross-sectional segment of the column which has a thickness dz and contains a certain amount of solute. The change of the amount of solute in this differential segment after passing dV volume of mobile phase can be written as

$$\frac{\partial C_M}{\partial z} \, dz \, dV \tag{5A.1}$$

The solute is, however, distributed between the two phases. The decrease of the amount of solute in the mobile phase is given by

$$-v_M \, dz \, \frac{\partial C_M}{\partial V} \, dV \tag{5A.2}$$

and that in the stationary phase by

$$-v_S \, dz \, \frac{\partial C_S}{\partial V} \, dV \tag{5A.3}$$

The principle of conservation of mass requires that the solute loss from the two phases in the infinitesimal segment be equal to the total loss of solute. This is expressed by the equation:

$$\frac{\partial C_M}{\partial z} + v_M \frac{\partial C_M}{\partial V} + v_S \frac{\partial C_S}{\partial V} = 0 \tag{5A.4}$$

which is often called the de Vault equation and can be considered the basic equation of chromatography. It can be solved easily when linear elution exists (K independent of sample concentration).

Substituting K into Eq. 5A.4, we obtain.

$$\frac{\partial C_M}{\partial z} + (v_M + v_S K) \frac{\partial C_M}{\partial V} = 0 \tag{5A.5}$$

This equation has the general solution

$$C_M = \varphi[V - z(v_M + v_S K)] \tag{5A.6}$$

where φ is an arbitrary function, determined by the initial conditions, that is, by the way in which the same is introduced at the column inlet. When the sample is injected as a narrow plug, the assumption of an initial delta function distribution is reasonable. The delta function is defined by

$$\delta(y) = 0 \quad \text{if } y \neq 0 \text{ and } \int_{+\infty}^{\infty} \delta(y) = 1 \tag{5A.7}$$

With this equation, Eq. 5A.6 becomes

$$C_M = \delta[V - z(v_M + v_S K)] \tag{5A.8}$$

which has nonzero value only for zero arguments. Hence

$$V = z(v_M + v_S K) \tag{5A.9}$$

The solute emerges at the end of the column, where z is equal to the length of the column L and V is equal to the retention volume V_R, which is obtained as

$$V_R = L(v_M + v_S K) \tag{5A.10}$$

Equation 5A.10 can be rewritten as

$$V_R = V_M + V_S K \tag{5A.11}$$

which is equivalent to Eq. 5.6.

Six

SEPARATION CHARACTERISTICS OF INDIVIDUAL METHODS

Preceding chapters have reviewed the fundamental aspects of separation, and following chapters will provide a detailed treatment of the individual separation methods and their application. The present short chapter serves as a bridge between these two major parts of the book. Here we list and discuss briefly some of those characteristics of a particular separation method that make it suitable or unsuitable for a given application. Our treatment draws upon the basic principles discussed in Chapters 2 to 5 and attempts to put them into perspective and to emphasize their practical significance. Thus this chapter can serve as a practical survey to select the most appropriate method to solve a given separation problem, as well as a preview to the contents of Chapters 7 to 18. Moreover, Chapter 19 will describe separation schemes involving two or more methods, based on the characteristics outlined in this chapter.

So far we have examined separation with regard to the underlying physico-chemical phenomena and possible operational modes. These have enabled us to organize separation methods from the fundamental point of view. In practice, however, one is usually confronted with a given separation goal. Therefore it is of considerable importance to classify the various separation techniques according to the goal involved. Because of the practice-oriented nature of such classification, and the fact that each particular separation problem involves specific considerations, the difficulties are great. Nevertheless, there are a few important features of the various methods that are relevant to the separation goal and can serve as the basis for such a classification. The most significant of these features, which are discussed in this chapter, are as follows:

Adaptability.
Load capacity.
Fraction capacity.
Selectivity.
Speed and convenience.

Let us take a brief look at each of these points, keeping in mind that our main interest lies in analytical- and laboratory-scale separations. No claim is made that the following discussion is complete, because the peculiarities of the separation goals and processes make the characteristics listed above somewhat interdependent.

In Table 6.1 a number of methods for separating chemical compounds (cf. Tables 1.1 and 1.2) are compared with respect to the above features. The list of methods is not complete, but it covers the more important separation techniques. The significance of the individual process characteristics and their relationship to the achievement of a given separation goal are covered in the rest of this chapter.

6.1 Adaptability

We define the adaptability of a separation method as the ability of the technique to be used with certain types of samples with regard to their physico-chemical properties. Materials within the range of conventional separation methods can be classified as atomic, molecular, macromolecular, or particulate. Molecular and macromolecular compounds are of major interest to us. Because molecular compounds can be subdivided into volatile and non-volatile materials, we obtain the following three major classes: volatile, nonvolatile, and macromolecular. Table 6.1 shows the adaptability of various separation processes according to these classes. Many separation methods, such as distillation, gas chromatography, or mass spectrometry, involve gases or evaporated samples; hence their adaptability is restricted to volatile compounds (or compounds that can be conveniently made volatile).*
Macromolecules not only are nonvolatile but also possess some properties (i.e., high molecular weight and low diffusivity) that necessitate the use of separation processes having appropriate adaptability, such as gel chroma-

---

* Of course the volatility requirements differ for the three methods.

tography or ultracentrifugation. In between are lower-molecular weight substances having polar and/or ionic functions, which may make these molecules thermally liable so that they cannot be volatilized. Under such circumstances the method selected has to be adaptable to the non-volatile nature of the sample; for example, liquid chromatography or fractional crystallization might be used.

In practice, the adaptability of a particular separation process with regard to the separation goal can easily be recognized, since in most cases an individual method has been developed and used for the separation of one particular type of sample. In some instances, particularly in analytical work, it is advisable to make nonvolatile molecules volatile by derivatization. Separation methods that have adaptability to volatile samples can then be employed. These are often superior with respect to speed, sensitivity, and convenience to techniques that are adaptable to nonvolatile or macro-molecular samples.

6.2 Load Capacity

By load capacity we mean simply the *maximum* quantity of mixture that can be separated with a given efficiency by a particular process. Load capacity is rarely an intrinsic property of the process but depends rather on the particular procedure and the equipment employed. Yet within the range of practical applications we can roughly characterize the various separation methods by the possible load capacity. Distillation and crystallization can be carried out easily on the production scale, but they are less suitable for the separation of a few micrograms in analytical work. Conversely, mass spectrometry and chromatography do not readily lend themselves to production-scale separations.

Table 6.1 indicates typical load capacity ranges for the various separation methods. This information is useful in selecting the proper technique to meet a particular separation goal.

6.3 Fraction Capacity

We define the fraction capacity of a separation process as the maximum number of components that can be separated in a single operation. Many separation methods, such as single-stage extraction, crystallization, and clathration, have a fraction capacity equal to 2. Processes suitable for the separation of multicomponent samples into individual components can have much greater fraction capacities, depending on their efficiency.

In chromatography, fraction capacity is equivalent to peak capacity,

Table 6.1. Characterization of Individual Separation Methods

Method	Adaptability			Selectivity[a]					Fraction Capacity[b]	Load Capacity[c]	Speed[a]	Con-venience[e]
	Volatile	Non-volatile	Macro-molecular	Mol. Weight	Functional Group			Isomer				
					Phys.	Chem.	Shape					
Distillation	++			++	+				10	g-kg	−	−
Gas-liquid chromatography	++	++	++	++	+				100+	mg-g	++	+
Extraction		++	++	+	++	++			2	g-kg	++	++
Liquid-liquid chromatography												
Column	+	++	+	++	++	++			10–20	mg-g	+	+
Paper		++	+	+	++	++			10	μg–mg	+	++
Fractional crystallization		++			+		++		2	g-kg	−	−
Precipitation		++	+			++			2	g-kg	++	++
Ion-exchange chromatography		++	+		++	++			10–100	mg-g	+	+
Liquid-solid chromatography												
Column	+	++		++	++	+	+[f]	++[f]	10–100	mg-g	+	+
Thin layer		++		++	++	+	+[f]	++[f]	10–50	mg-g	++	++

Method						Amount[b,c]			
Gas-solid chromatography	+	++	++	+	+	10–50	mg	++	+
Adsorptive bubble methods (foam)	++	++	+	+		2	mg	−	−
Molecular sieves	++	++	+			2	g	−	+
Exclusion chromatography	+	++	++	+		5–10	mg–g	+	+
Clathration	++	++	++	++	+	2	mg–g	−	−
Dialysis	+	++	++	+		2	mg–g	−	+
Osmosis	+	++	++	++		2	mg–kg	−	+
Ultrafiltration	+	++	++	+	+	2	mg–g	+	+
Electrophoresis		++	++	++		10–50	mg	++	−
Ultracentrifugation		++	++	+		5	mg	+	−
Zone melting	++	+	++	+		2	g–kg	−	−
Thermal diffusion	++	+	+			2–5	g	−	−
Mass spectrometry	++	++	++	++	+	1000+	µg	++	+

[a] ++, marked; +, significant (see Section 6.1).
[b] Rough values which can vary markedly.
[c] Maximum values.
[d] ++, fast; +, intermediate; −, slow.
[e] ++, very convenient, simple apparatus; −, inconvenient and/or complex apparatus.
[f] Can be quite significant, but usually much less marked than functional group selectivity.

which is defined in Section 5.52 as the maximum number of separable bands or compounds yielded by a chromatographic system for a given set of experimental conditions. Fraction capacity or peak capacity can be defined in similar fashion for such separation methods as electrophoresis and mass spectrometry, where a series of separated bands is developed during separation. In chromatography, electrophoresis, and mass spectrometry the fraction capacity can be relatively high, normally between 10 and 100. Fraction capacity is related to the efficiency of a technique for separating sample mixtures in a single step; hence it can serve also as a measure of the efficiency of the separation methods mentioned above.

Table 6.1 classifies the various separation methods according to fraction capacity. These values are in most cases only rough ones, varying widely with experimental conditions. In general, fraction capacity can be increased by allowing experimental conditions to vary during a separation run. Thus an increase in pot temperature during distillation, a change of pH during electrophoresis, or a variation in solvent composition during chromatography can result in a higher fraction capacity than would be otherwise obtained.

6.4   Selectivity

The word selectivity refers to the intrinsic capability of a given separation method to distinguish between two components. Selectivity is related to the fundamental physicochemical phenomena underlying a separation; it will receive the greatest attention in our discussion. In the case of equilibrium processes (Chapters 7 to 15), selectivity is measured by the separation factor α, which was defined in Section 2.13. For the nonequilibrium processes (Chapters 16 to 18) the ratio of migration rates for two components in question yields a corresponding expression for selectivity, for example, the ratio of ion mobilities in electrophoresis. Thus the "selectivity" of a given separation method is rooted in the ability of that method to discriminate on the basis of one or more properties of the components at the molecular level. For example, distillation usually exhibits *molecular weight selectivity*, and as a result volatile compounds that differ significantly in molecular weight can be separated readily by distillation. On the other hand, ion-exchange processes have marked *acid* or *base selectivity*. Therefore, by using such methods, acids or bases can be separated readily from neutral compounds, and individual acids or bases can be separated from each other on the basis of differences in their pK_a values.

We will discuss various types of separation selectivity in terms of the molecular properties that enable us to separate the individual components of a given sample. When these properties all fall into the same selectivity

Table 6.2. Various Types of Separation Selectivity

Type of Selectivity	Examples of Separable Mixtures
Molecular weight or size	Homologs within a given series; polymers and biopolymers (proteins or nucleic acids of different molecular weights).
Functional group	
Physical	Molecules with different, nonreacting functional groups, e.g., R—CO$_2$—R, R—OH ^{a}; molecules with varying numbers of such groups, e.g., monoesters, diesters, triesters.
Chemical	Acids or bases of different pK values; olefins that form silver complexes of different dissociation constants; ions, one of which forms an insoluble precipitate.
Shape	Butane–isobutane; n-hexane–cyclohexane; m-xylene–p-xylene.
Isomer	2-Methyl, 3-methyl, and 4-methyl pyridines; *cis* and *trans* olefins.
Optical	LD-Amino acids; LD-Dopa.
Biochemical	Enzymes; steroids; nucleic acids.

^{a} R is an alkyl or aromatic substituent group.

category, a separation method is selected that offers the appropriate type of separation selectivity, as shown in Table 6.2. Often, however, the molecular properties of sample components cover various selectivity categories. Then either a suitable method that possesses all the required types of selectivity must be found, or several methods have to be used in sequence in order to accomplish the separation of all components (see Chapter 19). It should be noted that the classifications in Table 6.2 are at best rough and may be occasionally misleading. Whereas a few techniques employ heat as the sole separating agent, most of the methods can be used with a variety of separating agents (e.g., solvents in extraction or stationary phases in chromatography), which can exhibit different types of selectivity. As a result, the selectivity of a technique may vary within a wide range, depending on experimental conditions. The following discussion should be helpful in selecting not only the proper separation method but also the optimum conditions.

Molecular weight selectivity refers to the ability of a separation method to resolve compounds on the basis of molecular weight. Thus, in the separation of homologs, only molecular weight selectivity is involved. Many separation methods exhibit some degree of this type of selectivity, but this characteristic is most dominant in mass spectrometry, distillation, gas-liquid chromatography, and exclusion chromatography. With the exception of mass spectrometry, however, other types of selectivity always contribute to separation to some extent. This problem has already been discussed in Section 2.2, for the case of distillation, with respect to the relationship between molecular weight and boiling point. Molecular weight selectivity is often a significant factor in bringing about separation by such methods as extraction, liquid-liquid chromatography, gas-solid chromatography, and foam separations, but here other types of selectivity are of comparable importance, as shown in Table 6.1.

Functional group selectivity refers to the ability of a method to separate molecules that differ in the type or number of functional groups present. The selectivity can be *physical* or *chemical*, depending on the nature of interactions between the functional group and the separating medium. Interactions involving dispersion, dipole and/or hydrogen-bonding forces are considered physical, and their single arrayed effect is designated as "polarity," as discussed in Section 2.2. This term is widely used for comparison but is difficult to quantify. For example, hydrocarbons and halogen derivatives are less "polar," esters and ketones have intermediate "polarity," and alcohols and amines are highly "polar." On the other hand, acid-base equilibria, chemical complexation, and other reaction processes are classified as chemical interactions.

Physical selectivity is dominant, for instance, in liquid-solid chromatography because in this method the components are separated on the basis of their "polarities." Thus, liquid-solid chromatography often makes possible the separation of complex mixtures into groups of compounds that have the same type of functional group, as discussed in Chapter 13. This type of functional group selectivity is also marked in gas-solid chromatography, liquid-liquid chromatography, and extraction, but is accompanied by comparable molecular weight selectivity.

Chemical functional group selectivity is generally much more pronounced than physical selectivity. Therefore compounds having appropriately different functional groups can be easily separated, because physical interactions are weaker than chemical interactions and do not interfere significantly. Since the separation is based on the specificity and strength of chemical bonding, we can frequently estimate or even predict chemical selectivity (e.g., secondary chemical equilibria, Section 2.13). Acid-base equilibria, complex formation, and precipitation are the most commonly employed phenomena for utilizing chemical selectivity. For example, in separation methods involving ion exchange or extraction with acids or bases, components are separated according to their acid dissociation constants (pK values). Most olefins form complexes with silver ions, and this interaction is widely employed in various types of chromatography to bring about the separation of olefins from nonolefins and of individual olefins from each other. Chapter 9 deals in detail with such complexing interactions.

Shape selectivity expresses the ability of a given separation method to distinguish between two components on the basis of molecular geometry and shape. This type of selectivity usually plays a decisive role only when the molecular weights (i.e., the molecule sizes of the sample components) are similar. Some separation processes exhibit marked shape selectivity. Since lattice energy and the melting point of a crystalline solid constitute a function of molecular symmetry, the separation of xylenes by fractional crystallization is based on shape (i.e., symmetry) selectivity. Separations of n-alkanes from branched or cyclic alkanes of similar molecular weight, which utilize adsorption on molecular sieves or clathration with urea, involve another type of shape selectivity based on differences in the minimum cross-sectional area of the molecules. Many adsorption methods employing graphitized carbon black, alumina, or bentonite also exhibit shape selectivity, since planar molecules are preferentially adsorbed. Shape selectivity plays a role also in separations by exclusion chromatography and thermal diffusion. In order to take full advantage of shape selectivity, application of one of these methods is preferably preceded by a separation process having molecular weight selectivity.

Isomer selectivity may be considered as a particular case of functional group selectivity. It is involved in the separation of isomers that do not differ markedly from each other in terms of molecular shape or symmetry, but whose functional groups occupy different relative positions in a given molecule (e.g., substituted pyridines). Liquid-solid and gas-solid chromatography frequently exhibit pronounced isomer selectivity, as the functional groups fixed on the adsorbent surface can interact quite differently with those present in the molecules in such positional isomers.

Biochemical selectivity is based on highly specific interactions between biological molecules, one of which is usually a biopolymer. This type of selectivity is generally too large for multicomponent separations but offers unsurpassed efficiency in the isolation or purification of single biological substances. Immunoprecipitation based on antigen-antibody interactions, and affinity chromatography that utilizes biosorbents made by immobilization of biologically active molecules on solid supports, are processes that utilize biochemical selectivity. Affinity chromatography is discussed in Section 1.3.4.

Optical selectivity refers to the ability of a separation method to resolve racemic mixtures of optical antipods. Biochemical selectivity usually involves high optical selectivity as well. Isomer selectivity can also be used to separate optical isomers when the efficiency of the method is high, such as that of gas chromatography. The most commonly used separation techniques of high optical selectivity are extraction and precipitation with, respectively, an optically active solvent and precipitating reagent.

Other types of selectivity have less practical importance at present. As an example the separation of ortho- and parahydrogen by processes that exhibit magnetic moment selectivity may be noted in passing.

6.5 Speed and Convenience of Separation

Table 6.1 lists also an evaluation of the various separation processes in terms of speed and convenience. These features are related to the separation goal and are particularly important in laboratory work. Routine analysis is preferably carried out with a rapid, automated method, and the high cost of the necessary equipment may be tolerable. Routine separations usually have to be optimized, so that the actual time of separation may be a small fraction of the time required for the preliminary work. Nevertheless fast and convenient methods are always preferred unless the cost is prohibitive. Therefore a great deal of effort is spent on improving separation methods in this regard.

The speed and convenience of separation are interrelated with efficiency, fraction capacity, and load capacity. Greater fraction and load capacities can usually be obtained at a cost of longer time and lower efficiency.

Part Two

Methods Based on Phase and Distribution Equilibria

Seven

DISTILLATION

R. H. McCORMICK

7.1 Introduction

Distillation, a method used for the separation of the components of liquid mixtures, depends on the distribution of constituents between the liquid mixture and the vapor in equilibrium with the mixture. The two phases exist by formation of the vapor phase through partial evaporation of the liquid mixture. Each phase can be recovered separately with the more volatile components concentrated in the vapor, while the less volatile ones are in greater concentration in the liquid. The effectiveness of the separation is dependent on the physical properties of the components in the mixture, the equipment used, and the method of distillation.

A number of different distillation methods, or processes, are used in both laboratory and pilot-plant research, as well as in commercial production of chemicals and related products. The separation of liquid mixtures by distillation is one of the most important and widely used processes in the chemical industry today. The more important distillation methods are discussed in this chapter.

Distillation can be distinguished from evaporation in that it is the separation of a mixture in which all components are volatile, whereas in evaporation volatile components are separated from nonvolatile ones. For example, distillation would be used to separate a mixture of alcohol and water, but evaporation would be employed to separate water from an aqueous sodium chloride solution.

Conventional distillation is carried out by two methods. The first consists of producing vapor by boiling the liquid mixture to be separated, and then condensing and collecting these vapors without returning them to the still pot. This method is commonly referred to as *simple distillation*. The second method, and the one most widely used, is based on the return of a portion of the vapor condensate to the distillation unit under such conditions that this condensate is continuously and countercurrently in contact with the vapors. The liquid portion returned is referred to as *reflux*, and the method is called *fractional distillation* or *rectification*. This method, which provides a greater enrichment of the vapor in the more volatile component than can be produced in simple distillation, is usually performed in multiplate columns such as are shown in Fig. 4.4.

Fractional distillation can be carried out batchwise or continuously, whereas simple distillation is a batch process. As mentioned in Chapter 4, fractional distillations in the laboratory are usually done batchwise, mainly because the equipment is relatively small and the stream flow rates are low.

In *batch distillation* the only streams are the distillate product and reflux, both of which can be easily controlled automatically by the product take-off arrangement. On the other hand, in *continuous distillation* there are both distillate and bottom product feed and reflux streams, all of which must be controlled automatically. This is difficult to do at low flow rates. For pilot-plant-size equipment, in which flow rates are higher, continuous distillations can be done rather easily. For industrial uses, essentially all distillations are carried out by a continuous process.

In addition to the two conventional methods of distillation mentioned— simple and fractional—there are a number of more specialized types adaptable to particular mixtures, conditions, and applications. *Flash distillation* consists of vaporizing instantaneously and continuously a definite fraction of the liquid mixture in such a way that the total vapor produced is in equilibrium with the residual liquid. In contrast, in simple distillation the distillate product is removed differentially, and only at the time of removal is it in equilibrium with the residual liquid. The vapor and liquid fractions from the flash distillation are separated and have compositions quite different from those produced by any other distillation method. This method is ordinarily used for a mixture containing components with widely different boiling points.

Vacuum (reduced-pressure) *distillation* is ordinarily used to separate high-boiling mixtures or materials that decompose below their normal boiling points. Decreased pressure reduces these boiling-point temperatures.

Steam distillation is another special method utilized to separate high-boiling mixtures or to separate a material from a nonvolatile impurity. This method utilizes rather simple and inexpensive equipment. The boiling temperature of the mixture is reduced by vaporizing it into a stream of carrier vapor (steam), which upon condensation is immiscible with the original mixture and thus can be easily separated from it.

Azeotropic and *extractive distillations* are used for mixtures that are difficult to separate. In each of these distillation methods the relative volatility of the components in the mixture is altered by adding another substance. These methods find their principal use in the separation of mixtures whose components boil too close together for economic conventional fractional distillation.

Sublimation can be classed as a highly specialized distillation method, in that it is the direct vaporization of a solid without passing through a liquid phase. Ordinarily it is used as a purification method, and the desired product can be either the sublimed or the residual material.

7.2 Theory

7.21 VAPOR-LIQUID EQUILIBRIA

The equilibrium of phases can be defined as a state in which transfer of material in one direction and transfer in the reverse direction are taking place at the same rate, so that the concentrations in the two phases remain constant. This definition can be applied to the vapor and liquid phases in the distillation of a mixture. A limit to the changes in concentration, or the mass transfer, between phases is reached when the two phases come to equilibrium, because then a net transfer of material no longer takes place. Therefore, for a practical separation process *in which finite production is needed*, achievement of equilibrium may be undesirable, since the rate of mass transfer between phases is proportional to the driving force, which is the concentration difference or the departure from interphase equilibrium.

The controlling variables of equilibria are temperature, pressure, and the concentrations in the phases. The number of independent variables or degrees of freedom available for a specific case can be determined from Gibbs's phase rule, which was given as Eq. 2.11:

$$P + F = C + 2$$

where F = number of degrees of freedom, C = number of components, and P = number of phases. In a conventional distillation process there are only vapor and liquid phases, so that $P = 2$.

Although numerous laws have been derived to determine the vapor-liquid relationships for mixtures, in general these simple laws are not followed very closely. Raoult's law (Section 2.22), which assumes that the liquid phase is ideal, states that the partial pressure of any component is equal to the mole fraction of the component in the liquid times the vapor pressure of that component in the pure state at a given temperature:

$$P_A = X_A P_A^0 \quad \text{and} \quad P_B = X_B P_B^0 \tag{7.1}$$

and

$$P = P_A + P_B \tag{7.2}$$

where P = total pressure, P_A and P_B = partial pressures of components A and B, P_A^0 and P_B^0 = vapor pressures of components A and B, and X_A and X_B = mole fractions of components A and B. Diagrammatically this law is illustrated in Fig. 7.1. It relates the concentration and partial pressures for the components, A and B, and the total pressure for a mixture.

Raoult's law can frequently be applied to a component present in high concentration in a mixture, for example, P_A when X_A is high. Occasionally this law is followed over the entire concentration range of the component. A mixture of stereoisomers of a compound generally exhibits such ideal behavior.

Henry's law applies to a component present in low concentration in a mixture, for example, P_A when X_A is low. This law states that at any given temperature the partial pressure of a component is equal to its mole fraction in the liquid multiplied by a constant (Eq. 2.14):

$$P_A = H_A X_A \tag{7.3}$$

where H_A is the Henry's law constant.

It can be seen that Henry's and Raoult's laws are identical for ideal solutions, and that Henry's law constant H_A for a given component is the vapor pressure P_A^0.

Dalton's law applies to an ideal gas phase. This law states that the partial pressure P_A of any component in the vapor is equal to its mole fraction in the vapor Y_A times the total pressure P of the system:

$$P_A = Y_a P \tag{7.4}$$

Then equating Dalton's and Raoult's laws gives

$$Y_A P = P_A = X_A P_A^0$$

or

$$Y_A = \frac{X_A P_A^0}{P} \tag{7.5}$$

The equilibrium relationship between the mole fraction of a component in the vapor phase Y_A and the mole fraction of the same component in the

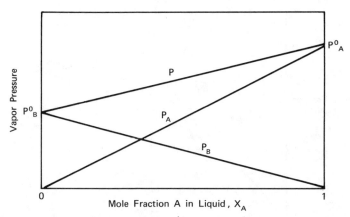

Figure 7.1.   Illustration of Raoult's law.

liquid phase X_A can be determined for a binary mixture for any given total pressure from known vapor pressure data. A plot of these relationships is an *equilibrium diagram* $(X - Y)$, such as is shown in Fig. 7.2.

Combining these two laws leads to the expression used as the definition of *relative volatility* α between components A and B (cf. separation factor, Eq. 2.22):

$$\alpha_{AB} = \frac{Y_A/Y_B}{X_A/X_B} \tag{7.6}$$

and

$$\alpha_{AB} = \frac{P_A{}^0}{P_B{}^0} \tag{7.7}$$

Equation 7.6 defines α in terms of the mole fractions of the components in the vapor and liquid phases at equilibrium. This relationship holds for both ideal and nonideal mixtures, whereas Eq. 7.7 is valid only for mixtures that obey Raoult's law.

The ratio of the vapor pressures of the pure components is frequently almost constant over moderate temperature and concentration ranges, and for such cases the relative volatility also remains essentially constant. Conversely, for nonideal mixtures the relative volatility usually varies widely with composition.

As mentioned earlier, few mixtures or solutions exhibit ideal vapor-liquid behavior. A number of equations have been developed to provide corrections for this nonideal behavior from theoretical considerations (10). A quantitative basis for anticipating solution behavior was discussed in Chapter 2.

A mixture whose total pressure is greater than that computed for ideality

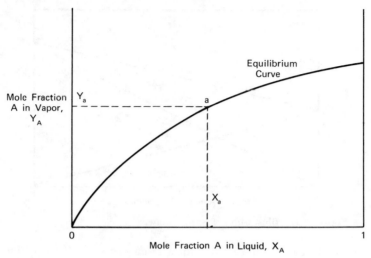

Figure 7.2. Vapor-liquid equilibrium diagram.

from Eq. 7.2 shows a positive deviation from ideality. Conversely, solutions with a lower pressure than that computed show negative deviations. Most solutions show positive deviations. These deviations are due to nonideal behavior in either the liquid or vapor phase or in both phases. They are dependent on the molecular structure of the compound, as well as on many other variables such as mixture composition and temperature.

These deviations from ideality can be determined experimentally for each component and then treated as *activity coefficients* γ by insertion into Eqs. 7.5 and 7.7 as multiplying factors (cf. separation factor in gas-liquid chromatography, Eq. 8.21):

$$PY_A = \gamma_A P_A{}^0 X_A$$

and

$$PY_B = \gamma_B P_B{}^0 X_B \tag{7.8}$$

or

$$\alpha_{AB} = \frac{\gamma_A P_A{}^0}{\gamma_B P_B{}^0} \tag{7.9}$$

Because of the nonideal behavior of most mixtures, equilibria are frequently determined in the laboratory to predict accurately the separation of binary and multicomponent mixtures. To obtain such equilibria, single- or multiple-stage equilibrium units are utilized in the laboratory. The relative volatility of the mixture, which is a measure of the separation effected between the vapor and liquid phases, is then determined from the concentrations of the components in the two phases. The total separation over such an equilibrium unit is referred to as the *enrichment factor* and is defined as

$$\text{Enrichment factor} = \alpha_{AB}^{n} \tag{7.10}$$

where n = the number of theoretical plates in the equilibrium unit.

This equation is another form of the Fenske equation (2). For single-stage units the relative volatility is then equal to the enrichment factor. It can be readily seen that for systems difficult to analyze or having a low relative volatility the single-stage unit may produce unreliable equilibria and relative volatility values. The use of multistage units produces much more accurate results because it reduces those sources of errors in the calculations due to the power factor n being greater than unity. In the multistage unit the relative volatility determined is an average value over the concentration range covered.

A typical procedure to determine such experimental data is to distil a mixture in a previously calibrated equilibrium unit without removing any distillate product; this corresponds to total reflux conditions. The exact number of theoretical plates that the unit contains is known from previous testing with a standard mixture. At equilibrium, samples of the vapor and liquid phases are removed for analysis.

Whether equilibrium data are obtained experimentally or calculated theoretically, they can frequently be best shown on diagrams (5, 9, 12). One such method is a plot of temperature as a function of concentration at a constant pressure, shown in Fig. 7.3. On this diagram T_A and T_B are the boiling points of components A and B, respectively. Component A is the more volatile. These points can be determined by experiment or from vapor pressure data for pure A and B. The upper curve is the saturated vapor line, and the lower curve is the saturated liquid line. At any point between these two curves a mixture of the vapor and liquid phases exists. Below the bottom curve only a liquid phase is present, while above the upper curve only vapor exists.

A liquid mixture with composition X_a has a boiling point T_1. The vapor in equilibrium with X_a has a composition Y_a. The boiling point of Y_a, after it is condensed, is T_2. The composition of any mixture lying on the saturated liquid line is likewise in equilibrium with the vapor on the saturated vapor line at the same temperature.

Another graphical method used to show the equilibria between the vapor and liquid phases of a binary mixture is the equilibrium curve, Fig. 7.2, which was defined earlier. Since this relationship is for a given pressure, a series of such equilibrium curves can be developed as the pressure is changed. On this diagram a is a point on the equilibrium curve where a liquid phase of composition X_a is in equilibrium with a vapor phase with composition Y_a.

Distillation

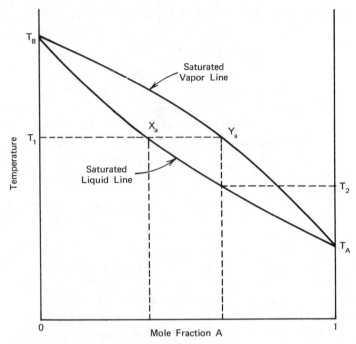

Figure 7.3. Boiling point-composition diagram.

Any other point could be similarly selected on this equilibrium curve. This curve can be experimentally established as previously described from data obtained in a single- or multiple-stage equilibrium unit, or from vapor pressure data if the mixture is assumed to behave ideally. In a multistage unit a number of points such as a on the equilibrium curve can be established in one run.

7.22 FRACTIONAL DISTILLATION

Fractional distillation was defined earlier in this chapter, and some aspects of it were discussed briefly in Chapter 4. A fractional distillation column can be thought of as a cascade of single-stage distillation units, stacked vertically and interconnected so that the vapor and liquid phases can move counter-currently through the assemblage of stages. Figure 4.4 shows such a column for continuous operation. In practice, particularly in the laboratory and pilot plant, packings are frequently used in these columns instead of the individual stages as shown.

Much of the theory and many of the equations used in fractional distillation are similar to those for simple distillation, particularly those involving the

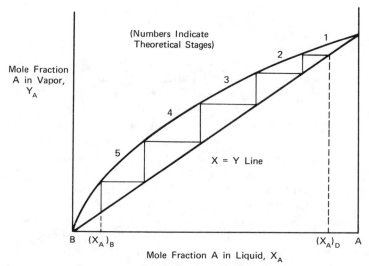

Figure 7.4. Graphical method for minimum theoretical plates at total reflux.

vapor-liquid relationships. On the other hand, fractional distillation is multistage, using reflux, and frequently it is a continuous process. In the design and operation of fractional distillation equipment there are many important factors to consider, including material and energy balances. Only a few of these are discussed in this chapter. Some of these factors can be calculated from the equilibria and the specifications for the end products. One of the most important characteristics of a distillation column is its separational ability. For reflux in a fractional distillation column, the Fenske equation relates the number of theoretical plates n and the equilibrium for the binary mixture, assuming a constant relative volatility α:

$$\left[\frac{Y_A}{Y_B}\right]_{Distillate} = \alpha^n \left[\frac{X_A}{X_B}\right]_{Bottoms} \tag{7.11}$$

If α and the equilibrium data are known for a binary mixture, it can be seen that the minimum number of theoretical plates needed to make a specified separation at total reflux can be calculated from this equation. On the other hand, Eq. 7.11 can be used also to determine experimentally the number of theoretical plates in a column from the analysis of the distillate and the bottom product, and the value of α for the test mixture.

Graphically, the use of this equation for a binary mixture can be represented on a conventional equilibrium diagram (e.g., Fig. 7.2) by stepping off the number of stages between the $X = Y$ line and the equilibrium curve from the distillate to the bottoms composition, as shown in Fig. 7.4 (5). Thus, for a separation that will produce a distillate with composition $(X_A)_D$ and a bottoms product with composition $(X_A)_B$, five theoretical plates are required at total reflux.

The number of theoretical plates required from the Fenske equation and from this graphical method is a minimum because of the conditions of total reflux. In actual practice, for distillation at a practical finite reflux ratio, about twice this number is required. This method of graphically calculating the number of theoretical plates can be used for either batch or continuous distillation columns, since at total reflux feed is not introduced and products are not removed.

In the graphical stepwise method just presented, the $X = Y$ line can be referred to as the *operating line* for the distillation. The operating line for a fractional distillation can be defined as a locus of points for the composition of the liquid phase leaving and the vapor phase entering a stage in a column. This operating line then shows the relationship between the concentrations in the two phases at any point in the column. In a typical distillation the vapor and liquid streams leaving a stage are at equilibrium, but those entering are not. At total reflux the concentrations of the liquid phase leaving and the vapor phase entering a stage are the same, since the compositions on each stage remain constant. In an actual distillation at a finite reflux ratio these are not equal; hence the operating line will lie somewhere between the $X = Y$ line and the equilibrium curve.

A complete presentation of the different methods available to analyze the operation of distillation columns is beyond the scope of this chapter. However, a diagram representing the McCabe-Thiele graphical stepwise method (6) for analysis of a continuous distillation operation (such as would be done for a column like that of Fig. 4.4) is shown in Fig. 7.5. For the case of a continuous column there are two operating lines, as shown. One represents the operation below the feed point, in the stripping section of the column. The other is for the enriching section, above the feed plate. Their intersection is at the feed composition. The slope of these lines is the liquid-to-vapor mole ratio in that part of the column, which for ease of calculation is usually assumed to be constant.

As mentioned before, distillation columns are operated at less than total reflux, with the reflux ratio dependent on the experimental conditions, the mixture to be separated, and the desired product. There is a lower limit, or minimum reflux ratio, at which a separation is not effected even with an infinite number of theoretical plates. From Fig. 7.5, as the reflux ratio is decreased, the point at which the two operating lines intersect will move toward the equilibrium curve along a line dictated by the condition and composition of the feed. When either or both of the operating lines just intersect the equilibrium curve, the number of stages, shown by steps on the diagram, becomes infinite. The reflux ratio corresponding to this situation is, by definition, the *minimum reflux ratio*.

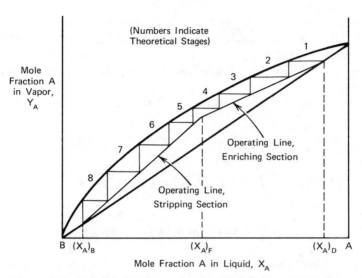

Figure 7.5. McCabe-Thiele graphical method for theoretical plates.

In batch distillation, the equipment is essentially the same as that for continuous distillation, that is, a reboiler, a packed or plate column, and a condenser. However, the operation is different. Instead of continuous feed to the column, a batch of liquid is charged to the reboiler and the system is allowed to reach equilibrium (steady state at total reflux). Distillate product is then continuously or intermittently removed, so that the column operates as an enriching section only. Batch distillation is a differential operation, in that the more volatile constituents are removed as distillate product; hence the concentrations in the reboiler and throughout the column are constantly changing as the distillation proceeds. On the other hand, the composition of the distillate product is dependent on the reflux ratio. If the reflux ratio is maintained approximately constant, as is usually the case, the concentration of the more volatile component in the distillate will decrease. In order to maintain a fixed distillate composition throughout a portion of the distillation, it is necessary to constantly increase the reflux ratio.

Another factor that is important in batch distillation is the liquid holdup in the column, as compared to the holdup in the reboiler. As this ratio decreases, the separation improves. The following equation (9) applies for a batch distillation at constant reflux ratio with negligible holdup (cf. Eq. 4.4):

$$\ln \frac{L_1}{L_2} = \int_{x_2}^{x_1} \frac{dX}{X_D - X} \tag{7.12}$$

where L_1 and L_2 = total moles of liquid in the reboiler at the start and finish of distillation, respectively; X, X_1, X_2 = concentrations of a component in the reboiler at any time, at the start of distillation, and at the end of distillation, respectively; X_D = concentration of a component in the distillate product. This equation is usually solved by graphical integration between the desired limits.

For the case of a constant, or fixed, distillate composition Eq. 7.12 becomes

$$\ln \frac{L_1}{L_2} = \ln \frac{X_D - X_2}{X_D - X_1}$$

or

$$(L_1 - L_2)X_D = L_1 X_1 - L_2 X_2 \tag{7.13}$$

7.3 Experimental Techniques

In the first part of this chapter, as well as in sections of Chapters 2 and 4, the theoretical aspects of distillation were covered. In the rest of this chapter experimental techniques employed in laboratory and pilot-plant distillation studies, as well as their application to industrial uses, are discussed.

The design of small laboratory distillation equipment often presents a number of problems not encountered in the design of larger units; therefore the conversion or scale-up of results must be done with care. One common problem is that heat (enthalpy) losses from small-diameter columns can represent an appreciable percentage of the heat loss in the vapor stream, so that the stream and heat flows measured may then be inaccurate and not representative. Another example is the difficulty in maintaining an accurate control of the reflux ratio or the splitting of the overhead condensate into reflux and product streams for small flows. On the other hand, high individual plate efficiency or, in the case of packed columns, a low "height equivalent to a theoretical plate" is frequently attained more easily in the laboratory, where operating conditions can be more accurately controlled.

As mentioned previously, much of the theory and many of the equations utilized in predicting vapor-liquid equilibria and the distillation performance of binary and multicomponent systems are based on ideal behavior in the vapor and liquid phases. In general, however, most mixtures do not follow ideality. Even though rather elaborate methods have been developed to make corrections for this nonideal behavior, experimental procedures must frequently be resorted to for accurate results.

In the area of vapor-liquid equilibria, accurate experimental data are used in the design of equipment and the prediction of the behavior of mixtures. The procedure for procuring such data is relatively simple if the proper equipment and analytical techniques are available. Basically, it amounts to

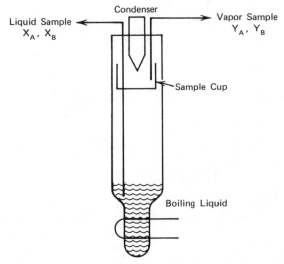

Figure 7.6. Single-plate vapor-liquid equilibrium unit.

determining the compositions of the vapor and liquid phases at equilibrium, for the conditions provided, in a unit with a predetermined number of theoretical distillation plates. A schematic drawing for a typical one-plate equilibrium unit is shown in Fig. 7.6. This unit would be entirely insulated to minimize heat losses. In such a unit the amount of the mixture in the vapor phase represents a small fraction of the total mixture in the unit. The mixture is refluxed, and samples of vapor and liquid are removed for analysis to determine the equilibria. The relative volatilities α of the components in the mixture are then calculated from Eq. 7.6.

Since the relative volatility α corresponds only to these compositions, to obtain a complete equilibrium diagram a number of different compositions must be used. This procedure can be utilized at atmospheric, as well as reduced and elevated pressures if the equipment is adaptable. Proper sampling of the phases and accurate analytical procedures are of utmost importance. By using multiplate equilibrium units, the accuracy of the results is enhanced and the number of determinations required for a knowledge of the complete equilibrium of a system is reduced.

Many examples can be cited to illustrate the importance and usefulness of accurate experimental equilibrium data that are not predictable from previous correlations or equations. One illustration is the equilibrium for a three-carbon-atom hydrocarbon system used for the purification of propylene to produce polypropylene polymer (4). To produce most polymers, compounds of extremely high purity are required, and in this case the relative volatilities α between propylene and the trace impurities are extremely small

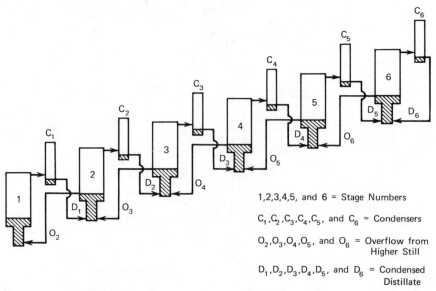

1,2,3,4,5, and 6 = Stage Numbers

$C_1, C_2, C_3, C_4, C_5,$ and C_6 = Condensers

$O_2, O_3, O_4, O_5,$ and O_6 = Overflow from Higher Still

$D_1, D_2, D_3, D_4, D_5,$ and D_6 = Condensed Distillate

Figure 7.7. Flow diagram for six-stage vapor-liquid equilibrium unit.

and vary with the composition of the mixture. Therefore accurate equilibria are required to design highly efficient fractional distillation columns. This research was done in a six-stage vapor-liquid equilibrium unit (7). The use of such units (8) makes it possible to attain a higher degree of accuracy on mixtures of low relative volatility. A diagrammatic sketch of this six-stage unit is shown in Fig. 7.7.

By employing such a unit, equilibrium data can be obtained for six different concentrations. Also, by using the compositions of the liquid from stage 1 and the vapor from stage 6, the enrichment factor over six stages, which is the relative volatility to the sixth power, α^6, can be calculated. This greatly improves the accuracy of the equilibrium and relative volatility α values.

7.31 CONVENTIONAL DISTILLATION

Conventional distillation includes both simple and fractional processes. Simple distillation is always a batchwise process, whereas fractional distillation can be batch or continuous. Only batch distillation operation will be covered here.

In a simple distillation the apparatus is usually a conventional laboratory distillation flask fitted with a condenser and a product receiver, as shown in Fig. 7.8. This is a differential type distillation in which none of the distillate is returned as reflux. For a simple distillation, in which entrainment or

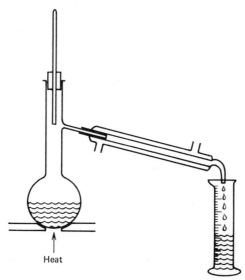

Figure 7.8. Simple distillation apparatus.

cooling of the vapor does not occur before entering the condenser, the vapor leaving the liquid at any time is in equilibrium with the liquid only at that particular time.

As a method of separation, simple distillation is not very effective. However, many such processes are employed in pilot-plant operation and in industry, especially for cases in which the components to be separated have widely different boiling points, and where sharp separations are not required. The practice frequently followed in simple distillations is to collect distillate fractions that represent various product purities, for example, successive 10% fractions of a charge. In commercial use, simple batch distillation has the advantage of easily installed, usually relatively inexpensive equipment. The equipment includes a steam-heated boiler, a condenser, and one or two tanks as receivers. Operation is relatively simple.

One of the most popular laboratory techniques for simple distillation, which also serves as a good example of the usefulness of this type of distillation, is the ASTM method published under the title "Distillation of Petroleum Products" (1). This is a standardized, simple distillation procedure for volatile petroleum products, by which their volatilities can be measured and compared with those of other samples from the resulting "boiling point versus percent distilled" curves. The equipment is similar to that shown in Fig. 7.8. A plot of temperature versus percent distilled for the simple distillation of a multicomponent petroleum fraction is shown in Fig. 7.9, along with

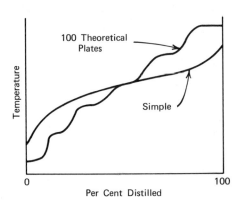

Figure 7.9. A comparison of simple and fractional distillation for a petroleum fraction.

a plot for the batch fractional distillation of the same fraction in a 100-theoretical-plate column. The boiling points of the components in this multicomponent petroleum fraction have moderate differences. Distillation in a 100-theoretical-plate column at a finite reflux ratio would then effect the separation between the components that is indicated by the plateaus in the curve. On the other hand, in simple distillation, reflux is not used, so that separation between individual components is at a minimum with no identifiable plateaus on the curve. As the number of components in such a mixture decreases and the boiling-point differences increase, these separations become more pronounced in both types of distillation. For example, for a binary mixture with boiling-point differences of 5°C for the components, distillation in a 100-theoretical-plate column would show a high percentage recovery of each component in high purity.

An example of simple distillation in pilot-plant and industrial use is the removal of small amounts of a volatile material, such as solvent, for final purification of a product.

7.32   FRACTIONAL DISTILLATION

Batch fractional distillation is one of the most widely used separation methods in the laboratory. This type of distillation is also a differential process, in that the distillate product is removed in fractions, and the equilibrium is constantly changing. The design and the resulting separating power of fractionating columns depend largely on their intended use. Their height may range from 1 to 50 ft, and their diameter from $\frac{1}{2}$ in. to 6 in.; and the number of theoretical plates may vary from a few to several hundred. These columns invariably contain a packed section (or the equivalent such as a spinning band) for contacting the vapor and liquid phases and usually are

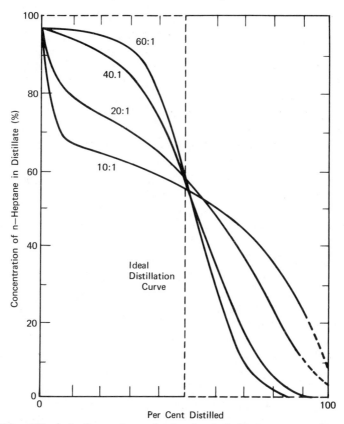

Figure 7.10. Effect of reflux ratio on the fractional distillation of a 50–50 mixture of *n*-heptane and methylcyclohexane. Column: 15-mm-i.d. glass tubing packed for 45 in. with $\frac{1}{16}$-inch-i.d. wire helius; 75 theoretical plates; throughput = 550 cc/hr; operating holdup = 50 cc.

not made up of individual stages. Many types of packing are available; these yield different operating characteristics, of which the most important are the number of theoretical plates and the pressure drop. The packing used also is somewhat dependent on the column diameter, so the packing must be selected to fit the column.

The use of batch laboratory fractional distillation varies widely, from rather easy separations requiring only a few theoretical plates and a low reflux ratio, to difficult separations necessitating a high number of theoretical plates and a high reflux ratio. The degree of separation obtained is largely dependent on the number of theoretical plates in the column and the reflux ratio used.

Figure 7.10 is a plot of distillate composition as a function of percentage of charge distilled to show the effect of reflux ratio for the fractionation of a 50–50 mixture of n-heptane and methylcyclohexane. The equilibria for this binary mixture are well established and are frequently used to determine the efficiency of fractional distillation columns. The relative volatility α for the mixture is constant at 1.07 over the complete concentration range. The boiling points at 760 mm of mercury are as follows: n-heptane $= 98.4°C$ and methylcyclohexane $= 100.8°C$. The separation for the 60:1 reflux ratio approaches that of total reflux. At 10:1 reflux ratio the separation shown approaches that of a simple distillation. This reflux ratio is lower than the minimum required to effect an efficient separation for this mixture. In making a routine laboratory distillation in this column on such a mixture, a reflux ratio of about 40:1 would be used, since it would allow a fairly good separation to be made in a reasonable length of time.

The effect of the number of theoretical plates in the column is not shown here, but it would be similar to that of the reflux ratio; that is, the degree of separation decreases rapidly as the number of plates decreases, especially as it approaches the minimum number. As discussed previously, there is a minimum number of theoretical plates and a minimum reflux ratio that can be calculated for each mixture. Operation below either of these levels will not produce the desired separation.

Before the development of chromatography and modern spectroscopic methods of analysis, batch fractional distillation was widely used in the analysis of complex multicomponent mixtures. The procedure usually consisted of a careful distillation in an efficient column to separate the mixture into various fractions. Thus the components were concentrated into selected boiling ranges that allowed a more accurate analysis from physical property information. Regardless of the complete analytical procedures used, distillation is still an excellent method to prepare large-volume fractions for subsequent analysis.

Probably the extreme example of the use of batch laboratory distillation is the preparation of high-purity compounds from mixtures that are otherwise practically inseparable, for example, hydrocarbons with a 0.5°C difference in the normal boiling points. Typical conditions for such a separation are as follows:

Column diameter	= 0.75 to 1.25 in.
Column height	= 12 to 15 ft
Number of theoretical plates	= 200 to 300
Charge	= 1 to 2 liters
Time for distillation	= 1 week

Fractional distillation columns with a number of theoretical plates comparable to that shown above are also employed to make separations industrially. Two examples are the separation of high-purity ethylbenzene from the xylene isomers for the production of the styrene polymer, and the separation of high-purity propylene from other three-carbon-atom hydrocarbons for the manufacture of propylene polymer. In both cases the products must be of high purity, that is, their preparation by fractional distillation requires very efficient columns.

7.4   Other Types of Distillation

Although conventional types of distillation are the most widely used in the laboratory, as well as in pilot plants and industry, they are not applicable to many separational problems that involve distillation. A number of other distillation processes will now be described.

7.41   VACUUM DISTILLATION

Distillation at reduced pressures has many different laboratory and industrial applications, covering a range in absolute pressures from several hundred to 0.001 mm of mercury. As might be expected, entirely different equipment is required for these pressure extremes. High-vacuum distillation frequently involves fractionations at absolute pressures of 1 to 50 mm of mercury. This pressure range is generally applicable to the use of fractional distillation in packed columns. Below a pressure of 1 mm of mercury, special types of distillation are normally required and these are usually single-stage operations. Above 50 mm of mercury, conventional stage type or packed columns can be employed.

Probably the main industrial use of vacuum distillation at pressures of 1 mm of mercury and higher is for the separation of high-boiling compounds and petroleum fractions. For example, petroleum fractions from the kerosene to the lube oil range are produced by fractionation at reduced pressures. A typical laboratory vacuum distillation is the purification of diethylhexylphthalate at an absolute pressure of 10 mm of mercury. At pressures below 1 mm of mercury, some of the main industrial uses are in the food, pharmaceuticals, and plasticizer industries whose products cannot withstand excessive temperatures. Operation at reduced pressures will effect a better separation of some mixtures because of the dependence of relative volatility α on pressure; for example, α increases as absolute pressure decreases.

As in other fractional distillations, good separation power is desirable for vacuum distillation equipment. In a packed column this would be referred to as a low value for the height equivalent to a theoretical plate (HETP).

The maximum operating throughput of a vacuum column is much less than at atmospheric pressure, because of the resulting increase in vapor volume and velocity at reduced pressures. The importance of heat losses is also increased. To maintain the reboiler at a minimum temperature and to enhance the operation of the column, a low pressure drop is very important. The main problems in designing and operating vacuum distillation columns are the result of these considerations. Therefore there are many designs for vacuum distillation equipment, largely dependent on the intended use.

A laboratory distillation column frequently used for absolute pressures of 20 to about 250 mm of mercury consists of a 24-in. high, 0.5- to 1.0-in.-i.d. tube packed with one of the conventional small-column packings. This column has the equivalent of about 20 theoretical plates, and a pressure drop in the range of 5 to 8 mm of mercury. It is fitted with auxiliary equipment necessary for vacuum operation and operates satisfactorily at these reduced pressures.

With careful operation, such a column can be used at pressures as low as 8 to 10 mm of mercury at the top of the column. However, for the best operation below an absolute pressure of 20 mm of mercury, the pressure drop over the column should be less than 5 mm of mercury. This usually means a shorter packed section or a less efficient packing, resulting in a less effective separation.

The spinning-band column provides another approach to laboratory-scale distillation. In this column the vapor and liquid streams are contacted by a spiral screen band rotated on a vertical shaft in the column. Vapor-liquid contact is enhanced by the spiral band, which throws both the liquid and the vapor onto the column wall. The wall is contacted by the edges of the spinning band, which contain teeth to brush the liquid from the wall.

Spinning-band columns can be operated at quite high throughputs with a very low pressure drop. The operating holdup of the column is quite low, making it possible to carry out a distillation on small samples. The separating power of the columns decreases as the throughput increases. The separations obtainable are considered good at lower throughputs and modest at higher levels. Because of the low pressure drop in the columns, they can be utilized quite satisfactorily at pressures as low as 1 mm of mercury. Since the columns are available in a variety of sizes, they are adaptable to many different applications. For example, the columns range in diameter from 6 to 23 mm and in height from 8 to 36 inches. Such columns have the equivalent of 10 to 25 theoretical plates.

Fractional distillation below 1 mm of mercury is difficult to carry out effectively and is usually done in a single-plate unit. For distillation at pressures in the range of 0.001 to 0.005 mm of mercury, molecular distillation can best be employed both in the laboratory and in industry (9). This is a specialized type of nonequilibrium low-pressure distillation, used principally in the preparation of heat-sensitive materials such as vitamins, high-boiling synthetic compounds, plasticizers, and fatty acids. When the absolute pressure is reduced to values used in molecular distillation, the mean free path of the molecules becomes very large, e.g., 1 cm. By placing the condensing surface at a distance of several centimeters from the evaporating liquid, few molecules will return to the liquid, but instead will be removed as product. The vigorous agitation or boiling present during ordinary distillations is absent in molecular distillation. In most units the liquid flows in a thin film over a solid surface, thus continually renewing the surface and keeping the liquid composition the same, but at the same time maintaining a low holdup of liquid. In conventional distillation of a mixture, the quantity of a constituent distilling is proportional to its partial pressure p_i. For molecular distillation this quantity is proportional to both the partial pressure and the molecular weight M, that is, $p_i/M^{0.5}$. Since substances of like molecular weight distil at about the same temperature, the separation obtained from molecular distillation is not widely different from that with equilibrant distillation. The unique feature of molecular distillation is the much lower degree of thermal exposure.

7.42 STEAM DISTILLATION

Steam distillation is used to separate mixtures at a temperature lower than the normal boiling points of their constituents. In this way it is similar to vacuum distillation and finds its main use in separating or purifying heat-sensitive materials. It is a simple distillation in which vaporization of the mixture is achieved either by continuously blowing live steam through the mixture or by boiling water and the mixture together. The steam then carries the material along with it overhead as vapor in the molecular ratio of the partial pressures, up to a maximum partial pressure equal to the vapor pressure of the material at the operating temperature. If equilibrium is established between the steam and the material, this partial pressure will equal its vapor pressure.

In this type of distillation the temperature of the material is reduced by vaporizing it into a stream of carrier vapor, usually steam. This mixed vapor of steam and vaporized material is then condensed. The liquid layers are not miscible and usually separate by gravity. Actually, any inert carrier gas can be used, but steam is normally the most economical and most convenient in the laboratory, as well as industrially. Most steam distillations are done batchwise.

Whether steam distillation is carried out in the laboratory or in an industrial plant, the equipment is essentially the same. In the laboratory the material to be distilled is charged to a still (usually a flask), which is equipped with an overhead vapor line, a condenser, and a condensate receiver to separate the two liquid phases. Steam is injected continuously into the bottom of the flask through a perforated tube to give maximum contact between the charge and the steam. If the conditions are such that liquid water is not condensed in the still, then all steam blown through the liquid passes out with the product. This occurs if the steam is sufficiently super-heated or if external heat is supplied so that none of the steam is condensed in the still. On the other hand, if the only heat supplied to the still is from the steam, some of it is condensed to raise the temperature of the still and its contents to the operating level, to supply the heat of vaporization of the material, and to compensate for heat losses. In this case a liquid phase of water collects in the still which increases in volume as the distillation progresses. In either case, when the sum of the partial pressures of the steam and the material reaches the total pressure, both substances pass overhead in the molecular ratio of their partial pressures. Nonvolatile impurities remain in the still. The mass ratio of steam to material in the vapor is

$$\frac{m_s}{m_A} = \frac{n_s M_s}{n_A M_A} = \frac{P_s M_s}{P_A M_A} \tag{7.14}$$

where m_s = mass flow rate of steam in vapor, m_A = mass flow rate of material in vapor, P_s = partial pressure of steam, P_A = partial pressure of material, M_s = molecular weight of steam, M_A = molecular weight of material, n_s = moles of steam in vapor, and n_A = moles of material in vapor.

It is possible to vary M_s and M_A, or n_s and n_A, independently by varying the rate of steam or its degree of superheat, so both the temperature and pressure of the distillation can be selected.

7.43   AZEOTROPIC AND EXTRACTIVE DISTILLATION

These two types of distillation are similar, since in each case a substance not normally present in the mixture is added to effect a better separation by altering the volatilities of the components and thus increasing the relative volatility between them. Therefore they will be discussed together, but the experimental methods will be differentiated. These types of distillation are frequently used to separate substances of dissimilar chemical nature that boil close together and form azeotropes, making the mixture impossible to separate by conventional distillation methods. In such cases the addition of another substance alters the relative volatility and therefore makes the separation possible.

Both of these types of distillation are done in fractionating columns. Azeotropic distillation can be carried out batchwise or continuously.

Although extractive distillation is usually not particularly adaptable to batch operation, it can be done in this way by the proper product and solvent recovery arrangement. Both types of distillation require removal of the added substance from the end products, so that auxiliary steps are needed after the distillation.

Azeotropic Distillation

The applications and different kinds of azeotropic distillation are quite numerous, and only some of the more typical ones will be mentioned.

Since in the majority of cases an azeotrope boils at a temperature lower than that of any of the pure components (minimum-boiling azeotrope), most of the added substance usually appears in the overhead product. In the most common type of azeotropic distillation the added substance forms an azeotrope with one of the feed components, and the separation is made between this azeotrope as an overhead product and the other feed components as a bottoms product. In some cases the overhead azeotrope may be of the ternary type. Another, somewhat different example of the use of azeotropic distillation is the separation of binary mixtures of close-boiling components having similar normal volatilities; by the addition of a substance these volatilities are altered with the formation of an azeotrope with each component. In this case the boiling points of the azeotropes produced are sufficiently different so that in a continuous distillation process one azeotrope will be the overhead product and the other the bottoms product. An example of such a process is the separation of a mixture of aromatic and paraffinic hydrocarbons by adding ethyl alcohol. The alcohol forms an azeotrope with both hydrocarbon types, but these are sufficiently different in boiling point that they can be separated by fractional distillation. A substance that forms a heterogeneous azeotrope with one or more of the original components, and forms two liquid layers upon condensation at ambient temperatures, may often be used to good advantage, since this usually eliminates a separation step for the azeotroping agent and the product.

The selection of an azeotroping agent is quite important to making an effective separation between the components, as well as in isolating the components from the agent. The following properties are desirable in an agent for the separation of multicomponent hydrocarbon mixtures and apply also to mixtures of other chemical compounds.

1. Its boiling point should be 10° to 40°C below that of the hydrocarbon mixture.

2. It should show an appreciable, positive deviation from ideality to form a minimum-boiling azeotrope with one or more of the hydrocarbon types in the mixture.

3. It should be completely miscible with the hydrocarbon mixture at the distillation and reflux temperatures.

4. It should be easily and cheaply separable from the hydrocarbons with which it forms azeotropes.

An azeotropic distillation in the laboratory can be done batchwise in a conventionally packed fractionating column. An excess of the required amount of azeotroping agent is added with the charge to the reboiler. The mixture is then distilled in the same manner as for a conventional fractional distillation of a mixture. The distillate product is the azeotropic mixture of the agent and the individual components in the charge. The composition of the product changes as the distillation proceeds since the separation taking place is between the different azeotropic mixtures, with the lowest-boiling one in the early fractions.

An example of such a distillation is the separation of a mixture of a paraffin and an aromatic hydrocarbon, as mentioned earlier, with ethyl alcohol as the azeotroping agent. The first distillate fractions contain the paraffin and ethyl alcohol, and as the distillation proceeds the aromatic and ethyl alcohol azeotrope are also taken overhead. The same type of distillation can be done on a multicomponent mixture. The ethyl alcohol can then be removed easily from the other components by washing out with water.

Extractive Distillation

As previously mentioned, in azeotropic distillation the added substance is usually much lower boiling than the components to be separated. On the other hand, in extractive distillation the added substance, or solvent, is usually relatively nonvolatile with respect to the components to be separated. It therefore appears in the bottoms product, from which it can be readily separated. As its name indicates, extractive distillation is a combination of extraction and distillation. For the extraction part, the solvent separates the components according to molecular type, while the distillation achieves the separation by boiling point or molecular size. The solvent enters at the top of the column and is mixed with the reflux. Since the solvent is considerably higher boiling than the other components, it is present principally as a liquid flowing countercurrent to the vapor. In this way it is in continuous contact with the vapor and constantly washes it. The solvent rate to the column controls its concentration in the extractive distillation.

If the components to be separated are not of the same type, as is usually the case, the greater the deviation from ideal behavior produced by the addition of the solvent, the greater will be the increase in the relative volatility. This increase is always greatest at the highest solvent concentration; hence to make an effective practical separation a fairly high concentration of the solvent is required (e.g., 70 to 80%). Therefore, heat loads and steam flows are greatly increased by its use, and the degree of separation or plate efficiency in the column may be decreased. For example, for a given separation with a fixed feed rate and reflux ratio, as the rate of introduction of solvent increases the number of plates required will decrease because of the higher relative volatility. On the other hand, this step will result in an increased heat load and may require a larger column diameter to satisfy the liquid flow, corresponding to a less efficient separation. If the solvent rate is decreased, however, the number of plates must be increased because of a lower relative volatility.

Two examples will serve to illustrate the effect of the solvent on the vapor-liquid equilibria of binary hydrocarbon mixtures. For a mixture of n-heptane (b.p. = 98.4°C) and methylcyclohexane (b.p. = 100.8°C) the normal relative volatility α is 1.07. By using aniline as an extractive distillation solvent at 85% concentration, this value is raised to 1.4, an increase that greatly enhances the ease of separation. For a mixture of toluene (b.p. = 110.6°C) and n-octane (b.p. = 125.7°C) the normal relative volatility is 1.4; by using aniline in 80% concentration, it is decreased to 0.6, thus effectively reversing the boiling points of the components.

Although extractive distillation is sometimes made in the laboratory, in most cases it is performed on a pilot-plant scale, mainly because the flow rates and heat balances can be better controlled on a larger-size unit. It is done in a conventional type of fractional distillation column, usually packed and operated continuously, with auxiliary equipment for solvent and product recovery. A typical pilot-plant setup might be to use a fractional distillation column with a diameter of 1 to 1.5 in., packed for 10 to 15 ft and having the equivalent to 50 to 75 theoretical plates.

A number of equipment arrangements are suitable for extractive distillation. A typical example is shown in Fig. 7.11 (9). Solvent is introduced to the column at point S, and the feed at point F. The section above point S removes solvent from the vapors passing up the column and is referred to as the solvent recovery section. In some cases this requires a separate unit. In the section between points S and F, referred to as the enriching section, the higher-boiling components are separated from the vapor phase. The section below the feed plate removes the low-boiling components from the liquid and is termed the stripping section. These three sections are characteristic of all continuous extractive distillation columns.

Since the solvent leaves the main column with the bottoms product, it must be separated and recycled. This is done in a separate solvent recovery column, as shown, from which the overhead is the high-boiling extraction product and the bottoms is the recovered solvent. This overhead product will normally contain small amounts of solvent, which must in turn be

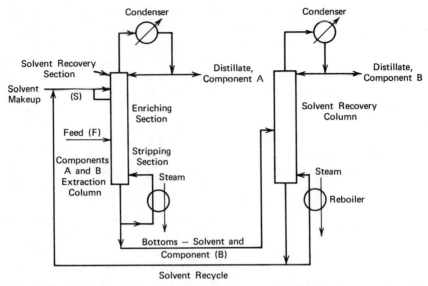

Figure 7.11.   Flowsheet for extractive distillation. Reprinted from Ref. 9, by courtesy of McGraw-Hill Book Company.

replaced as makeup solvent to the main column. This general equipment arrangement is used industrially for the separation of benzene and cyclohexane, using phenol as the extraction solvent.

As mentioned previously, extractive distillation can be done on a batchwise basis. In the flowsheet shown, the continuous feed is replaced by using a batch charge in the reboiler. The overhead from the solvent recovery column continuously flows back to the reboiler. In this way the only product is the distillate from the main column, as in a conventional batch fractional distillation. Such a setup is much more adaptable to a laboratory type study than is continuous operation.

7.44   SUBLIMATION

Sublimation is defined as a direct vaporization of a solid without the appearance of a liquid phase; hence it is essentially the distillation of a solid. As a complete process, the successive phase changes are as follows: solid → vapor → solid. Sublimation is another rather specialized form of distillation, used frequently for materials that cannot withstand heat and must be prepared at low temperatures.

There are many processes in which some form of sublimation is utilized as a separation step. Depending on the use, the process can be either continuous or batchwise. Simple sublimation is a process in which the vapor

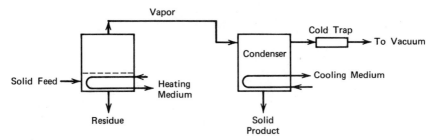

Figure 7.12. Simple vacuum sublimation. Reprinted from Ref. 9, by courtesy of McGraw-Hill Book Company.

phase consists essentially of the component being sublimed, with only small amounts of other gaseous components present. If the vapor pressure of the solid during vaporization is below atmospheric level, as is usual, sublimation must be performed at a reduced pressure. A schematic diagram of a continuous simple sublimation apparatus is shown in Fig. 7.12 (9).

Another type of frequently employed sublimation involves passing an inert noncondensing gas (the entrainer or carrier) over or through the product to be sublimed. This gas-product mixture is then cooled to condense the solid product from the gas. This process, sometimes referred to as *extractive sublimation*, is an operation similar to steam distillation in that in both cases the separation temperature is lowered by a carrier gas sweeping off vapors and transporting them into the overhead fraction.

If both components of a binary solid mixture are volatile, a batch fractional sublimation is possible by a simulation of stages to produce countercurrent flow. Ordinarily, however, this technique is not employed.

The sublimation process can best be illustrated on a temperature-pressure diagram, Fig. 7.13 (9), which includes the triple point *B* where all three phases (solid, liquid, and vapor) coexist in equilibrium. The sublimation curve is line *AB*, which is the locus of all points representing temperature and corresponding vapor pressure for the solid-vapor equilibria. The sublimation point of a material is defined as the temperature at which the vapor pressure of the solid phase is equal to the total pressure of the gas phase in contact with it, represented by line *AB*. This definition is analogous to that for the boiling point of a liquid or the vapor-liquid equilibrium line, represented by line *BC*.

In the use of sublimation as a purification process, either the product can be the sublimed material, in which the impurity remains as the residue, or the impurity can be the sublimable material, with the residue as the desired product. Freeze drying is a special case of sublimation in this latter category, in that water in the solid state is usually the volatile material to be removed from the product as vapor.

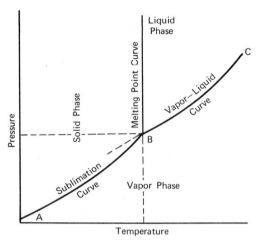

Figure 7.13. Three-phase diagram. Reprinted from Ref. 9, by courtesy of McGraw-Hill Book Company.

The equipment necessary to carry out sublimation in the laboratory or in industry is usually relatively simple, as suggested by Fig. 7.12. However, because of the nature and behavior of different solids and the various methods of handling required, the experimental procedure must be adapted to each specific case.

The limiting factors in carrying out sublimation are heat flow to sublime the solid material without melting; rate of change of solid to vapor phase, or sublimation rate; and rate of change of gas to solid phase, or condensation. The third factor is frequently encountered as a problem because of the accumulation of solids on the condenser, which results in poor heat transfer and the necessity for frequent removal of the solids.

7.45 SUMMARY OF DISTILLATION METHODS

The distillation methods discussed in this chapter and a typical industrial application for each are listed below.

Type of Distillation	Industrial Application
Simple	Removal of traces of solvent from the products of extraction.
Flash	In petroleum refining, to make rough separation between high- and low-boiling components in crude oil.
Fractional	Separation of ethylbenzene from xylenes for production of styrene polymer.

Type of Distillation	Industrial Application
Vacuum	Purification of 2-ethylhexyl alcohol at 10 mm of mercury for use in the manufacture of plasticizers.
Molecular	Distillation of high-boiling synthetic compounds, vitamins, and fatty acids.
Steam	Purification of nitrobenzene.
Azeotropic	Separation of water from acetic acid, using butyl acetate as azeotroping agent.
Extractive	Separation of benzene and cyclohexane, using phenol as the solvent.
Sublimation	Freeze-drying to preserve fruit juices, vegetables, meats, blood plasma, vitamins, and antibiotics.

References

1. American Society for Testing Materials (ASTM), *Annual Book of Standards*, D86-67, "Distillation of Petroleum Products."
2. M. R. Fenske, *Ind. Eng. Chem.*, **24**, 482 (1932).
3. G. H. Hanson, R. J. Hogan, W. J. Nelson, and M. R. Cines, *Ind. Eng. Chem.*, **44**, 604 (1952).
4. A. B. Hill, R. H. McCormick, Paul Barton, and M. R. Fenske, *A.I.Ch.E. J.*, **8**, 681 (1962).
5. Warren L. McCabe and Julian C. Smith, *Unit Operations of Chemical Engineering*, 2nd ed., McGraw-Hill, New York, 1967.
6. W. L. McCabe and E. W. Thiele, *Ind. Eng. Chem.*, **17**, 605 (1925).
7. R. H. McCormick, P. Barton, and M. R. Fenske, *A.I.Ch.E. J.*, **8**, 365 (1962).
8. C. C. Peiffer, R. H. McCormick, and M. R. Fenske, *Ind. Eng. Chem.*, *Process Design and Develop.*, **10**, 380 (1971).
9. J. H. Perry, *Chemical Engineers' Handbook*, 4th ed., McGraw-Hill, New York, 1963.
10. J. M. Prausnitz, *Molecular Thermodynamics of Fluid Phase Equilibria*, Prentice-Hall, Englewood Cliffs, N.J., 1969.
11. H. H. Reamer and B. H. Sage, *Ind. Eng. Chem.*, **43**, 1628 (1951).
12. Robert E. Treybal, *Mass Transfer Operations*, 2nd ed., McGraw-Hill, New York, 1968.

Selected Bibliography

C. H. Robinson and E. R. Gilliland, *Elements of Fractional Distillation*, 4th ed., McGraw-Hill, New York, 1950.

Arthur Rose and Elizabeth Rose, *Distillation Literature, Index and Abstracts, 1946–52*, Applied Science Laboratories, State College, Pa.

Arthur Rose and Elizabeth Rose, *Distillation Literature, Index and Abstracts, 1953–54*, Applied Science Laboratories, State College, Pa.

List of Symbols

H_A	Henry's law constant	*Subscripts*	
L	moles of liquid	A	component A
m	mass flow rate	B	component B
X	mole fraction in the liquid	D	distillate
Y	mole fraction in the vapor	s	steam
α	relative volatility		

Eight

GAS-LIQUID CHROMATOGRAPHY

Since the first publication concerning gas-liquid chromatography (GLC) in 1952 by James and Martin (1), the method has enjoyed a phenomenal growth and at present is one of the most often used and most powerful tools available for separation and analysis. Many reasons can be given for this growth, but the three major factors are speed, resolving power, and extreme sensitivity. Both speed and resolving power are illustrated in Fig. 5.1, which shows a chromatogram of the separation of 10 isomers of heptane in less than 60 sec on an open tubular (capillary) column. In addition, detection limits as low as $\sim 10^{-12}$ to 10^{-14} g can be obtained readily with ionization detectors.

Gas-liquid chromatography also has the advantages of simplicity and ease of automation.* Indeed, a large number of chromatographs can be coupled to a single computer for automated separation and analysis of many samples. In addition, gas chromatography (GC) has been coupled on-line to a mass spectrometer (MS), so that not only are components separated and quantified by GC, but their structures are also elucidated directly in the MS. This coupled arrangement has resulted in one of the most important methods available for the analysis of complex mixtures.

Gas-liquid chromatography consists of a mobile gas phase and a stationary liquid phase that is coated onto either a porous solid matrix (e.g., diatomaceous earth) or the walls of a capillary tube. Typically, the stationary phase has a sufficiently low vapor pressure (<0.5 mm) at the column

---

* Many of the characteristics of GLC apply to gas-solid chromatography as well (see Chapter 14).

211

temperature that it can be considered nonvolatile. The sample components, once eluted from the column, are most often detected by thermal conductivity or ionization of the gas (e.g., flame ionization). These ionization devices produce the highest sensitivities for GLC.

Although GLC is similar to distillation (Chapter 7) in that separation is based on vapor-liquid equilibrium, there are several major differences. First, separation in ordinary distillation is a result of differences in vapor pressure of the pure components, whereas in GLC differences in solubility of the components in the stationary phase influence separation, along with vapor pressure. In this sense GLC is more closely related to extractive distillation. As a result of this added degree of freedom over ordinary distillation, GLC has a much higher potential for selectivity. Second, the efficiency of the chromatographic process (per unit of column length) is greater than that of distillation, again indicating that GLC should be able to achieve more difficult separations. Counterbalancing these advantages, however, are the cost and difficulty of scaling up GLC to production-scale levels. Hence, for large-scale fractionations of easy-to-separate volatile mixtures, distillation continues to be the method of choice. It is at the analytical and preparative scales that GLC has its greatest value.

Gas-liquid chromatography can also be compared with liquid chromatography (LC) in columns (Chapters 10, 12, 13, and 15). The conditions of GLC are such that it is most applicable to nonelectrolytes that have molecular weights up to 300 (i.e., the components that have sufficient volatility for elution at the column temperature) and are thermally stable. However, it is possible in a number of cases to synthesize volatile derivatives of high-molecular-weight species or even of metals (see Section 8.35). Nevertheless LC finds primary applicability in the separation of high-molecular-weight, ionic, as well as thermally labile, species.

8.1 Theory

8.11 EQUILIBRIUM THEORY—RETENTION

Section 5.21 dealt in detail with retention and its relationship to equilibrium distribution in chromatography. Much of that discussion is directly applicable to GLC; however, the assumption was made that the pressure drop necessary for mobile-phase flow is linear. Since gases are compressible, this assumption is not valid for GLC. The nonlinear pressure drop means that the carrier gas velocity is a function of the axial position in the column. Let us examine this effect.

Darcy's law (Section 3.62) gives the relationship of velocity to pressure drop in a packed column:

$$v = -\frac{B_0}{\varepsilon \eta} \frac{dP}{dx} \tag{8.1}$$

Here v = mobile phase velocity, B_0 = specific permeability coefficient, ε = interparticle porosity (0.4 ± 0.03), η = mobile-phase viscosity, and dP/dx = pressure gradient. For a given column and a given flow rate, a simple extension of Boyle's law leads to the fact that the pressure-velocity product is constant. Thus

$$P_o v_o = Pv \tag{8.2}$$

where P_o and v_o are the pressure and velocity at the column outlet. Substitution of Eq. 8.2 into Eq. 8.1 results in

$$v_o P_o \int_0^x dx = -\frac{B_0}{\eta \varepsilon} \int_{P_o}^P P \, dP \tag{8.3}$$

When $x = 0$, $P = P_o$, and when $x = L$, $P = P_i$, where P_i is the inlet pressure and L is the column length. With these boundary conditions, Eq. 8.3 can be solved for v_o and then for v in Eq. 8.2 to give

$$\frac{x}{L} = \frac{P_i^2 - P^2}{P_i^2 - P_o^2} \tag{8.4}$$

This equation quantifies the change of pressure with position in the column. A similar equation can be derived for the velocity gradient in the column.

Figure 8.1 shows plots of the change of P/P_o and v/v_o with fractional distance down a GLC column, as a function of the pressure drop ratio P_i/P_o. Two points are worth noting on these diagrams. First, when P_i/P_o is small, the pressure and velocity do not change greatly from the beginning to the end of the column, and $\bar{v} \simeq v_o$, where $\bar{v}$ is average velocity. Second, at high P_i/P_o ratios, the pressure and velocity have their greatest changes near the end of the column, and a significant fraction of the column is operated at roughly constant velocity.

James and Martin (1) recognized this gas compressibility effect in their original GLC publication. They showed that the average column velocity $\bar{v}$ is related to the outlet velocity through a *correction factor j*:

$$\bar{v} = jv_o \tag{8.5}$$

where

$$j = \frac{3}{2} \frac{(P_i/P_o)^2 - 1}{(P_i/P_o)^3 - 1} \tag{8.6}$$

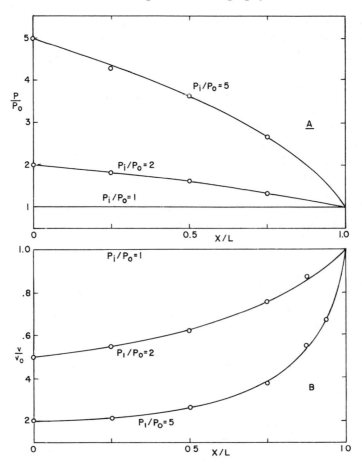

Figure 8.1. Plots of relative pressure P/P_0 and relative velocity v/v_0 as a function of fractional column position. P_o, $v_o =$ outlet pressure and velocity; P, $v =$ pressure and velocity at position X in the column.

Since the velocity and the flow rate are related to each other* (i.e., $F = \varepsilon v A$, where $A =$ the cross-sectional area of the column tube), it follows that $\bar{F} = jF_o$, where $\bar{F} =$ the average column flow rate and $F_o =$ the outlet flow rate.

The correct retention volume and retention time quantities of Section 5.21 can then be easily computed. Thus

$$t_R = \frac{L}{\bar{v}}\,(1 + k') = \frac{L}{jv_o}\,(1 + k') \qquad (8.7)$$

* The free cross-sectional area is assumed to be constant.

Typically, the flow rate of carrier gas is measured after the column with a soap-bubble flowmeter. In addition to the compressibility correction, two other factors must be taken into account with this flowmeter arrangement in order to obtain the average flow rate *in the column*. The first correction results from the fact that the flow rate is measured at room temperature T_0 rather than at column temperature T. The correction factor in this case is T/T_0. The second correction is for the vapor pressure of water P_w when using the soap-bubble meter: $(P_0 - P_w)/P_0$, where P_0 is the atmospheric pressure. The average flow rate $\bar{F}$ can be obtained from the measured flow rate F_m:*

$$\bar{F} = jF_m\left(\frac{T}{T_0}\right)\left(\frac{P_0 - P_w}{P_0}\right) \tag{8.8}$$

and this average flow rate should be used to obtain the retention volumes.

The net retention volume is related to the distribution constant by Eq. 5.7, $V_N = KV_S$, and therefore is a function of liquid loading. In order to facilitate the comparison of retention from column to column, V_N must be converted to an intensive parameter. This can be done by dividing V_N by W_S, that is, the net retention volume per gram of stationary phase or the *specific retention volume* V_g (ml/g):

$$V_g = \left(\frac{V_N}{W_S}\right)\left(\frac{273}{T}\right) \tag{8.9}$$

The specific retention volume in this case has been corrected to the reference temperature at $0°C$. Less often the specific retention volume at the column temperature is used. The relationship of V_g to the distribution constant is

$$V_g = \frac{K}{\rho_s}\left(\frac{273}{T}\right) \tag{8.10}$$

where the density of the solvent ρ_S is taken at temperature T.

We focus now on the distribution constant to gain deeper insight into the retention process. This is necessary for the proper selection of stationary liquid phases in order to achieve separation of specific mixtures, for the control of retention near the optimum capacity factor values (k') of 2 to 5, and for the identification of unknown components by relating molecular structure to retention. The discussion is presented on two levels. First we examine the relationship of retention to fundamental thermodynamic parameters in the light of the distribution laws presented in Chapter 2. Then we describe a semiempirical approach to the understanding of retention in GLC, based on a subdivision of intermolecular force contributions to the total solute-solvent interaction.

---

* Note that $F_o = F_m(T/T_0)[(P_o - P_w)/P_o]$.

The distribution constant is defined as

$$K = \frac{C_S}{C_M} = \frac{(n_S/V_S)}{(n_M/V_M)} \tag{8.11}$$

where C and n are, respectively, the concentration and number of moles of the volatile solute, and the subscripts S and M refer to the stationary liquid and mobile gas phases, respectively. Also, V_S and V_M are the total column volumes of the stationary and mobile phases, respectively. At the low pressures ordinarily encountered in GLC and with the inert gas helium, non-ideality of the gas phase contributes less than 1 to 3% to the distribution constant. Therefore ideal gas behavior can be assumed, so that

$$\frac{n_M}{V_M} = \frac{P}{RT} \tag{8.12}$$

where P is the partial pressure of the solute at the column temperature T.

Raoult's law (Section 2.22) can be used to relate the mole fraction of solute in the liquid phase X_S to the partial pressure P:

$$X_S = \frac{P}{\gamma P^0} \tag{8.13}$$

where γ is the activity coefficient, based on the pure liquid solute standard state, and P^0 is the saturation vapor pressure of the solute at the column temperature. At the low concentrations common for analytical GLC (linear elution conditions), the concentration of solute in any volume element of stationary phase can be assumed to be infinitely dilute, i.e. Henry's law is obeyed. Hence Eq. 8.13 can be written as

$$X_S \simeq \frac{n_S}{N_S} = \frac{P}{\gamma^\infty P^0} \tag{8.14}$$

where N_S is the moles of stationary phase, and γ^∞ is the solute activity coefficient at infinite dilution in S. Combination of Eqs. 8.12 and 8.14 with 8.11 results in

$$K = \frac{N_S RT}{\gamma^\infty P^0 V_S} = \frac{W_S RT}{\gamma^\infty P^0 M_S} \tag{8.15}$$

where M_S = molecular weight of the stationary phase. The specific retention volume can be found by combination of Eqs. 8.10 and 8.15:

$$V_g = \frac{273R}{\gamma^\infty P^0 M_S} \tag{8.16}$$

Equation 8.16 indicates that in a given solvent (M_S constant) there are two factors which determine the specific retention volume: one dependent only on the solute, P^0, and the other dependent on both the solute and the solvent, γ^∞.

We see from Eq. 8.16 that the larger the vapor pressure P^0, the shorter will be the retention time, other factors being equal. There are two general means to alter the vapor pressure of a solute—temperature and derivative formation. The Clausius-Clapeyron equation (Eq. 2.12a) states that $\ln P^0$ is proportional to $1/T$, the proportionality constant being equal to $\Delta H^V/R$, where ΔH^V is the heat of vaporization. Thus P^0 increases with temperature. Derivative formation is often used with nonvolatile species to produce a compound with sufficient volatility for elution in a reasonable time. An example is the elution of metals as volatile chelates, for instance, trifluoro-acetylacetone or hexafluoroacetylacetone derivatives (2).

We turn next to the activity coefficient γ^∞. This parameter is a measure of the influence of the solvent on the vapor-liquid equilibrium of the pure solute. As discussed in Chapter 2, the activity coefficient is related to the partial molar excess thermodynamic functions:

$$\frac{\Delta\mu^e}{RT} = \frac{\Delta h^e}{RT} - \frac{\Delta s^e}{R} = \ln\gamma^\infty \tag{8.17}$$

It is convenient to divide the activity coefficient into a temperature-dependent portion $\gamma^{\infty,t}$ and an athermal portion $\gamma^{\infty,a}$:

$$\ln\gamma^\infty = \ln\gamma^{\infty,t} + \ln\gamma^{\infty,a} \tag{8.18}$$

where $\gamma^{\infty,a}$ is related to Δs^e, and $\gamma^{\infty,t}$ is related to Δh^e.

The *athermal activity coefficient* is usually associated with statistical effects arising from size differences of the solute and solvent. This configurational entropy term was discussed in Section 2.23. The Flory-Huggins equation can be written as:

$$\ln\gamma^{\infty,a} = \ln\frac{\bar{V}_i}{\bar{V}_s} + \frac{\bar{V}_s - \bar{V}_i}{\bar{V}_s} \tag{8.19}$$

where $\bar{V}_i$ and $\bar{V}_s$ are the molar volumes of the solute and solvent, respectively. In ordinary solutions $\bar{V}_i \simeq \bar{V}_s$, and this entropy effect is not significant. In GLC, however, the requirements for volatility of the solute and nonvolatility of the solvent often make $\bar{V}_i \ll \bar{V}_s$. Under this condition we see that $\gamma^{\infty,a}$ will be less than unity. This creates a positive excess partial molar entropy and a negative excess chemical potential (Eq. 8.17). As a result, the configurational entropy term favors solution of the solute molecules in the large solvent molecules over an ideal solution (because of more disorder in the nonideal case); and, as seen from Eq. 8.16, the retention volume is increased over that for an ideal solution ($\gamma^\infty = 1$).

The *thermal activity coefficient* $\gamma^{\infty,t}$ is associated with differences in solute-solvent interactions over those that occur in an ideal solution. Although expressions can be formulated to approximate this factor in the case of a hydrocarbon dissolving in hydrocarbon solvents, the problem becomes much more difficult when specific forces (dipole-dipole, hydrogen bonding, etc.) occur. This is related, of course, to the fact that a full understanding of the solution process is not yet at hand. Semiempirical expressions and empirical behavior must therefore be employed.

Before turning to these topics, let us try to obtain a better understanding of $\gamma^{\infty,t}$. The value of this portion of the activity coefficient can be either greater or less than 1 (i.e., Δh^e can be positive or negative). This can be seen from the following approximation, taken from Eq. 2.42:

$$\ln \gamma^{\infty,t} = \frac{\bar{V}_i}{RT}\left(\frac{E_{ii} + E_{ss}}{2} - E_{is}\right) \tag{8.20}$$

where E_{ii}, E_{ss}, and E_{is} are the interaction potential energies between two solute molecules, two solvent molecules, and a solute and a solvent molecule, respectively. When $(E_{ii} + E_{ss})/2 > E_{is}$, $\gamma^{\infty,t} > 1$ (i.e., Δh^e is positive) and the specific retention volume is less than that for an ideal solution (Eq. 8.16). This is the usual case encountered in GLC. Since $\gamma^{\infty,t} > 1$ and $\gamma^{\infty,a} < 1$, the activity coefficient of the solute is typically close to unity, especially when strong specific solute-solvent interactions are not involved. If strong interactions occur, then $(E_{ii} + E_{ss})/2 < E_{is}$, and $\gamma^{\infty,t} < 1$ (i.e., Δh^e is negative). This enhanced solubility of the volatile solute will result in increased retention volumes.*

In chromatography, we are interested not only in V_g, but also in the relative retention volumes for solute pairs whose separation is required. We previously introduced the *separation factor* α (Eq. 2.22a) as the relative retention parameter and related it to the resolution R_s of two solutes (Eq.

---

* This discussion has dealt solely with gas-liquid partition. There are circumstances, however, under which other retention mechanisms, namely, gas-liquid, liquid-solid, and/or gas-solid adsorption, can contribute significantly to the retention volume. Mixed mechanisms have been discussed in detail (3). As noted in Chapter 5, a term for each distribution equilibrium must be added to the retention volume equation. In general, gas-liquid adsorption will occur when the solubility of the volatile component is low. As a rule, when $\gamma^{\infty} > 5$, interfacial adsorption is likely. Gas-solid adsorption takes place when the liquid stationary phase does not cover all the adsorption sites of the solid support (e.g., too light a liquid loading). Liquid-solid adsorption is independent of liquid loading but can be important when polar solutes and nonpolar stationary liquid phases are used. Although in some rare cases mixed mechanisms can improve separation, conditions are usually established to make gas-liquid partition the only measurable retention mechanism. The reason is that sample overloading and nonsymmetrical peaks readily occur when adsorption and partition take place simultaneously. For the purpose of this chapter, we shall continue to assume that gas-liquid distribution is the only mechanism of retention.

5.43). The relationship of the activity coefficient to the separation factor can be seen from either Eq. 8.15 or 8.16:

$$\alpha = \frac{V_{g_2}}{V_{g_1}} = \frac{K_2}{K_1} = \frac{\gamma_1^\infty P_1^{\,0}}{\gamma_2^\infty P_2^{\,0}} \tag{8.21}$$

where the subscripts 1 and 2 refer to the first and second components, respectively, eluted from the column.

We thus see that separation results from differences in a property of the solutes themselves (P^0) and a property of the solute and solvent (γ^∞). For closely related components whose α value is near unity, we may assume to a first approximation that $\gamma_1^{\infty,a} = \gamma_2^{\infty,a}$. Equation 8.21 can then be written as

$$\alpha = \left(\frac{\gamma_1^{\infty,t}}{\gamma_2^{\infty,t}}\right)\left(\frac{P_1^{\,0}}{P_2^{\,0}}\right) \tag{8.22}$$

The ratio of thermal activity coefficients is a measure of the differences in the interaction of the solutes with the solvent. The more the ratio ($\gamma_1^{\infty,t}/\gamma_2^{\infty,t}$) differs from unity, the more selective is the phase. Note, however, that the solvent can either enhance or diminish separation by the relationship of the $\gamma^{\infty,t}$ ratio to the P^0 ratio. Selectivity refers only to the influence on the $\gamma^{\infty,t}$ ratio, not α. Thus it can be seen that in a multicomponent-multifunctional group mixture the solvent may help the separation of some component pairs, but hinder the separation of others.

8.12   KOVATS' RETENTION INDEX

A variety of parameters have been suggested for reporting retention, the object in all cases being to present a parameter that is simple to use and that can be compared from column to column and from laboratory to laboratory. We have already commented on the value of using the intensive parameter, the specific retention volume, rather than the net retention volume. However, the determination of the amount of stationary liquid phase is subject to some error, which can affect column-to-column reproducibility. In addition, it does not give a description of the retention position of a component relative to other components. In practice, it has proved useful to report data in terms of relative retention, in which a standard component is assigned a net retention volume of unity, and all other components are reported relative to this solute. However, it is difficult to fix one standard suitable for many separations, so that published data often may not be directly used.

To overcome these problems, Kovats introduced the *retention index I* (4), defined as

$$I = 100\,\frac{\log V_{N,x} - \log V_{N,n}}{\log V_{N,n+1} - \log V_{N,n}} + 100n \tag{8.23}$$

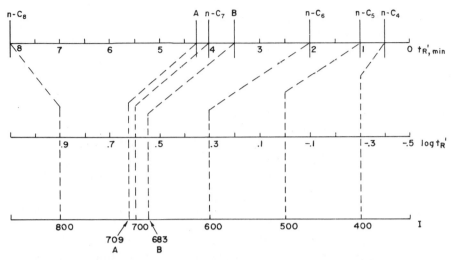

Figure 8.2. Representation to explain the meaning of the Kovats index I. t'_r = net retention time. From L. S. Ettre, *Anal. Chem.*, **36**, 31A (1964).

where $V_{N,X}$ is the net retention volume of the sample component, and $V_{N,n+1}$ and $V_{N,n}$ are the net retention volumes for the two normal alkanes with carbon numbers $n + 1$ and n which bracket the component. Figure 8.2 will serve to explain the principle of this index. In the top scale, the net retention times of seven solutes—five n-alkanes and two hypothetical components—are shown. Note that the relative retention of each n-alkane pair is constant, equal to 2.0 in this example. This constancy is characteristic of a homologous series (see Chapter 2) and requires that log t'_R be equally spaced for each solute in the series. The second scale, which sets the net retention times on a logarithmic basis, shows this. From Eq. 8.23 we see that each n-alkane is assigned a number, $100n$. We now change log t'_R into the retention index, which is linear in log t'_R as shown in the bottom scale. The two hypothetical solutes have I values of 683 and 709, respectively.

The retention index is seen to be descriptive of the positions of the sample components (e.g., A lies between n-heptane and n-octane). Two standards that bracket the solute are used. In addition, the change in I with temperature is generally smaller than with relative retentions and much smaller than with absolute retentions. Thus reproducibility is found to be better with this retention parameter. Finally, I can be a most useful qualitative tool, and many workers have attempted to relate structural effects to retention index increments; one obvious example is that two adjacent homologs in a series differ by 100 units. To aid in solute identification by gas chromatography, retention indices for a large number of compounds on a variety of stationary phases have been cataloged (5, 6).

8.13 STATIONARY LIQUID PHASE

We have seen that the selectivity of a stationary phase for the separation of a given mixture depends (to a first approximation) on the relative values of the thermal activity coefficients and thus on the molecular interactions that the vapor components undergo with the solvent. As a general rule, the ease in separating two components of equal vapor pressure follows this order: (a) compounds with different functional groups, (b) isomers with polar functional groups, and (c) isomers with no functional groups. Thus the separation of an alcohol from a ketone can be easily achieved using an electron-donor stationary liquid phase in order to retard the alcohol relative to the ketone via hydrogen bonding. On the other hand, a relatively difficult separation for GLC is that of the isomers (m- and p-xylene isomers) which contain no functional group. Chapter 2, which discusses interaction forces in detail, provides further consideration of these points.

Chromatographers have often tried to rank solvents according to their ability to retard polar compounds, that is, to classify solvents on the basis of polarity. Examination of Table 10.1 reveals that the solubility parameter might be used as a measure of polarity, in that its value generally increases with solutes that are called polar. In another polarity scale, introduced by Rohrschneider (7), the *polarity* $P*$ was defined as

$$P* = a\left(\log \frac{_2V_{R,p}}{_1V_{R,p}} - \log \frac{_2V_{R,u}}{_1V_{R,u}}\right) \tag{8.24}$$

where the subscripts 1 and 2 refer to butane and butadiene, respectively; the subscripts p and u refer to a given phase of polarity $P*$ and a standard nonpolar phase (squalane), respectively; and a is constant for adjustment of the scale to give $P* = 100$ for β,β'-oxydipropionitrile. Squalane was assigned $P* = 0$, and all other phases were ranked between these two solvents. Thus dibutyl oxalate was found to have $P* = 42$ and polyethylene glycol 600 $P* = 78$, values that are in the expected direction.

Polarity scales such as the above are useful in several instances. First, in multicomponent analysis it is not possible to isolate a given solute pair and to select a solvent that optimizes the activity coefficient ratio for this pair without influencing the relative retention of all other solute pairs. Hence we are forced to select solvents on a general basis of polarity when we have many components to separate. Second, well over 200 phases have been suggested for use in GLC. With a quantitative polarity scale, we are in a position to minimize the choice of solvents, for many phases will have essentially equivalent $P*$ values.

One problem with a single measure of polarity is that it cannot simultaneously take account of all interaction possibilities. The Rohrschneider scale is a measure of the ability of the stationary phase to induce a dipole moment in butadiene, and the scale is thus related to the dipole moment of the phase. However, consider the stationary phase 1,2,3-tris(2-cyanoethoxy)-propane, which, besides having a large dipole moment, is an electron donor. If two components with the same dipole moment, one having electron-donor capabilities and the other proton-donating abilities, are chromatographed on this phase, hydrogen bonding will substantially retard the latter component, and this fact is not predicted by the $P*$ scale.

Rohrschneider (8) has also developed an empirical solvent scale, based on a splitting of the total polarity into the various interaction forces. The scale is developed from ΔI, the *retention index difference* of a solute between a polar liquid phase and a standard nonpolar phase (squalane). To a first approximation ΔI represents the specific interaction forces (i.e., no dispersion forces) that occur on the polar phase. Then

$$\Delta I = \sum_i E_i \tag{8.25}$$

where E_i represents the interaction energy for dipole-dipole, induction, and other forces. Each energy term can be characterized by the product of two factors, one dependent only on the solute, a_i, and the other on the solvent, X_i:

$$\Delta I = \sum_i a_i X_i' \tag{8.26}$$

Knowledge of the various a's and X's permits the prediction of ΔI. The prediction of ΔI becomes better, the larger the number of terms in Eq. 8.26. In practice, it has been found that a five-term equation allows ΔI to be estimated within 10 retention index units for over 600 solute-solvent combinations. The several X's are obtained from the ΔI values of five standard components, chosen to account for various specific force interactions.

Table 8.1 lists the five solutes in the Rohrschneider scheme, as recommended by McReynolds (9). From the component solubility parameters, benzene is seen to account for induction interactions, butanol for both proton

Table 8.1. Solubility Parameter Values for the Five Standard Components used in Rohrschneider Solvent Polarity Scheme, as Modified by McReynolds

Solute	δ	δ_d	δ_o	δ_a	δ_b
Benzene	9.2	9.2	—	—	0.5
Butanol	9.8	7.4	2.2	4.0	4.0
Nitropropane	10.4	7.0	5.0	—	1.0
Pyridine	10.4	9.0	4.0	—	5.0
2-Pentanone	9.0	7.0	2.3	—	1.5

donating and proton accepting, nitropropane and 2-pentanone for dipole-dipole interactions, and pyridine for proton accepting.

Table 8.2 lists the Rohrschneider indices for 27 common liquid phases, along with the average polarities and the maximum operating temperatures (at which the vapor pressure of the stationary liquid phase is ~ 1 mm). From this table, it is clear that there are far too many solvents for GC. For example,

Table 8.2. Rohrschneider Constants for Common GC Liquid Phases,[a] as Modified by McReynolds[b]

Liquid Phase	Max. T (°C)	X'	Y'	Z'	U'	S'	P
Squalane	100	0	0	0	0	0	0
Apiezon L	250	32	22	15	32	42	29
SE-30	300	15	53	44	64	41	43
OV-1, methyl gum	350	16	55	44	65	42	44
OV-3, 10% phenyl	350	44	86	81	124	88	85
OV-7, 20% phenyl	350	69	113	111	171	128	118
Dioctyl sebacate	125	72	168	108	180	123	130
Dilauryl phthalate	—	79	158	120	192	158	141
Dinonyl phthalate	150	83	183	147	231	159	161
OV-17, 50% phenyl	375	119	158	162	243	202	177
Versamid 930	150	109	313	144	211	209	197
Trimer acid	150	94	271	163	182	328	218
OV-25, 75% phenyl	350	178	204	208	305	280	235
Polyphenyl ether (6 rings)	225	182	233	228	313	293	250
Ethylene glycol tetrachlorophthalate	200	307	345	318	428	466	373
Triton X-305	200	262	467	314	488	430	392
Carbowax 20M	225	322	536	368	572	510	462
Carbowax 4000	200	317	545	378	578	521	468
Reoplex 400 (polyester)	200	364	619	449	647	671	550
Carbowax 1540	175	371	639	453	666	641	554
Diglycerol	100	371	826	560	676	854	657
EGSS-X	200	484	710	585	831	778	678
Ethylene glycol phthalate	200	453	697	602	816	872	688
Diethylene glycol succinate	200	496	746	590	837	835	701
Tetrahydroxyethylenediamine	150	463	942	626	—	893	731
1,2,3,4,5,6-Hexakis(2-cyano-ethoxycyclohexane)	150	567	825	713	978	901	797
N,N-Bis(2-cyanoethyl)formamide	125	690	991	853	1110	1000	929

[a] $X' = \Delta I$, benzene; $Y' = \Delta I$, butanol; $Z' = \Delta I$, 2-pentanone; $U' = \Delta I$, nitropropane; $S' = \Delta I$, pyridine; $P = (X' + Y' + Z' + U' + S')/5$.

[b] W. O. McReynolds, *J. Chromatog. Sci.*, **8**, 685 (1970), reprinted by courtesy of the publisher.

in the cases of SE-30/OV-1 and Carbowax 4000/Carbowax 20M the indices for the two solvents of each pair are essentially equivalent. Other pairs are also closely similar. In fact some workers believe that 10 or fewer solvents are all that are required for GLC, even though over 200 have been recommended for use. The most polar phases are those in which multifunctional groups are present, such as N,N-bis(2-cyanoethyl)formamide. Note that for these very polar solvents all five indices are quite large, the large value for X' (benzene) being a result of the extensive dipole-induced dipole interaction. Finally, it is instructive to compare diethylene glycol succinate with diglycerol. Even though the average polarity is larger for the former solvent, Y' (butanol) is much greater for the diglycerol phase, indicative of the selective retention of proton donors on this phase. Note, on the other hand, that U' (pyridine) is much larger for the succinate phase. Another difference between these two solvents is the large difference in maximum operating temperatures.

Besides the standard GC phases, special liquid phases are often employed for specific separation purposes. One example is *liquid crystals*, used in the separation of positional isomers. Certain compounds, especially those with rodlike shapes and polar linkages at the ends of the molecule, form ordered liquid mesophases when melted; upon further heating, these convert to the isotropic liquid state. There are three types of mesophases—smectic, nematic, and cholesteric (10). As an example, the smectic phase consists of a two-dimensional orientation, in which the rod-shaped molecules are arranged in a parallel fashion in layers. The mechanism of separation of positional isomers can be understood from the separation of m- and p-dichlorobenzene. Since the para isomer is more elongated, it can align itself in the layers more readily than the meta isomer and thus be retained for a longer time. For further details, see Refs. 11 and 12.

A second type of selectivity involves the formation of *reversible complexes* between vapor molecules and nonvolatile components added to "inert" stationary phases. It is easy to show that the stoichiometric distribution ratio D of a single component and the separation factor of two components that form 1:1 complexes with the additive can be written as

$$D = K_0(1 + K_1 X_A) \tag{8.27}$$

and

$$\alpha = \frac{D_2}{D_1}\left(\frac{K_{02}}{K_{01}}\right)\left[\frac{1 + (K_1)_2 X_A}{1 + (K_1)_1 X_A}\right] \tag{8.28}$$

where the subscripts 1 and 2 refer to the first and second eluted components, respectively; K_0 is the distribution constant of the uncomplexed solute; K_1 is the formation constant of the 1:1 complex; and X_A is the mole fraction of the additive. From Eq. 8.28, we see that separation can be achieved if one component complexes with the additive and the other does not. For example, $AgNO_3$ has been used as an additive in polypropylene glycol to retard olefins in relation to saturated hydrocarbons. Olefins form a charge transfer complex with Ag^+, whereas saturated compounds are retained only by physical forces. The concentration of the additive, X_A, can be used to control the extent of olefin retardation. The use of complexation in GC is simply an example of secondary chemical equilibria (see Section 2.13). Further details on selective stationary phases can be found in Ref. 13.

8.14 QUALITATIVE ANALYSIS

The prediction of retention, as discussed in the preceding section, is of interest not only for the selection of a stationary liquid phase, but also for the identification of a solute. The use of GC as an identification tool has been extensively developed (14), with interesting relationships between retention and solute structure.

There are several structure-retention relationships that are useful for identification purposes. The basic assumptions for all of these were discussed in terms of the Martin equation in Chapter 2. First, the logarithm of the partition coefficient is linearly related to the carbon number in a homologous series. Thus a straight-line relationship between log retention volume and carbon number exists for a homologous series, and any component in this series can then be identified. A homologous series will also give a linear plot of log retention volume versus boiling point.

For homologous series, it follows that log retention on one column is linearly related to log retention on a second column. Since the retention index is a log retention parameter, the indices of a homologous series on two phases are then linearly related. Recently, Walraven et al. have been able to show rather subtle structural relationships between positional isomers (15). An example of the type of correlations is shown in Fig. 8.3, which is a plot of retention index on two different liquid phases for methyl esters of carboxylic acid. It is seen that all of the isomers with a given carbon number can be positioned on a straight line, the points on a particular line being only a function of the position of the carbomethoxy group. Plots such as the one in Fig. 8.3 illustrate the so-called "roofing tile effect." Another type of structural pattern can be realized when the retention index on a given column at one temperature is plotted against a second temperature. For homologous series, a plot of log retention versus boiling point is a straight line, and therefore the same type of linear behavior as shown in Fig. 8.3 is found when two temperatures are used.

It should be pointed out that the identification of a component from its GC retention can never be absolute. There is always the possibility that

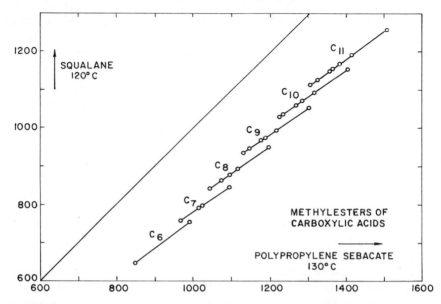

Figure 8.3. Plot of the relationship of the retention index of a series of methyl esters of carboxylic acids on two phases. The fine structure reveals a "roofing tile" effect. Reprinted from Ref. 15, by courtesy of Pergamon Press, Inc.

several components can elute at the same time. Of course, the more efficient the column is and the more precisely the retention is known, the more confident the identification. A more unique retention description of the solute can be obtained by chromatographing the solute on three or four liquid phases of different polarities.

At the present time, qualitative analysis is often achieved by coupling a second instrument to the gas chromatograph, for example, a mass spectrometer. In addition chemical reactions have been performed on solute peaks before and after the column for identification purposes. Nevertheless identification from retention time can still be important. First, it is inexpensive and simple to do. Second, if one has a mass spectrometer coupled to the gas chromatograph, the use of retention plus a mass spectrum can more positively identify the component.

8.15   COLUMN EFFICIENCY AND TYPE

Column efficiency for gas chromatography was discussed in some detail in Section 5.3. It was seen that the total variance for band broadening consisted of a series of independent mechanisms, each contributing a separate variance to the total and culminating in the van Deemter equation for plate height H:*

$$H = A + \frac{B}{\bar{v}} + C\bar{v} \qquad (8.29)$$

Here A represents the plate height contribution for eddy diffusion (Eq. 5.30), $B/\bar{v}$ for axial diffusion (Eq. 5.18), $C\bar{v}$ for mobile and stationary-phase resistance to mass transfer effects (Eqs. 5.31 and 5.34), and $\bar{v}$ is average velocity. As shown in Fig. 5.11, a hyperbola results from a plot of Eq. 8.29, with a velocity at which H is a minimum. Equation 8.29 can be solved to find this velocity (i.e., $dH/d\bar{v} = 0$):

$$\bar{v}_{\text{opt}} = \sqrt{B/C} \qquad (8.30)$$

The smaller the C value, the greater will be $\bar{v}_{\text{opt}}$, resulting in faster analysis. In practice, it is found that the best velocity, in terms of rapid analysis, occurs at roughly twice the optimum velocity, because $H/\bar{v}$ (see Eq. 5.46) continues to decrease up to this point. Beyond this velocity $H/\bar{v}$ becomes constant (the C term dominates), and no gain in analysis time is achieved. In addition, the pressure drop is proportional to velocity, so that it is actually detrimental to increase velocity beyond the point at which the H versus $\bar{v}$ plot approximates linearity. By using columns with low C values, it is possible to achieve very rapid analyses.

We turn next to the expressions relating H to velocity for the two main types of column—*open tubular* and *packed*. For the open tubular column an exact expression can be formulated:

$$H = \frac{2D_g}{\bar{v}} + \frac{1 + 6k' + 11k'^2}{24(1 + k')^2} \frac{r^2}{D_g} \bar{v} + \frac{k'}{6(1 + k')^2} \frac{d_f^2}{D_l} \bar{v} \qquad (8.31)$$

where D_g and D_l are the gas and liquid diffusion coefficients, respectively; r is the radius of the tube; and d_f is the thickness of the stationary liquid-phase layer coated on the column walls. Note that the A term does not occur in this equation. Because layer thicknesses are small, the last term on the right-hand side of Eq. 8.31 can usually be neglected; that is, at velocities above $\bar{v}_{\text{opt}}$, the mobile-phase resistance to mass transfer term predominates. Thus, for good efficiency and speed characteristics, the tube diameter should be kept narrow (i.e., capillary tubing ~ 0.25 mm). In addition, H is secondarily related to k', with small values resulting in better efficiency; however, optimum k' values are determined on the basis of the resolution equation (Eq. 5.44), rather than solely from the effect of k' on H.

* Strictly speaking, the outlet velocity should be used with the B and C_M terms in Eqs. 8.29, 8.31, and 8.32. However, for purposes of simplicity, we shall use the time-average linear velocity $\bar{v}$. The differences are not great until high velocities are attained.

The big advantage of open tubular columns is that the permeability is high, and long columns and/or high mobile-phase velocities can be used. It is possible to employ columns of 100 meters or longer, with several hundred thousand theoretical plates. Such columns have also been successfully used in the resolution of optical isomers, in which the stationary liquid phase is optically active (see Fig. 8.6).

Open tubular columns have a disadvantage, however, in that the small amount of stationary liquid phase results in low sample capacity and, at times, small k' values. The low sample capacity means that very small sample sizes ($\sim 10^{-2}\,\mu l$) must be used to prevent column overloading (i.e., asymmetrical peaks). Usually the injected mixture must be split before entering the column in order to achieve sufficiently low sample sizes. Small k' values mean that the number of effective plates will be substantially less than the total number of theoretical plates (see Section 5.42).

To increase the capacity of the column, *support-coated open tubular* and *packed capillary columns* have been introduced. In the former case, a thin layer of support is coated on the column walls, or alternatively the walls are roughened by chemical means (e.g., etching of glass). The increased surface area then permits a larger amount of stationary liquid phase to be retained in the column, without substantial loss in efficiency. The packed capillary (or irregular packed) column is one in which support particles are loosely packed into the capillary tube (often in the drawing process of glass capillaries). The tube-to-particle diameter ratio is often 5:1 or less. Under such conditions, the support cannot be densely packed, and such a loose structure gives a permeability intermediate between the values for a packed and an open tubular column.

Equation 8.29 is the general equation of H versus velocity for a packed column. In terms of column parameters we may write

$$H = 2\lambda d_p + \frac{2\gamma D_g}{\bar{v}} + q\,\frac{k'}{(1 + k')^2}\,\frac{d_f^2}{D_l}\,\bar{v} + \frac{\omega d_p^2 \bar{v}}{D_g}$$

$$= A + \frac{B}{\bar{v}} + C_S\bar{v} + C_M\bar{v} \tag{8.32}$$

where λ = packing factor (~ 0.5), γ = obstruction factor to correct D_g (~ 0.7), q = factor to take account of the configuration of the liquid phase (~ 0.67 for uniform films), and ω = packing factor function to correct for radial diffusion (~ 0.02 to 5). At the low reduced velocities in GC, the coupled form of the A and C_M terms (see Eq. 5.33) is not used. In addition, since diffusion in the gas phase is rapid, the effect on mass transfer diffusion in the stagnant mobile phase contained within the intraparticle void volume (Eq. 5.34) is negligible.

When heavily loaded columns are used (e.g., 30 to 40% w/w liquid coated on porous supports), the C term is determined by slow diffusion in the stationary liquid phase, with the mobile-phase mass transfer effect making a much smaller contribution. The function $k'/(1 + k')^2$ reaches a maximum (poorest efficiency) at $k' = 1$, and then decreases slowly with larger k' values. In addition, H depends on d_f^2 in this case. This means that, for any given loading, it is essential that the liquid be coated as uniformly as possible on the support; puddling will be detrimental to column efficiency. The functionality d_f^2 indicates that lower-loaded columns will be more efficient and have higher speed characteristics (i.e., lower C values). This is true up to a point; however, experimentally the C term is much less dependent on layer thicknesses below loadings of roughly 15% w/w.

As the liquid loadings become less, the C_M term becomes important, and on lightly loaded columns (ca. 5% w/w) it must be included. Since $H \sim d_p^2$, good efficiency and speed require small particle diameters. Of course, the pressure drop then has to be increased, as this is the price that must be paid for small particles. In practice, particles of ~ 100 to $150~\mu$ represent the best compromise between speed/efficiency and permeability. Pressure drops of roughly 3 to 4 atm are encountered for columns of 2 to 3 meters. For further details on all column types, see Refs. 16 to 18.

8.2 Experimental Gas-Liquid Chromatography

8.21 BASIC APPARATUS

Figure 8.4 shows a block diagram of the basic gas chromatographic apparatus. Here we briefly discuss the various components. Later, more detailed examinations of the column (Section 8.22) and detector (Section 8.23) are presented.

The carrier gas first passes through a flow controller for constant (and repeatable) flow settings. The sample is picked up by the gas in the injection

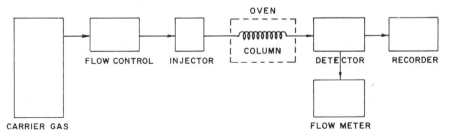

Figure 8.4. Block diagram of a basic gas chromatographic apparatus.

port and carried onto the column, where separation takes place. From the column the sample components travel to the detector, where the amount of sample is converted to an electrical signal. This signal is then amplified to drive a recorder, which displays the separation. Determination of the flow rate should be made after the dtector, when possible. Finally, if desired, samples can be collected after the detector (e.g., preparative-scale GC), if the latter is nondestructive of components. Alternatively, for a detector that decomposes sample components, a device to split the flow stream can be positioned before the dtector, with one of the streams leading to the collection device.

8.22   COLUMN DETAIL

As we have noted, the two types of columns commonly used in GLC are the packed and the open tubular (capillary). Let us first examine the packed column. The *solid support* that holds the liquid phase should ideally have the following properties: (*a*) chemical inertness; (*b*) spherical shape with narrow particle size distribution; and (*c*) open pore structure for rapid mass transfer. A variety of support materials have been used in GLC; however, none meet fully the above criteria. A complete review of this subject has been written by Ottenstein (19).

By far the most popular support is diatomaceous earth—Chromosorb P (pink) and W (white). In both cases the diatoms are fused together at around 900°C; however, for Chromosorb W a small amount of sodium carbonate flux is added in the firing process. Chromosorb P is a more active support than W, while Chromosorb W is considerably more fragile.

Activity of the solid support is undesirable for three reasons: (*a*) adsorption sites can act as catalysts for chemical reactions on the column; (*b*) peak tailing can result through overloading of the adsorption sites, even at the smallest sample sizes; and (*c*) the retention volumes can be influenced either through modification of the liquid coating the support or introduction of gas-solid or liquid-solid adsorption into the overall retention process. The influence of the solid support is most prevalent with polar solutes and non-polar solvents at low liquid loadings.

To decrease the influence of the support less active materials have often been used. For example, fluorine-containing polymers of low specific surface area (e.g., Teflon 6 or Fluoropak-80) have been employed in the separation of aqueous samples. Glass microbeads, with surface areas of 0.1 m^2/g have been used for separation of polar materials. This support is especially good for the separation of high-boiling materials, since lightly loaded columns (0.05 to 0.5% w/w) work best. It should be noted that the bulk density of the glass beads is 4 times greater than that of Chromosorb P, so that for the same weight of liquid phase there is a considerably smaller weight percentage of the stationary phase on the beads. Etched glass beads provide a more uniform coating; however, the activity of these beads is greater than that of the simple glass beads.

For diatomaceous earth and other silica materials, the major activity arises from the silanol groups (Si—OH) on the surface. These groups are partially deactivated by the stationary phase. Above 5% w/w loading on Chromosorb P, for example, little or no support influence is observed in typical analytical separations, especially when using polar solvents. In order to reduce the influence of the support at low loadings, the adsorption sites can be covered by adding a small amount of a polar liquid (e.g., a phosphate) to the stationary liquid or mixing a small amount of water vapor with helium in the carrier-gas phase. In addition, surfactants can be added to the stationary phase in small amounts to help spread nonpolar phases on the polar supports.

The most common procedure for deactivation of the support is chemical removal of the adsorption sites by silanization. For example, hexamethyl-disilizane (HMDS) reacts with silica as follows:

$$
\begin{array}{ccc}
\overset{|}{-}\!\!\underset{|}{\text{Si}}\!\!-\text{OH} & & \overset{|}{-}\!\!\underset{|}{\text{Si}}\!\!-\text{O}\!\!-\text{Si(CH}_3)_3 \\
\text{O} \quad + \ (\text{CH}_3)_3\text{Si}\!\!-\!\text{NH}\!\!-\!\text{Si(CH}_3)_3 \ \rightleftharpoons & \text{O} & + \ \text{NH}_3 \\
\overset{|}{-}\!\!\underset{|}{\text{Si}}\!\!-\text{OH} & & \overset{|}{-}\!\!\underset{|}{\text{Si}}\!\!-\text{O}\!\!-\text{Si(CH}_3)_3 \\
\end{array}
$$

In this case the hydrophilic silica surface is converted to a hydrophobic one. Other silanization reagents are trimethylchlorosilane (TMCS) and dimethyl-dichlorosilane (DMCS). Commercially available supports can be obtained in the silanized form.

Open tubular columns are much more difficult to prepare and maintain stable than are packed columns. Tubes are typically composed of either stainless steel or glass. The glass capillaries can be drawn into a compact helical form by using a specially designed commercial drawing machine. Before coating, the capillaries are washed successively with a variety of polar and nonpolar solvents to remove impurities.

How well a solvent coats a particular tubular material is a function of the critical surface tension of wetting and the structure of the wall (e.g., smooth bore versus roughened). For glass, the critical surface tension for wetting is 29.6 dynes/cm at 25°C. This means that nonpolar solvents (e.g., squalane) will coat uniformly, but polar solvents (e.g., cyanoethyoxypropanes) will puddle, exposing glass for adsorption. Stainless steel is somewhat better than glass, but polar phases are still difficult to coat. To aid in coating with polar solvents, a small amount of surface-active agent is added to the solvent, or the inner walls are roughened. Etching, by means of sodium hydroxide solutions at 100°C, has been employed for this purpose. For full experimental details see Refs. 20 and 21.

To increase the amount of stationary liquid phase, porous-layer open tubular columns have also been developed. In this case an emulsion of solid particles (e.g., graphitized carbon black or ferric oxide) is used to deposit the support (no liquid) on the column walls. The stationary liquid is then coated in the typical manner, or alternatively the porous-layer column is used directly for GSC. The advantage of these columns is that sample size and k' values are larger than with simple open tubular columns, with little if any diminution in speed or efficiency. This type of column is discussed in more detail by Halasz and Heine (17).

8.23   DETECTORS

After the column the detector is the most important component of the gas chromatograph. In comparison to liquid chromatography, detection of effluent composition in GC is simpler and in general more sensitive. The latter advantage arises from the following factor. Molecules are more easily ionized in the gas phase than in the liquid state, and the properties of ions can often be measured at higher sensitivity than the properties of neutral molecules.

The most common detectors are the thermal conductivity cell, flame ionization detector, and electron capture detector. We shall briefly describe each of these and then discuss several of their characteristics that are important in GC detection.

The *thermal conductivity cell* has the advantages of simplicity, reliability, universality, and nondestructibility of sample components. The last advantage means that the effluent can be collected after passing through the detector cell, a feature that might be desired in preparative-scale GC. The negative feature of this detector is that its sensitivity is considerably less than that of ionization detectors. Thus, for open tubular columns, the thermal conductivity cell is often not sufficiently sensitive; however, for classical packed and preparative-scale columns, it is more than adequate.

The principle of detection is based on the removal of heat from a hot wire or thermistor by the gas (thermal conduction). The change in temperature of the hot wire is measured as a change in resistance, the wire being one arm of a Wheatstone bridge circuit. In practice, another arm of the bridge contains a second thermal conductivity cell through which the carrier gas alone flows. When no sample components are eluting from the column, carrier gas then flows through both cells, and the bridge is balanced. When a sample component elutes from the column, however, the thermal conductivity of the gas mixture is different from that of the carrier gas alone. The temperature (and thus the resistance) of the sample cell is different from that of the reference cell, resulting in an out-of-balance condition in the bridge, which causes a displacement on the recorder.

For best sensitivity the thermal conductivity of the carrier gas should be as different as possible from that of the sample components. The values of some typical gases (in cal/cm sec °C) are as follows:

Hydrogen	41.6×10^{-5}
Helium	34.8×10^{-5}
Nitrogen	5.81×10^{-5}
Organics	$1–4 \times 10^{-5}$

On this basis H_2 is the preferred carrier, but the danger of explosion and the tendency for reduction of the oxides that make up thermistors give the edge to helium. Note that a major loss in sensitivity occurs if N_2 is used.

It is difficult to give an exact figure for the detection limits of this type of detector since it is quite dependent on operating conditions (e.g., current in hot wire, sample components); however, an order of magnitude is 10^{-6} to 10^{-7} mole/cc for a 1-mV response on a recorder. Note that the thermal conductivity cell responds to the concentration of vapor.

The *flame ionization detector* can be considered as universal for organic compounds (vapors containing CH groups), but it does not respond to common inorganics (e.g., H_2O, NH_3, N_2, He, CO_2). The latter property allows more freedom in the selection of the carrier gas than is possible with the thermal conductivity detector and permits the straightforward analysis of aqueous solutions. Typically, the ionization takes place in a hydrogen flame emanating from a metal jet. The conductivity of the flame is measured by grounding the metal jet and positioning a collector electrode several millimeters above or around the side of the flame, to which is applied ~ -300 V. Positive ions are accelerated to the collector electrode, and electrons are grounded in the jet. The small current (ca. 10^{-11} A baseline) is amplified by a high-input-impedance electrometer, and the signal is then displayed on a recorder.

The detector responds to the mass flow rate of a vapor, that is, grams per second. The limit of detection is typically quoted at 10^{-14} g/sec, which is equivalent to 10^{-9} M of —CH_2— groupings at a flow rate of 60 ml/min. If we consider a molecule with 10 methylene groups, the limit of detection is then 10^{-10} M, or roughly 3 orders of magnitude more sensitive than the thermal conductivity detector. In addition to sensitivity, the flame ionization detector is also known for its linearity, which extends over at least 4 orders of magnitude. Indeed, the detector is usable over 7 orders of magnitude.

The *electron capture detector* is highly sensitive and selective for vapors containing high-electron-affinity elements (e.g., halogens) and electron-capturing groups (e.g., polyaromatics). In operation, a radioactive short-range beta emitter, consisting of ^3H or ^{63}Ni (for high temperatures), ionizes the carrier gas (e.g., N_2) to produce slow electrons, which under a potential of 50 V between two electrodes results in a current of roughly 10^{-9} A. When an electron-capturing vapor enters the detector cell, there is a drop in the current due to the absorption of some of the slow-moving electrons by the vapor. This change in current can be related to the amount of sample present (the concentration) after suitable calibration.

The detector finds its greatest use in the analysis of trace concentrations of pesticides, compounds that are heavily halogenated and may contain phosphorus. To illustrate the sensitivities possible, Table 8.3 lists the responses for some typical pesticides and other compounds. Note that the response seems to increase exponentially with the number of electron-capturing groups or elements. The influence of temperature on relative response factors is great (a factor of 4 to 5 per 100°C, either positive or negative), so that the values in the table should be used only as orders of magnitude.

Further details on GC detectors can be found in Refs. 22 to 25.

Table 8.3. Relative Sensitivities for Pesticides and Other Compounds in the Electron Capture Detector[a]

Compound	Sensitivity (approx.)
Hexane	10^{-7} M
Chlorobenzene	10^{-9} M
n-Butyl chloride	10^{-9} M
Dichloroethane	10^{-11} M
Benzophenone	10^{-12} M
Malathion	8×10^{-13} M
DDT	8×10^{-14} M
Heptachlor	5×10^{-14} M
Dieldrin, Endrin	2×10^{-14} M
Aldrin	10^{-14} M
Chloroform	10^{-14} M
Dibromoethane	10^{-14} M
Carbon tetrachloride	6×10^{-16} M

[a] Recalculated from G. Zweig and J. M. Devine, *Recent Advances in Gas Chromatography*, I. I. Domsky and J. A. Perry, eds., Marcel Dekker, New York, 1971. Original data from Varian Aerograph Technical Bulletin, 106–63 (1963).

8.24 QUANTITATIVE ANALYSIS

In quantitative analysis it is necessary to relate the chromatographic signal to the amount of each sample component present upon injection. Both peak height and area have been used as a measure of the quantity of material. For rapidly eluting peaks, in which the band widths are quite narrow, peak height has been found to be less prone to error. For all other cases, peak area has proved superior.

As implied in Section 8.23, two fundamentally different classes of detectors are used in column chromatography (GC or LC) (26): (a) *concentration sensitive* (e.g., thermal conductivity, ultraviolet), and (b) *mass-flow-sensitive* (e.g., flame ionization). In general, type *a* detectors are nondestructive of the sample components, whereas in type *b*, which has a response proportional to the mass flow rate, chemical reaction occurs. The signals for the two types of detectors can be expressed as

$$S = R_c C \tag{8.33}$$

and

$$S = R_m C \cdot F = R_m (w/t) \tag{8.34}$$

where C = concentration, F = flow rate, w = weight, t = time, and R_c and R_m are the response factors for the concentration- and mass-flow-sensitive detectors, respectively. From Eq. 8.34 it can be seen that the signal in the mass-flow-sensitive detector is proportional to the flow rate: the higher the flow rate, the larger the signal.

The area of a peak is simply

$$A = \int R \, dt \tag{8.35}$$

Now, since $w = \int C \, dV$ and $dV = F \, dt$ (V = volume of gas containing sample), it is easy to show for a concentration-sensitive detector that

$$A = \frac{R_c \theta w}{F} \tag{8.36}$$

where θ is a constant to account for chart speed and recorder response. For mass-flow-sensitive detectors, we have

$$A = R_m \theta w \tag{8.37}$$

It is seen that the peak area is independent of the flow rate for the carrier gas with the latter type of detector but is inversely proportional to flow with the former type. Thus careful control of flow rate is required for quantitative analysis with concentration-sensitive detectors.

With either detector, we see from Eqs. 8.36 and 8.37 that we must determine the peak area and the response factor. Peak area can be determined by a manual procedure, mechanical integrator, electronic integrator, or computer. There are three basic manual methods: (a) cut out and weigh the peak from the recorder trace; (b) calculate the area of the triangle from the tangents to the inflection points of the Gaussians; and (c) use a planimeter. The precision for these three approaches is roughly 3 to 4%. Of course, manual methods are the least expensive.

The most common type of mechanical integrator is the "ball and disk," which is attached directly to a strip-chart recorder. A ball is put into contact with a disk revolving at a constant rate. The position of the ball on the disk is proportional to the position of the recorder pen with the zero point at the center of the disk. The ball revolves at a rate proportional to its position on the disk, with the number of revolutions related to the peak area. The precision of this method is roughly 1%.

The electronic integrator is based on digital electronics. The chromatographic signal is converted to electrical pulses, the frequency of which is proportional to the voltage, and the total number of pulses is counted. The precision of this method is roughly 0.5%. Alternatively, the signal can be put through an analog-digital converter for digitization. The signal is then integrated in either a dedicated on-line computer or an off-line computer. In addition to providing good precision with peak area determinations, the computer can also correct for baseline noise and drift, resolve overlapping peaks electronically, and automate the operation of the gas chromatograph. The full power of the computer in conjunction with gas chromatography is only now beginning to be understood (27). Undoubtedly many new developments can be expected in the next few years.

Quantitative analysis also requires knowledge of the response factor. Since the full relationship between sensitivity and molecular structure is often unknown, it is necessary to determine these response factors experimentally. Although absolute values can be obtained, this is often tedious and unnecessary. Experimentally it is possible to determine the relative response factors, a procedure in which the response factor for a standard component is set equal to unity.

In practice a common procedure for quantitative analysis is the addition to the injected mixture of a known weight proportion of a pure substance whose retention is close to (but does not overlap) the retention of the sample components. This internal standard can then serve as the component on which the relative response factors are based. In addition, this method permits determination of absolute weight proportions for the individual components, regardless of the injection reproducibility or the loss of nonvolatile components. Errors due to slight changes in flow rate and detector nonlinearity are also minimized.

8.25 ANCILLARY TECHNIQUES

The power of GLC as an analytical tool has been greatly enhanced by coupling it to other analytical or chemical techniques (28, 29). These ancillary techniques expand the range of analytical GLC to nonvolatile solutes, as well as provide information impossible to obtain by individual analytical methods.

The ancillary techniques can be classified as (*a*) *chemical reaction* (reaction gas chromatography), or (*b*) *chromatographic detection*. In either case the equipment can come before or after the chromatographic column. In category *a*, most reactions are made before the column, with the products swept onto the column for separation and subsequent analysis. We can subdivide these reactions into three groups: (1) microreaction techniques to study catalytic behavior, (2) pyrolysis to study nonvolatile compounds, and (3) reactions to elucidate structure. Postcolumn reactions are usually employed to obtain enhanced quantitative information about the separated components.

In the second category of ancillary techniques, an instrumental method of analysis is either substituted for or added to a standard GLC detector (see Section 8.23). In the simplest mode, the sample components are collected after the column and then analyzed by an analytical method (e.g., infrared spectroscopy, NMR). On the other hand, the effluent from the column can be coupled directly to the ancillary system, as occurs in GC-mass spectroscopy. In general a coupling method is more convenient than the two-step method of collection and analysis.

These ancillary systems are used in the identification of sample components, since retention data always contain some uncertainty. A full discussion of identification techniques is given in Ref. 28.

Several ancillary techniques are worthy of special note. *Pyrolysis gas chromatography* has been often applied to the analysis of polymers, especially those that readily decompose to volatile products, such as poly(vinyl chloride) or poly(methyl methacrylate). Although the pyrolysis products can be first trapped and subsequently separated and analyzed by GC, most often the products are swept immediately onto the column. In terms of decomposition reproducibility, it is best to flash-heat the sample, that is, rapidly bring it up to the pyrolysis temperature. Recently, this has been accomplished using a Curie point pyrolyzer (30), in which a current is induced in a wire of ferromagnetic material. The wire rapidly reaches the temperature at which it ceases to be ferromagnetic (the Curie point) and remains at this temperature. The heating takes a matter of milliseconds, and the temperature is quite reproducible. In addition to direct thermal heating, high-voltage discharge or mercury-sensitized photolysis has been employed to achieve pyrolysis.

A wide variety of precolumn methods have been ingeniously developed to obtain qualitative structural information. Reactions typical for specific functional groups can provide information on the polar portions of a molecule. In one case, volatile products are formed which are then separated and analyzed by gas chromatography. A particularly good example is *carbon skeleton chromatography* (31). In this method, compounds are passed in a stream of hydrogen over a catalytic bed (e.g., 1% w/w Pd on a solid support) at an elevated temperature. Chemical reactions characteristic of specific functional groups take place (e.g., C=O, C—O, C—S, C—Cl are cleaved to the C—H), and unsaturated groups are saturated by hydrogenation. The resulting product is then the saturated hydrocarbon (i.e., the carbon skeleton) of the molecule, which can be subsequently analyzed. Since most qualitative analytical techniques provide information regarding the functional groups of a molecule, carbon skeleton chromatography can be particularly useful (e.g., steroids, long-chain carboxylic acids). In the second case of reaction gas chromatography, specific compound types are selectively subtracted from the sample for identification purposes. Either physical (e.g., adsorption) or chemical reactions are used to produce products of low volatility; a wide variety of reactions have been employed. Details are given in Ref. 31.

The coupling of a gas chromatograph with a mass spectrometer has evolved into one of the most important tools currently available. The gas chromatograph performs the separation, and the structure of each component is ascertained by its mass spectrum. Quantitative analysis can be performed either by using an ordinary chromatographic detector or by determining the total ion current at a given mass peak. The effluent from the gas chromatograph at 1 atm pressure must be reduced to 10^{-5} torr in the ion source of the mass spectrometer for successful operation. This can be accomplished by rapid removal of the carrier gas in the interface coupling the two systems. Fortunately, helium, the usual GC carrier gas, is much lighter than the sample components, so that separation can readily be achieved by means, for example, of helium diffusion through porous membranes (e.g., fritted glass, porous stainless steel). To accelerate the removal of helium a vacuum is applied on the outside of the membrane.

8.26 TEMPERATURE PROGRAMMING AND PREPARATIVE-SCALE CHROMATOGRAPHY

In Section 5.5 the basics of temperature programming were described. It will be recalled that this technique is applied to overcome the limited peak capacity of the column and the "general elution problem" when the sample contains many components having widely different boiling points. Temperature programming is particularly powerful in the analysis of complex mixtures, such as occur in petroleum or food and flavor analysis (see Fig. 8.7).

Dual-column temperature programming is a technique often employed to extend the usable temperature range of the stationary liquid phase. In single-column operation, above a certain temperature the vapor pressure of the stationary phase is sufficiently large that bleeding takes place. This results in an increase in the baseline displayed on the recorder, thus limiting sensitivity. If, however, a second, completely identical column is matched to the separation column, then bleeding should occur to the same extent in both columns. By passing only carrier gas through the second column and by bucking the signal (baseline) from this column to that from the analytical one, it is possible to extend the upper temperature limit by at least 25°C. This extension can be important, especially for the elution of high-boiling components of a sample mixture. For further details on the theory and practice of temperature-programming GC, see Harris and Habgood (32).

Preparative-scale gas chromatography has found use in the preparation of relatively pure components that have difficult-to-resolve impurities (33). Whereas typical sample sizes in analytical-scale GC are approximately in microliters, injection volumes of milliliters or greater are common in preparative-scale GC. To prevent overloading, it is necessary to use columns with a great deal more packing than is required for analytical-scale work. This can be achieved by employing columns of wide diameter (e.g., 1 to 2 cm) or of longer length (e.g., 10 to 15 meters). In general, wide columns result in some loss in efficiency; however, with care these losses can be minimized.

Collection of samples after the column has often been a problem. Aerosol formation is common, especially when components travel from the hot column to the cold trapping system. Frequently, collection is more efficient when the temperature of the trap is close to room temperature, rather than much lower. Trap designs have taken many forms, but the general principle is to have the vapor come into contact with a large area of a cold surface. Glass beads placed in the trap often improve collection efficiency.

8.3 Typical Applications

By now an enormous number of separation problems have been solved by GLC, and various books dealing with specific areas of application have been written (14, 34, 35, 36). Given this situation, we have no hope of covering in detail the many applications that have been made. Rather, we shall cite typical examples. The discussion will be on the basis of sample type. Further details can be found in Ref. 37.

8.31 HYDROCARBONS

Often the low-molecular-weight hydrocarbons (up to *n*-butane) occur in mixtures with the permanent gases. These samples are most conveniently separated by GSC (see Chapter 14). As the molecular weight (and boiling point) increase, however, GLC has been found to be superior. The hydrocarbons can easily be resolved with a wide variety of liquid phases. On nonpolar phases elution is in the order of boiling point, whereas on polar phases the elution order may change somewhat with, in general, more rapid retention (because of low solubility). As mentioned previously, Fig. 5.1 illustrates the separation of 10 isomers of *n*-heptane on a nonpolar phase.

Olefins are also typically separated in the order of boiling point on non-polar phases. A particularly good separation of the C_2 to C_7 alkanes on a 300-ft capillary column coated with silicone fluid 96 is shown in Fig. 8.5. The separation of low-molecular-weight olefins from saturated hydrocarbons is carried out most conveniently by using a stationary phase of silver nitrate doped in a polar phase such as propylene glycol. The olefins are retarded by the formation of a charge transfer complex between the π clouds and the Ag^+ ions. This is a particularly good example of secondary chemical equilibria (see Chapter 2).

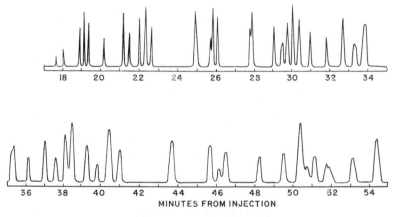

Figure 8.5. Gas chromatogram of master blend of C_2 to C_7 alkenes. Column: glass capillary, 0.012 in. × 300 ft, coated with silicone fluid 96; T = 23 to 25°C. Reprinted from Ref. 40, by courtesy of the American Chemical Society.

8.32 FATTY ACIDS AND ESTERS

In the original paper by James and Martin (1), fatty acids up to C_{12} were analyzed. On nonpolar phases, tailing results, probably because of adsorption on the solid support. However, when 10% w/w stearic acid is added, good symmetrical peaks are obtained. The acid serves the dual purpose of decreasing solid support adsorption by the low-molecular-weight fatty acids and diminishing the extent of dimerization. Although free fatty acids up to C_{20} have been analyzed by GLC, most often methyl esters are separated in order to enhance volatility and minimize tailing.

8.33 AMINO ACIDS

Analysis of amino acids by GLC requires formation of volatile derivatives. Typically the COOH group is esterified with either a methyl or n-butyl group, and the NH_2 group is esterified with the trifluoroacetyl group. More than 20 amino acid derivatives have been analyzed in one chromatogram in a standard procedure (38). It is possible to resolve the amino acids into their optically active isomers, using an optically active stationary liquid phase. Figure 8.6 shows a typical separation from the work of Gil-Av.

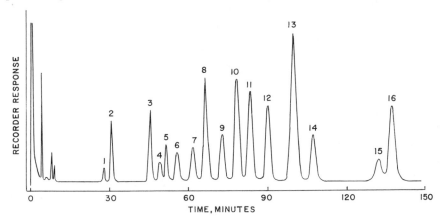

Figure 8.6. N-TFA-isopropyl esters of relatively volatile amino acids on a column (500 ft. × 0.02 in.) of N-TFA-L-valyl-L-valine cyclohexyl ester. $T = 110°C$. *1, 2* = DL-alanine; *3, 4* = DL-valine; *5* = glycine; *6, 7* = DL-threonine; *8, 9* = DL-alloisoleucene; *10* = L-isoleucine; *11* = β-alanine; *12* = D-leucine; *13* = L-leucine + D-serine; *14* = L-serine; *15, 16* = DL-proline. Reprinted from Ref. 41, by courtesy of the publishers of the *J. Chromatog. Science.*

8.34 STEROIDS

Much effort has gone into the analysis of steroids by GLC (36, 39). These solutes can be directly separated on lightly loaded packed columns with non-polar phases. However, derivatives are often made (e.g., trimethylsilyl ethers, methyl esters). In body fluids steroids occur mainly as conjugates (sulfates and glucuronides), and before GLC analysis the conjugates must be hydrolyzed, either with acid or with an enzyme, to produce the free steroids. When proper precautions are taken, reliable quantitative analysis can be obtained, even though two chemical steps—hydrolysis and derivatization—are required.

8.35 MISCELLANEOUS APPLICATIONS

Gas-liquid chromatography is a very powerful tool in pesticide analysis. Since most of these compounds possess halogen atoms in their chemical structures, the electron capture detector is found to be both selective and highly sensitive. For example, it is possible to measure picogram quantities of pesticides, such as lindane, heptachlor, and DDT. Liquid phases of moderate polarity are most often selected for these separations.

For food and flavor analysis, GLC has found many applications. Here, one usually is faced with the analysis of a complex mixture in which most, if not all, components are unknown. Such cases require large peak capacity column systems, that is, temperature programming with high-resolution capillary columns. Figure 8.7 shows a typical chromatogram of a cheese extract.

Gas-liquid chromatography has been applied to the separation and analysis of a surprisingly large number of inorganic substances. Initial work in this field involved the analysis of volatile metal halides, special precautions being necessary because of the reactive character of these compounds. For

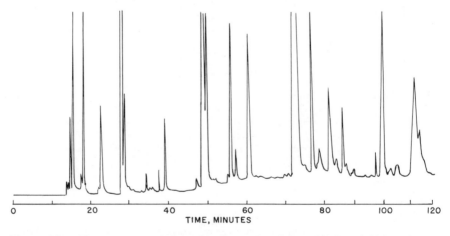

Figure 8.7. Chromatogram of Roquefort cheese. Conditions: 500 ft × 0.02 in. column; Dowfax 9N15; temperature programmed from 25°C to 125°C at 2°C/min; sample size 10 microliters of cheese oil. Reprinted from Ref. 42, by courtesy of the publishers of the *J. Chromatog. Science.*

less volatile halides, fused salts consisting of eutectic mixtures of inorganic ionic halides have proved useful as stationary phases. More recently workers have formed volatile derivatives of metals for straightforward chromatographic separation and analysis. Metal chelates made from trifluoroacetylacetone and hexafluoroacetylacetone are most often used in this area. Metals of high atomic number, including uranium, have been successfully chromatographed. The high sensitivity of the GLC detectors makes this method particularly suitable to trace analysis. Further details on this subject can be found in Ref. 2.

References

1. A. T. James and A. J. P. Martin, *Biochem. J.*, **50**, 679 (1952).
2. R. W. Moshier and R. E. Sievers, *Gas Chromatography of Metal Chelates*, Pergamon Press, London, 1965.
3. (a) D. E. Martire, in *Progress in Gas Chromatography*, J. H. Purnell, ed., Interscience, New York, 1968; (b) J. R. Conder, D. C. Locke, and J. H. Purnell, *J. Phys. Chem.*, **73**, 700 (1969); (c) P. Urone, Y. Takahashi, and G. H. Kennedy, *J. Phys. Chem.*, **74**, 2326 (1970); (d) B. L. Karger, P. A. Sewell, R. C. Castells, and A. Hartkopf, *J. Colloid Interface Sci.*, **35**, 328 (1971).
4. E. sz. Kovats, in *Advances in Chromatography*, Vol. I, J. C. Giddings and R. A. Keller, eds., Marcel Dekker, New York, 1965.
5. *Compilation of Gas Chromatographic Data*, O. E. Schupp and J. S. Lewis, eds., American Society for Testing and Materials, Philadelphia, 1967.
6. W. O. McReynolds, *Gas Chromatographic Retention Data*, Preston Publishing Co., Evanston, Ill., 1967.
7. L. Rohrschneider, in *Advances in Chromatography*, Vol. IV, J. C. Giddings and R. A. Keller, eds., Marcel Dekker, New York, 1968.
8. L. Rohrschneider, *J. Chromatog.*, **22**, 6 (1966).
9. W. O. McReynolds, *J. Chromatog. Sci.*, **8**, 685 (1970).
10. G. W. Gray, *Molecular Structure and the Properties of Liquid Crystals*, Academic Press, New York, 1962.
11. M. J. S. Dewar and J. P. Schroeder, *J. Am. Chem. Soc.*, **86**, 5235 (1964).
12. H. Kelker, in *Advances in Chromatography*, Vol. IV, J. C. Giddings and R. A. Keller, eds., Marcel Dekker, New York, 1968.
13. S. H. Langer and R. J. Sheehan, in *Progress in Gas Chromatography*, J. H. Purnell, ed., Interscience, New York, 1968.
14. D. A. Leathard and B. C. Shurlock, *Identification Techniques in Gas Chromatography*, Interscience, New York, 1970.
15. J. H. Walraven, A. W. Ladon, and A. I. M. Keulemans, *Chromatographia*, **1**, 195 (1968).
16. J. C. Giddings, *Dynamics of Chromatography: Principles and Theory*, Marcel Dekker, New York, 1965.
17. I. Halasz and E. Heine, in *Progress in Gas Chromatography*, J. H. Purnell, ed., Interscience, New York, 1968.
18. G. Guiochon, in *Advances in Chromatography*, Vol. VIII, J. C. Giddings and R. A. Keller, eds., Marcel Dekker, New York, 1969.

19. D. M. Ottenstein, in *Advances in Chromatography*, Vol. III, J. C. Giddings and R. A. Keller, eds., Marcel Dekker, New York, 1966.

20. L. S. Ettre, *Open Tubular Columns in Gas Chromatography*, Plenum, New York, 1965.

21. J. M. d'Aubigne, C. Landault, and G. Guiochon, *Chromatographia*, **4**, 309 (1971).

22. J. E. Lovelock, *Anal. Chem.*, **52**, 162 (1961).

23. B. J. Gudzinowicz, in *The Practice of Gas Chromatography*, A. Zlatkis and L. S. Ettre, eds., Interscience, New York, 1967.

24. A. B. Littlewood, *Gas Chromatography*, 2nd ed., Academic Press, New York, 1970, chaps. 8–11.

25. M. Krejci and M. Dressler, *Chromatog. Rev.*, **13**, 1 (1970).

26. I. Halasz, *Anal. Chem.*, **36**, 1428 (1964).

27. J. E. Oberholtzer and L. B. Rogers, *Anal. Chem.*, **41**, 1234 (1969).

27a. S. N. Chesler and S. P. Cram, *Anal. Chem.*, **43**, 1922 (1971).

27b. M. Goedert and G. Guiochon, *Chromatographia*, **6**, 37 (1973).

28. L. S. Ettre and W. H. McFadden, eds., *Ancillary Techniques of Gas Chromatography*, Interscience, New York, 1969.

29. V. G. Berezkin, *Analytical Reaction Gas Chromatography*, L. S. Ettre, transl., Plenum, New York, 1968.

30. C. Buhler and W. Simon, *J. Chromatog. Sci.*, **8**, 323 (1970).

31. M. Beroza and R. A. Coad, in *The Practice of Gas Chromatography*, A. Zlatkis and L. S. Ettre, eds., Interscience, New York, 1967.

32. W. E. Harris and H. W. Habgood, *Programmed Temperature Gas Chromatography*, Interscience, New York, 1966.

33. A. Zlatkis and V. Pretorius, eds., *Preparative Gas Chromatography*, Interscience, New York, 1971.

34. B. J. Gudzinowicz, *Gas Chromatographic Analysis of Drugs and Pesticides*, Marcel Dekker, New York, 1967.

35. R. W. Moshier and R. E. Sievers, *Gas Chromatography of Metal Chelates*, Pergamon Press, Oxford, 1965.

36. H. H. Wotiz and S. J. Clark, *Gas Chromatography in the Analysis of Steroid Hormones*, Plenum, New York, 1966.

37. L. E. H. Knapman, ed., *Gas Chromatography Abstracts*, Institute of Petroleum, London, yearly.

38. C. W. Gehrke and D. L. Stalling, in *Separation Techniques in Chemistry and Biochemistry*, R. A. Keller, ed., Marcel Dekker, New York, 1967.

39. E. C. Horning and M. G. Horning, *J. Chromatog. Sci.*, **9**, 129 (1971).

40. A. G. Polgar, J. J. Holst, and S. Groennings, *Anal. Chem.*, **34**, 1266 (1962).

41. S. Nakaparksin, P. Birrell, E. Gil-Av, and J. Oro, *J. Chromatog. Sci.*, **8**, 177 (1970).

42. H. M. Liebich, D. R. Douglas, E. Bayer, and A. Zlatkis, *J. Chromatog. Sci.*, **8**, 351 (1970).

Selected Bibliography

E. R. Adlard, R. Stock, and B. T. Whitman, in *Comprehensive Analytical Chemistry*, Vol. IIB: *Physical Separation Methods*, C. L. Wilson, D. W. Wilson, and C. R. N. Strouts, eds., Elsevier, New York, 1968, pp. 55–212.

L. S. Ettre and A. Zlatkis, eds., *The Practice of Gas Chromatography*, Interscience, New York, 1967.

L. S. Ettre, *Open Tubular Columns in Gas Chromatography*, Plenum, New York, 1965.

A. B. Littlewood, *Gas Chromatography*, 2nd ed., Academic Press, New York, 1970.

S. Dal Nogare and R. S. Juvet, Jr., *Gas-Liquid Chromatography: Theory and Practice*, Interscience, New York, 1962.

J. H. Purnell, *Gas Chromatography*, John Wiley, New York, 1962.

J. H. Purnell, ed., *Progress in Gas Chromatography*, Interscience, New York, 1968.

A. Zlatkis and V. Pretorius, eds., *Preparative Gas Chromatography*, Interscience, New York, 1971.

R. Kaiser, *Gas Phase Chromatography*, Vols. I, II, and III, P. H. Scott, transl., Butterworths, Washington, 1963.

H. P. Burchfield and E. E. Storrs, *Biochemical Applications of Gas Chromatography*, Academic Press, New York, 1962.

C. E. H. Knapman, ed., *Gas Chromatography Abstracts*, Institute of Petroleum, London, yearly.

List of Symbols

a	correction factor	ΔI	retention index difference
a_i	solute dependent factor determining E_i	j	correction factor to calculate the average flow velocity in the column
A	cross-sectional area of the column tube	K_i	complex formation constant
B_0	specific permeability coefficient	K_0	distribution constant of the uncomplexed solute
D	stoichiometric distribution ratio	M_s	molecular weight of stationary phase
D_g	diffusion coefficient of solute in the gas phase	N_s	moles of liquid stationary phase in the column
D_l	diffusion coefficient of solute in the liquid phase	P^*	polarity
E_i	interaction energy	P^0	saturation vapor pressure of solute at column temperature
E_{ii}	interaction potential between two solute molecules		
E_{is}	interaction potential energy between solute and solvent molecules	P_w	vapor pressure of water
		Q	constant used in peak area determination
		R_c	response factor for concentration-sensitive detector
E_{ss}	interaction potential energy between two solvent molecules		
		R_m	response factor for mass-sensitive detector
$\bar{F}$	average flow rate in the column	Δs^e	partial molar excess entropy
F_m	measured flow rate	T_0	ambient temperature
F_o	outlet flow rate	$\bar{v}$	average flow velocity in the column
Δh^e	partial molar excess enthalpy		

v_0	flow velocity at the outlet of the column	x_i	solvent dependent factor determining E_i
V_g	specific retention volume	X_s	mole fraction of solute in the liquid stationary phase
$\bar{V}_i$	molar volume of solute		
$V_{N,n},$	net retention volumes of		
$V_{N,n+1}$	two normal alkanes with carbon numbers n and $n+1$, respectively	γ^∞	solute activity coefficient at infinite dilution
$_1V_{R,p},$	retention volumes of	$\gamma^{\infty,a}$	the athermal portion of the activity coefficient
$_1V_{R,u}$	butane as measured with a polar and an unpolar stationary phase, respectively.	$\gamma^{\infty,t}$	the temperature-dependent portion of the activity coefficient
$_2V_{R,p},$	retention volumes of	λ	packing factor
$_2V_{R,u}$	butadiene as measured with a polar and an unpolar stationary phase, respectively	$\Delta\mu^e$	partial molar excess chemical potential
		ρ_S	density of the liquid stationary phase
W	weight	ω	packing factor for radial diffusion

Nine

SOLVENT EXTRACTION

H. FREISER

9.1 Introduction

Solvent extraction is a method of separation based on the transfer of a solute from one solvent into another, essentially immiscible solvent when the two solvents are brought into contact. The reason for the large popularity of solvent extraction lies in the speed, ease, and convenience of the technique. The separations are clean because the relatively small interfacial area between the two liquid phases avoids any effects analogous to the undesirable coprecipitation phenomena that plague most precipitation separations. In most cases a simple separatory funnel is all the apparatus required. The extraction step usually requires only several minutes to carry out, and the procedures are applicable to both trace and macro levels. Many sensitive organic and biochemical materials can be separated by extraction procedures with much less danger of decomposition than is present with other separation processes.

A further important advantage of solvent extraction methods lies in the convenience of subsequent analysis of the extracted species. Thus, if the extracted species is colored, as is the case with many chelates, spectrophotometric methods can be employed. Alternatively, the solution may be aspirated for atomic absorption or flame spectroscopic analysis. If radiotracers are used, radioactive counting techniques can be employed.

One objective measure of the continued popularity of solvent extraction in analytical chemistry lies in the wealth of literature continuing to appear on the subject. For example, the biennial review articles (1) that the author has helped to prepare summarize the work of approximately 500 papers a year. The reader is urged to look either at these or at the three major works listed in the selected bibliography at the end of the chapter.

One of the unique features of solvent extraction, particularly for metal ions, is the large variation in distribution ratios and separation factors made possible by controlling the chemical parameters of the system. Thus, although most simple metal salts are highly soluble in aqueous media and relatively insoluble in organic solvents, it is possible to employ a variety of chemical approaches to incorporate the metal in an organic-soluble species. The importance of such transformations can be gauged by their use in the classification of metal extraction systems (Section 9.3). Consideration of the extractability of neutral compounds such as elemental iodine, osmium tetroxide, and most organic substances involves chemical and quasi-chemical interactions such as hydrogen-bonding, donor-acceptor, and dipole-dipole interactions (see Section 2.2) that are more subtle but nevertheless play a determining role in these extraction systems. In addition, organic acids and bases are readily transformed by pH variation to charged species whose extractabilities are dramatically altered from those of the parent compounds. Further, complex formation such as that of alkenes with silver ion or of amines with transition-metal ions provides another important chemical means of changing the characteristics of organic compounds and thereby improving extraction selectivity (see Section 2.13).

9.2   Process of Extraction

Despite the great profusion of extraction systems, it is possible to describe every extraction by the following simple, three-part scheme.

1. Distribution of the Extractable Species. Any neutral species will distribute between two essentially immiscible solvents at constant temperature in such a manner that the ratio of the concentrations (strictly speaking, the activities) of the species in the two solvents remains constant (called K_D, the *distribution constant*) provided that there is no chemical interaction involving the species in either phase. This statement of the *Nernst distribution law* (see Section 2.13) indicates the mathematical simplicity of the extraction of any species, but ignores the complexity of relating the extractability of a given species to its molecular structural parameters or those of the solvent pair that affect it. In extraction systems involving hydrogen bonding between the solute and one or more of the solvents, it is possible to qualitatively evaluate the effect of structural changes on the course of the distribution processes. Similarly, when other more or less well-defined chemical forces come into play, their effect on distribution can be evaluated at least qualitatively.

For extraction systems in which specific chemical forces are not active, the classic principle of "like dissolving like" is of great help in predicting relative solubility and extractability. This principle is incorporated in the *Hildebrand theory* of regular solutions, which shows that solubility increases as the values of the solubility parameter δ of the solute and solvent approach one another (see Section 2.2).

By means of this approach (Eq. 2.8b and Eq. 2.44) the following expressions can be derived:

$$2.3RT \log K_D = \bar{V}_S[(\delta_S - \delta_1)^2 - (\delta_2 - \delta_S)^2] \tag{9.1}$$

where $\bar{V}_S$ is the molar volume of the distributing solute, δ_S is its solubility parameter, and δ_1 and δ_2 are the solubility parameters of the pair of immiscible solvents. Although one expects this relation to be of maximum assistance in organic extraction systems in which specific chemical or associative forces are at a minimum, Eq. 9.1 has proved useful in a number of inorganic extraction systems (both chelates and ion pairs) as well. The choice of the optimum organic solvent for an extraction of particular interest still cannot be made on a purely scientific basis but involves a great deal of empirical evaluation.

If, as is the case in most inorganic (metal) and many organic extraction systems, the extractable species is involved in chemical interactions in either phase or both of them, the ratio of the analytical concentrations (called D, *the distribution ratio*) obviously varies with experimental conditions. This does not represent a failure of the Nernst law, however, because, if the equilibrium expressions of all the participating reactions are included, it is possible to derive an expression based on the Nernst law relating D to the experimental parameters that quantitatively describe the extraction behavior of the system (Section 2.13).

2. Chemical Interactions in the Aqueous Phase. A major distinction between organic and inorganic extractions lies in the extent to which the formation of an uncharged extractable species depends on chemical interactions in the aqueous phase. Most organic compounds are uncharged, although those containing acidic or basic functional groups can undergo proton-transfer reactions that result in charged species, such as $RCOO^-$ and RNH_3^+. Such species have obviously different solubility characteristics in the aqueous and organic phases, and their formation must be accounted for in calculating D. Hence many organic and biochemical extractions can be carried out by means of pH control.

Metal salts are generally soluble in aqueous media not only because of the high dielectric constant of water, which readily permits dissociation of ionic species, but also, more importantly, because the basic character of water results in the solvation of metal ions, which gives these ions a solvent sheath that reduces electrostatic interaction and makes them more "solvent-like." The role of the complex-forming metal extraction agents is essentially to supplant the coordinated water from around the metal ion to give a species that is more likely to be compatible with organic solvents. So important is the formation of an undissociated complex in metal extractions that a classification of such extractions is based on the nature of the complex (Section 9.3).

3. Chemical Interactions in the Organic Phase. Chemical interactions of the extracted species in the organic phase lower its activity and hence move the overall equilibrium in the direction of greater extractability unless, of course, the reaction product is more soluble in the aqueous phase. Among organic compounds, dimerization of carboxylic acids in non-oxygen-containing solvents results in higher D values. Ion-association complexes (see Section 9.3) are dipolar and tend to polymerize in low dielectric constant solvents. Another important category of organic phase reaction is that of mixed ligand, or adduct, complexes of metal chelates whose formation can increase D values by several orders of magnitude.

9.3   Classification of Inorganic Extraction Systems

Fundamentally, there are two broad categories of neutral extractable complexes: those that involve chemical bonding or coordination, and those that form by essentially electrostatic forces or ion association. In the following seven categories, the first three describe various kinds of neutral coordination complexes, the fourth deals with ion-association complexes, and the last three involve complexes that combine attributes of some of the first four categories.

9.31   PRIMARY SYSTEMS

a. Simple (Monodentate) Coordination Complexes

These are formed by the combination of cationic metal ions, such as $Hg(II)$, $Ge(IV)$, and $As(III)$, with anionic monodentate ligands (e.g., halide anions), giving neutral complexes that are extractable in hydrocarbon (e.g., C_6H_6) and chlorinated hydrocarbon (e.g., $CHCl_3$) solvents.

b. Heteropoly Acids

These represent a class of coordination complexes in which the central ion is itself complex rather than monatomic, such as phosphomolybdic acid,

$H_3PO_4 \cdot 12MoO_3$. Heteropoly acids are highly solvated by hydrogen bonding; therefore their extraction requires the use of oxygenated solvents.

c. Chelate (Polydentate Coordination) Complexes

These are formed by the bonding of a metal ion by ligands which can occupy at least two coordination sites, resulting in a cyclic compound. When the charges of the metal ion and the ligands match, as with Fe(III) and three 8-hydroxyquinolinate ions, a neutral chelate results, such as Fe(8-hydroxyquinolinate)$_3$, which is often much more soluble in organic solvents than in aqueous media and is therefore of great interest in extraction procedures. Some of the most widely used chelating agents are listed in Table 9.1.

d. Simple Ion-Association Complexes

Large and poorly hydrated ions tend to associate to form neutral compounds that are soluble in water to only a small extent but are relatively soluble in organic solvents, particularly when one of the ions has organic character. Thus tetraalkylammonium ions (R_4N^+) will form benzene, amyl alcohol, and chloroform-soluble salts with large inorganic anions, such as perchlorate, thiocyanate, hexafluorophosphate, and triiodide, and such organic anions as tetraphenylborate and alkyl phenolates. Large, singly charged cations (e.g., Rb^+, Cs^+) form nitrobenzene-extractable salts with some of the anions mentioned (I_3^-, PF_6^-, alkyl phenolate anions).

9.32 MIXED EXTRACTION SYSTEMS

In addition to the primary types of extraction systems listed in Section 9.31, in which the extractable complex was formed by either coordination (monodentate, polydentate, or chelate, heteropoly acid) or ion pairing, there are several other important types in which the formation of the extractable complex involves a combination of these factors.

a. Mixed Simple Coordination and Ion-Pair Systems

Quite a number of metal cations form negatively charged complexes with monodentate anionic ligands, such as the halides, thiocyanate, and oxyanions. These complexes when paired with suitable cations are extractable in organic solvents. One such cation formed in acidic solution is the hydrated hydronium ion, $[H(H_2O)]_3O^+$, whose requirement of further hydrogen bonding for stabilization necessitates the use of oxygen-containing solvents. Thus extraction of Fe(III) from 6 M HCl as $[H_9O_4^+, FeCl_4^-]$, or that of Zn(II) from 1 M HCl as $(H_9O_4^+)_2ZnCl_4{}^{2-}$, is quantitative when ethers, alcohols, ketones, or esters are employed, but negligible if hydrocarbons or chlorinated hydrocarbons are used. A more stable type of cation, such as a

Table 9.1.   Selected Chelate Extraction Systems

A. Bidentate Chelating Agents

  a. Four-membered ring systems

   1. *Dialkyldithiocarbamates* (DDTC). The water-soluble species, Na^+, $(C_2H_5)_2NCSS^-$, will react with most of the heavy- and transition-metal ions to give chelates, for example, $Pb(DDTC)_2$, which are readily extractable into C_6H_6, $CHCl_3$, $CH_3COOC_2H_5$. To overcome the instability of the reagent to acid and to improve selectivity as well, an organic solution of the Zn or Pb chelate is sometimes used in what is termed metal-exchange extraction. Only metals that form more stable complexes (e.g., Cu, Ag) will extract. Thus in a widely used method Cu^{2+} is extracted from an 0.1 M HCl solution using a carbon tetrachloride solution of the Zn chelate of diethyldithiocarbamic acid.

   2. *Dialkyldithiophosphates* $[(RO)_2PSSH)]$. These compounds, which, like DDTC, bond metals through two S atoms, are much more resistant to decomposition in strong acid solution.

  b. Five-membered ring systems

   1. *Cupferron* (*ammonium salt of N-Nitrosylphenylhydroxylamine*) **1**. This water-soluble compound forms extractable complexes with a large number of metal ions but must be used in chilled solution to avoid decomposition. It was named with the expectation that the reagent would be selective for Cu and Fe, but, depending on the pH of the solution, cupferron reacts with many different metal ions. The more recently developed *N*-benzoyl-*N*-phenylhydroxylamine (**2**) behaves similarly but is more stable.

1   2

   2. *8-Quinolinol* (*oxine*) (**3**). Oxine (Ox) is generally used in $CHCl_3$ or benzene solutions and reacts with almost 60 metal ions to form complexes of such types as $CuOx_2$, $ZnOx_2 \cdot HOx$, $UO_2(Ox)_2$. Its chelates with nontransition metals are highly fluorescent. 2-Methyloxine (the 2-position is adjacent to the N atom) is of interest in separating metal traces from Al^{3+}, since steric hindrance prevents the formation of an extractable neutral Al complex.

3

3. 8-Mercaptoquinoline (thioxine) (**4**). This reagent is more selective than oxine, reacting only with the heavy-metal ions and the more electronegative of the transition-metal ions, but not with the alkaline earths or Al^{3+}. The reagent forms complexes that have higher proton displacement constants, K_{PD} (defined as the equilibrium constant of the reaction $M^{n+} + nHL \rightleftharpoons ML_n + nH^+$), than does oxine, which permits extraction from more acidic solutions. Although thioxine is readily oxidized, the sodium salt is very stable.

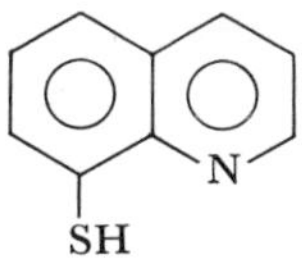

4

4. Dithizone (diphenylthiocarbazone) (**5**). Dithizone, a highly conjugated complexing agent, forms metal complexes having molar absorptivities of the order of 40,000 with heavy- and transition-metal ions, providing the basis for very sensitive spectrophotometric methods for these metal ions. For example, the extraction of Hg(II) from acid solution into dilute dithizone solutions in chloroform can be used to detect submicrogram quantities because its molar absorptivity at 500 nm is about 70,000.

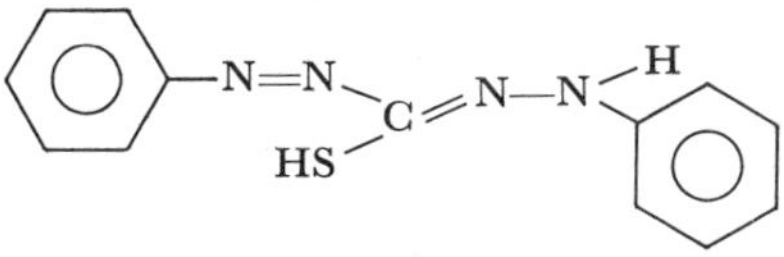

5

c. Six-membered ring systems

1. Thenoyltrifluoroacetone (TTA) (**6**). This β-diketone, like others in its class forms extractable chelates with over 60 metal ions. It first came into wide use in connection with separating the metallic (principally lanthanide) products of uranium fission, as it is fairly radiation resistant.

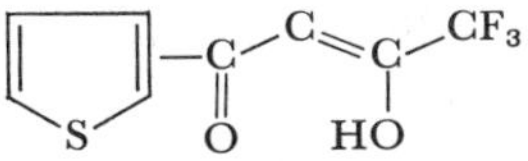

6

d. Larger ring systems

1. Di(2-ethylhexyl)phosphoric acid (DEHPA)
This dialkylphosphoric acid exists largely as a hydrogen-bonded dimer (**7**) in non-oxygen-containing organic solvents and forms extractable chelates with

lanthanides and actinides in which one of the protons of the dimer is replaced by the metal ion.

$$(C_8H_{17}O)_2P \underset{OHO}{\overset{OHO}{<}} \overset{}{\underset{}{>}} P(OC_{18}H_{17})_2$$

7

B. Polydentate Chelating Systems

Pyridylazonaphthol (PAN) (**8**) is a fairly general extractant, but glyoxal bis(2-hydroxylanil) (**9**) is unusual in forming an extractable Ca complex.

8   **9**

tetrasubstituted ammonium, phosphonium, or arsonium cation (R_4N^+, R_4P^+, R_4As^+), permits ion-pair extraction of $FeCl_4^-$, $ZnCl_4^{2-}$, MnO_4^-, and so forth into hydrocarbon solvents. In this latter category, cations of triphenyl-methane dyes not only are suitable for such extractions but also provide the basis for colorimetric determination of the extracted metal ion.

b. Mixed Chelation and Ion-Pairing Systems

If a neutral chelating agent such as dipyridyl or phenanthroline reacts with a metal ion, the resulting chelate is positively charged (e.g., $Fephen_3^{2+}$). Such large cations readily pair with suitable anions, such as ClO_4^- or $(C_6H_5)_4B^-$, to give extractable species. Analogously, if an anionic chelating agent forms a negatively charged chelate, then pairing with a suitable cation such as $(C_4H_9)_4N^+$ can bring about extraction. Anionic oxalato and even ethylene-diaminetetracetic acid (EDTA) chelates can be extracted using high-molecular-weight quaternary ammonium ions. The highly colored anionic complexes of metallochromic indicators such as Eriochrome Black T can be extracted and used in colorimetric analysis.

c. Mixed Ligand Chelates

Metal ions whose coordination number is more than twice their electrical charge react with bidentate ligands to form chelates that are termed co-ordinatively unsaturated. In such chelates, for example, Mg(oxinate)$_2$, the coordination sites not occupied by the chelating agent are filled by water, which results in a rather poorly extractable complex. Improved extraction results when these water molecules are replaced by organic ligands such as alcohols, ketones, esters, or amines. Extraction systems using combinations of chelating agents and auxiliary ligands can be so much better than either one alone that they are referred to as "synergic extractions." Extraction of Th(IV) with TTA is greatly enhanced in the presence of tributyl phosphate (TBP) because of the formation of the mixed ligand chelate, $Th(TTA)_4 \cdot TBP$.

9.4 Fundamental Extraction Parameters

9.41 DISTRIBUTION RATIO

The ratio of the total stoichiometric concentrations of a distributing substance in the organic and aqueous phases, C_o/C_w, called the distribution ratio D, is generally used as a measure of the extent of its extraction (see Section 2.13). The value of D is constant (equal to K_D) when the distributing substance does not react chemically in either phase but otherwise (and most often) is a function of specific experimental conditions.

9.42 FRACTION EXTRACTED

The fraction extracted, θ, is related to D by the following (see Eq. 2.19),

$$\theta = \frac{C_o V_o}{C_o V_o + C_w V_w} = \frac{DV}{1 + DV} \tag{9.2}$$

where C_o and C_w represent the organic and aqueous phase (total) concentrations, respectively, and V is the phase volume ratio, V_o/V_w.

Equation 9.2 demonstrates the possibility of increasing the extent of extraction with a given D value by increasing the phase volume ratio (see Fig. 2.7). If, instead of a single batch extraction, a second or third extraction is carried out on the same aqueous solution by successive portions of organic solvent such that V remains the same, the additional fractions extracted are $\theta(1 - \theta)$ and $\theta(1 - \theta)^2$, respectively. The fraction remaining in the aqueous phase following n successive extractions is $(1 - \theta)^{n-1}$.

9.43 SEPARATION FACTOR

If two substances A and B are present in a solution in an initial concentration ratio C_A/C_B, then upon extraction their concentration ratio in the organic phase will be $C_A \theta_A / C_B \theta_B$, where θ_A and θ_B are the corresponding fractions extracted. The ratio θ_A/θ_B, which is the factor by which the initial concentration ratio is changed by the separation, is a measure of separation. A corollary measure of separation which represents the change in the ratio of concentrations remaining in the aqueous phase is $(1 - \theta_A)/(1 - \theta_B)$.

Two substances whose distribution ratios differ by a constant factor will be most efficiently separated if the product $D_A D_B$ is unity (Eq. 2.23). Consider the case of a pair of substances whose distribution ratios are 10^3 and 10^1, respectively. If these substances were present in equal quantity, a single extraction would remove 99.9% of the first and 90% of the second. A much more efficient extraction would be obtained if, using the same factor of 100 between the distribution ratios, the two distribution ratios were 10^{+1} and 10^{-1}. In this case the respective fractions extracted would be 90% and 10%.

9.5   Methods of Extraction

9.51   BATCH EXTRACTION

When experimental conditions can be adjusted so that the fraction extracted is 0.99 or more ($VD \geq 100$), then a single, or batch, extraction suffices to place the bulk of a desired substance in the extract. Even with VD at 10, carrying out the batch extraction twice will transfer 99% of the substance to the organic phase.

If a desirable separation criterion for a pair of substances is that at least 99% of one and no more than 1% of the second is extracted is one extraction step, then $VD_1 \geq 100$ and $VD_2 \geq 0.01$.

9.52   CONTINUOUS EXTRACTION

For substances whose VD values are relatively small, even multiple-batch extraction cannot conveniently or economically (too much organic solvent is required) be used. Continuous extraction using volatile solvents can be carried out in an apparatus in which the solvent distilled from an extract collection flask is condensed, contacted with the aqueous phase, and returned to the extract collection flask in a continuous fashion. Apparatus for solvents that may be either heavier or lighter than water is available.

9.53   COUNTERCURRENT DISTRIBUTION

A special multiple-contact extraction is needed to separate two substances whose D values, even under optimum conditions, are very similar. In principle, countercurrent distribution (CCD) can be carried out in a series of separatory funnels, each containing an identical lower phase. The mixture is introduced into the first portion of upper phase in the first funnel. After equilibration, the upper phase is transferred to the second funnel and a new portion of upper phase (devoid of sample) is introduced into the first funnel. After both funnels are equilibrated, the upper phase of each is moved on to the next funnel and a fresh portion of upper phase is again added to the first funnel. This process is repeated for at least as many times as there are funnels, the upper phases being collected as "elution fractions." Lyman Craig introduced CCD in order to study an analog of partition chromatography, but it was found useful in its own right (see Section 4.8). With automated CCD equipment several hundred transfers can be conveniently accomplished, permitting the separation of two solutes whose θ_1/θ_2 is less than 2.

The relation of the distribution ratio D of a solute in a CCD process to the concentration in the various separatory funnels or stages can be shown to be given by following binomial expansion:

$$[\theta + (1 - \theta)]^n = 1 \tag{9.3}$$

where θ, the fraction extracted, is $DV/(1 + DV)$, as shown in Eq. 9.2, and n is the number of stages used in the CCD process.

The fraction $T_{n,r}$ of the solute present in the rth stage for n transfers can be calculated from

$$T_{n,r} = \frac{n!}{r! \, (n - r)!} \frac{(DV)^r}{(1 + DV)^n} \tag{9.4}$$

A modification of CCD useful for laboratory purposes involves using a small number of separatory funnels (e.g., three), together with a larger number of transfers of upper-phase portions (e.g., eight), and collecting the first five upper-phase portions as the product fraction. This will serve to quantitatively separate a substance with a D of 10 from one whose D is 0.1.

9.6   Quantitative Treatment of Extraction Equilibria

It is useful to develop equations relating the experimental quantity D to the equilibrium constants of the contributing chemical reactions, as well as to the experimental concentration variables. Examination of such equations serves to elucidate the role of each factor in the extraction process (Section 2.13).

9.61   EXTRACTION OF ORGANIC ACIDS AND BASES

The extraction of 8-quinolinol (HQ) between chloroform and water illustrates an organic extraction that is also of interest in metal-ion extractions.

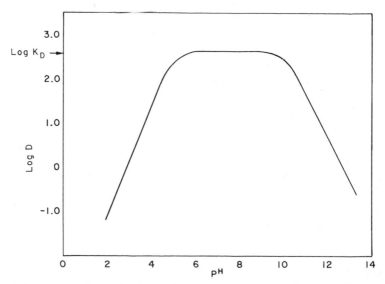

Figure 9.1.   Distribution of 8-quinolinol between chloroform and water at 25°C.

In aqueous media, three forms of 8-quinolinol can exist: H_2Q^+, in which the nitrogen is protonated, HQ, and Q^-. The acid-base equilibria that must be taken into account are

$$K_{a_1} = \frac{[H^+][HQ]}{[H_2Q^+]} \quad \text{and} \quad K_{a_2} = \frac{[H^+][Q^-]}{[HQ]}$$

In the chloroform phase only the neutral HQ is present. Hence the distribution ratio in this case is

$$D = \frac{[HQ]_o}{[H_2Q^+] + [HQ] + [Q^-]}$$

which, when the equilibrium expressions for acid dissociation and distribution are substituted, becomes

$$D = \frac{K_D K_{a_1}[H^+]}{[H^+]^2 + K_{a_1}[H^+] + K_{a_1}K_{a_2}}$$

From this equation or, more particularly, from a plot of log D versus pH (Fig. 9.1), one can see that there are essentially three pH regions, corresponding to those in which one of the three forms (H_2Q^+, HQ, and Q^-) predominates in the aqueous form. Hence, in the pH range between pK_1 (5.0) and pK_2 (9.9) where HQ is the major species, D for 8-quinolinol remains essentially constant at 400 ($= K_D$). Even though H_2Q^+ predominates at lower pH values and Q^- at higher pH values, it can be seen that more than

90% ($D \geqslant 10$) of the compound remains in the organic phase within the larger range of pH 3.4 to 11.5.

When the extraction behavior of 8-quinolinol, an amphiprotic substance, is compared with that of a base such as quinoline and of an acid such as phenol, it may be seen that D varies with pH at low but not at high values for the base, whereas for the acid the reverse is true. Therefore control of pH serves to separate by extraction acids and bases from each other and from neutral substances.

9.62 EXTRACTION OF METAL CHELATES

A large number of chelating agents which may be represented generally as HL (weak monoprotic acids) form uncharged metal chelates that are soluble in organic solvents and are therefore extractable. Such reagents, which include acetylacetone, dimethylglyoxime, and diphenylthiocarbazone, are generally dissolved in organic solvents immisicible with water.

Figure 9.2 illustrates the equilibria involved in a typical chelate extraction.

The definition of D, the metal distribution ratio, as the ratio of total metal-ion concentrations in the organic and aqueous phases, is seen to be

$$D \equiv \frac{C_{M(o)}}{C_M} = \frac{[ML_m]_o}{[M^{n+}]/\alpha_M} \tag{9.5}$$

where α_M represents the fraction of all of the metal-containing species in the aqueous phase that is in the M^{n+} form. Use of α_M is a simple and convenient way to account for all of the many side reactions involving the metal ion, such as hydroxy or other anion complex formation as well as masking reactions.

With appropriate substitutions of the equilibrium expressions involved in Fig. 9.2, then

$$D = \frac{K_{D_C} K_f K_a^n}{K_{D_R}^n} \frac{[HL]_o^n}{[H^+]^n} \alpha_M \tag{9.6}$$

Equation 9.6 shows how the extent of metal-ion extraction increases with organic-phase reagent concentration and with the pH in the aqueous phase.

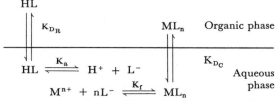

Figure 9.2. Illustration of various equilibria involved in metal-ion extraction. The different equilibrium constants are defined.

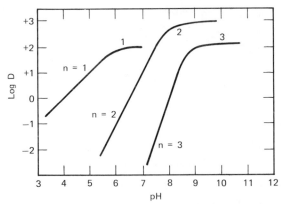

Figure 9.3. The solvent extraction of a series of metal ions with a chelating agent. Here n = number of hydrogen ions released from the chelating agent per metal ion.

The course of solvent extraction of a series of metal ions with a given chelating agent can be conveniently represented in Fig. 9.3, where the variation of log D with pH for three metal ions at constant reagent concentration is shown. As predicted by Eq. 9.6, each of the plots is seen to have a linear portion whose slope is equal to the charge of the metal ion (strictly speaking, the slope is equal to the number of hydrogen ions released per metal ion in the overall extraction equation). Each curve also has a horizontal portion of maximum log D, which is reached at $D = K_{D_C}$ because of the variation of α_M with pH. In this horizontal portion of the curve the intrinsic solubility characteristics of the chelate (as reflected by K_{D_C} rather than the formation equilibria) limit the extent of the extraction. From the curves it may be seen that the more highly charged the metal cation is, the narrower is the pH range required for complete extraction. For most analytical purposes extraction may be considered to be complete at the 99% level. This corresponds to a value of log $D = 2$ (assuming a phase volume ratio of unity). Since we can usually ignore extractions of less than 1%, the practical limits of extraction of a particular metal are as follows:

$$-2 < \log D < +2$$

By examining extraction data for various chelating agents represented in the form of curves, as in Fig. 9.4, it is possible to predict which metals may be quantitatively separated from each other by one extraction as well as the pH range in which this separation should be carried out. Complete separation of the two metals I and II in Fig. 9.4 will be possible if there is a pH, pH_A, where log $D = -2$ for metal II and log $D = +2$ for metal I. From the figure it is obvious that the minimum separation between the two curves must be $4/n$ where n is the valence of the metal ion. This relationship does

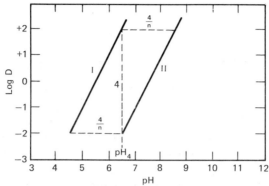

Figure 9.4. Extraction curves of two metals having the same valence, n.

not hold, however, if the two metals under consideration have extraction curves whose slopes are not the same. In general, for a series of extraction curves that may have different slopes the distance separating each curve must be described in terms of some arbitrarily chosen value of D. Usually the value $\log D = 0$, which corresponds to half extraction, is chosen, and the pH at which this occurs is designated as $pH_{1/2}$.

masking agents, is accounted for in Eq.9.6 by α_M, which can be shown to be

$$\alpha_M{}^{-1} = \sum_i \alpha_i{}^{-1} - (i - 1) \tag{9.7}$$

where α_i is the value of α_M that would obtain with only one side reaction considered. It follows, therefore, that in solutions involving a number of side reactions giving different α_i values only the reaction giving the lowest α_i value, that is, the reaction giving rise to the most stable (highest K_f) metal complex, contributes to the overall α_M value. Thus, in a solution at pH 10 in the presence of both NH_3 and EDTA, copper can form $CuOH^+$, $Cu(NH_3)_4^{2+}$, and CuY^{2-}, but α_{Cu} can be calculated by considering only the CuY^{2-} formation because it is the strongest of the three complexes.

The α_M values for individual complexing agents can be calculated with the help of the following equation:

$$\alpha_M = [1 + K_{f_1}\alpha_L C_L + K_{f_1}K_{f_2}(\alpha_L C_L)^2 + \cdots + K_{f_1}K_{f_2}\cdots K_{f_n}(\alpha_L C_L)^n]^{-1} \tag{9.8}$$

where K_{f_1} is the stepwise complex formation constant, C_L the total concentration of the uncomplexed reagent, and α_L the fraction of the reagent in the active form (a function of the pH and pK values of the reagent).

From the foregoing discussion it is easy to see how solvent extraction measurements can be used to study various solution equilibria involving metal ions. By determining the variation of D over a range of reaction conditions, the equilibrium constants of chelate formation and of hydroxy and other complex formation for quite a number of metal ions have been evaluated.

9.63 ION-ASSOCIATION EXTRACTION SYSTEMS

Ion-pair formation can also result in an uncharged species having solubility characteristics suitable for its extraction into organic solvents. Reactions involving ion-pair formation are widely used for the extraction of metal ions.* The metal may be incorporated into a very large ion containing bulky organic groups, or it may pair (or associate) with an oppositely charged ion that is large. For example, $[\text{Fe(phenanthroline)}_3^{++}, 2\text{ClO}_4^-]$, $[(\text{C}_4\text{H}_9)_4\text{N}^+, \text{FeCl}_4^-]$, and $[(\text{C}_6\text{H}_5)_4\text{As}^+, \text{MnO}_4^-]$ represent extractable ion pairs of analytical utility.

Since many ion-association extractions take place from aqueous solutions of relatively large ionic strengths, a region in which great differences exist between concentrations and activities, it is very difficult to describe quantitatively the behavior of such systems in terms of simple equilibrium expressions. Allowing for the uncertainty in the values of the appropriate formation constants, however, it is possible to derive expressions that are at least qualitatively useful. As an illustration consider the extraction of Zn^{2+} from an HCl solution, using a benzene solution of a high-molecular-weight amine such as tribenzylamine, $(\text{C}_6\text{H}_5\text{CH}_2)_3\text{N}$, symbolized by R_3N.

The scheme in Fig. 9.5 shows that ion-pair complexes, unlike chelate complexes, are capable of dimerizing or associating to an even greater extent in the organic phase. It has been assumed that sufficient HCl is present in the aqueous phase to permit the neglect of unprotonated tribenzylamine.

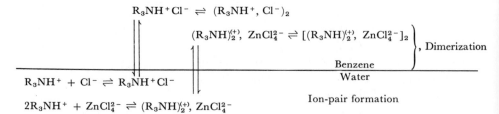

Figure 9.5.   Illustration of various equilibria unvolved in ion-pair extraction.

* Unfortunately the term "liquid ion exchange," originally introduced as an advertising gimmick, has found some use among chemists as an alternative name for ion-association extraction systems. The trouble with this name is that it implies a mechanism which has not been demonstrated to be valid and, even worse, it can conceivably be used for all chemical reactions that are not oxidation-reductions. That is to say, any acid-base reaction or metal complex formation reaction can be considered to be a metathetical, or ion-exchange, reaction. Any category that is so all-inclusive becomes meaningless.

For this system the distribution ratio of zinc is given by:

$$D = \frac{[(R_3NH^+)_2, ZnCl_4^{2-}]_0 + 2[((R_3NH^+)_2, ZnCl_4^{2-})_2]_0}{\sum\limits_{i=2}^{4}[ZnCl_i^{(2-i)+}] + [(R_3NH^+)_2, ZnCl_4^{2-}]} \quad (9.9)$$

which formally resembles the expression shown earlier for the distribution of acetic acid between water and benzene.

By the incorporation of the various equilibrium expressions as indicated in Fig. 9.5, the distribution ratio D may be expressed in terms of the total zinc concentration present, the total tribenzylamine concentration present, and the chloride-ion concentration.

Selected Bibliography

1. G. H. Morrison and H. Freiser, *Solvent Extraction in Analytical Chemistry*, John Wiley, New York, 1957. See also *Biennial Reviews in Analytical Chemistry*, from 1958 to 1968.

2. J. Stary, *Solvent Extraction of Metal Chelates*, Pergamon Press, London, 1964.

3. Y. Marcus and A. S. Kertes, *Ion Exchange and Solvent Extraction of Metal Complexes*, Wiley-Interscience, New York, 1969.

List of Symbols

K_a	ionization constant	α_i	the value of α_M with only one side reaction
K_D	distribution constant		
K_{D_C}	distribution constant of chelate	α_M	fraction of all metal containing species in the aqueous phase
K_{D_R}	distribution constant of chelating agent	θ	fraction extracted
K_F	formation constant of chelate		Subscripts
M	metal	A	component A
n	number of hydrogens released from a chelating agent per metal ion	B	component B
		M	metal
		o	organic phase
L	ligand	s	solvent
V	phase volume ratio	w	aqueous phase

Ten

LIQUID-LIQUID CHROMATOGRAPHY

In liquid-liquid chromatography (LLC), both the mobile and stationary phases are liquids. The underlying equilibrium process is therefore the same as in extraction: a distribution or partitioning of sample between two immiscible liquid phases. For this reason LLC is often referred to as "liquid partition" or simply "partition" chromatography. Liquid-liquid chromatography bears to extraction a relationship similar to the one that gas-liquid chromatography (GLC) has to distillation, and from this we can immediately infer many of its special features. The separation selectivity of LLC will be generally similar to that obtainable with extraction, since the same two liquid phases can be used in either procedure. This means that similar kinds of samples can be handled by extraction and LLC.

The differences between extraction and LLC are more or less the same as those that distinguish distillation from GLC: large samples are more readily handled by extraction, while LLC is capable of faster, more difficult separations. Often LLC is better suited for analytical applications, whereas extraction dominates the field of preparative- and production-scale separations. Compared to GLC, LLC is uniquely suited for high-boiling or thermally unstable samples, since the separation of these compounds by LLC can be accomplished at room temperature. Separation selectivity in GLC and LLC often differs markedly, so that many separations can be effected by one procedure but not the other (as is true of extraction versus distillation). The slower rate of sample diffusion in liquids than in gases makes LLC slower and/or less efficient than GLC under comparable conditions.

Experimentally, LLC can be an extremely simple procedure. Two basically different techniques exist: LLC in columns and open-bed or "paper" chromatography. Liquid-liquid chromatography in columns (e.g., Fig. 10.1*a*) is carried out in essentially the same way as GLC, except that the mobile phase is a liquid instead of a gas. A cylindrical column is packed with a finely divided, porous solid (the "support") that has been coated with stationary phase (liquid *S*). Sample is introduced at one end of the column, and the mobile phase (liquid *M*) is passed over the sample and through the column. The differing migration rates of the various sample components as they move through the column result in their separation. Each sample band can be collected as it leaves the column, without the special techniques necessary for sample recovery in GLC. In its simplest form LLC in columns requires nothing more than a packed glass tube, but more versatile units (e.g., Fig. 10.1*c*) can be as complex as those used in GLC.

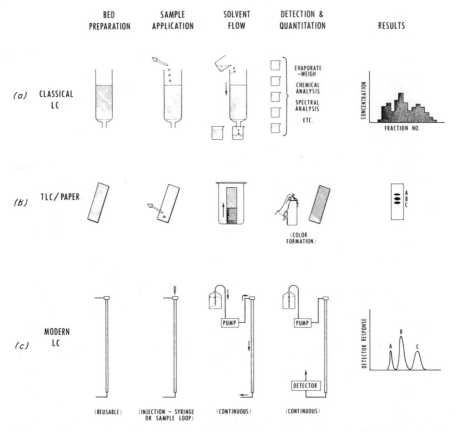

Figure 10.1. Various forms of liquid-liquid chromatography: (*a*) classical column chromatography, (*b*) open-bed or paper chromatography, (*c*) modern column chromatography.

Paper chromatography takes its name from the fact that the chromatographic bed consists of a sheet of absorbent paper (e.g., filter paper) that has been impregnated with liquid S (see Fig. 10.1b). The sample is applied ("spotted") at one end of the sheet, which is inserted into a container of liquid M. Then M moves across the sample and along the sheet by capillary wetting. In this manner the sample is separated, just as in column chromatography. The essential difference in open-bed (i.e., paper) chromatography is that separated components need not be washed or eluted from the bed.

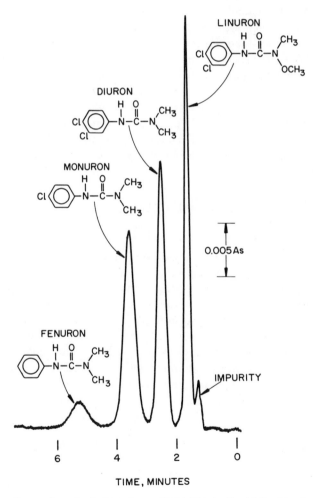

Figure 10.2. Separation of substituted urea herbicides by liquid-liquid column chromatography. Reprinted from Ref. 1.

At the end of separation, sample bands can be made visible on the bed (if necessary) by the use of color-forming reactions. Quantitation can in turn be achieved by a variety of measuring techniques.

Liquid-liquid chromatography in columns was first introduced by Martin and Synge in 1941, after unsuccessful attempts at separating certain amino acid derivatives by countercurrent distribution (Section 4.81). This discovery was followed by the development of modern paper chromatography in 1944 by Consden, Gordon, and Martin. During the next 15 years these two procedures gradually replaced adsorption chromatography in many of its original applications, and LLC in columns or on paper was eventually applied to almost every type of sample: amino acids and peptides, carbohydrates, natural products, synthetic-organic mixtures, inorganic compounds, and so on. After 1960 LLC was in turn replaced in some of these applications by other chromatographic methods, notably GLC and thin-layer chromatography (Chapter 13). More recently the development of so-called modern liquid chromatography (1), featuring special column packings and fully automated equipment, has resulted in renewed interest in LLC. The separation of a mixture of herbicides by modern LLC is shown in Fig. 10.2. Here four major components are resolved within 6 min. For a general review of the LLC field, see Refs. 1 to 7.

10.1   Theory

Any LLC theory must deal with certain practical questions. How do we select the mobile- and stationary-phase liquids (M and S) for a given separation? How do other experimental variables, such as temperature, solvent flow rate, and stationary-phase support, affect separation? To what extent can we relate band migration rates to the molecular structure of the migrating compound, in order to identify unknown bands from their positions in the chromatogram? These and other questions can be related to the fundamental quantities α, k_2', and N of Eq. 5.43. The separation factor α and capacity factor k' are determined by the equilibrium that exists in a given chromatographic system. These equilibrium processes are essentially similar in corresponding column and open-bed systems, and they can be discussed as a single subject (Section 10.11). Bed efficiency N and mobile-phase flow in the bed, on the other hand, are sufficiently different in columns and open beds to merit separate attention (Sections 10.12 and 10.13).

10.11   EQUILIBRIUM THEORY

Dependence of K_i on the Solubility Parameter δ

The selection of the right two phases for an LLC separation is a major decision in the design of an adequate system. We will look closely, therefore, at this aspect of separation. The main requirements of the two liquids S and M (stationary and mobile phases, respectively) are as follows:

1. Immiscibility of the two phases.
2. Sample k' values in the right range (roughly $1 < k' < 20$, with an optimum k' value of 2 to 5).
3. Sufficiently large separation factors α for all pairs of sample components.

For a general discussion of the thermodynamics of solution as it relates to LLC equilibria, see Refs. 8 and 9. For a review of LLC thermodynamics specifically, see Refs. 10 to 14.

The fundamental quantity of interest under linear elution conditions is the distribution constant $K_i = (i)_S/(i)_M$ of a solute i between phases S and M. We can relate K_i to parameters of more direct chromatographic interest: separation factors α for solutes i and j ($\alpha \equiv K_i/K_j$), capacity factors k' (equal to $K_i V_S/V_M$), and band migration rates (Section 5.2). In turn, K_i can be expressed as a function of the activity coefficients γ_S and γ_M of i in phases S and M (pure liquids as standard states; Eq. 2.8b):

$$K_i = \frac{\gamma_M}{\gamma_S} \tag{10.1}$$

Combination of Eqs. 10.1 and 2.44 gives K_i in terms of the molar volume of i ($\bar{V}_i$) and the solubility parameters of i (δ_i), S (δ_S), and M (δ_M):

$$\ln K_i = \frac{\bar{V}_i[(\delta_i - \delta_M)^2 - (\delta_i - \delta_S)^2]}{RT} \tag{10.2}$$

The solubility parameter δ of a substance (Section 2.22) is roughly equivalent to what chemists refer to as its "polarity," and these two terms are used more or less interchangeably; polar compounds have large values of δ. For $K_i = 1$, Eq. 10.2 yields $\delta_i = (\delta_M + \delta_S)/2$, that is, the polarity of i is intermediate between the values for S and M. As the polarity of i becomes more like that of S, K_i becomes larger. Similarly, as the polarity of i becomes more like that of M, K_i becomes smaller.

According to Eq. 10.2, K_i and relative band migration can be controlled by changes in the polarity of either phase. This is illustrated in Fig. 10.3 for K_i as a function of $\delta_M - \delta_i$ and $\delta_S - \delta_i$, assuming that $\bar{V}_i = 100$ ml and $T = 25°$. The heavy diagonal lines bound the region for $4 \leqslant (\delta_S - \delta_M) \leqslant 17$. Most practical systems fall within this region. Thus phase immiscibility generally requires $\delta_S - \delta_M > 4$, while the combination of very polar phases (e.g., water) and nonpolar phases (hydrocarbons) leads to maximum values of $\delta_S - \delta_M < 17$. Similarly the solute solubility parameter δ_i usually lies between the extreme values set by δ_M and δ_S.

Maximum resolution per unit time requires k' values in the range of 2 to

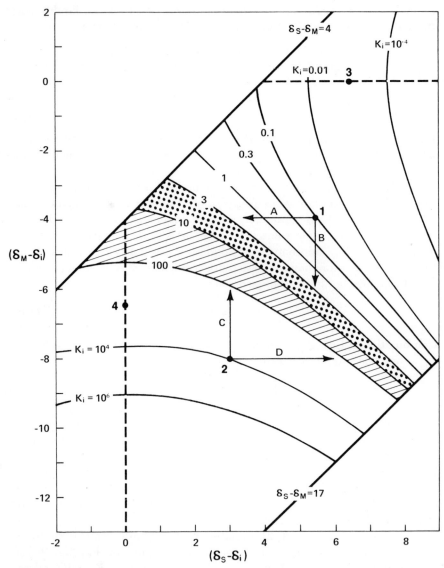

Figure 10.3. Solute retention as a function of the composition of mobile and stationary phases in normal LLC K_i as a function of $(\delta_M - \delta_i)$ and $(\delta_S - \delta_i)$, for $\overline{V}_i = 100$ and $T = 25°$ (K_i values indicated on curves).

5 (Section 5.42). Similarly V_S is generally less than V_M. Therefore K_i values should fall approximately in the shaded region of Fig. 10.3 ($3 \leqslant K_i \leqslant 10$) for optimum separation. Further reduction in V_S/V_M (less S on the support permits even larger values of K_i for optimum values of k' (cross-hatched region of Fig. 10.3). When K_i is too small (e.g., point 1 of Fig. 10.3), it can be increased by decreasing the polarity (i.e., value of δ) of either the mobile phase (arrow B) or the stationary phase (arrow A). Similarly, when K_i is too large (point 2 of Fig. 10.3), the polarity of one phase or both of them should be increased (arrows C and/or D).

Relatively nonpolar compounds (e.g., hydrocarbons) cannot be successfully separated in normal LLC systems (i.e., S more polar than M). This can be seen by reference to Fig. 10.3. For hydrocarbon samples $\delta_M - \delta_i$ cannot be made much less than zero, and the requirement that $\delta_S - \delta_M > 4$ then means that $K_i \approx 0$ (e.g., point 3). So-called *reverse-phase* systems provide a solution to this problem. Here the normally polar stationary phase and nonpolar mobile phase are interchanged, and all K_i values are then replaced by their reciprocals (e.g., point 4 in Fig. 10.3). Large K_i values for nonpolar samples result, and these may be reduced (if necessary) into a more desirable range of values by decreasing the amount of stationary phase on the support.

Effect of Specific Intermolecular Interactions

Equation 10.2 and Fig. 10.3 are useful in illustrating the control of k' by variation of the liquids S and M in LLC. Similarly values of δ for the mobile and stationary phases (see Table 10.1) provide a rough basis for the control of sample migration rates (an increase in δ_S or δ_M normally increases sample migration; a decrease in either of these parameters decreases migration*). More reliable prediction of k' as a function of solvent composition requires attention to the different kinds of intermolecular interactions (Section 2.21).† This becomes particularly important when we are interested in changing the selectivity of a particular LLC system (i.e., the variation of α with solvent composition).

The limitations of Eq. 10.2 can be seen by comparing the three liquids CS_2, butyronitrile, and 2-hexanol. Any one of these might form the mobile phase in an LLC system with water as stationary phase. The solubility parameters δ of the three solvents are about equal (10.0 to 10.2), and Eq. 10.2 then predicts that the LLC systems CS_2–H_2O, butyronitrile–H_2O, and 2-hexanol–H_2O would be essentially equivalent. This, however, is not the case. Consider the various types of solute-solvent interactions (Section 2.21)

---

* Except for reverse-phase operation, where the opposite is true.

† Note that Eq. 10.2 is derived on the assumption that only nonpolar solvents and solutes are involved.

that are possible in these three mobile phases. In CS_2, dipole orientation and hydrogen-bonding interactions are essentially absent. In butyronitrile, hydrogen-bonding interactions are largely precluded, occurring only in the case of strong proton-donor solutes i. In 2-hexanol, hydrogen bonding can occur with any solute i that exhibits moderate proton-acceptor or proton-donor properties. As a result, pronounced differences in separation selectivity would be expected for these three LLC systems. Proton-acceptor solutes should be preferentially retained in the mobile phase of the 2-hexanol–H_2O system, relative to the other two systems. Solutes with very polar groups (i.e., those possessing a large dipole moment) should be preferentially retained in the mobile phase of the butyronitrile–H_2O system.

The estimation of K_i as a function of solvent composition in LLC can be greatly improved by classifying solvents in terms of their ability to enter into different types of intermolecular interactions. Such a system, based on an expanded version of the Hildebrand solubility parameter treatment, has been described recently (14, 15). Parameters δ_d, δ_o, δ_a, and δ_b are defined for any solvent.* The values of these parameters for a given solvent characterize its ability to participate in interactions of various types. Solvents with large δ_d values exhibit increased dispersion interactions with solutes that also have large δ_d values. Solvents and solutes which possess large δ_o values likewise show strong dipole interactions. Strong proton-donor solvents (large δ_a values) undergo strong hydrogen bonding with strong proton-acceptor solutes (large δ_b values), and vice versa. The classification of some common solvents in this fashion is shown in Table 10.1.

The selection of liquids S and M for an LLC separation can be approached as follows, on the basis of our analysis of Eq. 10.2, Fig. 10.3, and the data of Table 10.1. Sample k' values are first adjusted into approximately the right range (2 to 5) by varying the polarity δ of S and/or M (i.e., suitable changes in the composition of either phase). If at this point α values for solutes of interest are not sufficiently different from 1, Table 10.1 is consulted for other solvents of similar polarity (similar values) but different values of δ_d, δ_o, δ_a, and/or δ_b. The substitution of such solvents into the LLC system is likely to provide roughly similar K_i values, with simultaneous change in α values (because of changing solute-solvent interactions). This is illustrated by a hypothetical example: two LLC systems A–B and C–B, where A and C are mobile-phase solvents of similar polarity (B is the stationary-phase solvent), and a given pair of solutes 1 and 2:

System	Mobile Phase (A or C) Properties					Solute K_i Values		
	δ	δ_d	δ_o	δ_a	δ_b	1	2	α
A–B	10	10	0	0	0	5.0	5.0	1.00
C–B	10	7	3	4	4	4.0	6.0	1.50

*An induction parameter δ_{i_n} must also be considered, but its effect approximately cancels in LLC systems.

Table 10.1. Solvent Properties of Chromatographic Interest

Solvent	I δ	II δ_d	III δ_o	IV δ_a	V δ_b
Perfluoroalkanes	6.0	6.0	0	0	1
CFCl$_2$—CF$_3$	6.2	5.9	1.5	0	0
Isooctane	7.0	7.0	0	0	0
Diisopropyl ether	7.0	6.9	0.5	0	0.5
n-Pentane	7.1	7.1	0	0	0
CCl$_3$—CF$_3$	7.1	6.8	1.5	0	0.5
n-Hexane	7.3	7.3	0	0	0
n-Heptane	7.4	7.4	0	0	0
Diethyl ether	7.4	6.7	2	0	2
Triethylamine	7.5	7.5	0	0	3.5
Cyclopentane	8.1	8.1	0	0	0
Cyclohexane	8.2	8.2	0	0	0
Propyl chloride	8.3	7.3	3	0	0
CCl$_4$	8.6	8.6	0	0	0.5
Diethyl sulfide	8.6	8.2	2	0	0.5
Ethyl acetate	8.6	7.0	3	0	2
Propylamine	8.7	7.3	4	0.5	5
Ethyl bromide	8.8	7.8	3	0	0
m-Xylene	8.8	8.8	0	0	0.5
Toluene	8.9	8.9	0	0	0.5
CHCl$_3$	9.1	8.1	3	0	0.5
Tetrahydrofuran	9.9	7.6	4	0	3
Methyl acetate	9.2	6.8	4.5	0	2
Benzene	9.2	9.2	0	0	0.5
Perchloroethylene	9.3	9.3	0	0	0.5
Acetone	9.4	6.8	5	0	2.5
CH$_2$Cl$_2$	9.6	6.4	5.5	0	0.5
Chlorobenzene	9.6	9.2	2	0	0.5
Anisole	9.7	9.1	2.5	0	2
1,2-Dichloroethane	9.7	8.2	4	0	0
Methyl benzoate	9.8	9.2	2.5	0	1
Dioxane	9.8	7.8	4	0	3
Methyl iodide	9.9	9.3	2	0	0.5
Bromobenzene	9.9	9.6	1.5	0	0.5
CS$_2$	10.0	10.0	0	0	0.5
Propanol	10.2	7.2	2.5	4	4
Pyridine	10.4	9.0	4	0	5
Benzonitrile	10.7	9.2	3.5	0	1.5
Nitromethane	11.0	7.3	8	0	1
Nitrobenzene	11.1	9.5	4	0	0.5
Ethanol	11.2	6.8	4.0	5	5

(continued)

Table 10.1. (*continued*)

Solvent	I δ	II δ_d	III δ_o	IV δ_a	V δ_b
Phenol	11.4	9.5			
Dimethylformamide	11.5	7.9			
Acetonitrile	11.8	6.5	8	0	2.5
Methylene iodide	11.9	11.3	1	0	0.5
Acetic acid	12.4	7.0			
Dimethylsulfoxide	12.8	8.4	7.5	0	5
Methanol	12.9	6.2	5	5	5
1,3-Dicyanopropane	13.0	8.0	8	0	3
Propylene carbonate	13.3				
Ethanolamine	13.5	8.3	Large	Large	Large
Ethylene glycol	14.7	8.0	Large	Large	Large
Formamide	17.9	8.3	Large	Large	Large
Water	21	6.3	Large	Large	Large

Solvents of similar polarity δ which can be substituted in this fashion, keeping k' values approximately constant while varying α, have been referred to as *isopartive* or (in adsorption chromatography) *equieluotropic*. We will use the latter term.

It should be noted that solvent mixtures (as opposed to pure solvents) provide an almost unlimited variety of properties of chromatographic interest. Generally the properties of a solvent mixture will be intermediate between those of the components of the mixture.* Values of the parameters δ, δ_d, δ_o, δ_a, and δ_b for a solvent mixture are generally equal to the arithmetic average for each component of the mixture, weighted according to its volume fraction in the mixture. Thus, for 40%v acetonitrile–benzene, δ is equal to (0.4 × 12.1) + (0.6 × 9.1), or 10.3. The polarity of a solvent mixture can be adjusted continuously by varying the proportions of its components, thus permitting precise control over the value of k' for a given solute.

The problem of incompletely resolved sample bands in a given LLC system ($\alpha \approx 1$) can be attacked more directly when dealing with known compounds. For example, assume that two unresolved bands i and j are an alcohol and an ester. An increase in the basicity (δ_b value) of either phase will cause preferential retention of the alcohol in that phase (via stronger hydrogen bonding) and a resulting change in α.

---

* We assume here that the two phases S and M are immiscible. When one of the two components of a binary mixture M is significantly soluble in the other phase S, a change in composition of M will simultaneously affect the composition of S. This effect can be used to advantage in some cases (see, e.g., Ref. 15a).

Use can also be made of selective, reversible chemical reactions for improved separation factors (see Section 9.62). For example, acids and bases can be made to migrate more or less rapidly by a suitable change in the pH of either phase (usually the polar phase), and other compound types can be subjected to complexation reactions, for example, silver-ion complexation of olefins, borate complexation of *vic*-diols, or bisulfate reaction of aldehydes and ketones. Where one or both of a pair of unresolved compounds can undergo reversible reaction with some reagent X, addition of X to one of the two phases (S or M) is likely to favor the separation of the compounds in question. For further discussion of these possibilities, see Section 9.62.

Yet another approach to the separation of unresolved band pairs via a change in α is to increase the polarity *difference* of phases S and M, while holding average polarity constant. If K_i values for two solutes are slightly different and are larger than 1, Eq. 10.2 predicts that such a change in solvent compositions will increase α. However, this also results in a large increase in the k' values of more strongly retained solutes, while the k' values for weakly retained solutes simultaneously approach zero. In other words, the total number of solutes that can fall in a desirable range of k' values (e.g., 1 to 20) is decreased by increasing the difference in polarities of S and M. Thus, by increasing the polarity difference of two phases, more favorable α values are likely to result, but fewer compounds can simultaneously migrate with favorable k' values. Another result of increasing $\delta_S - \delta_M$ is increased *molecular weight selectivity*, that is, improved separation of adjacent homologs. Thus, for water as stationary phase (see, e.g., Ref. 3, p. 115), the change in K that occurs for addition of a methylene group to a given molecule *increases* as the polarity of M *decreases*.

An inviting goal for chromatography is the a priori calculation of solute K_i values in a given chromatographic system as a function of solute molecular structure. This offers the possibility of confirming the structure of an unknown sample band by comparison of experimental and calculated band migration rates (Eq. 5.5 or 5.10). Such tests can be made more stringent by obtaining migration data for the compound in question in several different LLC systems, and by studying the migration of *derivatives* of the compound as well. Such an approach to structure elucidation has the apparent advantage of simplicity and small sample size requirements, and it has in fact been applied in the steroid field, using paper chromatography (4).

The calculation of K_i values for a given structure in LLC is usually based on Martin's relationship (Eq. 2.48). Each possible structural group X in a molecule is assumed to contribute an increment $\Delta \log K_i$ to $\log K_i$, where $\Delta \log K_i$ is constant for a given group X (in a specific LLC system). Addition of $\Delta \log K_i$ values for every group X in the molecule thus gives a calculated value for $\log K_i$. In the case of open-bed or paper chromatography, definition of the function $R_M = \log (1/R_f - 1)$ plus Eq. 5.10 yields

$$R_M = \log \left(\frac{V_S}{V_M}\right) + \log K_i \qquad (10.3)$$

Addition of the group X to a compound i leads to a change in K_i (to K_{ix}), such that

$$\log K_{ix} = \log K_i + \Delta \log K_i$$

for a given pair of phases M and S. If ΔR_{MX} for the group X in the system M–S is defined equal to $\Delta \log K_i$, then it is seen that

$$(R_M)_{ix} = (R_M)_i + \Delta R_{MX} \qquad (10.4)$$

Thus, given values of ΔR_{MX} for the common groups X (e.g., methyl, hydroxy, amino), we can calculate R_M and R_f values for a whole series of derivatives of any compound i whose R_f value has been measured. In this way a limited number of chromatographic data (R_f values for a few compounds) can be used to predict the migration of a much larger group of compounds. For examples of this approach in LLC, see Refs. 4 and 16.

Other Aspects of Equilibrium Theory

The two liquid phases in LLC must generally be immiscible (for an important exception to this rule, see Section 10.4). According to Eq. 2.43, immiscibility is favored for a large difference in δ values for the two phases, and a large molar volume for each of the two liquids. As we will see (Section 10.12), high-molecular-weight mobile phases are objectionable because of their excessive viscosities. Consequently the combination of low-boiling (nonviscous) mobile phases and higher-boiling or polymeric stationary phases is fairly common. Table 10.2 summarizes a number of LLC systems that have been found useful in paper and modern column chromatography.

A complication which we have so far ignored is the possibility that the stationary-phase *support* can affect K_i values. Such so-called support effects can lead to significantly different K_i values from those observed in the equilibrium distribution of i between the two unsupported phases. In the case of cellulose as support (e.g., in paper chromatography) it is known that water as stationary phase interacts strongly with the cellulose molecules, effectively yielding a concentrated cellulose solution (rather than water). As a result it is possible to carry out separations on cellulose with a single solvent (rather than the normal two). (For water as solvent, a water–cellulose solution constitutes the stationary phase, and pure water forms the mobile phase.) Other support effects arise from the adsorption of sample on the support, ion-exchange reactions between the sample and the support, and so forth. It has also been suggested (16a) that *adsorption* of solute at the liquid-liquid interface between M and S can affect K_i values.

Table 10.2. Some Useful Systems for Liquid-Liquid Chromatography

Stationary Phase	Mobile Phase	Comments
Water	n-Butanol–acetic acid–water (4/1/5)	Paper (3)
Water	Phenol	Paper (3)
Formamide	CHCl₃–benzene–cyclohexane^a	Paper (3)
Dimethylformamide	Cyclohexane	Paper (3)
Kerosene^b	70% Isopropanol–water	Paper (3)
Liquid paraffin^b	Dimethylformamide–methanol–water (10/10/1)	Paper (3)
β,β′-Oxydipropionitrile ⎫	Cyclopentane, hexane, heptane, isooctane	Columns (1)
Carbowax 600 ⎬	Same, modified with up to 10% chloroform, dichloromethane,	Columns (1)
Triethylene glycol ⎭	tetrahydrofuran, dioxane	
Ethylene glycol	Di-n-butyl ether	Columns (1)
H₂O–ethylene glycol	Hexane–CCl₄	Columns (1)
Chloroform–cyclohexane–nitromethane	(ternary mixture, two phases)	Columns (1)
Ethylenediamine	Hexane	Columns (1)
Hydrocarbon polymer	Aqueous methanol	Columns (1)

^a Polarity decreases in going from solutions rich in CHCl₃ to those rich in cyclohexane.
^b Reverse-phase LLC.

10.12 SEPARATION EFFICIENCY OF COLUMNS

As in the case of GLC (Section 8.2), our analysis of separation efficiency for LLC in columns begins with the variation of H (height equivalent of a theoretical plate, L/N) with the velocity v of the mobile phase. The same fundamental equation (Eq. 5.36) applies in both GLC and LLC. Substituting Eq. 5.28, 5.33, and 5.35 into Eq. 5.36 gives a relationship of the following form:

$$H = \frac{B}{v} + \left(\frac{1}{A} + \frac{1}{C_M v}\right)^{-1} + C_S v \qquad (10.5)$$

Here B, A, C_M, and C_S are constants that were discussed in Sections 8.15 and 5.3.* It is of special interest to compare and contrast the significance of Eq. 10.5 for the separation by liquid versus gas chromatography. Much of what we say in this connection will apply to all column liquid chromatographic procedures (i.e., ion exchange, adsorption, and exclusion, as well as LLC).

* The terms B/v, A, $C_M v$, and $C_S v$ are due, respectively, to longitudinal diffusion, eddy diffusion, mobile-phase mass transfer, and stationary-phase mass transfer.

The fundamental difference between liquid and gas chromatography, as regards Eq. 10.5, is the great difference in sample diffusion rates in liquids and gases (Section 3.3). Sample diffusion coefficients (D_M) in the mobile phase are typically 10^4 to 10^6 times smaller in liquid chromatography than in gas chromatography. Recalling that the longitudinal diffusion parameter B is proportional to D_M (Eq. 5.18), and the mobile-phase mass transfer parameter C_M is proportional to $1/D_M$ (Eq. 5.29), we see that the term B/v will be much smaller in liquid chromatography than in gas chromatography, whereas the term $C_M v$ will be much larger. In gas chromatography, where C_M is usually small, Eq. 10.5 reduces to the van Deemter equation (e.g., Fig. 10.4a):

$$H = A + \frac{B}{v} + C_S v \qquad (5.37)$$

In GLC plots of H versus v usually follow Eq. 5.37 closely. In liquid chromatography the term B/v is generally negligibly small (particularly in well-designed separations), whereas $C_M v$ is never small. Consequently for liquid chromatography Eq. 10.5 reduces to*

$$H = \left(\frac{1}{A} + \frac{1}{C_M v} \right)^{-1} + C_S v \qquad (10.5a)$$

The general form of this relationship is illustrated in Fig. 10.4b. At low values of v, H approaches zero (at *very* small values of v the term B/v becomes significant, and H then increases with decrease in v). At large values of v, Eq. 10.5a reduces to $H = A + C_S v$. The mathematical form of this relationship may be better appreciated, perhaps, in terms of an equivalent empirical relationship that is broadly applicable (19):

$$H = D v^n \qquad (10.5b)$$

Here D and n are constants for a given system (column, solvents, solute), and $0.2 \leqslant n \leqslant 1.0$ (usually $n \approx 0.4$). Thus a major difference between gas and liquid chromatography is the differing form of the H versus v relationship. This is illustrated in Fig. 10.4 (solid curves). At high mobile-phase velocities, H varies linearly with v in gas chromatography (Fig. 10.4a) and is approximately proportional to v. In liquid chromatography (Fig. 10.4b) H tends to level off at high values of v.† A second important difference in corresponding gas and liquid chromatographic systems involves separation

---

* Note that C_S is the sum of stagnant mobile-phase plus stationary-phase mass transfer terms (Section 5.3).

† Here we assume moderately large values of v at the upper end of the range (e.g., 5 cm/sec); that is, Fig. 10.4b applies to so-called high-speed or high-performance liquid chromatography (see Ref. 1).

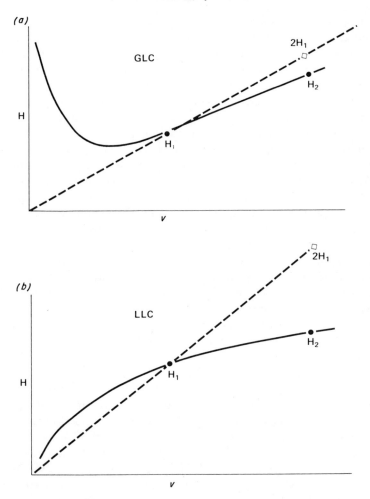

Figure 10.4. Dependence of plate height H on mobile-phase velocity μ in gas chromatography (*a*) versus liquid chromatography (*b*).

speed and efficiency. Liquid chromatography (at comparable particle diameters) is typically less efficient and/or slower. Again this is the result of slower diffusion of sample in the mobile phase and resulting larger values of C_M in liquid chromatography. Despite the generally lower efficiency of liquid chromatography, however, better separation can often be achieved because of greater selectivity.

These fundamental differences in liquid and gas chromatography have led to corresponding differences in separation conditions. The adverse effect on H of slow sample diffusion can be overcome by a reduction in v, and liquid chromatography is typically carried out at lower mobile-phase velocities than are used in gas chromatography. Lowered efficiency in liquid chromatography is thus partially overcome by increasing the time of separation. Smaller particle diameters d_p of the stationary-phase support offer improved separation efficiency in both gas and liquid chromatography, since A, C_M, and (for porous particles) C_S all decrease with d_p (Section 5.3). However, smaller particles require higher column pressures, since bed permeability decreases with $d_p{}^2$ (Eq. 3.62). In gas chromatography high column pressures are generally not used for several reasons. No such limitations exists for liquid chromatography, so that optimized separations generally feature lower values of d_p (e.g., 5 to 50 μ) and much greater column pressures (e.g., 1000 to 5000 psi) than are feasible in gas chromatography.

Separation in gas chromatography is usually carried out at values of v close to the value for minimum H. Some advantage is gained by further increase in v, but continued increase soon becomes counterproductive (Section 5.43). This is not the case in liquid chromatography, however, where it is generally desirable to use the highest solvent velocities possible for a given separation time, as illustrated by the data of Fig. 10.4b. Here H_1 and H_2 are plate heights at the corresponding solvent velocities $v_1 = \frac{1}{2}v_2$. Separation time can be held constant by simultaneously doubling solvent velocity and column length. The corresponding plate numbers of the resulting separations N_1 (for L_1 and H_1) and N_2 (for $2L_1$ and H_2) are, then, $N_1 = L_1/H_1$ and $N_2 = 2L_1/H_2$, and the condition for $N_2 > N_1$ is $H_2 < 2H_1$. From Fig. 10.4b it is seen that this condition is generally met in liquid chromatography (because the plot of H versus v is convex to the solvent velocity axis), so that greater column efficiency (plate numbers) per unit time are obtained for longer columns plus correspondingly increased solvent velocities. In gas chromatography (see Fig. 10.4a), H_2 approaches $2H_1$ past the minimum in the curve, and little increase in efficiency per unit time results with further increase in v.

The advantage of working at the highest possible solvent velocities in liquid chromatography is limited by the pressure required to achieve these velocities. In the example given above, where column length and v were doubled simultaneously, a corresponding fourfold increase in column pressure drop would be required. Ultimately, at very high pressures, the ability of the column packing to withstand crushing and compaction must also be considered. Typically, however, performance is limited by the pressure that can be achieved by the pumping system.

Another important difference between gas and liquid chromatography involves the relative ease of column packing. In general, as particle size d_p is decreased, it becomes more difficult to obtain reproducible columns of high efficiency.* The lower particle sizes used in liquid chromatography therefore require a proportionately greater effort in preparing efficient packed columns, in comparison to gas chromatography.

In the case of porous support particles, the small values of D_M encountered in liquid chromatography result in a large contribution to H from mass transfer in the stagnant mobile phase within the particle (Eq. 5.35) and to large values of C_S. This loss in column efficiency can be substantially reduced by using so-called *pellicular* supports or surface-coated beads (1) (also called porous-layer beads). These supports consist of solid spherical beads (usually silica) with a thin porous surface on which the stationary phase is coated. The thickness of the stagnant mobile phase is then only a small fraction of the particle diameter, and stagnant mobile-phase contributions to H are much reduced (C_S is much smaller). Pellicular supports can provide much higher column efficiencies, therefore, than porous supports of similar size and permeability. The disadvantage of pellicular supports is that their capacity (amount of sample that can be charged before overloading) is generally much smaller than that of comparable porous particles. This essentially rules out the use of pellicular supports for preparative separations, and detectors of high sensitivity are required in analytical separations. Recently, workers have begun using columns with very small porous particle diameters (e.g., 5 to 10 μ). Under proper packing conditions, columns of high performance and good sample capacity are obtained (see Table 10.3). To overcome the particle agglomeration problem the column is packed either by a slurry method or by tamping with a close-fitting rod within the column tube.

Optimum separation conditions in liquid versus gas chromatography can be summarized as follows:

1. Generally lower mobile-phase velocities, but higher velocities *relative* to the value of v for minimum H.

2. Smaller particle sizes (e.g., 5 to 50 μ versus 100 μ and higher in gas chromatography), and more difficult column preparation.

3. Higher column pressures.

Sample diffusion coefficients D_M increase proportionately with the viscosity η of the mobile phase (Section 3.32), so that H is larger for more viscous mobile phases (cf. Eq. 10.5a and Section 5.3). The permeability of the column also decreases with increasing mobile-phase viscosity. For a given maximum column pressure, separation in liquid chromatography therefore becomes slower and/or less efficient as η increases. The viscosity of the stationary phase is less important in this regard, because its thickness is generally much smaller than that of the mobile phase.

* With small particles there is a greater tendency for particle agglomeration (because of the greater importance of surface effects), which tends to disturb the regular packing structure within the column.

Table 10.3. Relative Performance (Efficiency) of Various Columns for Liquid-Liquid Chromatography, Compared with Figures for Gas-Liquid and Paper Chromatography

Columns	Effective Plates per Sec[a]
Liquid-liquid chromatography[b]	
0.8% ODPN[c] on Zipax (pellicular)	3.8
1.1% ODPN[c] on Corasil-I (pellicular)	1.9
2.2% ODPN[c] on Corasil-II (pellicular)	1.2
30% ODPN[c] on Silica Microspheres (dp = 5 to 6 μ)	20[f]
Gas-liquid chromatography[d]	
Conventional packed	10
Coated open tubular (capillary)	25
Paper chromatography[e]	0.1

[a] For liquid-liquid columns, assumes 15-min separation time and column pressure of 1500 psi.
[b] 25 to 50-μ particles (18).
[c] β,β-Oxydipropionitrile.
[d] Ref. 17.
[e] Assumes recent papers (or cellulose TLC) equivalent to TLC.
[f] Ref. 35.

The relative performance of different liquid-liquid chromatographic supports (measured as effective plates per second) is compared in Table 10.3, along with corresponding values for typical GLC systems and for paper chromatography. It is seen that LLC in columns can be much more efficient (or yield shorter separation times) than paper chromatography, and that LLC columns using the pellicular supports approach the performance of classical packed GLC columns. Properly packed columns of 5 to 6 μ porous silica bead on the other hand provide a performance close to that of open tubular columns. We can anticipate further developments with small particles in LLC and LSC. For further details on the efficiency of LLC columns, see Refs. 7, 18–22, 34, 35, 36.

10.13   MOBILE-PHASE FLOW AND SEPARATION EFFIENCY IN PAPER CHROMATOGRAPHY

The paper that forms the chromatographic bed in paper chromatography is composed of matted cellulose fibers. The stationary phase is absorbed within the fibers, and the mobile phase moves through the spaces between fibers. In the simplest model of mobile-phase or solvent flow in paper, the interfiber spaces can be approximated by capillaries of uniform, constant diameter d.

The driving force for migration of solvent through the paper is capillary filling. The resulting pressure at the solvent front is the same as that derived in the usual manner for the height of capillary rise as a function of solvent surface tension γ: $\Delta P = 4\gamma \cos \theta/d$. The contact angle θ between the wetting solvent and the side of the capillary is approximately zero. The rate of flow of the solvent through the capillary is given by Poiseuille's equation (Eq. 3.64) $v = C'd^2 \Delta P/\eta z$, where C' is a constant, η is the solvent viscosity, and z is the distance that the solvent front has migrated through the paper. Combining these two relationships, and expressing v as dz/dt (t is time), gives (see Eq. 3.73):

$$\frac{dz}{dt} = \frac{C'd^2(4\gamma/d)}{\eta z}$$

$$= \frac{4C'(\gamma d/\eta)}{z}$$

Rearrangement of this expression, followed by integration (with t at the start of separation set equal to zero), gives

$$z^2 = 2C'\left(\frac{\gamma d}{\eta}\right)t \tag{10.6}$$

According to Eq. 10.6, the distance traveled by the mobile phase along the paper bed increases as the square root of time. The time required for the mobile phase to migrate a certain distance z along the paper sheet increases with the viscosity of the mobile phase and decreases with its surface tension. Coarser papers (larger d) give shorter running times. For a more detailed discussion, see Ref. 23.

A limitation of the above model is that it does not explain the fact that in actual separations the mobile phase is unevenly distributed along the bed behind the advancing solvent front; that is, V_M/V_S (the ratio of mobile- to stationary-phase volumes) is not constant along the bed behind the solvent front. Instead V_M/V_S has a maximum value at the front of the bed ($z = 0$), and falls off to zero at the solvent front (Fig. 10.5). This is due to the presence within the paper of capillaries of differing diameter, whereas we assumed constant capillary diameters in the derivation of Eq. 10.6. The smallest capillaries are preferentially wetted by advancing solvent, while the main solvent flow occurs within the larger capillaries.

The main practical consequence of these *solvent concentration gradients* along the bed (i.e., variation of V_M/V_S in the direction of solvent flow) is that Eq. 5.10 does not accurately describe band migration rates (R_f values) as a function of K_i and V_M/V_S, because Eq. 5.10 assumes that V_M/V_S is constant. To a first approximation it is found that the effect of these mobile-phase concentration gradients is to change Eq. 5.10 as follows (see Ref. 24):

$$R_F = \frac{1}{\xi[1 + (A_S/A_M)K_i]} \qquad (10.7)$$

Here ξ is an empirical constant that varies from about 1.1 to 1.4. In other words, experimental R_f values are generally lower than values predicted in the absence of solvent concentration gradients.

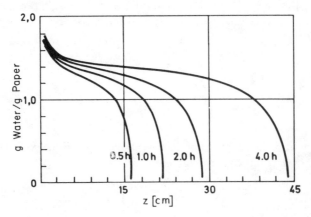

Figure 10.5. Distribution of solvent (concentration profile) as a function of time and distance z between immersion line (solvent level) and point of observation. Reprinted from Ref. 34, by courtesy of Elsevir Publishing Co.

The theory of bed efficiency in paper chromatography is similar to that for LLC in columns, but somewhat more complex (see Refs. 23 and 25). Thus in paper chromatography Eq. 10.5 must be integrated over the entire separation, since v changes with time. In paper chromatography we have much less control over mobile-phase velocity than is possible in column chromatography. Coarse and fine papers can provide changes in separation time (and average values of v) by a factor of 2 to 4. Wicks of special geometry can be used to feed the mobile phase to the paper for slower rates of separation. However, paper length L and solvent velocity v cannot be varied over a very wide range, as in column chromatography, and this limits the relative speed and efficiency of paper chromatography (but see Ref. 25). The best open-bed (i.e., paper) chromatographic procedures normally provide no more than a few hundred effective theoretical plates for separation times of one to several hours. Liquid column chromatography, on the other hand, can yield several thousand effective plates in separation times of an hour or less. Alternatively, separations equivalent to those achieved by paper chromatography can be duplicated in columns in much shorter times.

Typical papers for chromatography have been poorly designed for separation efficiency (large N), so that separation times were long and results were relatively poor. More recently, improved papers have been introduced, along with the technique of cellulose thin-layer chromatography (see Section 10.4). These new materials provide substantially faster, better separations by open-bed LLC.

10.2 Column Chromatography

In choosing between column and paper chromatography, several factors must be considered. Under optimum conditions columns can be many times more efficient and/or faster than paper. Quantitative analysis is more convenient and accurate with columns. Because of the low capacity of paper chromatography, preparative separations are best handled on columns. The advantages of paper chromatography include simple equipment and technique, easy variation of experimental conditions to fit the requirements of different samples, and the ability to carry out two-dimensional separations (Section 5.53).

Columns are generally preferred for repetitive quantitative analyses of a given type, whereas paper chromatography is favored for the qualitative analysis of unknown mixtures and conditions. In this connection a given separation on paper can usually be repeated on a column under similar experimental conditions (i.e., same two liquid phases).

10.21 EQUIPMENT

The basic equipment requirements for all liquid column chromatography methods (liquid-liquid, liquid-solid, ion-exchange, and exclusion) are similar to those for GLC: a mobile-phase drive unit, sample introduction system, column, and detector. In addition, fraction collectors are often used in liquid column chromatography. Two extremes in equipment are still common: very simple (usually homemade) units, such as those shown in Fig. 10.6, and more complex commercial units featuring semiautomatic operation and rapid, highly efficient separation (as in Fig. 10.1c; for a review, see Refs. 26 and 27).

The units of Fig. 10.6 are intended for use at low pressures (often less than 10 psi), solvent being driven through the column by means of gravity or compressed gas. The columns are constructed of glass and Teflon. Small eluate fractions are collected and analyzed manually for sample bands, as illustrated in Fig. 10.1a.

In the case of semiautomated units for column chromatography (Fig. 10.1c) the mobile phase is driven under high pressure (e.g., 500 to 5000 psi)

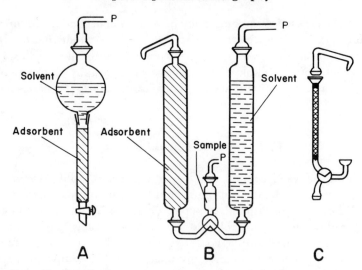

Figure 10.6. Simple (low-pressure) equipment for liquid chromatography.

through the column by means of a pump, at carefully controllable flow rates. Two general types of pumps are used: constant-volume flow (motor-driven pistons or syringes), and constant pressure (gas displacement). Constant-volume pumps are usually preferable, because they allow more accurate control of solvent flow rates through the column. This is particularly important when the composition of the solvent changes during the separation, as in gradient elution. A constant flow of solvent is important for repeatable retention times (and accurate qualitative analysis) and accurate quantitation. Peak heights and, especially, peak areas vary with flow rate.

Three types of sample introduction are used in semiautomated liquid chromatography units:

1. Syringe injection, most frequently through a septum.
2. Sample loop injection, through a multiport valve.
3. Stop-flow injection.

The use of syringes and loops is more convenient and generally preferable. Syringe injection permits easier variation of sample size, but sample loops are more convenient and give more precision. The main advantage of stop-flow injection is that it does not require elaborate equipment: the solvent flow is simply stopped, a fitting at the column inlet is removed, and the sample is applied directly to the column (at low pressure); the fitting is then replaced and solvent flow resumed.

Because of the slow diffusion of samples in the liquid phase, the holdup volumes of the sample introduction unit must be quite small. This is true also of all column connections, the detector, and connecting lines between the various components of the chromatographic unit. It should be emphasized that these *extra-column* contributions to band broadening are much more important in liquid chromatography than in GLC; this means that the volumes of detector, sample inlet system, and so forth in a liquid chromatography unit must be quite small.

Columns used in liquid chromatography are of glass or metal. Small-bore, heavy-wall glass tubing is useful for pressures below 1000 psi, and stainless steel tubing is suitable for any pressures desired. A variety of special fittings exist for connecting columns to inlet and outlet lines.

Several types of detectors are used in liquid chromatography, none of which is universally satisfactory. These are usually less sensitive than the detectors available for GLC. *Ultraviolet* (UV) *photometers* and *differential refractometers* are among the most popular detectors at present, and each is available with the very small cell volumes (e.g., 10 μl) required for high-performance liquid-chromatography. Ultraviolet detectors are extremely sensitive (e.g., 0.00005 optical density unit, equivalent to 1ppb in favorable cases), a feature that makes this type preferable for aromatic and other UV-absorbing sample constituents. Unlike refractometers, UV detectors can also be used with gradient elution. The best refractometers have a sensitivity (10^{-7} refractive index unit, equivalent to about 1 ppm) which is marginal for many LC applications, but most types of sample are detectable by refractive index. In other words, refractometers can be considered as *universal* detectors, whereas UV detectors are *specific*. Refractometers have found widespread use in exclusion chromatography (Chapter 15), where high sensitivity is not required.

Two other detectors have also been used widely for liquid chromatography, although to a lesser extent than the above detectors. *Conventional GLC detectors* (e.g., flame ionization units) can be coupled to units by means of mechanical transfer devices (moving wires, chains, belts, etc.) that accept eluate from the column, evaporate the mobile phase, and deliver the residual sample bands to the detector. Present devices of this type can be used only for relatively nonvolatile samples, and have a sensitivity between that of the UV detector and that of the refractometer. The *microadsorption detector* is based on a thermistor measurement of the heat of adsorption and desorption of the sample as it passes through a small cell packed with an adsorbent. This detector possesses intermediate sensitivity and responds to all solute bands, but gives a non-Gaussian chromatogram.

Other detectors include those based on the measurements of electrical conductivity, polarographic reaction, and infrared absorption, as well as other properties. None of these detectors is widely used.

Commercial fraction collectors are available which collect eluate on the basis of a certain volume or number of drops per fraction, or a certain number of fractions per unit time. With the increasing use of continuous detectors, fraction collectors are employed mainly for preparative separations.

10.22   COLUMN PREPARATION

Column preparation begins with the selection of a support for the stationary phase. Porous silica, kieselguhr (also called Celite or Hyflo-Supercel), and powdered cellulose have been widely used in the past. Efficient and/or rapid separations are provided by the pellicular supports (surface-coated glass beads; see Ref. 1). With the exception of cellulose, these supports are usually coated with the stationary phase S before packing the column. This is best done by evaporating a slurry of the support plus a solution of S in a low-boiling solvent. Porous supports give good efficiencies with stationary-phase loadings of 5 to 10% or less, depending on the surface area of the support. Maximum loadings for pellicular supports are usually 1 to 2% or less.

In the past, columns have often been packed by pouring a slurry of the coated support plus the mobile phase (saturated with S) into the column. The column is then compacted by gravity, by pressured flow of the mobile phase through the column, or by tamping the column packing with a per-forated disk that fits the column walls closely. A column can also be packed dry by pouring the coated support into the column in small increments, followed by tapping of the column for maximum settling. Dry column packing is usually preferred for support particles larger than 40 μ in diameter. For a detailed discussion of this topic see Ref. 1. High pressure packing of a slurry of the support is found useful for very small particles (e.g., 5 to 10 μ).

Reverse-phase separations require special consideration, since hydrophilic supports such as cellulose or silica do not hold a nonpolar stationary phase in the presence of a polar mobile phase. Reaction of the surface groups of a silica (as with a trialkylchlorosilane) can yield a hydrophobic material (see Section 8.22) suitable as a support for reverse-phase separation. Such supports can also be used directly, without addition of a stationary phase, the alkyl groups bonded to the silica surface serve as the nonpolar phase (see Section 10.4).

10.23   SEPARATION CONDITIONS AND TECHNIQUES

The number of theoretical plates N required for a given separation, and the time of separation, are key considerations in the design of a chromatographic system (Section 5.43). At a given mobile-phase velocity v, N is proportional to column length L. The time of separation is equal to t_0 (the retention time of a nonretained solute) times $(k' + 1)$ for the most strongly retained sample component, Eq. 5.3 ($k' + 1$ is generally 3 to 6 for a two-component sample mixture, and not more than 20 for multicomponent samples). Figure 10.7 shows how L and t_0 are interrelated for various particle sizes and a given pressure ($P = 1000$ psi) and solvent viscosity ($\eta = 0.3$ cP). Values of t_0 for other pressures and solvent viscosities can be obtained by multiplying these values by $(1000/0.3)\eta/P$. We have noted that N increases for smaller values of d_p, but a lower limit on particle size (5 to 20 μ) is set by the difficulty of obtaining a narrow size range and by other problems. Typical LLC separations are carried out with 2 to 4-mm.-i.d. columns, 50 to 100 cm in length, using particles that are 20 to 50 μ in diameter.

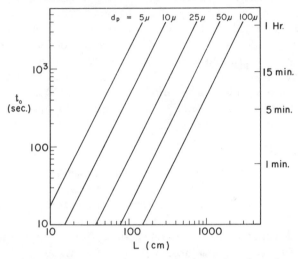

Figure 10.7. Separation time as a function of column length and support particle diameter d_p for column pressure = 1000 psi and solvent viscosity = 0.3 cP.

Separation begins with the introduction of sample to the column, usually as a solution in the mobile phase.* The total sample volume should be small, so as to avoid further band broadening beyond that produced during separation. The quantity of sample that can be separated on a given column varies; 0.1 to 1 mg of sample per gram of stationary phase is typical. In analytical separations column overloading results in a dependence of sample migration rates on sample size (Section 5.1). In preparative separations sample size is limited mainly by the deterioration of resolution that occurs for large samples. Column overloading can also result in dissolution (by sample) of stationary phase from the front of the column and eventual loss in column efficiency.

* The solvent used to dissolve the sample should not be stronger (i.e., give smaller k' values) than the mobile phase, since this leads to poor separation.

In addition to the requirements that the mobile phase be as nonviscous as possible, and that the combination of S plus M give suitable k' and α values, certain practical considerations should be kept in mind. The mobile phase must not interfere with the detection of eluted sample bands or (in preparative separations) their eventual recovery. The mobile phase must also be a reasonably good solvent for the sample, but this is usually the case if S and M provide desirable migration rates for the sample. Chemically stable, inexpensive solvents are of course preferred.

In high-efficiency (high-pressure) separations, the mobile phase should be degassed before use (by application of heat and/or vacuum). Dissolved oxygen can lead to reaction and deterioration of the stationary phase, as well as alteration of sample components. Column temperature should be controlled within 0.1° in order to prevent changes in the amount of stationary phase on the support. Under these conditions (constant temperature, no oxygen), column life should average 3 to 6 months or longer.

Temperature programming and gradient elution (Section 5.53) are not used in LLC, because of the difficulty in maintaining the stationary phase within the column.* Flow programming (Section 5.53) can be of some use in speeding up the elution of more strongly retained bands, while providing improved resolution for initially eluting bands.

10.3   Paper Chromatography

The experimental aspects of open-bed or paper chromatography (PC) are deceptively simple. A minimum of equipment and technique is sufficient for the successful completion of many separations, but the capability of PC can be expanded by attention to further details. Here we attempt to cover only the more important practical aspects; for a detailed review, see Refs. 2 to 6 and 29.

10.31   EQUIPMENT

The only essential piece of equipment in PC is the developing chamber. In its simplest form this consists of an enclosed (usually glass) container with an airtight lid. Its function is to hold the developing solvent and paper sheet during the separation, usually in an atmosphere that has been equilibrated with both stationary and mobile phases. For one-dimensional PC, either ascending or descending development can be carried out in simple units of

---

* An exception is the use of gradient elution with bonded-phase supports (9); see Section 10.4.

the type shown in Fig. 10.8. Ascending development with the unit of Fig. 10.8a is quite simple. Descending development (as in Fig. 10.8b) is more often used, however, because it is faster and more suitable for long paper sheets (which give higher efficiencies). The unit of Fig. 10.8b features a solvent trough at the top of the chamber, with an antisiphon rod at its side. For two-dimensional PC, ascending development can be used as in Fig. 10.8c. After the initial development, which provides partial separation of the sample along a vertical line (see Fig. 5.19), the sheet is uncoiled, dried, and recoiled perpendicularly to the first coil. The various sample bands are then arranged around the bottom of the coil, and the second development provides further separation of each band, just as in the first development. *Radial PC* can be carried out by spotting the sample in the center of the paper and then adding solvent dropwise directly to the sample spot. No special equipment is required.

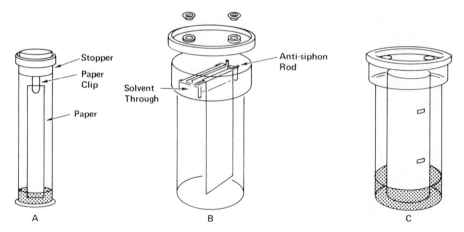

Figure 10.8. Developing chambers for paper chromatography: (*a*) ascending development, (*b*) descending development, (*c*) two-dimensional development.

More complex developing chambers have been described for the simultaneous development of several sheets, for more rapid equilibration of the development chamber, and for other refinements. Other equipment associated with the use of PC is indicated in the following sections and requires little comment.

10.32 PREPARATION OF THE BED

The paper used in PC is normally a commercial product, obtained from carefully selected short cotton fibers. The final paper is 99% α-cellulose with low metal content. Papers for PC come in slow, standard, and fast grades, corresponding to fine, medium, and coarse papers. For typical conditions (e.g., 35-cm paper length, 4/1/5 butanol–acetic acid–water) running times on standard papers are 10 to 20 hr. Fast papers (4 to 6 hr) give poorer resolution but faster separation; they are used for less difficult separations. Slow papers (20 to 50 hr) permit improved resolution but are not used to much extent. Continuous or multiple development (see Section 10.33) is generally preferred for difficult separations. Whatman No. 1 and Schleicher-Schuell No. 20436 are examples of standard papers that have been used widely in the past. These papers are in the process of being superseded by newer products that give improved separations in short development times (1 to 2 hr); see also Section 10.5 for alternative bed materials. Paper sheets for PC are kept in a controlled-humidity atmosphere before use, in order to maintain the starting water content within narrow limits. Stationary phases other than water can be added to the paper by impregnation (dipping) and subsequent removal of excess liquid by blotting (e.g., formamide and propylene glycol for normal PC; silicone oil, saturated hydrocarbons, and glycerides for reversed-phase PC). A completely wet paper will not imbibe mobile phase, as required for separation. Volatile, polar stationary phases (e.g., methanol) can be absorbed by the paper from a saturated atmosphere, and acetylated papers will absorb volatile, nonpolar stationary phases (e.g., benzene) for reversed-phase work.

A variety of special papers are available for certain applications. "Heavy" (thick) papers provide increased capacity for preparative separations, although preparative work is normally handled better in columns. Ion-exchange papers are available that have an ion-exchange resin incorporated into them. The resulting separations are intermediate between those obtained by normal PC and by ion-exchange chromatography (see Chapter 12).

10.33 TECHNIQUE

The commonest form of paper chromatography is descending, one-dimensional development as in Fig. 10.8b. Improved separation of complex, multicomponent samples can be obtained with two-dimensional PC (Section 5.53), shown in Fig. 10.8c. Greater resolution of two (or more) overlapping bands is possible with the techniques of multiple or continuous PC (see below). Radial PC, mentioned in Section 10.31, allows more rapid separation and is used with easily separable samples or for preliminary separations aimed at selecting a solvent of the right strength. For the separation of just two sample components, maximum resolution in PC is provided when the average $k' = 2$ (i.e., $R_f = \frac{1}{3}$).

Separation by PC begins with the application of the sample to the sheet. From 2 to 4 μl of a 1 to 2% sample solution is normally applied in one-dimensional PC (up to 12 μl in two-dimensional PC), followed by drying the area around the point of sample application. The total quantity of sample should not exceed 500 μg. Too large a sample leads to poor separation and variable band migration rates, as in column separations, and too small a sample may result in bands that are difficult to detect. The sample must be applied to the same relative position on the sheet, if band R_F values are to be reproducible, because of the solvent concentration gradient (varying A_M/A_S) along the sheet (Section 10.13).

After sample application and drying, the sheet is next introduced into the development chamber and equilibrated with both mobile and stationary phases (which are normally placed in the bottom of chambers such as that of Fig. 10.8b) for 1 to 3 hr (some workers prefer up to 24 hr equilibration). Equilibration of the paper before separation yields more reproducible R_f values.

After equilibration of the paper, development of the sheet is begun by insertion of one edge into the mobile phase (or addition of mobile phase to the solvent trough). The mobile phase should be pre-equilibrated with stationary phase. Close temperature control during separation is also required for reproducible R_F values. It has been stated that volatile solvents (either mobile or stationary phase) give poorer reproducibility, but this must be weighed against the better efficiency associated with volatile, nonviscous solvents. With volatile solvents, careful equilibration is required if band tailing is to be avoided.

After separation and subsequent drying of the sheet, the separated bands can be detected in a variety of ways. Most commonly the sheet is sprayed with a solution that reacts with individual sample components to give colored products. A large variety of such solutions have been described (see, e.g., Ref. 3) for different types of compounds. Aromatic compounds can be detected by exposure of the sheet to ultraviolet radiation in the dark. Fluorescent bands appear directly, and other aromatic compounds can be seen by fluorescent quenching if a fluorescing material is first added to the paper. Quantitative analysis via PC can be carried out if the location of individual bands can be controlled within narrow limits. The bands are cut from the paper and extracted with a suitable solvent. The resulting sample solutions are then analyzed by any suitable technique. Alternatively, direct quantitation of the original chromatogram can be achieved by densitometry, in either the reflection or the transmission mode.

Multiple development and continuous development are special techniques for increasing the resolution of two adjacent bands that are not adequately separated in normal development. The objective of these techniques is to simultaneously increase the distance of band migration (and therefore increase N) and increase k' (and therefore increase $k'/k' + 1$; see Eq. 5.43). In each case the solvent is selected to give k' values (for the two bands in question) significantly larger than the normally optimum value of 2, for example, $k' = 10$. In multiple development the sample is first separated in the normal manner; then the sheet is dried, and the separation repeated. This process is repeated some number of times for optimum resolution (see Table 10.4). In continuous development provision is made for solvent evaporation from the end of the sheet, thereby permitting continuous flow of solvent through the sheet. In this way the two bands of interest may be moved across the entire sheet, separation being terminated just before they arrive at the end of the sheet. The combination of increased migration distance and capacity factors can lead to an increase in resolution by a factor of almost 3, relative to normal PC separation; however, this is achieved at the expense of longer separation times and also involves a more complicated procedure.

Table 10.4. Maximum Resolution in Multiple-Development Paper Chromatography (or TLC)

| Number of Developments | Optimum R_f | | Relative Resolution[c] |
	Single Development[a]	Total[b]	
1	0.33	0.33	1.00
2	0.29	0.50	1.31
3	0.26	0.61	1.52
5	0.22	0.71	1.73
10	0.16	0.82	2.00

[a] Position of band after first development.
[b] Position of band at completion of multiple development.
[c] R_s divided by R_s for single development ($R_f = 0.33$).

10.4 Some Related Methods

In addition to normal column and paper chromatography as described in Sections 10.2 and 10.3, there have been a number of recent innovations in the field of LLC. Thin-layer chromatography (TLC) with powdered cellulose as stationary phase has been in use for several years and has displaced PC from many of its previous applications. Thin-layer chromatography, which normally uses an adsorbent as stationary phase (Section 13.22), is a form of open-bed chromatography in which the powdered support (cellulose in this case) is attached as a coherent film or layer to an impermeable plate or backing. Cellulose TLC is essentially equivalent to chromatographic paper in terms of band migration and separation factors, as shown by the two separations of Fig. 10.9. The same two solvent systems were used in each of these two-dimensional separations of a mixture of nuclei acid bases. However, the TLC separation is better with respect to band sharpness and resolution and is 4 to 5 times faster. Increased band sharpness in TLC also results in

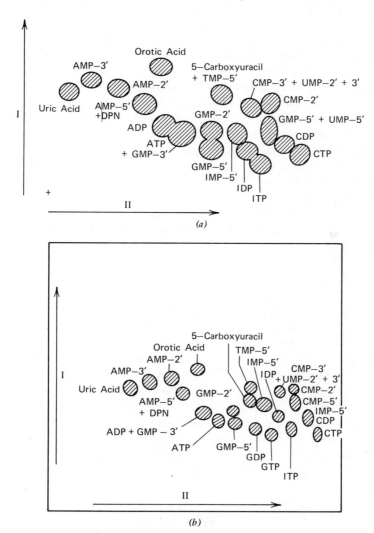

(a)

(b)

Figure 10.9. Comparison of paper chromatography (*a*) and cellulose TLC (*b*) in the separation of nucleotides. Note the sharper separation in (*b*). Reprinted from Ref. 28, by courtesy of Marcel Dekker, Inc.

significantly lower sample detection limits. The superiority of cellulose TLC to PC is apparently the result of differences in support particle shape: roughly spherical in the case of powdered cellulose, long fibers in PC. Recent improvements in chromatographic papers have reduced the margin of superiority of cellulose TLC over PC, however.

Reversed-phase TLC is another LLC technique that has found wide application. Here a nonpolar liquid (normally a hydrocarbon, such as undecane) is coated onto a TLC plate of silica or kieselguhr. Separation of the sample then proceeds as in normal TLC, using a more polar solvent as the mobile phase. This technique also provides more rapid and efficient separations than can be obtained by conventional PC.

A variety of *bonded-phase* supports, carrying both nonpolar and polar phases S, are available for LLC (e.g., Durapak® by Waters Associates, DuPont's Permaphase®) (1). The bonded-phase supports are similar in performance to conventional coated supports but are more convenient to use. Thus the support does not have to be coated before use, and the bonded phase cannot be washed from the support (thus avoiding the need for equilibration of the mobile phase with S before use, and permitting the application of gradient elution).

Another recent innovation is intermediate between GLC and LLC in columns; so-called *supercritical fluid* chromatography (30). This involves the use of a mobile phase above its critical temperature, for example, n-pentane or isopropanol above 200°. These supercritical fluids behave somewhat like liquids with respect to their equilibrium properties, so that very-high-boiling materials can be readily eluted. However their viscosities are lower by a factor of 10 or more than those of normal liquids at temperatures below their boiling points. This results in significant improvement in separation efficiency, with column pressure drops that are an order-of-magnitude smaller (relative to normal LLC). Supercritical fluid chromatography appears to complement both GLC and LLC in some of their present applications. This technique is limited, however, by the need for somewhat more complex equipment and higher than ambient temperatures, although a few mobile phases (e.g., ammonia) can be used at room temperatures. Jentoft and Gouw (31) were able to separate individual polystyrene oligomers containing from 1 to 32 styrene groups in about 1 hr, using supercritical fluid chromatography and a bonded-phase support. See also the discussion of Section 13.3.

10.5   Applications

For many years the major application of LLC has been the separation of hydrophilic samples of intermediate molecular weight and low volatility.

The first successful analyses of complex mixtures of amino acids and peptides were based on two-dimensional PC, and similar analyses are now carried out by cellulose TLC. A similar situation prevails for the nuclei acid bases and their derivatives (e.g., Fig. 10.9). The carbohydrates have long been handled successfully by PC and cellulose column chromatography. More recently TLC with kieselguhr or Celite has been used for similar separations; the low surface area of these supports makes it likely that these separations take place by partition rather than adsorption (cf. normal TLC; Section 13.22). Phenol derivatives such as synthetic mixtures, the flavonoid pigments, anthocyanins, and tannins have been separated primarily by PC or related column methods. Less polar, hydrophobic samples (e.g., lipids, steroids) were formerly separated to a large extent by PC and partition chromatography

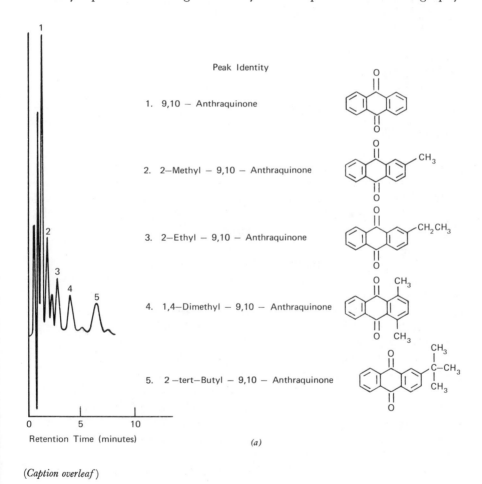

Peak Identity

1. 9,10 — Anthraquinone

2. 2—Methyl — 9,10 — Anthraquinone

3. 2—Ethyl — 9,10 — Anthraquinone

4. 1,4—Dimethyl — 9,10 — Anthraquinone

5. 2 —tert—Butyl — 9,10 — Anthraquinone

Retention Time (minutes)

(a)

(Caption overleaf)

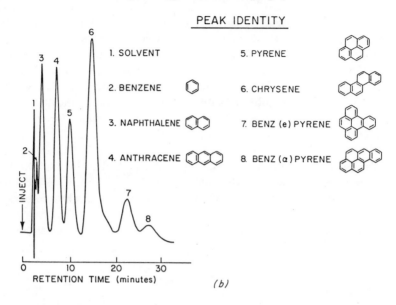

Figure 10.10. Reversed-phase separations by liquid-liquid column chromatography. Columns packed with hydrocarbon polymer on Zipax support in each case. Reprinted from Ref. 1.

in columns, but these older procedures have been largely replaced by TLC (adsorption) techniques (Section 13.22). However, the recent improvement in liquid-liquid column chromatography has led to its application to a wide range of less polar substances (e.g., Fig. 10.10). Figures 10.2 and 10.11 illustrate some typical applications of modern LLC in columns (i.e., more polar samples. Finally, Fig. 10.12a shows a spectacular separation of 5 hydroxylated compounds using porous microspheres (5 to 6 μ) (34). A second example using small particle sizes is shown in Fig. 10.12b (35). For further examples see Ref. 27.

Considerable attention has been devoted to the separation of inorganic mixtures by PC, for both qualitative and quantitative analysis (see, e.g., Ref. 32). The analyses for various cations and anions have been relatively successful, particularly in combination with chemical procedures. These methods have been largely superseded, however, by modern spectroscopic procedures—neutron activation analysis, flame emission and absorption, X-ray fluorescence, and the like. Liquid-liquid chromatography still plays an important role in the analysis of ionic molecular species, for example, mixtures of the phosphoric acid oligomers (H_3PO_4, $H_2P_2O_7$, metal complexes, etc.).

Other applications of LLC are described in Refs. 2 to 6 and 33.

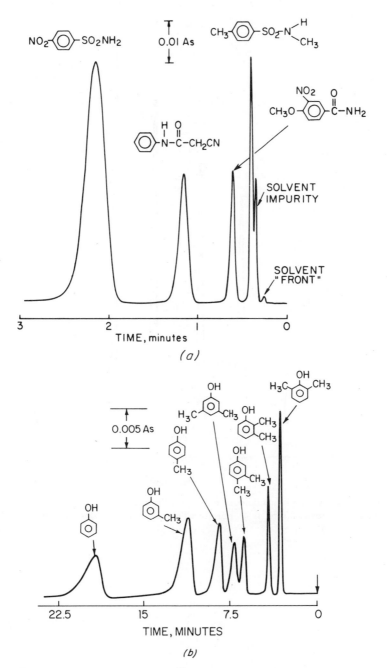

Figure 10.11. Separation of polar samples by liquid-liquid column chromatography: (*a*) bonded "nitrile" polymer on Zipax, (*b*) bonded "ether" polymer on Zipax. Reprinted from Ref. 1.

299

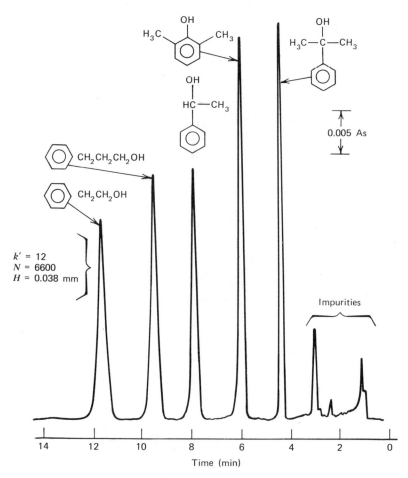

Figure 10.12a. High performance liquid-liquid chromatographic separation of hydroxylated substances using 5 to 6 μ porous silica particles. Stationary phase: β,β'-oxydipropionitrile. Mobile phase: hexane. Reprinted from Ref. 35 by permission of the publisher of *J. Chromatog. Science.*

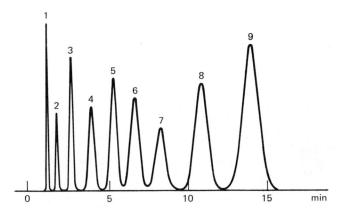

Figure 10.12b. Rapid separation of phenols by liquid-liquid chromatography using 5 to 15 μ Kieselguhr. Stationary phase: 1,2,3-tris (2-cyanoethoxy) propane. Mobile phase: 2,3,4-trimethylpentane. Reprinted from Ref. 36 by permission of the publisher.

References

1. J. J. Kirkland, in *Modern Practice of Liquid Chromatography*, J. J. Kirkland, ed., Wiley-Interscience, New York, 1971, Chap. 5.
2. R. J. Block, E. L. Durrum, and G. Zweig, *A Manual of Paper Chromatography and Paper Electrophoresis*, 2nd ed., Academic Press, New York, 1958.
3. I. M. Hais and K. Macek, *Paper Chromatography*, Academic Press, New York, 1963.
4. I. E. Bush, *The Chromatography of Steroids*, Pergamon Press, New York, 1961.
5. C. J. O. R. Morris and P. Morris, *Separation Methods in Biochemistry*, Interscience, New York, 1964.
6. R. Neher, *Steroid Chromatography*, Elsevier, Amsterdam, 1964.
7. J. F. K. Huber, "Liquid Chromatography and Columns," in *Comprehensive Analytical Chemistry*, Vol. 2B, C. L. Wilson and D. W. Wilson, eds., Elsevier, Amsterdam, 1968.
8. J. H. Hildebrand and R. L. Scott, *Regular Solutions*, Prentice-Hall, Englewood Cliffs, N. J., 1962; *The Solubility of Non-electrolytes*, 3rd ed., Dover Publications, New York, 1964.
9. J. H. Hildebrand, J. M. Prausnitz, and R. L. Scott, *Regular and Related Solutions*, Van Nostrand Reinhold, New York, 1970.
10. E. Soczewinski, *Advan. Chromatog.*, **5** (1966).
11. J. A. Thoma, *Advan. Chromatog.*, **6** (1967).
12. D. C. Locke, *Advan. Chromatog.*, **8** (1969).
13. D. E. Martire and D. C. Locke, *Anal. Chem.*, **43**, 68 (1971).
14. L. R. Snyder, in *Modern Practice of Liquid Chromatography*, J. J. Kirkland, ed., Wiley-Interscience, New York, 1971, Chap. 4.
15. R. A. Keller, B. L. Karger, and L. R. Snyder, in *Gas Chromatography, 1970*, R. Stock and S. G. Perry, eds., Institute of Petroleum, London, 1971.
15a. J. F. K. Huber, *J. Chromatog. Sci.*, **9**, 72 (1971).
16. J. Green and D. McHale, *Advan. Chromatog.*, **2** (1964).
16a. D. Locke, *J. Gas Chromatog.*, **5**, 202 (1967).
17. B. L. Karger, in *Modern Practice of Liquid Chromatography*, J. J. Kirkland, ed., Wiley-Interscience, New York, 1971, Chap. 1.

18. L. R. Snyder, in *Gas Chromatography, 1970*, R. Stock and S. G. Perry, eds., Institute of Petroleum, London, 1971.

19. L. R. Snyder, *J. Chromatog. Sci.*, **7**, 352 (1969).

20. J. H. Knox and M. Saleem, *J. Chromatog. Sci.*, **7**, 614, 745 (1969).

21. A. Zlatkis, ed., *Advances in Chromatography, 1969*, Preston Technical Abstracts Co., Evanston, Ill., 1969.

22. A. Zlatkis, ed., *Advances in Chromatography, 1970*, Preston Technical Abstracts Co., Evanston, Ill., 1970.

23. G. H. Stewart, *Advan. Chromatog.*, **1** (1963).

24. L. R. Snyder, *Principles of Adsorption Chromatography*, Marcel Dekker, New York, 1968, pp. 135, 365–369.

25. D. L. Saunders and L. R. Snyder, *J. Chromatog. Sci.*, **8**, 706 (1970).

26. L. R. Snyder, in *Chromatography*, E. Heftmann, ed., Reinhold, New York, 1967, Chap. 5.

27. J. J. Kirkland, ed., *Modern Practice of Liquid Chromatography*, Wiley-Interscience, New York, 1971.

28. G. Pataki, *Advan. Chromatog.*, **7**, 47 (1968).

29. V. C. Weaver, *Advan. Chromatog.*, **7**, 87 (1968).

30. S. T. Sie and G. W. A. Rijnders, *Separation Sci.*, **2**, 729 (1967).

31. R. E. Jentoff and T. H. Gouw, *J. Chromatog. Sci.*, **8**, 138 (1970).

32. G. Nickless, *Advan. Chromatog.*, **7**, 121 (1968).

33. E. Heftmann, ed., *Chromatography*, Reinhold, New York, 1967.

34. J. C. Giddings, *et al.*, *J. Chromatog.*, **3**, 239 (1960).

35. J. J. Kirkland, *J. Chromatog. Sci.*, **10**, 593 (1972).

36. J. F. K. Huber, 5th International Symposium, Column Chromatography, Lausanne, 1969, published as supplement to *Chimia*, 24 (1970).

List of Symbols

A	cross-sectional area		γ	surface tension
C'	constant in Eq. 10.6		ζ	constant in Eq. 10.7
d	capillary diameter			
D	constant in Eq. 10.5b			
n	constant in Eq. 10.5b		*Subscripts*	
Q	contact angle		i and j	sample components
X	a given group in a molecule		$a, b, d, o,$ and i_n	pertain to various solubility parameters
Z	distance		M	mobile phase
Δ	incremental quantity		S	stationary phase

Eleven

CRYSTALLIZATION

W. R. WILCOX

11.1 Introduction

Crystallization is one of the oldest separation techniques utilized by man. Centuries ago seafaring men were producing drinking water by slow melting of sea ice (56). One century ago wax was separated from oil by chilling the oil and squeezing it in fabric bags. Today almost all chemicals sold in solid form are produced by crystallization. In spite of its ancient origins, crystallization remains under active investigation and development.

Crystallization possesses some advantages over competitive methods. Separations can often be accomplished at low temperatures, minimizing decomposition of heat-sensitive compounds and contamination by attack of the container. Crystallization may be carried out at any scale from milligram to industrial. Some compounds can be separated by crystallization that cannot be separated practically in any other way, for example, close-boiling isomers having nearly identical vapor pressures. Even optical isomers can be separated (46). Germanium and gallium are ultrapurified by crystallization for semiconductor device applications. p-Xylene is separated from its isomers on an industrial scale by crystallization. Crystallization separates organics from sulfur at a low temperature, thus avoiding the contamination from the crucible walls observed in the usual thermal decomposition methods (18). Although radioactive components in neutron-irradiated organics can be separated by gas-liquid chromatography, the handling of the solid samples

Table 11.1. Examples of Purification by Crystallization

Crystallization Method Employed	Special Features	Material	Impurity	Measure	Impurity Content		Ref.
					Before	After	
Vertical directional solidification	6 passes with decanting at 1.5 cm/hr	Hexafluorobenzene	Native impurities	wt %	1.68	<0.01	20
Horizontal directional solidification	1 pass at 0.07 cm/hr	Lead	Silver	wt %	1.0	0.07	24
Vertical directional solidification	Czochralski method—pulled from melt at 0.03 cm/hr	α-d-Propoxyphene	α-l-Propoxyphene	mole %	3.1	0.05	46
Horizontal zone melting	Ultrasonic stirring, 1 pass at 7.5 cm/hr	Naphthalene	Azobenzene	%	0.2	0.06	1
Vertical zone melting	60 passes at 2.5 cm/hr	Propionamide	Acetamide	%	0.02	2×10^{-6}	24
Zone melting	1 pass at <40 cm/hr	KCl	Ba^{2+}	ppm	78	10	56
Electron beam floating zone melting	Pressure $<4 \times 10^{-9}$ torr, 6 passes at 10 cm/hr	Vanadium	Oxygen	ppm	153	16	7

Progressive freezing followed by slow melting with drainage	2 Fractionations	Benzoic acid	Native impurities	mole %	0.09	0.002	39
Progressive freezing followed by melting with drainage	Industrial scale	p-Dichlorobenzene	Isomers, etc.	mole %	3.1	0.05	46
Partial freezing, partial melting	Contact with cold brine to crystallize, separate in centrifuge, rinse with warm brine; industrial scale	Benzene	Natural impurities	Freezing point (°C)	3.4	5.4	56
Countercurrent crystallization: freezing staircase method	1 stage at 0.47 cm/hr	Naphthalene	Oil red dye	mole %	0.04	10^{-6}	38
Column crystallization	Rotating helix	Benzene	Thiophene	S content, %	1.88	0.99	8
Column crystallization	6 theoretical plates	Sulfur	Bitumens	%	0.03	2×10^{-5}	18
Column crystallization	Industrial scale (60,000 metric tons/year)	p-Xylene	Isomers	%	22	>99	56

is in most cases unsatisfactory (19). Crystallization methods, on the other hand, do not have this disadvantage. Further examples are given in Table 11.1.

In addition to separation and purification, crystallization is also useful for concentrating solutions and suspensions by solvent removal. Since high temperatures are not involved, volatile components are not lost. Possible applications are (*a*) analyses of trace quantities, and (*b*) industrial production of food concentrates.

11.11   CLASSIFICATION OF CRYSTALLIZATION METHODS

Crystallization is a generic term for any process in which a crystalline solid is formed. It is also used to denote a process in which discrete crystals or clumps of crystals are produced in a liquid, usually containing a solvent. When a massive layer of solid is produced from a melt containing no added solvent, the process is termed *solidification. Fractional crystallization* and *fractional solidification* indicate that the original mixture is separated into fractions or that a material is purified.

There are a host of related crystallization operations. *Precipitation* denotes rapid formation of small crystals or amorphous solid particles, usually by addition of a substance forming one or more insoluble compounds. In *adductive crystallization* one component of a mixture is preferentially removed by incorporation into crystals formed from an added compound. *Sublimation* denotes a vapor-solid transformation (Section 7.44) which is also used for separations (26). In *chemical vapor deposition* gaseous components react to form the solid. *Electrodeposition* (electrorefining) is useful for purifying metals, and electrochemical reactions are potentially useful for other separations as well. Sublimation, chemical vapor deposition, and electrodeposition are not discussed further here, although the general principles of crystallization apply to these as well.

11.2   Crystallization Phenomena

11.21   NUCLEATION

Consider a solution of a compound whose solubility increases with temperature T, as in Fig. 11.1. Imagine that this solution is initially at point A, that is, T is above the equilibrium temperature T_e. The solution is now slowly cooled along the dashed line. Crystals do not form until the solution is supercooled below T_e. This supercooling may be expressed in several ways,

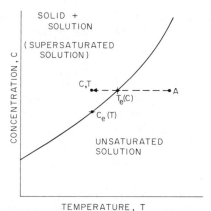

Figure 11.1. Solubility curve.

as shown in Table 11.2. Alternatively, attention may be focused on the excess concentration above the solubility C_e at the solution temperature T. This supersaturation may similarly be expressed in several ways (Table 11.2). If the solubility curve (liquidus) can be approximated by a straight line of slope $m = |dT/dC|$, then supersaturation and supercooling are related by

$$\Delta T = m \, \Delta C \tag{11.1}$$

There are three types of nucleation: homogeneous, heterogeneous, and crystal breeding or secondary nucleation. In *homogeneous nucleation* a small number of molecules first come together to form a submicroscopic solid, called an *embryo*. Molecules are added to and subtracted from the embryo in a random fashion. As the embryo grows, its free energy at first increases because of increasing surface energy. Thus only a few embryos out of many continue to grow. Eventually a critical embryo size is reached beyond which the free energy decreases with increasing size, because the surface-to-volume

Table 11.2. Expressions for Supersaturation and Supercooling
(See Fig. 11.1)

Name	Expression
Supercooling or undercooling	$\Delta T = T_e - T \geqslant 0$
Reduced temperature	$T_r = T/T_e \leqslant 1$
Fractional undercooling	$\Delta T/T_e = 1 - T_r \geqslant 0$
Supersaturation	$\Delta C = C - C_e \geqslant 0$
Saturation ratio or supersaturation ratio	$\alpha = C/C_e \geqslant 1$
Fractional supersaturation	$\sigma = \Delta C/C_e = \alpha - 1 \geqslant 0$

ratio has diminished sufficiently to overcome the influence of the surface energy. Beyond this critical size the crystal grows spontaneously and is then called a *nucleus*. The critical embryo size decreases rapidly as the supercooling increases. The nucleation rate depends so strongly on critical size, and hence on supercooling, that there is effectively a critical supercooling for nucleation beyond which nucleation is very rapid and below which it is negligibly slow. At very high supercoolings, transport in melts is so slow that nucleation again becomes negligible and amorphous solids may be formed.

In nearly all real systems, homogeneous nucleation as described above does not occur. Nucleation usually occurs on very small particles (motes) or on the container walls. Less supercooling is required for *heterogeneous nucleation* than for homogeneous nucleation. It is predicted that the ability of a foreign particle to cause nucleation increases as (1) the surface energy between the particle and the crystal embryo approaches the surface energy between the particle and the solution; (2) the size of the particle increases; and (3) the roughness of the particle increases. In practice it is exceedingly difficult to remove all nucleating particles and surfaces from a fluid.

A few materials nucleate with such difficulty that procedures in addition to supercooling are required to initiate crystallization. Nucleation may sometimes be induced mechanically (dynamic nucleation) by friction, by shock waves, by ultrasonic vibration, and by high-speed fluid motion. Cavitation appears to be necessary for dynamic nucleation, as well as a minimum supercooling (21). An electric field increases the nucleation of some materials (17). Foreign particles are sometimes deliberately added, as in cloud seeding. When all else fails, seed crystals of the same compound must be added.

Often the ease of nucleation is observed to depend strongly on the previous history of the fluid. For example, heating to higher temperatures frequently increases the supercooling necessary for spontaneous nucleation. It is thought that heterogeneous nucleation sites are destroyed or dissolved during heating.

Nucleation of new crystals is accelerated by the presence of existing crystals (44). This phenomenon is called *crystal breeding* or *secondary nucleation* and is thought to be the primary nucleation process in stirred suspensions of crystals. Secondary nucleii can come from a crystalline dust washed from the surface of the seed when it is first introduced into the solution (*initial* or *dust breeding*), from break up of an agglomerated polycrystalline seed, or from breakoff of dendritic or needle protuberances (*dendritic breeding*). Crystals may be fractured into pieces by collision with one another or with a stirrer or container walls, or by ultrasonic agitation (*collision* or *attrition breeding*) (33). Nucleii are also produced by collisions even when no visible particles are removed from the seed crystal and no damage to the seed crystal is apparent (*contact nucleation*) (29). Pressure on the crystal by another crystal or by a hard object similarly produces nucleii. At low supersaturations, the number of $MgSO_4 \cdot 7H_2O$ nucleii produced is proportional (1) to the energy of impact or to the pressure, (2) to the supersaturation, and (3) to the 0.6 power of the contact area (at constant total impact energy) (29). Approximately 15 sec is required between contacts for the number of nucleii to reach the maximum. The number of nucleii increases as the microscopic roughness of the surface increases and as the angle between the contacting surfaces increases. This author believes that submicroscopic pieces are broken off at atomically large steps, irregularities, and protuberances.

When an impurity both inhibits nucleation and is strongly incorporated into the crystal ($K_{eff} \gg 1$), fluid adjacent to a growing crystal is depleted into the crystal, fluid adjacent to a growing crystal is depleted in that impurity and nucleation occurs there (*impurity concentration gradient nucleation*) (11). One might expect similar behaviour when an impurity enhances nucleation and is strongly rejected by the growing crystal.

11.22 GROWTH KINETICS

The rate of growth of an existing crystal is determined by interfacial processes and by transport processes. We discuss the interfacial processes first and the transport processes in Section 11.43. Many crystals grow with a surface that is atomically flat except for the presence of a few steps (Fig. 11.2). Molecules that strike the interface and are adsorbed are not immobilized until they

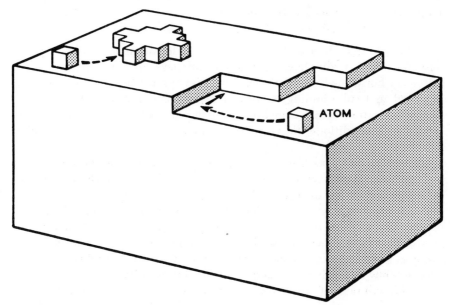

Figure 11.2. Growth of a perfect crystal. An incomplete layer and a two-dimensional nucleus are shown.

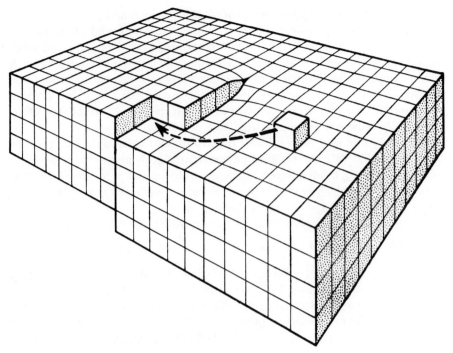

Figure 11.3.   Growth on a surface intersecting a screw dislocation. (A spiral ramp, or growth spiral, rapidly forms.)

migrate to a step. On a perfect crystal, molecules continue to add to a step until the layer is complete. In this way all steps grow themselves out of existence. For growth to continue, a new layer must be formed by several adsorbed molecules joining to form a two-dimensional nucleus. Thus the rate of formation of two-dimensional nucleii often limits the growth rate. Under such conditions the growth rate increases rapidly as supercooling increases.

In reality, most crystals are not perfect but contain a special type of imperfection called a *screw dislocation* (Fig. 11.3). Because of the presence of the screw dislocation the growth layer is never completed but instead forms into a spiral ramp with the dislocation at the center (14, 44). Thus it is unnecessary to form new layers, and growth is continuous. Although the growth rate again increases with supercooling, the dependence is not as strong as when two-dimensional nucleation limits growth. Cracks and foreign particles may similarly act as layer sources to form spirals.

The growth rates of different crystal faces are not equal. These differences in rate determine the crystal habit (shape), with the slowest-growing faces being the largest. The relative growth rates of different faces often change with supersaturation, so that the habit depends on supersaturation. Even if all faces of a crystal are crystallographically identical, they frequently do not grow at the same rate, because the number and/or size of screw dislocations varies from face to face. Thus, even in a group of crystals grown from the same solution, the crystal habit may vary widely.

In agitated crystal suspensions the McCabe ΔL law is often valid. This law states that "all geometrically similar crystals of the same material suspended in the same solution grow at the same rate, if the growth is measured as the increase in length of geometrically corresponding distances on all the crystals." In other words, the growth rate of a particular face is independent of crystal size. Although such behavior is usually observed. several exceptions have been noted, generally showing more rapid growth of larger crystals.

*Twinned crystals** occasionally occur and exhibit other growth mechanisms. Twinning may lead to formation of a re-entrant angle where the two portions meet. The re-entrant angle serves as a continuous source of growth steps. (Re-entrant angles and accelerated growth sometimes also occur when two crystals come into intimate contact and grow together.) Twinned KBr crystals growing from aqueous solutions form a surface with an angled step resembling a screw dislocation (54). This pseudo-screw dislocation acts as a source of growth steps, which coalesce to form waves moving down the sides of the crystal.

The surfaces of some growing crystals are not smooth as in the foregoing description but are rough on the atomic scale. This appears to occur primarily during melt growth of materials having low entropy of fusion (some metals and the plastic-crystal organics composed of nearly spherical molecules) (28). Atoms or molecules can add to a rough interface without the presence of a step. Since stepwise growth does not occur, the crystal does not form facets (planar faces). Normally faceted crystals tend to become nonfaceted when grown in a sufficiently steep temperature gradient (53).

11.3 Separation by Crystallization

The driving force for separation by conversion to a solid is the difference in equilibrium composition between the solid and the fluid. It should not be concluded from this, however, that crystals are necessarily in equilibrium with the fluid from which they are formed. In fact true equilibrium is seldom attained in a crystallization operation. The reason is that mass transfer is very slow in solids because of the lack of convection and the very small

---

* A twinned crystal consists of two or more parts oriented with respect to one another according to well-defind crystallographic laws.

diffusion coefficients. In the extreme, the solid is removed as effectively from the fluid as is vapor from a liquid during a batch evaporation. Thus the Rayleigh equation is often obeyed; for segregation of an impurity this is written as*

$$C_s = K_{eff}C_0(1 - g)^{K_{eff}(\rho_f/\rho_s) - 1} \tag{11.2}$$

Here C_s is the concentration of impurity in the solid at the point where mass fraction g of the fluid has solidified, C_0 is the original impurity concentration in the fluid, K_{eff} is the effective distribution coefficient between the solid and the bulk fluid (C_s/C_f), and ρ_f and ρ_s are fluid and solid densities. We see later that mass transfer in the fluid and adsorption on the surface of the growing solid lead to nonequilibrium between the solid and the fluid. Even after crystallization ceases, the approach to equilibrium is slow and becomes increasingly negligible as the crystal size increases.

11.31 CRYSTALLIZATION FROM SOLUTION

Crystallization from solution finds wide application to the separation and purification of inorganic salts and organic compounds. In a mixture the solubilities of the various components in a particular solvent differ. This permits separation of one component from a solution by crystallization. As crystallization proceeds, however, another compound also becomes saturated and begins to crystallize as well. Thus the amount of crystallization must be limited, or the purity of the product is reduced. Separation of more than one compound from the mixture requires special measures such as alternate crystallizations from two solvents.

In crystallization from solution, discrete crystals or clumps of crystals are produced either by evaporation of solvent or by temperature changes. Vacuum evaporation both removes solvent and cools the solution. Deliberate temperature gradients are usually not imposed. If only one crystal is present, the optimal separation is usually obtained with slow growth, rapid stirring, and high solubility. Since solubility increases with temperature, operation at as high a temperature as possible is desirable. In most practical crystallizations, however, many crystals are growing simultaneously. For reasons that are explored later, separation (or purification) is a function of crystal size. The maximum separation is sometimes obtained for very small crystals, sometimes for very large crystals, and sometimes for intermediate-size crystals. Since nucleation and growth both depend on supercooling, it is seen

---

* Valid for constant K_{eff}, partial mixing of the fluid, and no diffusion in the solid. For solution growth some authors define compositions and distribution coefficients in terms of ratios, such as moles of impurity/moles of solute or impurity/solute + impurity. Expressions differing slightly from Eq. 11.2 are then found.

that the average crystal size depends on supercooling. For many systems larger crystals are produced by slow growth (low supercooling), whereas for other systems rapid growth produces larger crystals. Stirring facilitates the mass transfer in the fluid and usually improves the separation, but may also enhance nucleation. Certain additives have a beneficial effect on some crystallizations by inhibition of nucleation and by modifying the habit of the crystals, but other impurities are deleterious.

It may be seen from the foregoing that optimal crystallization conditions really cannot be predicted. If the same separation is to be performed many times, the optimal conditions should be determined experimentally. If a one-time separation is desired, it is recommended that crystallization be carried out slowly with gentle stirring at as high a temperature as practical.

The kinetics of crystallization varies widely from material to material. Advantage may sometimes be taken of this fact to separate two compounds, both present in concentrations near saturation. Thus an aqueous solution saturated with borax and potassium chloride crystallizes only KCl on rapid cooling. The borax remains behind as a supersaturated solution, and the KCl crystals can be removed before the slower borax crystallization starts. Sometimes seeds of one component can be added to cause preferential crystallization of that component before others can nucleate appreciably.

When impurities are rejected during crystallization, they are concentrated in the mother liquor, especially in the fluid adjacent to the crystals. The impurity content both in the mother liquor and in the crystals increases as crystallization proceeds (Eq. 11.2). The bulk of the mother liquor may be removed either by filtration or by decantation with possible assistance by centrifugation. Some mother liquor remains on the surface of the crystals and is usually removed by rinsing. Clean solvent is sometimes used for rinsing; or, if the crystals melt at a reasonable temperature, a portion of them may be melted and used for this purpose.

The ability to filter and rinse crystals depends strongly on crystal size and shape. Larger crystals are much better from this standpoint. Crystal habit may be strongly influenced by certain additives, but it should be noted that additives which modify habit are incorporated into the crystal, often markedly so (12). Thus habit may be greatly improved by a certain additive, but one must be willing to tolerate the presence of the impurity in the product.

11.32 PRECIPITATION

Precipitation is actually a special type of crystallization from solution in which an insoluble compound is produced by addition of a reagent to a solution containing a mixture, for example, of ions (22). Thus it is used to separate one or more of these ions from the mixture. As a consequence of the low solubilities and the mixing of solutions, enormous supersaturations are momentarily produced. Nucleation is rapid and microscopic crystals result. Such minute particles are usually undesirable, both because the potential separation is not fully realized and because they are difficult to filter. Particle size may often be increased by adding the precipitating reagent (1) in a more dilute form, (2) at a slower rate, (3) with more or less agitation, (4) at a higher temperature, or (5) by release in a slow, homogeneous chemical reaction. Once the particles are formed, their size can be increased by Ostwald ripening, whereby the precipitate and the mother liquor are held at an elevated temperature for a long period. Because of their larger surface energy, smaller particles have an increased solubility and tend to dissolve while larger crystals grow at their expense.

Postprecipitation is the slow precipitation of a second insoluble substance after the formation of the primary precipitate. There are several possible mechanisms for this behavior. Nucleation of the secondary precipitate may either be slow or be dependent on heterogeneous nucleation on the primary precipitate. Contamination by postprecipitation increases with time, reaching a constant value when the concentration of the precipitating impurity reaches its solubility product.

11.33 FRACTIONAL SOLIDIFICATION

In recent years the applications for fractional solidification have increased enormously. Most materials that melt without decomposition and with low to moderate vapor pressures have been purified by zone melting and related techniques. The list includes organics, salts, metals, semiconductors, and water.

Fractional solidification has the advantage over fractional crystallization in that no solvent is required. Solvents are considered to be impurities for many applications, and some contamination by solvent is almost unavoidable in crystallization. The disadvantage of fractional solidification is that higher temperatures are required than when a solvent is employed. Not only is it difficult to heat to the melting point of some refractory materials, but also the melts themselves are so reactive that suitable containers are not available. (We shall see shortly, however, that containers are not always necessary.)

Fractional solidification will now be considered in more detail. If a melt is slowly cooled without imposing a temperature gradient, many discrete crystals are usually formed. Sometimes the entire melt is frozen and then slowly melted with drainage. Consider, for example, the freezing of saline water. Ice crystals are surrounded by brine, which is also frozen when the temperature is lowered sufficiently. Upon raising the temperature, the brine melts first and drains away. The surfaces of the ice crystals melt next and rinse the remaining brine from the crystals. Many industrial processes have been based on such process of crystallization, melting, and rinsing (see Section 11.35). The nearest approximation for laboratory purifications has been slow freezing of organics or aqueous solutions in a large bottle or flask immersed in a cooling bath (39). Freezing progresses from the outside in. Usually the remaining melt is poured out. Subsequent partial melting is accomplished, for example, by placing a small heat source, such as a light bulb, in the interior from which the melt is drained.

Progressive freezing may be controlled by placing the melt in a boat or cylindrical container and moving the container slowly out of a heater and into a cooler. We see later that a sufficiently large temperature gradient leads to a smooth freezing interface without formation of discrete crystals. Such controlled solidification produces a larger separation than formation of discrete crystals, although partial melting is now of no advantage. The concentration profile obeys Eq. 11.2.

Continuous and multistaged progressive freezing methods have been devised for both laboratory and for industrial use (56).

11.34 ZONE MELTING

In crystallization, precipitation, and progressive solidification one may not always attain the desired purification in a single state. It is then necessary to repeat the process, that is, to redissolve the crystals or to remelt the progressively frozen solid (after discarding one end or the other) (see Chapter 4). If yield is important the multistaged processes may be arranged in a crosscurrent scheme. However, this is laborious, and the extensive handling makes chance contamination more likely. For fusible substances, zone melting has proved to be very effective for multistaged solidification. As shown in Fig. 11.4, only a narrow zone is melted and moved through the solid. Most impurities are preferentially carried along with the zone, while a few move in the reverse direction. (When several impurities with unknown distribution coefficients are present, it is advisable to discard both ends of the zone-refined ingot.)

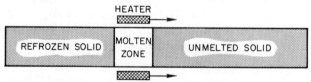

Figure 11.4. Zone melting with a single zone. More rapid separation is obtained by forming many molten zones in the solid. Reprinted from Ref. 56, by courtesy of Marcel Dekker, Inc.

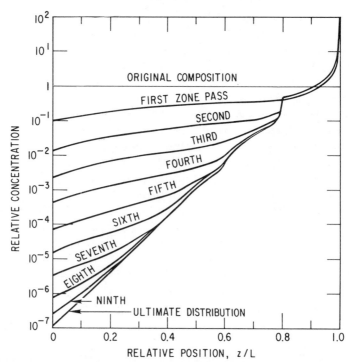

Figure 11.5. Calculated concentration profiles for zone melting with $K_{\mathrm{eff}} = 0.1$, $l/L = 0.2$, and $\rho_s = \rho_f$. Reprinted from Ref. 56, by courtesy of Marcel Dekker, Inc. See also Ref. 13.

The concentration profile for a single-zone pass of a long rod is given approximately by*

$$\frac{C_s}{C_0} = 1 - (1 - K_{\mathrm{eff}}) \exp\left(- K_{\mathrm{eff}} \frac{z}{l} \frac{\rho_s}{\rho_f}\right) \tag{11.3}$$

where C_s is the impurity concentration in the solid at distance z from the front of the ingot, and l is the zone length.

Many molten zones can be moved through the solid not only sequentially but also simultaneously without additional handling. Examples of concentration profiles are shown in Fig. 11.5. After approximately $2(L/l) + 1$ zone passes, little further separation occurs (L = length of solid rod). The ultimate concentration profile is given approximately by

$$\frac{C_s}{C_0} = Ae^{Bz} \tag{11.4}$$

* Exact equations are available but are much more complex. See Ref. 13 for an example.

where the constants A and B are given by

$$K_{\text{eff}} = \frac{Bl}{e^{Bl} - 1} \tag{11.5}$$

and

$$A = \frac{BL}{e^{BL} - 1} \tag{11.6}$$

The ultimate separation increases as the zone size becomes smaller and as K_{eff} deviates from unity.

The rate of separation increases as the number of zones present at one time increases. Thus the optimum condition is to have as many zones present as possible. The number of zones is limited by the minimum size for a zone, which is about equal to the container diameter because of heat transfer limitations. Under some experimental conditions only a single zone can be generated in the ingot at any time. With a single zone the maximum rate of separation is attained by making the first zone as large as possible; this means progressive freezing. Subsequent zones are gradually reduced in size to attain the maximum possible ultimate separation.

The density of the liquid ρ_f may be either greater or less than the density of the solid ρ_s. For a material with $\rho_s > \rho_f$ in a cylindrical container, the container may be ruptured during formation of a zone unless allowance is made for expansion. With organics this may be accomplished by starting the zone at a Teflon plug, which the expanding melt can force down the tube. A moment's reflection reveals that the density difference causes each zone to move a small amount of material relative to the container. In a horizontal open boat this may be avoided by tilting the boat. It is sometimes advantageous to tilt the boat sufficiently to cause a portion of each zone to overflow when it reaches the end of the boat. This constitutes an automatic cropping of what is usually the most impure portion of the ingot and thus increases the possible purification, although the yield is reduced.

As mentioned earlier, suitable zone-melting containers are not available for some high-melting metals and salts. Even moderately high-melting materials (~ 500 to $1500°C$) may be contaminated by containers. The floating-zone-melting technique was devised to avoid these problems. The rod is held vertically, and the zone is held by surface tension. Magnetic suspension of metallic melts has been proposed, as well as growth in the microgravity environment of an earth satellite. The zone may be heated by an electron beam (7), by induction heating, or by light from an infrared lamp or an arc. Although volatile materials have been "float zoned," this is not generally practical. The evaporation rate of slightly volatile materials (vapor pressure $\sim 10^{-4}$ to 10^{-1} torr) may be controlled by operation in

1 atm or more of inert gas. Nonvolatile materials are most advantageously zoned in a vacuum, since this permits additional purification by evaporation of volatile impurities. In fact, purification of refractory metals by electron-beam floating-zone melting generally takes place more by outgassing than by segregation in the solid-liquid transformation.

A variety of continuous zone-refining schemes have also been devised (56).

11.35   COLUMN CRYSTALLIZATION

In column crystallization crystals are moved countercurrent to the melt, at least during a portion of the cycle. In one technique crystals are forced down in a vertical column by means of a rotating helix (8, 56). The bottom of the column is maintained at a slightly higher temperature than the top. In one industrial method crystals are formed by chilling and are fed into the top of a column (56). A heater melts the crystals at the bottom, while either a piston at the top or a pulse pump at the bottom periodically forces melt up countercurrent to the crystals. Impure melt is taken off at the top of the column, and purified melt is removed from the bottom. In another method the melt is placed into a toroid (doughnut shape) (38). Stationary cold fingers on alternate sides of the tube cause islands of solid to form. Upon rotation the fluid is forced countercurrent to the solid.

Column crystallization may be operated in either batch or continuous modes. Separation appears to be much more rapid than with zone melting, although the apparatus required is more complex and successful operation is more difficult to achieve.

11.4   Phenomena Influencing Separation

In theoretical treatments of fractional crystallization the effective distribution coefficient K_{eff} is usually taken as constant for simplicity. In fact, however, K_{eff} varies somewhat during a given crystallization process. Some of the many phenomena influencing K_{eff} are discussed in the following sections.

11.41   SOLID-LIQUID EQUILIBRIA

A wide variety of solid-liquid phase equilibria are encountered, including complete solubility of both phases, compound formation, and limited liquid miscibility. Metal phase diagrams are covered in Refs. 5 and 23, salts in Refs. 6 and 34, organics in Ref. 6, and solubilities in solvents in Refs. 6, 35, and 43. Some variety of the limited solid solubility phase diagram (Fig. 11.6) is often encountered (see Section 2.12). This eutectic diagram

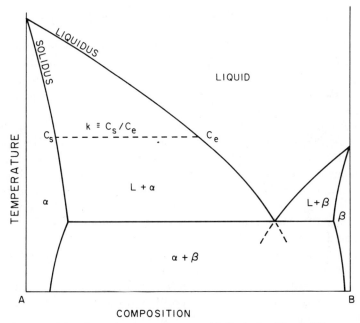

Figure 11.6. Eutectic phase diagram with limited solid solubility.

is really the superposition of two liquid solubility curves, one for solid A in solvent B and one for solid B in solvent A. The intersection is the eutectic point. As a point of interest, if one phase nucleates with difficulty, it is actually possible to follow the solubility curve of the other component for some distance below the eutectic point.

The solidus of most published phase diagrams is in error, for several fundamental reasons. Most significantly, equilibrium between solid and liquid is difficult to obtain, as indicated previously. Phase diagrams showing no solid solubility are most frequently wrong. Most solid-liquid phase diagrams are determined by thermal analysis, which is incapable of revealing low solid solubility. Indeed, formation of a crystal followed by chemical analysis appears to be the best means for determining the solidus.

It has occasionally been claimed on the basis of phase diagrams showing no solid solubility that total purification is possible by crystallization. The possible existence of zero solid solubility has frequently been debated, but cannot be proved or disproved by thermodynamic arguments. Whenever sufficiently sensitive methods of analyses have been employed, some solid solubility has always been revealed. It seems reasonable that, if two components are sufficiently similar to exhibit liquid miscibility, then they will also exhibit some solid solubility. It is very unlikely, therefore that total purification can be obtained by any crystallization operation, although the solid solubility may be so small ($K < 0.01$) that a satisfactory separation in one step is possible.

Generally, solid solubility increases with increasing temperature. Thus, for example, $NaNO_3$ and KNO_3 form a continuous series of solid solutions at high temperatures, so that a single solidification of their molten mixtures brings about only a small separation. However, at room temperature the solid solubility is negligible; hence crystallization from an aqueous solution can effect almost total separation. Solid solubility also tends to increase as the size, shape, charge, and other characteristics of the impurity approach those of the host crystal. When more than one impurity is present, strong interactions may occur between two impurities either in the fluid or in the solid to cause a significant change in the equilibrium distribution constant K. Benefit can sometimes be taken of such interactions by adding a "getter" or "scavenger," which drastically reduces K for the impurity. For example, metal contamination could be removed from anthracene by zone melting only when 0.1 wt % of the complex agent 7,7,7,8-tetracyanoquinodimethane was added (42).

In systems with a minimum liquidus temperature, such as a eutectic system, the fluid approaches the composition at the minimum as crystallization proceeds. Thus the presence of a minimum represents a limitation on the amount of separation or yield that can be achieved. When the minimum is a eutectic, however, separate crystals of the components are produced, although the crystals are small and thoroughly mixed. These crystals can be separated by various techniques, such as electrostatic deflection while falling (4), electrophoresis, or flotation. Alternatively, the composition of a binary eutectic can be shifted by addition of a solvent (Fig. 11.7). Since

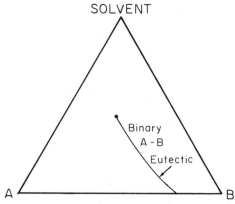

Figure 11.7. Ternary-phase diagram illustrating shift of binary eutectic composition by addition of solvent.

this shift depends on the amount of solvent added, crystallizations may be carried out alternately, first at one solvent concentration and then at another so as to ultimately separate both components from the mixture.

The equilibrium may be shifted even more strongly by addition of a material forming a compound or adduct preferentially with one component. Precipitation of an ion by production of a low-solubility compound has already been discussed. Organics may similarly be separated. For example, fully halogenated methanes, such as CCl_4 form molecular compounds with various aromatic compounds. In these addition complexes the added material crystallizes with a structure containing open spaces that preferentially incorporate one compound for steric and chemical reasons. Thus stoichiometric compositions are not always observed. Examples of adducts are naphthalene, phenol, and fatty acids in desoxycholic acid; quinol clathrate complexes with H_2S, SO_2, methanol, argon, krypton, and xenon; the nickel–cyanide–ammonia complex with benzene; gas hydrates (useful for desalination); urea complexes with straight-chain or nearly straight-chain compounds; and thiourea complexes with highly branched hydrocarbons. Such adducts can be decomposed by heating and/or dissolution in a suitable solvent (see Section 15.3).

11.42 INTERFACIAL DEPARTURES FROM EQUILIBRIUM

As noted earlier, equilibrium is seldom attained in crystallization. Some impurities adsorb strongly on the surfaces of the crystal. During growth these impurities are covered by layers before they can escape. At high growth rates, the concentration of impurity in the solid approaches that of the adsorbed layer. Since the number of surface sites for adsorption is limited, these adsorption effects are much stronger at low impurity concentrations. Thus, for example, when KCl crystallizes slowly from water containing $PbCl_2$, $K_{eff} = 2700$ for solutions containing 10^{-8} mole $PbCl_2$/mole KCl, 28 for 10^{-5} mole $PbCl_2$/mole KCl, and 1 for 10^{-3} mole $PbCl_2$/mole KCl (10). For ionic crystals the adsorption of anions or cations may depend strongly on pH, as well as on the concentrations of other anions and cations. For very small crystals ($\leqslant 1000$ Å), such as might result from precipitation, the surface molecules make up a large proportion of the molecules of the crystal. Thus adsorbed species will be present in large amounts, even without trapping by growth.

Adsorbed impurities may also inhibit the growth of a crystal face, particularly if the adsorbed species is much larger than the native crystalline species or carries a different charge. The degree of adsorption and inhibition varies from face to face, so that habit changes are produced. The concentration of trapped impurity likewise may vary markedly from one face to another of a different type.

Aside from surface adsorption effects, one can also imagine that the ease of incorporation of different species would be different, so that k would depend on the growth rate. This would be particularly likely, for example, with organics, where odd-shaped molecules must be oriented properly to fit into the crystal lattice. Experimental indications (3, 16) of these kinetic departures from equilibrium have been attributed (57) to dendritic trapping (p. 329).

11.43 MASS TRANSFER IN CRYSTAL GROWTH

There are two basic mass transfer problems to consider. (1) In growth from solution or vapor, solute must be transported from the bulk fluid to the crystal surface. This mass transfer influences the growth kinetics and in some cases may be limiting. (2) If an impurity is being rejected by the growing crystal in solution or melt growth, it must diffuse back into the fluid. Thus the concentration of the compound being incorporated into the crystal is decreased at the crystal surface. The concentration of a rejected impurity, or of solvent, is increased at the crystal surface over the bulk value. Both transport problems may be treated by the same mathematical methods. The stagnant film model for mass transport is employed and gives the correct results except for very high crystallization rates.

If we regard the crystal surface as the plane of reference, then it is important to realize that crystal growth generates a net flow of fluid toward the crystal surface. We call this the *crystallization flow*; it is related to the rate of crystallization V_c by the partial molar volumes in the solid and fluid (58). The influence of the crystallization flow has frequently been incorrectly assumed to be negligible in solution growth.

If we first consider solute transport in solution growth, the growth rate is found to be

$$V_c = \frac{D}{\delta} \frac{\rho_{sf}}{\rho_s} \ln \left[\frac{C_s(\rho_{sf}/\rho_s) - C_i}{C_s(\rho_{sf}/\rho_s) - C_f} \right]$$

$$\approx \frac{D}{\delta} \frac{\rho_{sf}}{\rho_s} \left[\frac{C_f - C_i}{C_s(\rho_{sf}/\rho_s) - C_i} \right]$$

(11.7)

for $\delta V_c \rho_s / D \rho_{sf} < 0.1$. Here D is the diffusion coefficient of the solute in the fluid, δ is the film thickness (which decreases as stirring is increased), C_s is the molar concentration in the solid of the same species used for D, C_f is the solute fluid concentration, C_i is the solute concentration in the fluid at the interface, and ρ_{sf} is the density the solid would have if its components had

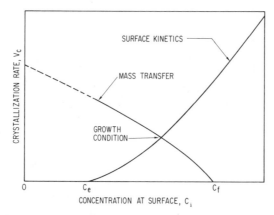

Figure 11.8. Mass transfer kinetics and interface attachment kinetics as a function of interfacial concentration in the fluid. At steady state the two rates must be identical. Reprinted from Ref. 55, by courtesy of Marcel Dekker, Inc.

the same partial molar volumes that they have in the fluid (58). When crystallization flow is incorrectly deleted, the growth rate is found to be given by the second expression, except that C_i no longer appears in the denominator. When interfacial processes are rapid compared to mass transport, then $C_i = C_e$, the equilibrium solubility at the growth temperature. In general, however, this is not true and C_i is unknown. To find the growth rate we must equate the mass transport rate to the interface kinetics, both of which depend on C_i as shown in Fig. 11.8. The intersection of the curves gives C_i and V_c. Changes in stirring move the mass transfer curve up and down with one end fixed at $C_i = C_f$ (zero growth). A near-vertical mass transfer curve represents surface kinetics control of the rate. Similarly, a near-vertical surface kinetics curve (fixed at C_e) represents mass transfer control of the growth rate.

For impurity segregation we assume that the growth rate is known and calculate the effective distribution coefficient (58):

$$K_{\text{eff}} = \frac{C_s}{C_f} = \frac{K_i}{K_i(\rho_{sf}/\rho_s) + [1 - K_i(\rho_{sf}/\rho_s)] \exp\left[-(\delta V_c/D)(\rho_s/\rho_{sf})\right]} \quad (11.8)$$

where $K_i = C_s/C_i$ is the interfacial distribution coefficient. When there is equilibrium at the interface, $K_i = K$ (the equilibrium distribution constant), and K_{eff} varies with freezing rate as shown in Fig. 11.9. Under such conditions it can be shown that the freezing rate producing the most rapid purification at the beginning of zone refining is given by $V_c \approx (D/\delta)(\rho_{sf}/\rho_s)_r$. If $\rho_s = \rho_{sf}$, $D = 10^{-5}$ cm²/sec, and $\delta = 0.4$ mm, then the optimum zone travel rate is ~ 1 cm/hr. To approach the maximum ultimate separation $K_{\text{eff}} \approx K$, that is, the travel rate of the last few zones should be reduced.

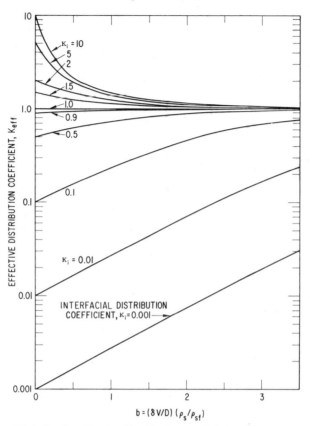

Figure 11.9. Variation in effective distribution coefficient with crystallization rate V_c and interfacial distribution coefficient K_i. Reprinted from Ref. 56, by courtesy of Marcel Dekker, Inc.

In general, however, we do not have equilibrium at the crystal surface, so that $K_i \neq K$. For example, if surface adsorption is significant, then K_i may become large and > 1 at high values of V_c even though $K < 1$, so that K_{eff} may show a maximum versus V_c, although this has apparently never been reported. When there is not equilibrium at the interface, more insight into impurity segregation can be gained by writing the result as

$$K_{\text{eff}} = \left(\frac{\rho_s}{\rho_{sf}}\right) \frac{1 - (C_i/C_f) \exp\left[-(\delta V_c/D)(\rho_s/\rho_{sf})\right]}{1 - \exp\left[-(\delta V_c/D)(\rho_s/\rho_{sf})\right]} \qquad (11.9)$$

We may also write, by use of the definitions of K_i and K_{eff} that

$$K_{\text{eff}} = K_i \frac{C_i}{C_f} \qquad (11.0)$$

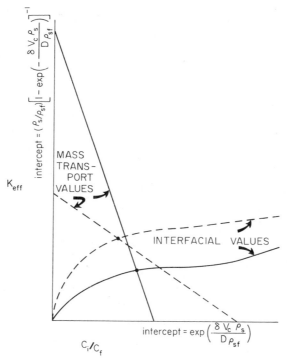

The y-axis is labeled K_{eff}, with $\text{intercept} = (\rho_s/\rho_{sf})\left[1 - \exp\left(-\dfrac{\delta V_c \rho_s}{D \rho_{sf}}\right)\right]^{-1}$

The x-axis is labeled C_i/C_f, with $\text{intercept} = \exp\left(\dfrac{\delta V_c \rho_s}{D \rho_{sf}}\right)$

MASS TRANS- PORT VALUES

INTERFACIAL VALUES

Figure 11.10. Effective distribution coefficient K_{eff} as a function of the interfacial concentration C_i (for constant bulk fluid concentration C_f). The observed distribution coefficient is given by the intersection of the curve for interfacial processes with that for mass transport. ——— $(V_c)_1$, ------ $(V_c)_2 = 3(V_c)_1$.

for interfacial departures from equilibrium. For a given growth rate, K_i is a function of C_i. Thus both Eq. 11.9 for mass transport and Eq. 11.10 for interfacial process are plotted in Fig. 11.10. (If equilibrium were attained at the interface, the curve for Eq. 11.10 would be a straight line through the origin of slope K, assuming constant K.) The predicted value of K_{eff} is given by the intersection of the mass transfer and the interfacial curves. An increase in freezing rate V_c causes the mass transfer curve to fall at the $C_i = 0$ intercept and to move to the right at the $K_{eff} = 0$ intercept. As V_c increases further, the $C_i = 0$ intercept approaches unity, while the $K_{eff} = 0$ intercept increases more and more rapidly. If adsorption is entirely responsible for deviations from equilibrium at the interface, then the interface curve also increases with increasing V_c and approaches a limiting curve for which C_s equals the surface adsorbed concentration. If kinetic departures from equilibrium occur, then the interfacial curve may shift either up or down without bounds with increasing V_c. If V_c is kept constant and δ is altered by changing the stirring, then the mass transfer line shifts as before, while the interface curve is unchanged. Thus changes in V_c and δ do not produce the same changes in segregation.

11.44 CONVECTION

We have just seen that stirring has a strong effect on crystal growth kinetics and on separation. In a gravitational or centrifugal field some stirring is caused by variations in density in the fluid. This is called *free convection* or *natural convection*. The density variations arise from temperature gradients due to heat transfer and to concentration variations due to the crystal growth itself. In zone melting, free convection is often the primary stirring mode. If the density differences are due primarily to temperature differences, convection is more vigorous at the upper interface because a density inversion occurs there. When this is true, and when no bubble is present, the freezing interface should be on top, that is, the zone should be moved down. If the density difference is due primarily to concentration differences, either upward or downward motion may be best. In ultrapurification, concentration differences are small, so that downward motion is best unless bubbles are present. Free convection may be considerably enhanced over that due to the gravitational field by winding the zone-melting tube into a helix and rotating rapidly about the axis (2).

Free convection in horizontal zone melting or solidification also leads to a phenomenon known as *gravity segregation*, in which the composition of the solid varies with vertical position. This arises as follows. The convection current either rises or falls at the vertical interface, depending on the density gradient. As the fluid moves past the interface, crystal growth takes place and changes the composition of the fluid. Thus the fluid composition varies with position, causing the solid composition to vary as well. The solidification behavior of certain eutectic mixtures may be related. It has been found that solidification of a eutectic in which one solid phase is much more dense results in that phase being concentrated at the bottom (56). Repeated melting (without stirring) and solidification enhance the effect. It is possible that separate crystals are formed, permitting the denser crystals to sink.

Convection may also be brought about by mechanical stirring, which may take the form of a rotated paddle, propeller, or magnetically coupled rod. The container may be rotated alternately in both directions. Slow rotation in one direction is effective if the containing tube is horizontal and a gas space is present over the melt. A gas may be bubbled through a liquid. With electrically conducting fluids, electromagnetic stirring may be employed. This may consist either of a rotating magnetic field or passage of a current through the melt in the presence of a magnetic field that varies along the melt. Ultrasonic agitation is sometimes employed, although nucleation may be increased thereby (1, 30). Gases are more soluble in melts than in solids, so that bubbles are generated at the crystal surface, especially during the first zone-melting pass. These bubbles tend to gather at the upper interface, where they influence mass transfer. A large gas layer may permit drainage from a polycrystalline feed or preferential transport from a mixture with one or more volatile components. Gas may be incorporated in the solid as a hollow core.

The absence of deliberate stirring and a gravitational field do not guarantee the absence of convection. If a gas-liquid interface exists, surface-driven flow is generated as follows. Surface tension depends on concentration and on temperature. As temperature increases, surface tension decreases, while ionic salts raise the surface tension of water. When surface tension varies, a flow is generated from areas of low surface tension to areas of high surface tension. Thus, when bubbles are near the interface, vigorous circulatory or oscillatory convection may result. These bubbles also experience a net force in the direction of decreasing surface tension, which may drive them either to or away from the crystal.

In solution growth, changes in stirring cause changes in growth rate through the effect on mass transport, as indicated by Fig. (11.7). In controlled solidification, changes in stirring cause changes in heat transfer which also produce fluctuations in the growth rate. It has been found in many crystal growth processes that irregular free convection occurs. It appears that free convection is laminar at low density gradients and turbulent at high gradients. At intermediate gradients in closed systems, oscillatory and fluctuating convection frequently occur, resulting in a fluctuating growth rate and a fluctuating composition. The separation is considerably less than would be predicted from Fig. 11.9 for the same average growth rate V_c without fluctuations (27). Compositional fluctuations are also produced by any mechanical or thermal fluctuating process during crystal growth. Thus, for example, a vapor space between the molten zone and the solid in zone melting of a volatile organic leads to drops forming on the solid and dripping back into the melt to cause temperature fluctuations (49).

Irregular free convection as described above can be eliminated in controlled solidification by lowering the temperature gradients in the melt, by lowering the temperature difference between heaters and coolers, or by installing heat shields and insulation around surfaces losing heat. Convection in conducting melts can be controlled by superimposing a transverse magnetic field of moderate strength.

11.45 CRACKING

In controlled solidification, thermal stresses are generated during cooling, both within the solid itself and between the solid and any container. If growth is slow enough, plastic deformation can take place to relieve the stress. In many organics under ordinary zone refining conditions ($V_c \approx$ 1 cm/hr), cracks are formed in the solid and impure melt is sucked into the cracks for distances of up to several centimeters. Similarly, the solid sometimes pulls away from the container wall and allows melt to flow into the gap.

11.46 DRAINAGE AND RIPENING

In zone melting there is not only a freezing interface, but a melting interface as well. Although the freezing interface has the primary influence on separation, the melting interface also has an effect. In the first pass, the zone traverses a solid rod which is usually formed by casting. The grains of the solid are separated by lower-melting material. Thus, as the zone approaches, this intergranular material melts first and can flow into the zone, especially if the melting interface is located above the melt and separated by a gas space (52). In addition, the large crystallites can grow at the expense of the small (Ostwald ripening). It is not uncommon during the first zone pass to observe relatively large crystals falling from the melting interface to be incorporated in the freezing interface without ever fully melting. The tendency of the system to exhibit such behavior depends on the crystals having a larger density than the melt and is greater for starting materials of lower purity.

11.47 INTERFACE BREAKDOWN DURING CONTROLLED SOLIDIFICATION

Consider a planar solid-liquid interface being forced to move by movement of a superimposed temperature gradient. Because of segregation, the melt composition varies as one moves away from the interface. Since the freezing point depends on composition, the freezing point will increase as one moves into the melt. If the freezing point increases faster than the actual temperature, the supercooling increases with distance from the interface. This is known as *constitutional supercooling* and leads to breakdown of the interface (40). A measure of the intensity of the constitutional supercooling is the gradient of the difference between the actual temperature T and the freezing point T_e. On the basis of these elementary considerations we expect a stable interface only for

$$G_s = \left[\frac{d(T_e - T)}{dy} \right]_0 \leqslant 0 \qquad (11.11)$$

and an increasingly unstable interface as G_s increases. It may be shown that the tendency of the interface to break down increases as

1. the imposed temperature gradient $G = (dT/dy)_0$ decreases;

2. the absolute value of the liquidus slope $|m|$ increases;
3. the freezing rate V_c increases;
4. the impurity content of the melt increases;
5. the separation between solidus and liquidus increases (k deviates from unity);
6. stirring decreases (δ increases).

The variables under the operator's control are G, V_c, and δ.

For nonfaceted interfaces the transition from planar to unstable interface is found experimentally to occur at $G_s = 0$. As the supercooling gradient increases, the interface at first develops pox, next elongated grooved cells, and then hexagonal grooved cells. As supercooling increases further, the grooves become deeper and rods or unbranched dendrites are formed. Eventually side branches develop on these rods and the growth becomes fully dendritic (36). Sometimes nucleation occurs in front of the interface, probably on foreign particles (32). In dendrite growth, dendrite arms may break off to form new crystals in the melt.

The presence of grooves reduces the separation because mass transfer is ineffective, particularly in deep grooves. Thus, in eutectic systems, it is common to find a second phase forming in the grooves even at low impurity concentrations. The presence of dendrites reduces the separation still further (Fig. 11.11). In fractional solidification of organic eutectic systems under dendritic growth conditions, the concentration profiles correspond to Eq. 11.2, with K_{eff} varying with V_c according to Eq. 11.8 (15). The value of K_i calculated therefrom did not represent equilibrium behavior but was related to the "surface void fraction" for the dendrites. The meaning of "surface void fraction" is unclear, although we can imagine that melt is effectively trapped after the first side branch forms on the dendrites. Thus the fraction of the cross-sectional area occupied by melt at that plane represents the fraction of the original melt trapped.

The elementary constitutional supercooling criterion for interface stability is no longer sufficient when faceted growth occurs. When anisotropic growth kinetics and surface energy are accounted for, the interface stability depends strongly on the orientation of the crystal. A planar interface may be greatly stabilized if it corresponds to a slow growth facet and lies on or near an isotherm. On the other hand, if slow growth planes lie at large angles to an isotherm, a planar interface tends to break down and form facets of these planes. Impurity trapping then takes place at corners and edges where these facets meet. Dendritic growth of strongly faceted materials may appear quite different from the regular treelike dendrites formed by nonfaceting materials. Needles, blades, parallel sheets, and other types may be formed. Extensive melt trapping occurs as with normal dendrites.

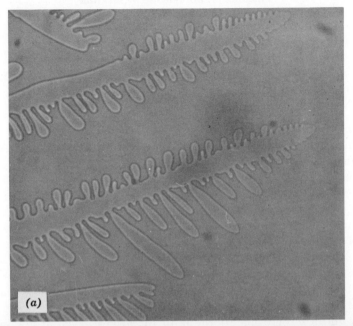

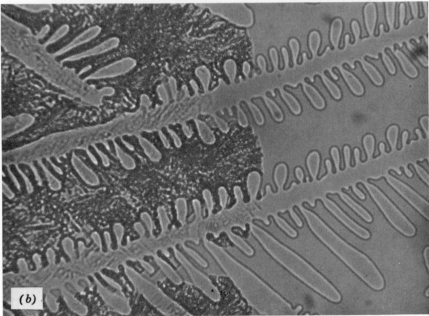

Figure 11.11. Camphor dendrites growing from a mixture of 88 wt % camphor and 12 wt % anthracene at 5.4 cm/hr: (*a*) dendrite tips, (*b*) eutectic solidifying between dendrites. Reprinted from Ref. 52, by courtesy of Marcel Dekker, Inc.

11.48   INCLUSION FORMATION AND REMOVAL

Constitutional supercooling always occurs in solution growth of faceted crystals in a nearly isothermal environment. Because solute is being removed at the crystal surface, the crystallization temperature T_e increases as one moves into the solution, so that with constant T the value of G_s is always positive. Thus it is very difficult to grow crystals from a solution without the interface breaking down to trap solvent (50). See, for example, Fig. 11.12. That faceted crystals can be grown is a tribute to the stabilizing influence of interfacial processes, in which layers move across a face much faster than new layers are formed. Theoretical analysis predicts that the shape of a suspended crystal is stable up to a critical size and unstable above this size. This critical size decreases as the growth rate increases and as the stirring rate decreases. Experimentally it is found that inclusions (trapped solvent) resulting from instabilities become more common as crystal size increases.

Trapping of solvent may be understood by considering the influence of mass transfer on growth mechanisms. The edges and corners of a faceted crystal project further into the solution and hence experience a slightly higher solute concentration than the centers of the faces. If growth did not occur by steps, this concentration difference would be sufficient to cause accelerated growth at the corners. As it is, screw dislocations control the rate of generation of new layers, and the screw dislocations are generally

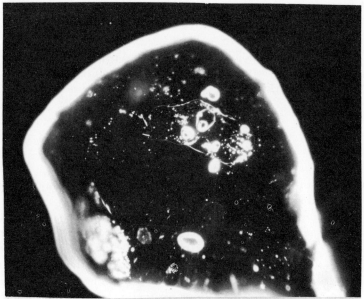

Figure 11.12.   Aqueous and bubble inclusions in AgNO₃ reagent crystal.

located near the centers of the faces. Under such conditions the faces should be stable and inclusions should not form. However, as the faces grow larger, or growth occurs more rapidly, the concentration difference between centers and corners increases. Eventually the supersaturation at the corners becomes sufficient to cause new layers to form there by two-dimensional nucleation. Although growth is smooth at the corners and no inclusions form there, the steps, as they move away from the corners and readily trap solution, pile up to form waves. When the slope of a wave becomes sufficiently steep, a new facet forms that is inclined to the face on which the wave is moving. Step movement along the face nearly ceases at that point, and the new facet becomes larger. Eventually the supersaturation at the point becomes sufficient to permit new layers to bridge over and form an inclusion. Needless to say, many variations are found on this basic theme. The point is that solution is almost invariably occluded in crystal growth. The occluded solvent contains impurities, usually in higher concentrations than are present in the bulk mother liquor because of rejection of impurities by the growing crystal.

Inclusions are sometimes introduced by spontaneous cracking, followed by healing of the crack to form a veil of tiny inclusions. Although this process has been observed fortuitously in growing sugar crystals, cracks and veils often appear to be associated in some crystals, thus suggesting a connection. The origin of the stress causing cracking is unknown, but is probably related to small variations in composition.

The solvent itself may be considered as an impurity in some crystals. The traditional method for removing solvent has been to heat the crystals, preferably above the boiling point of the solvent and perhaps also in a vacuum. Although this removes surface moisture, most inclusions are left intact. The solvent in the inclusions must reach the surface in order to escape. If sufficient pressure is generated in heating, some of the larger inclusions may crack the crystals. Even this may not be enough to remove solvent from these inclusions, for solvent often flows into a crack only to heal and form a veil (50). Small inclusions expand through plastic deformation of the crystal and vaporize. Upon cooling, the solvent remains in the enlarged inclusion at reduced pressure. Thus constancy of weight does not indicate removal of all solvent, as often assumed.

Solvent inclusions can be removed from crystals by placing them in a temperature gradient (50). Solute tends to dissolve on the hot side of the inclusion and to crystallize out on the cool side (assuming that solubility increases with temperature). This results in movement of the inclusion toward the heat source. The rate of movement is found to increase with increasing temperature gradient, decreasing slope of the liquidus $|m|$, increasing solubility, and increasing size of the inclusion. Generally the travel rate increases rapidly as the average temperature is raised.

Solvent inclusions containing gas bubbles tend to break up, with the bubble and a portion of the solvent moving in the "wrong" direction, that is, away from the heat source (51). This is caused by evaporation combined with a surface-driven flow. Solvent evaporates from the hot side of the bubble and condenses on the cool side. It leaves behind solute on the hot side and dissolves fresh solute on the cold side. Solution is returned to the hot side by surface-driven flow. Air bubbles are introduced into inclusions when they reach the crystal surface, and a portion of the solvent escapes. Therefore trapped solvent in crystals (and chemicals) is best removed by placing the crystals in a temperature gradient in a vacuum.

Inclusion movement as described above also occurs in fractional solidification (52). Trapped melt moves through the solid toward the bulk melt. If the trapped melt contains bubbles and has a volatile component, it may instead move further into the solid, with the limiting position being the eutectic isotherm. In zone melting these phenomena may act either to reduce or to enhance the separation, depending on the direction in which the inclusions move and on whether this occurs at the melting or the freezing interface.

Solid particles have also been moved through a solid (ice) by means of a temperature gradient (25). The mechanism is thought to involve a film of liquid (water) on the surface of the particles, which explains the strong increase in rate as the freezing point is approached.

11.49 PUSHING OF FOREIGN PARTICLES

It has been known for many years that a growing crystal can exert large forces on foreign surfaces. Foreign particles are pushed by a freezing interface so long as the freezing rate is below a critical value. The critical freezing rate depends strongly (and unpredictably, at present) on the particle-material system. Thus total separation is achieved at sufficiently low freezing rates, which, however, must be determined experimentally. In solution growth, particles are not pushed as readily and under stagnant conditions may block access of solute to the crystal and create inclusions thereby.

11.5 Related Processes

11.51 FIELD FREEZING

A separation may be generated by passage of large electrical currents through a conducting melt. This is called *electromigration* or *electrodiffusion* and is thought to be caused by differences in the collision cross section for electrons. When this procedure is combined with progressive freezing or zone melting, the separation is increased if the current is in the proper direction (56).

The beneficial effects are reduced, however by increased stirring of the melt, since stirring overpowers the segregating influence of the current. The detrimental effects of stirring may be eliminated by freezing at electrodes separated by a diaphragm or capillary (47). In this way, for example, eutectic mixtures may be separated.

11.52 ZONE CHROMATOGRAPHY

Consider the addition of a multicomponent mixture to a solvent. This solution is then placed at the beginning of a long column of frozen solvent, and molten zones are passed through the column, beginning at the end containing the added mixture. Since each component of the mixture has a different distribution coefficient, the mixture will be separated by passage of many zones through the column. Such a technique is called zone chromatography and appears to be useful primarily for analytical separations (37).

11.53 COPRECIPITATION

Coprecipitation has been defined as the contamination of a precipitate by substances that are normally soluble under the conditions of precipitation. Although this may be due to solid solubility or to occlusion of mother liquor, the primary cause is surface adsorption. This phenomenon can be exploited to concentrate or separate constituents present in very small concentrations. The primary precipitate acts as a carrier or collector for removal of the microcomponent from solution. Hydrous oxides of aluminum, iron, and other transition metals have proved most effective for general scavenging action. It is not unusual for a single precipitation to remove 90% or more of a trace constituent at dilutions of 1:109.

Acknowledgment

This work was supported by the Joint Services Electronics Program (U.S. Army, U.S. Navy, and U.S. Air Force) under Grant No. AFOSR–69–1622A.

References

1. O. V. Abramov, I. I. Teumin, V. A. Filonenko, and G. I. Éskin, *Soviet Phys.–Acoust.*, **13**, 141 (1967).
2. E. L. Anderson, *Chem. Ind.*, 1615 (1966).
3. J. C. Baker and J. W. Cahn, *Acta Met.*, **17**, 575 (1969).
4. I. M. le Baron and G. L. Samsel, U.S. Pat. 3,022,889 (Feb. 1972); through *Chem. Abstr.*, **57**, 2026 (1962).
5. A. W. Bamforth, *Industrial Crystallization*, Leonard-Hill, New York, 1965.

6. J. Bartels et al., eds., *Landolt-Bornstein Zahlenwerte und Functionen*, 6th ed., Springer-Verlag, Berlin, 1960.
7. H. A. Beale and R. J. Arsenault, in *Proceedings of the Fourth International Conference on Electron and Ion Beam Technology*, Electrochem. Soc., N.Y. (1970).
8. W. D. Betts, J. W. Freeman, and D. McNeil, *J. Appl. Chem.*, **17**, 180 (1968).
9. G. D. Botsaris, E. G. Denk, G. S. Ersan, D. S. Kirwan, G. Margolis, M. O'Hara, R. C. Reid, and J. Tester, *Ind. Eng. Chem.*, **61**, 86 (Oct. 1969), 92 (Nov. 1969), 65 (Dec. 1969); *Annual Review of Crystallization*.
10. G. D. Botsaris, E. A. Mason, and R. C. Reid, *A.I.Ch.E. J.*, **13**, 764 (1967).
11. G. D. Botsaris, E. G. Denk, and J. O. Chua, *Chem. Eng. Prog. Symp. Ser.* (in press).
12. H. E. Buckley, *Crystal Growth*, John Wiley, New York, 1951.
13. L. Burris, Jr., C. H. Stockman, and I. G. Dillon, *Trans. AIME*, **203**, 1017 (1955).
14. W. K. Burton, N. Cabrera, and F. C. Frank, *Phil. Trans. Roy. Soc.* (*London*), **243**, 40 (1951).
15. C. S. Cheng, D. A. Irvin, and B. G. Kyle, *A.I.Ch.E. J.*, **13**, 739 (1967).
16. C. T. Cheng, "Rates of Growth of Crystals from Solutions," Ph.D. Thesis, University of California, Berkeley, UCRL-19518 (1969).
17. J. Cisse and G. F. Bolling, *J. Crystal Growth*, **7**, 37 (1970).
18. G. G. Devyatykh and M. F. Churbanov, *J. Appl. Chem. USSR*, **41**, 2267 (1968).
19. B. Diehn, F. S. Rowland, and A. P. Wolf, *Anal. Chem.*, **40**, 60 (1968).
20. F. D. Evans, M. Bogan, and R. Battino, *Anal. Chem.*, **40**, 224 (1968).
21. S. N. Gitlin and S. Lin, *J. Appl. Phys.*, **40**, 4761 (1969).
22. L. Gordon, *Precipitation from a Homogeneous Solution*, John Wiley, New York, 1959.
23. M. Hansen, *Constitution of Binary Alloys*, McGraw-Hill, New York, 1958.
24. A. Hellawell, *Trans. AIME*, **233**, 1516 (1965).
25. P. Hoekstra and R. D. Miller, *J. Colloid Interface Sci.*, **25**, 166 (1967).
26. C. A. Holden and H. S. Bryant, *Separation Sci.*, **4**, 1 (1969).
27. D. T. J. Hurle and E. Jakeman, *J. Crystal Growth*, **5**, 227 (1969).
28. K. A. Jackson, D. R. Uhlmann, and J. D. Hunt, *J. Crystal Growth*, **1**, 1 (1967).
29. R. T. Johnson, R. W. Rousseau, and W. L. McCabe, "Factors Affecting Contact Nucleation," paper presented at *A.I.Ch.E.* Meeting, Denver, Colorado (Sept. 1970).; also N. A. Cloutz and W. L. McCabe, *Chem. Eng. Prog. Symp. Ser. 67*, No. 110, 6 (1971).
30. A. P. Kapustin, *The Effects of Ultrasound on the Kinetics of Crystallization*, Plenum, New York, 1963.
31. E. V. Khamskii, *Crystallization from Solutions*, Plenum, New York, 1969.
32. G. Kvajic and V. Brajovic, *Can. J. Phys.*, **48**, 2188 (1970).
33. D. P. Lal, R. E. A. Mason, and R. F. Strickland-Constable, *J. Crystal Growth*, **5**, 1 (1969).
34. E. Levin, C. R. Robbins, and H. F. McMurdie, eds., *Phase Diagrams for Ceramists*, American Ceramic Society, Columbus, Ohio, 1964.
35. W. F. Linke, *Solubilities of Inorganic and Metal-Organic Compounds* (Seidell), 4th ed., Van Nostrand, Princeton, N. J., 1958.
36. L. R. Morris and W. C. Winegard, *J. Crystal Growth*, **6**, 61 (1969).
37. W. G. Pfann, *Zone Melting*, 2nd ed., John Wiley, New York, 1966.
38. C. P. Saylor, in *Purification of Inorganic and Organic Materials*, M. Zief, ed., Marcel Dekker, New York, 1969, p. 125.
39. F. W. Schwab and E. Wickers, *J. Res. Natl. Bur. Std.*, **32**, 253 (1944).
40. R. F. Sekerka, *J. Crystal Growth*, **3**, **4**, 71 (1968).
41. N. N. Sheftal, ed., *Growth of Crystals*, Vols. 1–7, Consultants Bureau, New York, 1959–1969.

42. G. J. Sloan, in *Molecular Crystals*, Gordon and Breach, New York, 1966, p. 161.
43. H. Stephen and T. Stephen, eds., *Solubilities of Inorganic and Organic Compounds*, Macmillan, New York, 1963.
44. R. F. Strickland-Constable, *Kinetics and Mechanism of Crystallization*, Academic Press, New York, 1968.
45. A. Szymanski and M. M. Labes, *Nature*, **220**, 159 (1968).
46. L. G. Tensmeyer, P. W. Landis, and F. J. Marshall, *J. Org. Chem.* **32**, 2901 (1967).
47. R. S. Wagner, C. E. Miller, and H. Brown, *Trans. AIME*, **236**, 554 (1966).
48. A. G. Walton, *The Formation and Properties of Precipitates*, Wiley-Interscience, New York, 1967.
49. W. R. Wilcox, *Separation Sci.*, **2**, 411 (1967).
50. W. R. Wilcox, *Ind. Eng. Chem.*, **60** (3), 13 (1968).
51. W. R. Wilcox, *Ind. Eng. Chem.*, **61** (3), 76 (1969).
52. W. R. Wilcox, *Separation Sci.*, **4**, 95 (1969).
53. W. R. Wilcox, *J. Crystal Growth*, **7**, 203 (1970).
54. W. R. Wilcox and A. Leon, *J. Crystal Growth*, **8,** 230 (1971).
55. W. R. Wilcox, in *Preparation and Properties of Solid State Materials*, Vol. 1, R. Lefever, ed., Marcel Dekker, New York (1971).
56. M. Zief and W. R. Wilcox, eds., *Fractional Solidification*, Marcel Dekker, New York, 1967.
57. D. D. Edie and D. J. Kirwan, *Ind. Eng. Chem. Fund.* (in press).
58. W. R. Wilcox, *J. Crystal Growth*, **12**, 93 (1972).

List of Symbols

A	constant in Eq. 11.4	V_c	crystal growth rate
B	constant in Eq. 11.4	Z	distance
g	mass fraction		
G	gradient of the difference between actual temperature and the freezing point in the Y direction	α	saturation or super-saturation ratio
		δ	film thickness
		ρ	density
		σ	fractional supersaturation
K_{eff}	effective distribution coefficient		Subscripts
		e	equilibrium
l	zone length	f	fluid
L	length of bar	i	interface
m	slope of the solubility curve	0	starting value
		s	solid

Twelve

ION-EXCHANGE SEPARATION
PROCESSES

H. L. ROTHBART

12.1 Introduction

Ion exchange involves the substitution of one ionic species for another without significant change in structure. In chromatographic systems in which the stationary phase is relatively rigid, the restriction against significant change in structure becomes important. Although "significant change" is sometimes difficult to define, it is apparent that substitutions at surfaces and in porous media often may be classified as ion-exchange phenomena. On the other hand, if potassium ion with a crystal radius of 1.33 Å were substituted for cesium ion, with a crystal radius of 1.67 Å, in a cesium chloride crystal, the structure would have significantly different crystallographic properties. Therefore this is not considered to be ion exchange. Most useful ion exchangers are porous materials containing ionogenic groups; that is, they have fixed ionic groups and potentially mobile counterions.

Utilization of ion exchange in separation schemes frequently involves techniques and instrumentation similar to those used in liquid-liquid chromatography (Chapter 10). However, the ion-exchange process is not limited to these applications and systems. It has been cited as an important process in the transport of ions in living cells and is significant in maintaining electrolyte balance. Cellular membranes apparently display some properties that are formally equivalent to those of ion-exchanger membranes. Transmission of nerve impulses has been theorized as an exchange process. Similarly, the mechanism and the electrical potential of the glass electrode have been described in ion-exchange terms.

Each of these fields has benefited by the stimulation provided by the others. Some major advances in ion-exchange theory have resulted from the pursuit of biophysical principles, and phenomena discovered while studying ion exchange have been important in the aforementioned fields (1, 2).

Most applications of this versatile technique have been carried out with ion-exchanger particles in packed beds such as chromatographic columns (3). Other arrangements such as incorporation of ionogenic groups in membranes, are also important (Chapter 16), but the major direction of this chapter is the consideration of column systems, primarily those in which solutes to be separated are present in small quantities compared to the stationary phase. Although exchangers have been utilized for catalysis and the preparation of molecular species through oxidation or reduction of starting materials (utilizing redox ion exchangers), these applications are not considered in this chapter.

12.2 Structure

12.21 INORGANIC MATERIALS

Naturally occurring sands that desalt brackish waters were known in classical times, although characterization of these occurred only in the twentieth century. Such common materials as the zeolites, which include some sodium aluminum silicates, and clays, such as montmorillonite, have ion-exchange properties. Zirconium phosphate, $Zr(HPO_4)_2 \cdot H_2O$, and hydroxyapatite, $Ca_5(PO_4)_3OH$, are among the inorganic exchangers produced in the laboratory. The former has a layered structure with planar Zr atoms bridged by phosphate groups between the planes. Interlayer cavities about 8 Å in height provide the sites for cation exchange. Hydroxyapatite has found important application in the chromatography of proteins and nucleic acids such as DNA (4). The mechanism of separation is probably surface exchange, since these macromolecules are too large to penetrate the pores of the stationary phase. The relatively rigid frameworks of these materials and their fixed pore sizes restrict the penetration of molecular species into the framework. Thus separations have been carried out on the basis of molecular size as well as ion exchange (see Section 15.2).

12.22 ORGANIC MATERIALS

Styrene–Divinylbenzene Copolymers

Resins based on these copolymers are the most widely used ion exchangers. They are commonly produced by suspension polymerization in which styrene and divinylbenzene are suspended in water as droplets. Addition of a catalyst such as benzoyl peroxide initiates polymerization. The resulting beads are porous, and swell but do not dissolve when placed in common hydrocarbon solvents. The degree of swelling depends on the amount of divinylbenzene cross-linking agent present. Typically, 8 to 12% of this compound is used. Much lower crosslinking results in soft materials, while higher quantities of divinylbenzene produce an excessively rigid structure. Treatment of the hydrocarbon-swollen beads with sulfuric or chlorosulfonic

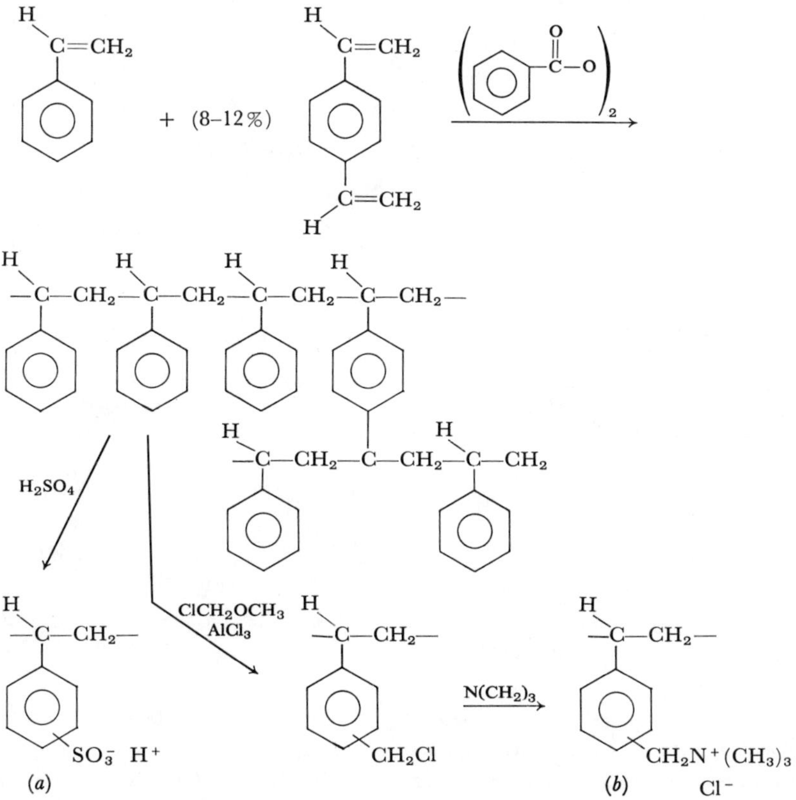

Figure 12.1. Preparation of styrene–divinylbenzene-based ion exchangers: (*a*) sulfonic acid cation exchanger, (*b*) quaternary ammonium chloride anion exchanger.

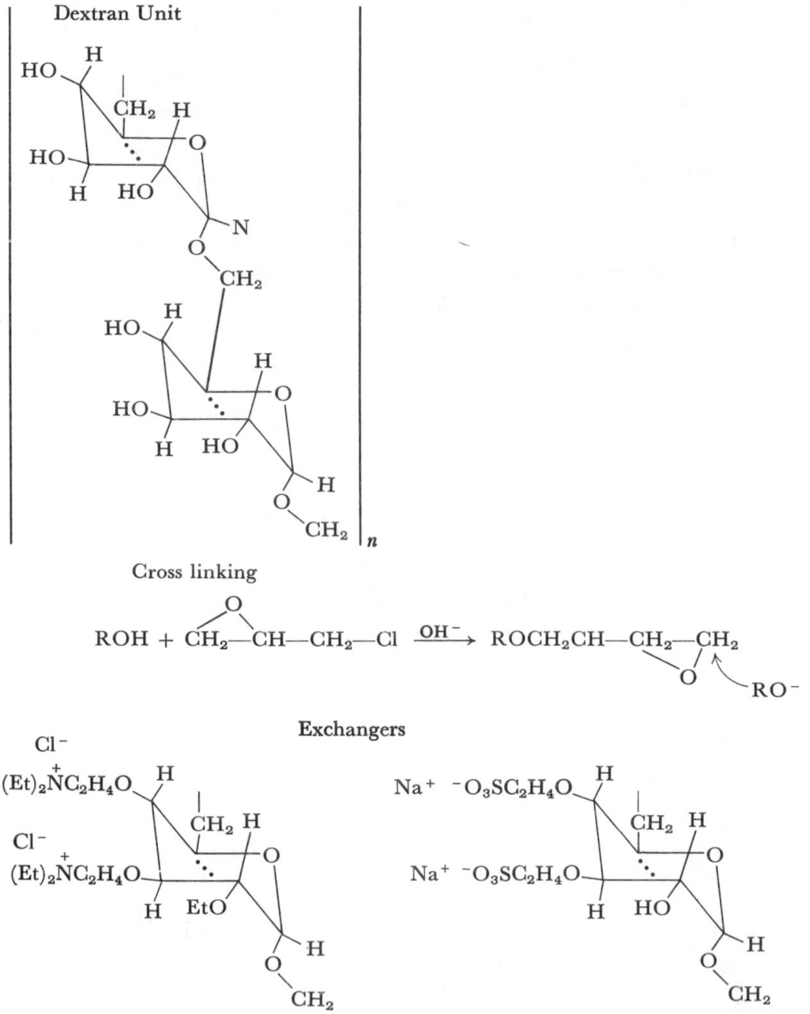

Dextran Unit

Cross linking

$$ROH + CH_2\!\!-\!\!CH\!\!-\!\!CH_2\!\!-\!\!Cl \xrightarrow{\ OH^-\ } ROCH_2CH\!\!-\!\!CH_2\!\!-\!\!CH_2$$

Exchangers

(a)   (b)

Figure 12.2. Preparation of dextran-based exchangers: (a) tertiary amine (weak base) anion exchanger, (b) Soudim-form cation exchanger.

acid results in a sulfonic acid cation exchanger. When the beads are treated with chloromethyl ether and then trimethylamine, the common quaternary ammonium chloride anion exchanger results (Fig. 12.1). The resins are treated to remove unpolymerized monomers, low polymers, and foreign counterions and are sold commercially. They swell slightly or not at all in organic solvents but markedly in aqueous solutions. The sulfonic acid resin is referred to as a strong-acid type, whereas the quaternary-ammonium resin in the hydroxide form acts as a strong base. Resins thus produced have fairly good mechanical stability. Strong-acid resins, their salts, and strong-base resins in their salt forms are relatively stable chemically and are among the most commonly used ion exchangers. Many other functional groups have also been incorporated, such as tertiary amines, carboxylic acids, and iminodiacetic acid $[N(CH_2COOH)_2]$ moieties, which result in weak-base, weak-acid, and chelating resins, respectively.

When copolymerization is carried out in the presence of an inert diluent that is a good solvent for the monomers, but not the polymer, resin beads with pools of diluent distributed throughout the matrix are obtained. When these are washed out, a macroporous structure results which is usually rigid because of the use of large amounts of cross-linking agent. Functionality is incorporated as described. The pores are much larger than in the usual exchangers, and these "macroreticular resins" have been finding increasing use in technological applications such as water purification.

Carbohydrate-Based Polymers

An important class of materials commonly used in biochemical and polymer studies is based on dextran gels prepared by the action of the microorganism *Leuconostoc mesenteroides* upon sucrose. The linear polymers are cross-linked with epichlorohydrin and converted to the commonly used diethylamino-ethyl (DEAE) and sulfonic acid derivatives (Fig. 12.2). Similar materials have been prepared directly by treatment of cellulose, and ion-exchange celluloses have been packed into columns as powders, cast onto plates for use in thin-layer chromatography, or, in preformed sheets, utilized in paper chromatographic work.

12.3 Properties of Ion Exchangers

12.31 CAPACITY

The number of ionogenic groups per unit weight or volume of exchanger is easily determined. For anion exchangers, chloride ions are readily exchanged for nitrate and the desorbed halide is easily determined with silver ion. In the case of cation exchangers, hydrogen ions may similarly be displaced by sodium ion in a salt solution and the resulting acid titrated with base. Alternatively, the hydrogen-form resin may be titrated directly with base. A typical titration curve is depicted in Fig. 12.3. The apparently great buffer capacity at the start of the titration is deceptive, since addition of strong acid

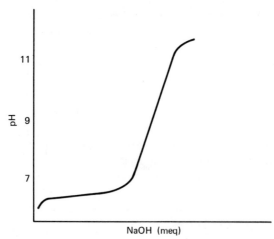

Figure 12.3. Direct titration of a cation exchanger with NaOH. The capacity of the exchanger can be calculated from the number of milliequivalents of base required to reach the equivalence point.

at this point would greatly decrease the pH of this system. The capacities of the usual polystyrene-based exchangers are on the order of 3 and 5 meq/g of resin (on a dry basis) for anion and cation exchangers, respectively.

12.32 SWELLING

Dry ion exchangers have a strong affinity for water. Use is made of this property, as molecular sieves are widely utilized for removing traces of water from organic solvents. Water diffuses into the porous exchangers, hydrating the fixed ions and counterions. In the early stages of water sorption, the volume of the system, water plus resin, is less than the sum of the volumes of the original materials. As subsequent water molecules penetrate the exchanger, the increase in volume is an additive function of the amount of water sorbed. Although linear polymers would eventually dissolve or form micelles, the presence of cross links limits the amount of water imbibed. The flexibility of these matrices allows swelling of the exchanger; however, swollen resins placed in contact with concentrated salt solutions shrink.

The amount of swelling depends on the degree of crosslinking and flexibility of the chains, and highly cross-linked resins swell least. Other important considerations are the concentration of the external solution and the ionic form of the resin. For example, the sequence of counter ions giving increasing resin volumes for most sulfonate type cation exchangers is $Cs^+ < Rb^+ < K^+ < Na^+ < Li^+$ and parallels the hydrated ionic volumes, that is, the hydrated ionic volume of lithium is greater than that of sodium. Swelling is similarly favored by high capacity of resin, polar solvents, and complete dissociation of ionogenic groups within the resin. For example, carboxylic acid resins do not swell appreciably in acid solution.

12.33 EQUILIBRIA

Uncharged Species

Thermodynamic description of resin swelling and sorption of undissociated solutes is formally equivalent to a system with a semipermeable membrane separating a solution containing a polyelectrolyte from another solution containing undissociated solute. The polyelectrolyte is unable to pass through the membrane, whereas all other species may be transported through this semipermeable barrier. Van't Hoff's law of osmotic pressure applies to these cases. Solute may pass into the polyelectrolyte solution, but equilibrium will be attained only if pressure is applied to the chamber containing the polymer. The difference in pressure applied to the two chambers is π, the osmotic pressure. In a resin the fixed ions are equivalent to the polyelectrolyte described, and the cross links, due to copolymerization of divinylbenzene, are the source of pressure on the solution within the exchanger. As species such as water diffuse into the resin the matrix is expanded until the force exerted by the cross links results in an osmotic pressure sufficient for equilibrium to be reached.

The chemical potentials of species in solutions corresponding to the one external to the resin were considered in Chapter 2, as in Eq. 2.3:

$$\mu_i{}^{aq} = \mu_i{}^{0,aq} + RT \ln a_i{}^{aq} \tag{2.3}$$

and succeeding expressions. In many studies the external solution is essentially incompressible and is generally near 1 atm in pressure; the effect of pressure on its chemical potential is thus usually neglected. Pressure, however, must be considered within the resin phase. The chemical potential of species within the resin, $\mu_i{}^r$, is a function of the internal pressure P and the molar concentration of the solute C:

$$d\mu_i{}^r = \left(\frac{\partial \mu_i{}^r}{\partial C}\right)_{T,P} dC + \left(\frac{\partial \mu_i{}^r}{\partial P}\right)_{T,C} dP \tag{12.1a}$$

$$\mu_i = v_i \, dP - s_i \, dT \tag{12.1b}$$

$$\mu_i{}^r = \mu_i{}^{0,r} + RT \ln a_i{}^r + v_i(P^r - P^{aq}) \tag{12.2}$$

and

$$\mu_i{}^r = \mu_i{}^{0,r} + RT \ln a_i{}^r + v_i\pi \tag{12.3}$$

All terms have their usual significance, while s_i and v_i are the partial molar entropy and the volume of species i. Expression 12.3 is useful for study of the equilibrium of an electrically neutral solute such as water or ethanol between the resin and an external solution.

Charged Species

Consider a resin with a typically univalent fixed functional group R and a counterion A of charge Z_A. When in equilibrium with a totally dissociated salt, $Y_{Z_A}A_{Z_Y}$, electroneutrality requires that the total concentration of positive charges within the resin equals the total concentration of negative charges within the resin:

$$[R] + |Z_Y| [Y^r] = |Z_A| [A^r] \tag{12.4}$$

The presence within the resin of small amounts of ion with the same charge sign as the fixed groups on the resin is termed *Donnan penetration*. In general, it is favored by high concentration of ions in the external solution, low cross-linking, and low capacity of resin. Donnan equilibrium also has been used to give a thermodynamic representation of ion exchange for electrolytes.

When charged species are to be considered, the electrochemical potentials, η of the species in each phase are equal at equilibrium, for example,

$$(\eta_i)^r = (\eta_i)^{aq} \tag{12.5}$$

The electrochemical potential is related to the chemical potential by

$$\eta_i = \mu_i + Z_i \mathscr{F} \phi \tag{12.6}$$

where $\mathscr{F}$ is the Faraday constant, and ϕ is the electrical potential acting on species i. The Donnan potential is the difference in electric potential between the ion exchanger ϕ^r and the solution ϕ^{aq}, and the chemical potential may be determined as in Eqs. 2.3 and 12.3. Reference states for ionic species in both resin and solution are taken usually as the infinitely dilute aqueous species. Thus

$$E_{\text{Donnan}} \equiv \phi^r - \phi^{aq} = \frac{1}{Z_i \mathscr{F}} \left(RT \ln \frac{a_i{}^{aq}}{a_i{}^r} - \pi v_i \right) \tag{12.7}$$

The partial molar volume of species i is assumed to be the same in aqueous or resin phase.

Consider the exchange of ion B, which is initially in the resin, by ion A:

$$|Z_B| A^{Z_A} + |Z_A|(|Z_B| R^{\pm 1} + B^{Z_B}) \rightleftharpoons |Z_B|(|Z_A| R^{\pm 1} + A^{Z_A}) + |Z_A| B^{Z_B}$$

The exchange reaction may be described in terms of a thermodynamic equilibrium constant:

$$K_B{}^{0,A} = \frac{(a_A{}^r)^{|Z_B|}(a_B{}^{aq})^{|Z_A|}}{(a_A{}^{aq})^{|Z_B|}(a_B{}^r)^{|Z_A|}} \tag{12.8}$$

a classical equilibrium constant where the molar concentrations of solute in both phases are used:

$$K_B{}^A = \frac{[A^r]^{|Z_B|}[B^{aq}]^{|Z_A|}}{[A^{aq}]^{|Z_B|}[B^r]^{|Z_A|}} \qquad (12.9)$$

and the Donnan potential:

$$E_{\text{Donnan}} = \frac{1}{|Z_A|\mathscr{F}} \left(RT \ln \frac{(a_A{}^{aq})}{(a_A{}^r)} - \pi v_A \right) = \frac{1}{|Z_B|\mathscr{F}} \left(RT \ln \frac{(a_B{}^{aq})}{(a_B{}^r)} - \pi v_B \right) \qquad (12.10)$$

The value of this potential is independent of whether ion A or B is chosen for consideration and is a measure of the relative abundance of the ions in the two phases. Thus

$$\ln \left[\left(\frac{a_B{}^{aq}}{a_B{}^r} \right)^{|Z_A|} \left(\frac{a_A{}^r}{a_A{}^{aq}} \right)^{|Z_B|} \right] = \ln K_B{}^{0,A} = \frac{\pi}{RT}(|Z_A|v_B - |Z_B|v_A) \qquad (12.11)$$

Selectivity of the resin for A over B is reflected in a positive value of $\ln K_B{}^{0,A}$. Equation 12.11 indicates that species of higher (absolute value) charge are favored by the resin over those of lower charge. Conversely, species of low partial molar volumes sorb preferentially, but it should be stressed that "partial molar volume" here refers to the hydrated species. Selectivity should be favored by large values of π, which occur when a high degree of cross linking is present. Examination of Section 12.32 indicates that swelling of cation exchangers increases when counterions are in the order from cesium to lithium. A related order prevails for selectivity, that is, cesium ion is favored. Similarly, barium ion is favored over strontium, and so on through beryllium ion. Typical selectivity for strong-acid cation exchangers is as follows: $Li^+ < H^+ < Na^+ < NH_4^+ < K^+ < Rb^+ < Cs^+ <$ Ag^+ and $Mg^{2+} < Zn^{2+} < Co^{2+} < Cu^{2+} < Cd^{2+} < Ni^{2+} < Ca^{2+} <$ $Sr^{2+} < Pb^{2+} < Ba^{2+}$.

The order for anion exchange is somewhat more complex, although iodide ion is most strongly held of the halides. A typical order for anion-exchange resins of the strong-base type is: $F^- \simeq OH^- < CH_3COO^- < HCOO^- <$ $H_2PO_4^- < HCO_3^- < Cl^- < NO_2^- < HSO_3^- < CN^- < Br^- < NO_3^- <$ $HSO_4^- < I^-$. Some of the complexity is due to formation of ion pairs in the resin phase or association, as in the case of acetate ion interacting with hydrogen ion to form dimerized undissociated acetic acid. The presence of tertiary and secondary amines as fixed groups in anion exchangers further complicates the picture. These sites can be protonated during washing of the resin and can act as a source of hydrogen ion, as well as a site for hydrogen bonding.

All of the foregoing trends are noted in the absence of special effects such as sieving, which is the partial or total exclusion of large ions due to limited pore size (see Chapter 15).

12.34 SELECTIVITY

The importance of this phenomenon to separations is so great that it must be considered in somewhat more detail. The approach given in the preceding sections, due mostly to Gregor (5), described selectivity in terms of the activities of solutes. Activity coefficients in electrolyte solutions can be calculated from the Debye-Hückel equation. Unfortunately, no such approach is available for species in the resin phase. Retention volume equations such as Eq. 5.6 require the use of equilibrium constants based on concentration (Eq. 12.9). Plots of these classical equilibrium constants against the fraction of the ionic groups in a resin with a specific counterion, χ_A, result in curves depicting a large variation in $K_B{}^A$ (Fig. 12.4). The reversal in sign of log $K_H{}^{Na}$ reflects an apparent reversal in selectivity for these two ions. This has led

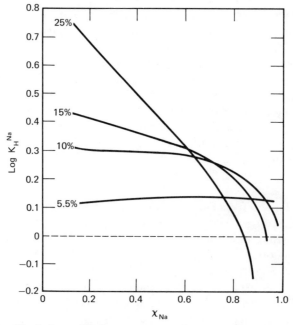

Figure 12.4. Classical equilibrium constants for styrene–divinylbenzene–sulfonate exchangers. The percentages indicate nominal cross linking, and the variation in selectivity for Na$^+$ and H$^+$ is noted as a function of fraction of resin in the Na$^+$ form, χ_{Na}. Reprinted from Ref. 6, by courtesy of The Chemical Society, London.

to the convention that discussions of selectivity should be referred to a resin that contains 50 equivalent per cent of the two ions being considered. When this is the case and the resin is in equilibrium with an extremely dilute solution of two counterions of the same charge, the thermodynamic and classical equilibrium constants are equal by convention. As the equivalent fraction of an ion in the resin approaches 1, the selectivity, on the concentration scale, for that ion decreases. If we assume for the present that $\ln K_B^A$ is proportional to $\pi(|Z_A|v_B - |Z_B|v_A)$, it is easy to understand why highly cross-linked resins display high selectivity. Furthermore, as the resin is converted to the ionic form of lower partial molar volume, π decreases, leading to a decrease in $\ln K_B^A$. This apparently is most important for resins with a high degree of cross linking.

Selectivity reversals are difficult to understand on the basis of this theory. It has been suggested that in a group such as the alkali-metal ions the most highly hydrated species tend to be stripped of some or all of their exterior water of hydration when the resin is under the stress of a large osmotic pressure. This should be most important for highly cross-linked resins. It has been proposed that the partial molar volumes of the unhydrated ions are the important parameters in such cases. Although transport and activity coefficient measurements tend to support the concept of hydrated ionic radii, there is no direct method for measuring these quantities, nor have "partially hydrated" species been determined in ion-exchange resins. The difficulties encountered in rationalizing some of the aforementioned phenomena with thermodynamic and classical equilibrium expressions have led to searches for other interpretations of ion-exchange selectivity.

One of the most interesting of these was developed for glass electrodes, which show specificity (ion-exchange selectivity) for alkali-metal ions (7). In this approach it was assumed that ionic species interacting with fixed groups on the exchanger matrix are stripped of water of hydration and the energy of attraction is purely Coulombic. When an ionic species is replaced by another, the "desorbed" ion gains water of hydration after the exchange and gives up free energy of hydration. The change in free energy for the conversion of a resin, initially the A form to the B form, is

$$\Delta G_A^{0,B} = -RT \ln K_A^{0,B} =$$

$$(\Delta G_{B,r}^{0,\text{Coulombic}} - \Delta G_{A,r}^{0,\text{Coulombic}}) + (\Delta G_A^{0,\text{hydration}} - \Delta G_B^{0,\text{hydration}}) \quad (12.12)$$

and

$$\Delta G_{B,r}^{0,\text{Coulombic}} \approx \frac{-332}{\rho_+ + \rho_-} \quad (12.13)$$

The terms have their usual significance, and $\rho_+ + \rho_-$ is the internuclear distance between the oppositely charged species in the absence of hydrating water molecules.

The free energy of hydration is roughly inversely proportional to the crystal radius of the exchanged ion. Eisenman (7) has made a careful study of the selectivity of glass electrodes for alkali-metal cations. When the anionic exchange site is large and dispersed, the difference in hydration energy between the two exchanging cations controls the overall free energy and thus the equilibrium constant. Conversely, small electrically dense exchange sites result in Coulombic control. Extrapolating this approach to porous exchangers, Eisenman and others have described sulfonate-ion groups as large exchange sites (8). For species of the same charge, if the crystal radius of ion A is smaller than that of ion B, the term $(\Delta G_A^{0,\text{hydration}} - \Delta G_{B,}^{0\ \text{hydration}})$ is negative, since the exothermic energy of hydration of species A is a larger negative number than that of B. In this type of case, the hydration energy controls the equilibrium. Thus the exchange of a resin initially in the A form for a larger (when unhydrated) B ion is a spontaneous process with a favorable free energy and equilibrium constant. This agrees with the observations that the resin favors the species of larger crystal radius, but the explanation differs from that presented before. In the interpretation of Gregor's work, selectivity was attributed to an affinity of the resin for the smallest hydrated species. In the interpretation of Eisenman's work, however, the affinity of the resin for the species of larger crystal radius is due to the relatively great (negative) energy of hydration of the species of smaller crystal

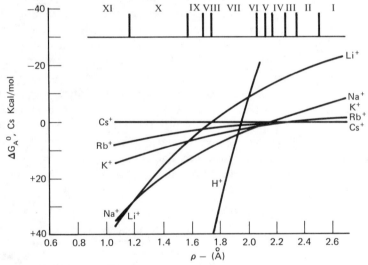

Figure 12.5. ΔG° for the production of a Cs$^+$-form resin from a resin initially in the A form (A = Li$^+$, Na$^+$, etc.) plotted against radius of the exchange site. The Roman numerals indicate 11 distinct selectivity orders at differing values of ρ^-, more than half of these orders have been experimentally observed (7).

radius. In the former case the explanation claims that the resin favors the species with the smaller hydrated radius, whereas in the latter case the solution favors the species with the smaller crystal radius, which might have a larger hydrated radius. Eisenman's approach has had an interesting degree of success.

In Fig. 12.5 a plot of $\Delta G°$ calculated as a function of the size of the unhydrated exchange site is depicted for alkali-metal cations, based on their crystal radii. For large exchange sites the predicted selectivity order is the same as that observed for sulfonic acid type resins. For small, electrically dense exchange sites, however, the selectivity order is reversed, that is, Li^+ is apparently favored over Na^+. This is the observed selectivity order for carboxylic acid resins, which in Eisenman's approach would be defined as small exchange sites. Phenomena observed for highly cross-linked resins have been explained on the basis of overlapping electron clouds of the fixed sites, yet no fully acceptable rationalization of selectivity reversals, which occur as the fraction of a counterion in the resin is changed, has been advanced. Although a number of other theories of selectivity have been published, the explanation of these phenomena awaits further investigation.

12.35   KINETICS

In ion exchange, as in the other forms of chromatography, kinetic processes are extremely important (Chapter 3). In addition to longitudinal transport of solute-containing eluent through the column, a number of other processes are also significant: (1) transport of species in the void volume between the resin particles, (2) diffusion of species through the relatively stationary film around the resin particles, (3) diffusion within the resin particles, and (4) exchange of ions at the fixed groups within the resin. Steps 2 and 3, called *film* and *particle diffusion*, are the slowest processes in ion exchange.

Ion-exchange kinetics is complex, and rigorous solutions have been made in only a few cases. Most attention has been paid to spherical beads and planar membranes, and much of the knowledge in this field is due to the work of Helfferich (1). He has shown for particle diffusion control that the half time for complete conversion may be given by

$$t_{1/2} = 0.0075 \frac{dp^2}{D^r} \tag{12.14}$$

Here dp is the bead diameter, and D^r is the diffusion coefficient within the resin. In this case the beads were in contact with a large excess of solution, in which the system was agitated so that they did not clump together. Since half time is inversely proportional to the rate of exchange, this rate is then proportional to the diffusion coefficient and inversely proportional to the square of the bead radius. In the case of film diffusion control, for a system similar to that previously described, half time for conversion of a resin initially in the A form to the B form may be calculated by

$$t_{1/2} = (0.17 + 0.064K_B{}^A) \frac{dp\delta q}{2D^{aq}[B]} \qquad (12.15)$$

Here the concentration of B in the external solution remains constant throughout the experiment. Film thickness is denoted by δ, the diffusion coefficient within this stationary layer of solution is D^{aq}, and q is the capacity of the resin in milliequivalents per milliliter of resin.

The question of whether film or particle diffusion is the rate-controlling step is readily answered in these nonchromatographic systems by setting the half times equal and rearranging. Remembering that the rate-determining step is the slowest process, that with the longest half time, we can express the control criterion as

$$\begin{pmatrix} \text{Particle} \\ \text{diffusion} \end{pmatrix} 1 \gg (6 + 2K_B{}^A) \frac{D^r\delta q}{dpD^{aq}[B]} \gg 1 \begin{pmatrix} \text{Film} \\ \text{diffusion} \end{pmatrix} \qquad (12.16)$$

A typical value for D^r/D^{aq} is about 0.1, q is often 3 meq/ml, and δ is about 0.001 cm in well-stirred solutions. In a system where $K_B{}^A = 1$, the product of dp and $[B]$ would have to be 2×10^{-3} for the system to lie in the intermediate range between clear-cut film and particle diffusion. A common bead size is 200 mesh, which reflects a diameter of 8×10^{-3} cm. In this system when $[B]$, with a charge of ± 1, is less than 0.1 M, film diffusion would occur. It should be remembered that these situations are related to, but are not equivalent to, columnar processes. Only rarely are large quantities of resin beads of identical diameter available. Even if such a condition were met, packing these into columns under the ideal condition of closest packing would result in each bead touching twelve others. Film diffusion through the complex figure would have somewhat different properties than that through a system with radial symmetry. In a similar manner the boundary conditions stated for particle diffusion, although useful in separation processes such as countercurrent ion exchange, are not often equivalent to those in chromatography. The overall pictures, however, show great similarity.

Diffusion of ions in the mobile phase was considered in Eq. 3.46. In this case the coupled movement of cations and anions results in an interdiffusion coefficient. Diffusion within the resin is at first sight more complex, since the counterions are mobile whereas the fixed ions are stationary and exclude mobile coions. Under this condition

$$|Z_A|[A^r] + |Z_B|[B^r] = q \qquad (12.17)$$

and

$$Z_A J_A = - Z_B J_B \qquad (12.18)$$

Equation 12.18 indicates that, when ion A moves in one direction, an equivalent amount of ion B moves in the opposite direction. This coupled diffusion reflects an absence of electrical current. The driving force for transport, as described in Chapter 3, is the gradient in chemical potential or, in the case of charged species, the gradient in electrochemical potential. Equation 3.1 described diffusion in the absence of an electrical potential. The existence of the Donnan potential, however, requires consideration of electrical transport. The flux of a species within a resin bead may be described by

$$J_A = - D_A{}^r \left(\text{grad } [A^r] + |Z_A|[A^r] \frac{\mathscr{F}}{RT} \text{ grad } \phi \right) \qquad (12.19)$$

Combination of Eqs. 12.17 through 12.19 results in

$$J_A = - \left[\frac{D_A{}^r D_B{}^r (Z_A{}^2 [A^r] + Z_B{}^2 [B^r])}{Z_A{}^2 [A^r] D_A{}^r + Z_B{}^2 [B^r] D_B{}^r} \right] \text{grad } [A^r] \qquad (12.20)$$

The term in brackets, the so-called interdiffusion constant, depends on the concentrations of A and B. Diffusion coefficients are of the same magnitude for most species within the resin phase. For ions of similar charge magnitude, it becomes apparent that the term in brackets reduces to $D_B{}^r$ when $[B^r] \ll [A^r]$. This means that the effective diffusion coefficient is that for the less concentrated species within the resin phase! This consideration is important in chromatographic systems where the eluent counter ion A, is in great excess to the ion being chromatographed, B. The diffusion coefficient is dependent on the latter. Furthermore, in the conversion of resin in the A form to the B form and, after completion, the reconversion of the latter to the A form, the initial rates are quite different.

12.4 Applications

12.41 REMOVAL OF SPECIFIC IONS

Use of many techniques is restricted by the presence of interfering counterions. Ion exchange becomes a valuable tool in the removal of these species. Sodium ion may be substituted for other cations in a solution by passing the solution through a bed containing cation exchanger in the sodium form. Similarly, nitrate ions may be substituted for interfering anions by passing the solution through an anion exchanger in the nitrate form.

One example of a frequently carried out technique is the preparation of tetramethylammonium hydroxide from tetramethylammonium chloride. This quaternary ammonium ion is desired in many potentiometric studies in which the effects of the complexing of a supporting electrolyte are to be minimized. It is desirable to convert the solution used as titrant to tetramethylammonium hydroxide, which is expensive and not always available at high purity. The procedure requires converting the bed of cation exchanger to the tetramethylammonium form by passing a solution of tetramethylammonium chloride through the bed. The bed is washed with water, and sodium hydroxide solution is passed through the bed. Sodium ions exchange with the quaternary ammonium ions, and tetramethylammonium hydroxide appears in the effluent ready for use. Because of the instability of anion exchangers in the hydroxide form, this procedure is far superior to utilizing an anion exchanger, converting it to the hydroxide form, and then passing the quaternary amine chloride through the anion exchanger. Furthermore, the low selectivity of resins for hydroxide ion results in a requirement for large quantities of base in order to prepare a resin that is totally in the hydroxide form.

12.42 DETERMINATION OF TOTAL SALT

Determination of the total number of equivalents in a salt solution is often a difficult task but is readily carried out by passing the solution through a cation exchanger in the hydrogen form. The acid that evolves is determined. Anion exchangers may be utilized, but their use in the hydroxide form involves a number of difficulties. Besides the aforementioned instability and difficulty of regeneration, they have lower capacities per unit volume than cation exchangers. Alkaline effluents have to be protected against atmospheric carbon dioxide, and insoluble hydroxides precipitate on the exchanger and clog it. The alternative is to use anion exchangers in the chloride form, but in utilizing this technique it is necessary to make sure that undissociated complexes do not cause errors.

An interesting alternative procedure useful for obtaining salt samples from remote areas involves passing a measured amount of fluid, such as ground waters, through small columns of exchanger contained in plastic. The sealed columns are readily transported back to the laboratory for study of individual components after they have been eluted from the resins.

12.43 DEIONIZATION

Among the most common uses for ion exchangers is the preparation of deionized water. Solutions to be treated are passed successively through a cation exchanger in the hydrogen form and an anion exchanger, usually a weak-base type, in the hydroxide form. The eluted solutions generally have low conductivity and may be useful for conductivity water, but they may contain traces of unremoved organic solutes. Similarly, decomposition of resins leaves traces of resin-breakdown products in the solutions. These solutions often have some surface activity, so that the deionized water, although useful in many applications, should not be employed for surface chemical studies without further treatment.

12.44 REMOVAL OF ORGANIC SPECIES

Ion-exchange resins tend to sorb some organic species and have some use for the removal of these materials. The removal of aldehydes may be carried out by converting anion exchangers to the bisulfite form. Solutions of aldehydes in ethanol may be purified by the formation of a bisulfite adduct of the aldehydes, leaving the ethanol unchanged. Weak acids may be removed by passage through weak-base anion exchangers in the hydroxide form. Use can be made of the sorption of organic (particularly benzenoid) species by the organic matrix of exchangers. Relatively large molecules are efficiently removed by macroreticular resins, which are used when the solvent is not sufficiently polar to swell ordinary exchange resins. The macroporous structure permits rapid mass transfer and facilitates sorption in continuous processes.

12.45 MICROSTANDARDS

Cation-exchange beads with rather homogeneous properties can be obtained in a wide range of particle sizes down to 0.001 mm in diameter. Since the size of individual beads can be measured with a calibrated microscope, one can make use of a single bead containing 10^{-12} equivalents of a cation such as sodium or calcium. Such beads, suspended in water and transferred by pipet, have been used as microstandards in mass regions where as yet there are no other reliable materials. For example, resin heterogeneity resulted in a 5% relative standard deviation in sodium-ion content at the 3-pg level (9).

12.5 Chromatography

12.51 SELECTIVITY AND ELUENT CONCENTRATION

Whenever a separation is to be attempted, a primary consideration is the relative selectivity of the separating tool for the solutes in question. In chromatographic systems the solutes are usually present in low concentrations. In ion exchange the resin is initially in equilibrium with an eluent that is relatively concentrated, and the resin counterion is one of the ions present in the eluent. Consider the exchange of ion A^{Z_A} for the eluent ion $El^{\pm 1}$ and the subsequent separation of ion A^{Z_A} from B^{Z_B} (let $Z_A = Z_B$ for simplicity), that is,

$$|Z_A|(R^{\pm 1} + El^{\pm 1}) + A^{Z_A} \rightleftharpoons (|Z_A|R^{\pm 1} + A^{Z_A}) + |Z_A|El^{\pm 1}$$

Then

$$K_{El}{}^{0,A} = \frac{(a_{El}{}^{aq})^{|Z_A|}(a_A{}^r)}{(a_{El}{}^r)^{|Z_A|}(a_A{}^{aq})} ; \qquad K_{El}{}^{0,B} = \frac{(a_{El}{}^{aq})^{|Z_B|}(a_B{}^r)}{(a_{El}{}^r)^{|Z_B|}(a_B{}^{aq})} \qquad (12.21)$$

The thermodynamic separation factor α^0 (see Eq. 2.22) is identical to the selectivity coefficient, since

$$\alpha^0 = \frac{K_{El}{}^{0,A}}{K_{El}{}^{0,B}} = K_B{}^{0,A} = \frac{(a_A{}^r)(a_B{}^{aq})}{(a_A{}^{aq})(a_B{}^r)} \qquad (Z_A = Z_B)$$

In chromatographic systems the pertinent terms are concentration-based equilibrium constants. For example, the retention equation utilizes a classical equilibrium term K^A (Chapter 5):

$$V_R{}^A = V_M + V_S K^A \qquad (5.6)$$

In the systems under question, this term may be related to the selectivity coefficient, $K_{El}{}^A$, since

$$K^A = \frac{[A^r]}{[A^M]} \qquad (2.1b)$$

and

$$K^A = K_{El}{}^A \frac{[El^r]^{|Z_A|}}{[El^{aq}]^{|Z_A|}} \qquad (12.22)$$

Under the conditions of chromatography, virtually all the sites on the resin have $El^{\pm 1}$ as counterions; thus $[El^r]^{|Z_A|}$ depends on the capacity of the resin and is essentially constant. Similarly, under these conditions, $K_{El}{}^A$ is invariant. Let the product of these two terms be $J_{El}{}^A$. Then

$$K^A = \frac{J_{El}{}^A}{[El^{aq}]^{|Z_A|}} \qquad (12.23)$$

and

$$\alpha = \frac{K^A}{K^B} = \frac{J_{El}{}^A}{J_{El}{}^B} = K_B{}^A \qquad (Z_A = Z_B) \qquad (12.24)$$

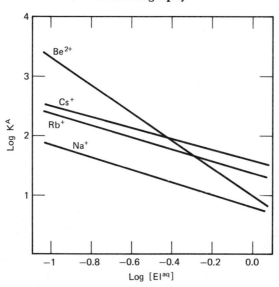

Figure 12.6. The effect of eluent concentration upon K^A. A hydrogen form cation exchanger with 12% divinylbenzene was utilized in conjunction with an HCl eluent. Reprinted from Ref. 10, by courtesy of the American Chemical Society.

An example of this approach is depicted in Fig. 12.6. Plots of log K^A against log $[El^{aq}]$ yield straight lines with slopes of $-|Z_A|$. For species of the same charge, the difference in log K^A is constant although $[El^{aq}]$ is varied. At low concentrations of eluent, the K values are large, leading to wide separation of peak centers. At higher concentrations of eluent, the retention volumes result in closer peak centers.

It is clear that for a column of given efficiency and dimensions an eluent concentration can be chosen that will provide the desired separation of solutes. An alternative approach is the utilization of stepwise changes in eluent concentration. An eluent concentration is chosen so that the first eluted peak has a desirable retention volume. Subsequent peaks, which would have excessively large retention volumes at this initial eluent concentration, are removed from the column by appropriate increases in eluent concentration. This technique has been modified by slowly increasing the eluent concentration during the course of a separation. The resulting procedure, called *gradient elution*, is formally equivalent to temperature programming in gas-liquid chromatography (see Chapter 5).

When the ions to be separated are of different charge, the situation is somewhat more complex but offers more options for separation. At low eluent concentrations K^A is usually higher for the higher-charged species (if all other factors are about the same; see Eq. 12.11). The slope of the log K^A

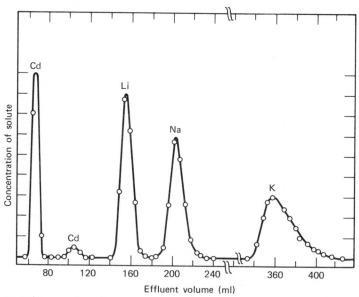

Figure 12.7. Separation of some cations with 0.7 M HCl as an eluent. About 5 hr was required for the separation. Reprinted from Ref. 11, by courtesy of Plenum Publishing Corporation.

versus log [El] line is steeper for the ion of greater charge. Somewhere in the graph a line for a species of high charge crosses the line for a species of lower charge. Thus, at low concentrations of eluent, the solute of lowest charge tends to be eluted first; at some intermediate concentration no separation is possible; and at a higher concentration the solute of greater charge may be eluted first. (Some anomalous effects, such as curvature of the lines at high eluent concentrations, have been encountered, but the aforementioned trends are usually observed.) During the course of stepwise or gradient elution, the solute of higher charge may move more slowly, then at the same rate, and finally faster through the column than a solute of lower charge. A classical separation of some cations is depicted in Fig. 12.7.

12.52 pH

The elution patterns of weak acids or their salts may be influenced markedly by changes in pH (see secondary chemical equilibria, Section 2.13). Weak bases or their salts are similarly affected, but strong acids, bases, and their salts are not influenced by pH. The dissociation of a weak acid may be described by

$$HA \rightleftharpoons H^+ + A^-; \qquad K_{eq} = \frac{[H^{aq}][A^{aq}]}{[HA^{aq}]} \qquad (12.25)$$

The anion is free to exchange with eluent anions in the resin, as in the case

$$A^- + (R^+ + El^-) \rightleftharpoons (R^+ + A^-) + El^- ; \qquad K_{E1}{}^A = \frac{[A^r][El^{aq}]}{[A^{aq}][El^r]}$$

Then

$$K^A = D = \frac{[A^r]}{[HA^{aq}] + [A^{aq}]} = \frac{K_{E1}{}^A[El^r]}{[El^{aq}]} \cdot \frac{K_{eq}}{K_{eq} + [H^{aq}]} \qquad (12.26)$$

In the formulation of Eq. 12.26, it was assumed that only the anion, and not the undissociated acid, penetrates the resin. Under the conditions of chromatography, the first two terms in the numerator are constant. When $K_{eq} \gg [H^{aq}]$, this equation reduces to the form of Eq. 12.23. Under these conditions, the acid is fully dissociated and the situation is equivalent to the exchange of salts of strong acids. When $[H^{aq}] \gg K_{eq}$—and in aqueous solution this indicates that $K_{eq} \ll 1$—the second part of Eq. 12.26 is very small. For all reasonable values of the first half of this equation, the resulting value of D nears zero under this condition. Equilibrium constants such as K_{eq} depend upon the acid HA, the temperature, and the solvent. Variation in [El] results in a small change in K_{eq}, which often can be neglected over the range in which [El] is varied in most chromatographic systems. In the intermediate range, where K_{eq} and $[H^{aq}]$ are of comparable magnitude, fixing pH results in the second half of Eq. 12.26 remaining essentially constant. Changes in [El] may be utilized to change D and hence the retention volumes. Similarly, pH may be varied to influence D. Many separations, such as that depicted in Fig. 5.2, utilize both pH and eluent concentration to control retention volumes.

Polyprotic acids (or bases) and their salts may be considered in similar terms. For the acid H_mA the dissociated species are commonly found in the resin phase, while these and the undissociated acid may be present in the external solution. Then

$$D = \frac{[H_{(m-1)}A^r] + [H_{(m-2)}A^r] + \cdots + [A^r]}{[H_mA^{aq}] + [H_{(m-1)}A^{aq}] + [H_{(m-2)}A^{aq}] + \cdots + [A^{aq}]} \qquad (12.27)$$

and

$$D = \frac{(K_{E1}{}^{H(m-1)A}K_1[H^{aq}]^{m-1}/[El^{aq}]) + (K_{E1}{}^{H(m-2)A}K_1K_2[H^{aq}]^{m-2}/[El^{aq}]^2) + (K_{E1}{}^A K_1 K_2 \cdots K_m[El^{aq}]^m)}{[H^{aq}]^m + K_1[H^{aq}]^{m-1} + K_1K_2[H^{aq}]^{m-2} + \cdots + K_1K_2 \cdots K_m} \qquad (12.28)$$

The terms K_1, K_2, and so on indicate the first, second, and higher dissociation constants. At virtually any pH only one or two species are present, resulting in three to five terms in equations analogous to Eq. 12.27. Consider the homologous series of polyphosphoric acids, the salts of which are important in detergents and fertilizers. The acids have the following structure:

$$HO-\overset{\overset{O}{\parallel}}{\underset{\underset{H}{|}}{P}}-\left(O-\overset{\overset{O}{\parallel}}{\underset{\underset{H}{|}}{P}}- \right)_n OH$$

When $n = 2$, the species are referred to with the prefix *tripoly*. (The monomer is referred to using the prefix *ortho*, and the dimer is called *pyrophosphate*.) Another group of related compounds consists of cyclic species, with the ends joined through oxygen bridges. These are referred to with the modifier *meta*, for example, *trimetaphosphate ion*. The equilibrium constants are such that one proton is dissociated from each phosphate group at pH 5 in aqueous solution. Thus

$$D = \frac{J_{El}{}^{H_2A}}{[El^{aq}]^{n+1}} \tag{12.29}$$

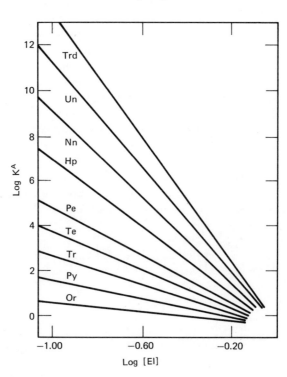

Figure 12.8. Variation of log K^A with log [El] for the linear phosphates. A KCl eluent at pH 5.0 was utilized in conjunction with an anion exchanger. Reprinted from Ref. 12, by courtesy of Pergamon Publishing Co.

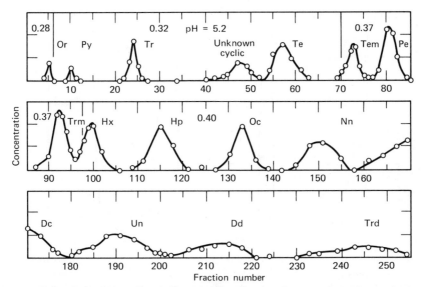

Figure 12.9. Separation of some linear and cyclic phosphates on an anion exchanger. The molarity of the KCl eluent is shown on the figure; pH was 5.0 except where noted, and fraction volumes were 26 ml. Reprinted from Ref. 12, by courtesy of Pergamon Publishing Co.

Plots of log D against log $[El^{aq}]$ result in straight lines with slopes of $-(n + 1)$ (Figure 12.8), and separations can be devised as described in Section 12.51. A separation of these phosphates is depicted in Fig. 12.9. Although other species will have different pH dependence, the trends are usually similar.

12.53 COMPLEXING

Complexing agents may be used to influence retention volumes and hence separation (see Section 2.13). The aqueous rare-earth ions are difficult to separate because of their similar charge-to-size ratios. This result of the lanthanide contraction leads to similar ion-exchange distribution coefficients. However, the formation constants of these ions with a variety of complexing agents are sufficiently different so that separations are practicable. The pertinent equilibria can be written as:

$$L^- + M^{+Z_M} \rightleftharpoons ML^{+(Z_M+1)}; \qquad K_f = \frac{[ML^{aq}]}{[L^{aq}][M^{aq}]} \qquad (12.30)$$

$$D = \frac{[M^r] + [ML^r]}{[M^{aq}] + [ML^{aq}]} = \frac{[M^r]}{[M^{aq}]} \cdot \frac{1 + K_f[L^r]}{1 + K_f[L^{aq}]} = \frac{J_{E1}{}^M}{[El^{aq}]^{Z_M}} \cdot \frac{1 + K_f[L^r]}{1 + K_f[L^{aq}]}$$

$$(12.31)$$

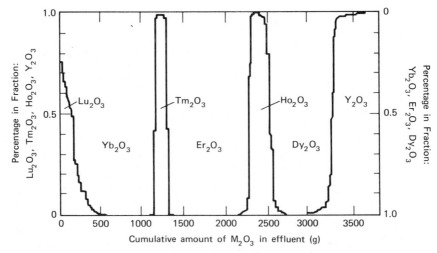

Fig. 12.10. Preparative separation of some rare-earth ions as their citrate complexes on a cation exchanger. Odd-numbered rare earths are plotted upward; even-numbered species, downwards. Concentration is expressed in terms of the oxides of the species recovered. Reprinted from Ref. 13, by courtesy of Academic Press, Inc.

The negatively charged ligand is excluded from cation exchangers by the fixed charges. Thus the numerator of the second part of Eq. 12.31 is smaller than the denominator. Complexing in this fashion leads to a reduction in D and retention volume. In a series such as the lanthanides, the species with the smallest unhydrated crystal radii have the highest K_f and are most affected by complexing agents. An example of the well-known preparative separation of these materials is shown in Fig. 12.10. Retention volumes may be influenced simultaneously by use of a noncomplexing eluent. Control of pH must also be provided, since the concentration of ligand ions is dependent on this parameter through the equilibrium

$$HL \rightleftharpoons H^+ + L^-; \qquad K = \frac{[H^{aq}][L^{aq}]}{[HL^{aq}]} \qquad (12.32)$$

Strong ligands that are polyprotic or cations that can be complexed by a number of anionic species may result in complexes with negative charge. These are excluded from cation exchangers and have effective D values of zero. They may, however, be sorbed on anion exchangers.

Kraus and his coworkers have carried out extensive investigations of the effects of complexing agents on the elution behavior of ionic forms of the elements. They popularized the use of "periodic tables" such as Fig. 12.11. The symbol D_V is equal to the retention volume V_R divided by the bed volume

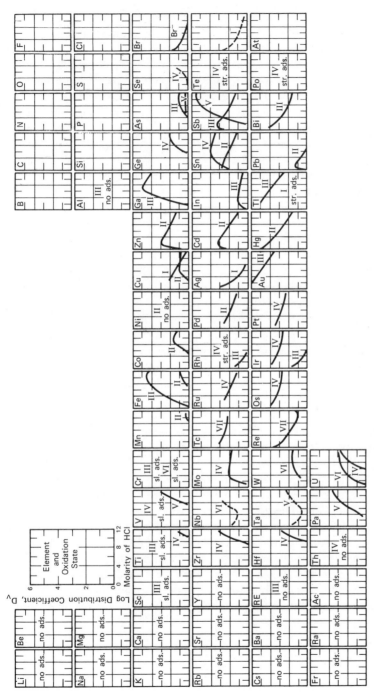

Figure 12.11. Distribution coefficients for the elements on an anion exchanger as a function of the HCl eluent concentration: no ads. = no adsorption, $0.1 < M$ HCl < 12; sl. ads. = slight adsorption in 12 M HCl $(0.3 \leq D \leq 1)$; str. ads. = strong adsorption, $D_v \gg 1$. Reprinted from Ref. 14, by courtesy of United Nations Press.

$(V_M + V_S)$ and is proportional to D. Five types of behavior are noted: (1) no adsorption by the anion exchanger, (2) sorption increasing with complexing agent concentration, (3) sorption decreasing with complexing agent concentration, (4) systems displaying a maximum in D_V, and (5) systems displaying a relative minimum. The first case relates to species that do not form chloride complexes. Case 2 reflects systems with relatively low formation constants for production of the complex anion in which a considerable concentration of HCl is required before the exchangeable species is produced. In case 3, which is often a special form of case 4, the ion usually displays a large formation constant which results in anionic species at low chloride concentrations. Increasing chloride concentration tends to displace the solute from the resin, since the chloride ion is the eluent. Examination of distribution coefficient curves indicates a maximum at a low acid molarity. Other cases with more obvious maxima involve a low degree of complexing at low ligand concentration. Increasing this concentration leads to anions capable of ion exchange with the resin. At high concentration of ligand, the ligand acting as eluent removes the complex from the resin, as in the previous case. Sometimes species of higher negative charge are formed; at these high ligand concentrations, they tend to be removed from the resin with the factor $[El^{aq}]^{|Z_A|}$. This is analogous to the discussion in Section 12.51. In case 5, the systems display minima but this phenomenon is not well understood.

Electrically neutral species that are able to complex ions may be separated by the exchange process. A well-known example involves the separation of glycols, including some sugars, through the charged species that result from their borate adducts. An anion exchanger in the borate form and a buffered sodium borate eluent were utilized (15). "Ligand exchange" is the name given to the approach in which separations of ligands are produced by their interaction with central ions sorbed by a resin. The eluent contains another ligand that competes for central ions with the ligands to be separated. An example is the separation of amino acids on a cation exchanger in the zinc or cadmium form (16). The acids compete with ammonia in the eluent for sites on the cations. Removal of metal ions from the resin is compensated for by adding these species to the eluent. Another approach has involved the use of chelating resins that bind metal ions strongly, reducing the need for replacement ions in the eluent. Use of metal-ion orbitals in this binding reduces the number of available sites and hence the resin capacity, but this term is usually larger than the capacity of ion exchangers in ordinary separations.

12.54   NONEXCHANGE APPLICATIONS

Ion-exchange resins have been utilized as stationary phases for the separation of nonelectrolytes. Sometimes species may be resolved by elution with

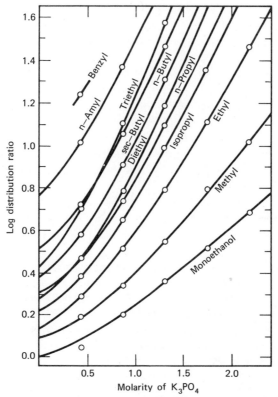

Figure 12.12. Salting out of some amines by K_3PO_4 onto a cation exchanger in the K^+ form. Reprinted from Ref. 17, by courtesy of Elsevier Publishing Co.

water, although greater selectivity can be achieved by the use of aqueous salt solutions. The solute interacts with the hydrocarbon matrix of the resin as well as with the ionic species and solvent molecules in both phases. In contrast to exchange separations, increases in eluent concentration result in increases in distribution coefficients (Fig. 12.12). This removal of solute from the aqueous phase by salt is reminiscent of "salting out" methods for the precipitation of nonelectrolytes or enhanced extraction by organic solvents of limited aqueous solubility. The mathematical form of the classical salting-out equation is similar to the following expression:

$$\log K^A = \log K_0{}^A + \kappa \qquad (12.33)$$

which has been used to describe, phenomenologically, K^A in salting-out chromatography. Here K^A is the observed distribution ratio while $K_0{}^A$ is the distribution ratio with pure water, κ is a constant for a particular species and resin in a stated ionic form, and M is the molarity of salt. Some curvature noted in Fig. 12.12 is due to acid-base equilibria of the amine and potassium

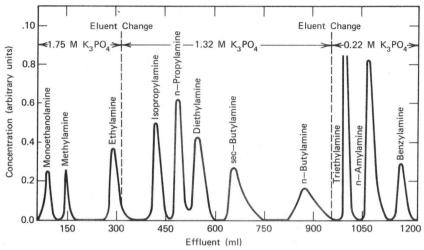

Figure 12.13. Separation of some amines by salting-out chromatography on a cation exchanger. Reprinted from Ref. 17, by courtesy of Elsevier Publishing Co.

phosphate eluent. When the concentration of salt in the eluent is high enough to fix pH, though M is varied, straight lines are observed. Other solutes less prone to acid-base equilibria show rather linear behavior. Separations involve the use of high initial concentrations of salt solutions as eluents, followed by gradient or stepwise reductions in concentration. A considerable number of compounds have been separated by this approach; one group is depicted in Fig. 12.13.

Solutes that show high affinity for the resin may be eluted by the use of organic solvents. Among the well-known examples of this approach are the separation of n-amyl through n-nonyl alcohol with aqueous acetic acid (18) and the separation of a variety of mono- and disaccharides with aqueous ethanol (19).

This and the ion-exchange approach are particularly useful for compounds of low volatility, when derivatization is troublesome and/or the solutes are thermally unstable but compatible with resin systems.

12.55 CROSS LINKING

The significance of cross linking in selectivity has been discussed and is widely recognized; its importance in chromatographic separations, however, has not always been considered. Producers of styrene-based ion exchangers use commercial divinylbenzene, which contains positional isomers of the cross-linking agent as well as ethyl styrene. The reported value for cross linking reflects the actual mole percentage of divinylbenzene isomers present

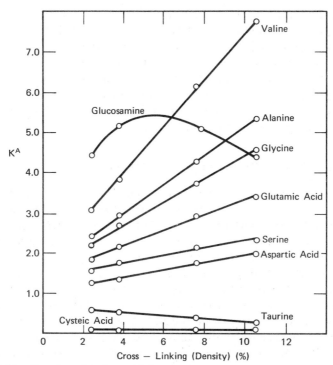

Figure 12.14. Sorption of amino acids by a cation exchanger, in the sodium form, as a function of cross linking as determined by density. Reprinted from Ref. 21, by courtesy of the American Chemical Society.

and is called the *nominal cross linking*. However, not all of the beads in a batch have the same nominal cross linking. Furthermore, incorporation of divinylbenzene into the growing bead occurs at a rate faster than that for styrene. Beads thus formed are rich in cross linking at their centers and have relatively low concentrations of cross links at their peripheries. Similarly, beads formed early in the polymerization are relatively highly cross linked.

Fortunately, resin density is roughly proportional to the percentage of cross linking, and this value may readily be determined. A convenient technique utilizes a series of solutions of sodium tungstate with specific gravities from 1.200 to 1.400. A small sample of resin placed in each solution results in some fraction of the beads being suspended at the top of the solution, in the solution, or deposited at the bottom of the container (20). Thus the degree of heterogeneity of a particular batch of resin may be determined, or if large quantities are utilized the technique may serve for the fractionation of the resin.

The importance of the cross-linking parameter is depicted in Fig. 12.14. Many commercial resins vary $\pm 0.5\%$ in the degree of cross linking. This could be significant for the separation of the amino acid alanine from gluco-samine if the cross linking were 9 rather than a desirable 8.5 or 9.5. Indeed a reversal in order of elution would occur over this range of cross linking. Variation in this parameter has been cited as an important factor in irre-producibility of separations when different batches of the same resin are used. When separations are to be repeated and are carried out in many different laboratories, it is essential that enough information be given to reproduce accurately the pertinent resin conditions.

12.56   HIGH PERFORMANCE CHROMATOGRAPHY

Important separations are often carried out repetitively; in such cases it is desirable to optimize conditions. A significant example is the separation of the amino acids. When Moore and Stein (22) published their cation-exchange procedure, utilizing buffered sodium citrate eluents, it was hailed as an out-standing achievement, although about 3 days was required for the separation. Hamilton (23) and his coworkers improved this procedure, adapted it for clinical use, and increased the sensitivity of the system. As a result, complete amino acid analyses have been carried out on as little as 0.1 ml of serum. They made a careful study of the effect of pH and eluent concentration on the distribution coefficients of the solutes, utilizing a resin with cross-linking homogeneity, chosen as described in Section 12.55. Hamilton et al. showed that peak dispersion σ, in volume units, is directly proportional to column cross-sectional area, while pressure drop across the column is inversely pro-portional to cross-sectional area, temperature, and bead radius, but is proportional to column length and flow rate of eluent. The last term is extremely important, since the time required for an elution is inversely proportional to the flow rate for a particular set of column and solute con-ditions.

Under elution chromatographic conditions the flux of species B within the resin bead was shown to be approximated by the expression

$$J_{B^r} \simeq \frac{15}{dp^2} D^r([B^r]^* - [B^r]) \tag{12.34}$$

where $[B^r]^*$ is the concentration of ion B at the resin surface, and $[B^r]$ is the mean concentration of this species within the exchanger. This is apparent-ly true when σ of the eluted peak is greater than 0.1 dp^2/D^r (24). Particle size and diffusion coefficients within the resin phase are thus important in chromatographic work, as well as in the batch systems described in Section 12.35.

Resin beads have been fractionated carefully to achieve homogeneity of cross linking and particle size, and velocity studies have been carried out as

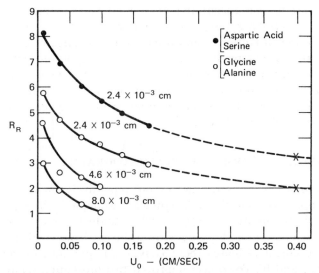

Figure 12.15. Resolution of amino acid pairs ($R_R = 4R_s$) as a function of flow rate. Average particle diameters are indicated for each curve. Reprinted from Ref. 25, by courtesy of the American Chemical Society.

depicted in Fig. 12.15. It is apparent that resolution at any velocity is greater for smaller beads. This is in agreement with conclusions drawn from Eq. 5.34; that is, the plate height contribution due to the stationary phase is directly proportional to the square of the particle diameter. An example of the separation of amino acids attained by Hamilton was given in Fig. 5.2.

Utilization of small particle diameters and high flow rates, resulting in short elution times, has led to large pressure drops across columns. These, in turn, have necessitated the development of pumps capable of maintaining constant flow rates at the high pressures required. An especially rapid separation, in which 7 to 14-μ beads of cation exchanger and a flow rate of 80 ml/hr were utilized, is shown in Fig. 12.16. The desired column conditions required a pressure of 4600 psi but resulted in a separation equivalent in time scale to those usually encountered in gas-liquid chromatography.

A related approach has involved the use of thin layers of ion exchangers coated over inert supports. The utility of this approach can be predicted from Eq. 5.28:

$$H_S = \frac{qk' \, d^2 v}{D_S(1 + k')^2} \tag{5.28}$$

It is apparent that the plate height contribution from the stationary phase is proportional to the square of the depth d of the layer. One approach to improving efficiency in this manner was shown by Horvath et al. (27), who

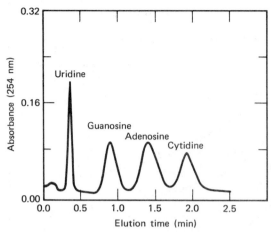

Figure 12.16. Separation of some nucleosides by cation-exchange chromatography. Reprinted from Ref. 26, by courtesy of Elsevier Publishing Co.

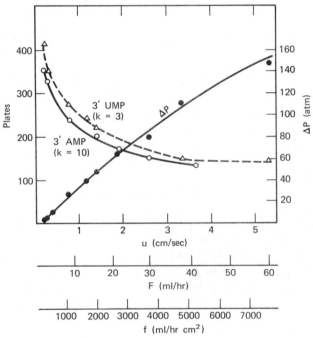

Figure 12.17. Change in number of plates and pressure drop as flow rate is changed for two nucleotides on a pellicular anion exchanger. 3'UMP = uridine-3'-phosphate, 3'AMP = adenosine-3'-phosphate. Reprinted from Ref. 27, by courtesy of the American Chemical Society.

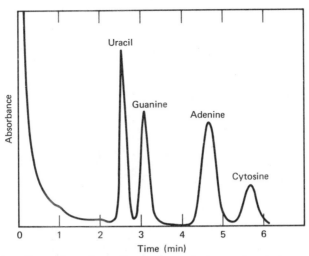

Figure 12.18. Separation of uracil, guanine, adenine, and cytosine on a pellicular cation exchanger. Reprinted from Ref. 28, by courtesy of the American Chemical Society.

coated glass beads with styrene and divinylbenzene and polymerized a thin layer around the glass core. Functionality was produced as described in Section 12.2. The "pellicular exchangers" thus obtained were utilized for the separation of a variety of charged solutes of biological interest. Figure 12.17 depicts the decrease in the number of plates as the flow rate and pressure drop across the column were increased. Figure 12.18 represents the separa-

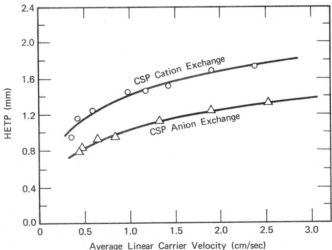

Figure 12.19. Change in HETP (H) as flow rate is varied for uridine-5'-monophosphate on a CSP cation exchanger and 2-aminobenzimidazole on a CSP anion exchanger. Reprinted from Ref. 29, by courtesy of the publishers of *J. Chromatog. Science.*

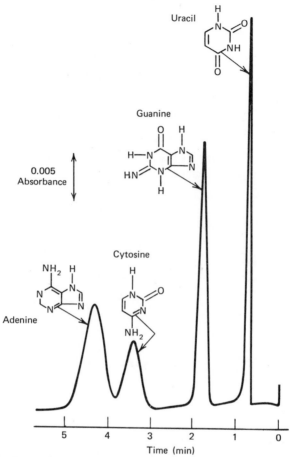

Figure 12.20. Separation of uracil, guanine, cytosine, and adenine on a CSP cation exchanger. Reprinted from Ref. 30, by courtesy of the publishers of the *J. Chromatog. Science*.

tion of some base components of RNA, achieved with a pellicular cation exchanger. Use of this sort of exchanger with relatively large particles allows rapid separations without extremely large pressure drops; however, pellicular exchangers have lower capacities than pure resin beads of similar radius, so that small amounts of sample are required.

Another approach based on similar theoretical reasoning was developed by Kirkland (29), who introduced the use of fluoropolymers with sulfonic acid or tetraalkylammonium chloride functional groups. These consist of a porous crust which surrounds a siliceous core. The materials, dubbed controlled surface porosity (CSP) exchangers, apparently have high performance characteristics, as depicted in Fig. 12.19. They have been used for many separations, one of which is shown in Fig. 12.20.

Obstacles to high performance in liquid chromatographic systems, such as inadequate fluid-pumping devices and low-efficiency stationary phases, appear to have been overcome in principle. However, the requirement for continuous monitoring of effluents has been met only partially, although many advances have been made in recent years and refractive index and ultraviolet absorption monitors with high performance characteristics are now available. These monitoring devices are used widely but are thought to be of limited utility when compared with the thermal conductivity and flame ionization detectors used in gas chromatography. It is hoped that future developments in the area of monitoring devices will lead to high-performance instruments of broader utility than those now available.

12.57 ION EXCHANGE MEMBRANES

These materials have become extremely important in large-scale separations. The emphasis is on deionization and removal of small ions from solutions containing macroions. Little or no effort is made to utilize resin selectivity. The membranes are polymeric materials, often formed on nylon gauze, and contain fixed groups and mobile counterions. In the system shown in Fig. 12.21, cations can pass through cation exchangers, but anions are excluded by the fixed charges on the membrane. The converse applies to

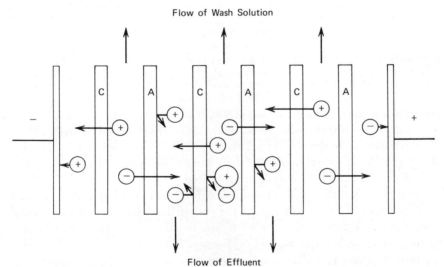

Figure 12.21. Membrane system for deionization of solvents with countercurrent flow.

anion-exchange membranes. In this manner, alternate compartments concentrate or deionize the solution. Countercurrent flow through the system provides for the collection of deionized liquid at one end and concentrated solution at the other. Large ions, as indicated in the central compartment, cannot penetrate the pores of the membrane; this effect is of use in the continuous demineralization of proteins. In another application, one ion may be exchanged for another in solution. Consider a system with two anion-exchange membranes and three solution compartments. Anions in the central compartment are transported toward the anode. These anions are replaced by a desired species from the cathode compartment. This exchange is made with no change in the equivalent concentration in the central compartment.

The use of ion-exchange membranes in various separation processes is discussed in Chapter 16 on membrane separations. Such applications appear to be so promising that predictions have been made that the use of ion-exchange membranes (in fuel cells, in desalinization of water, and in membrane electrodes) will eventually far exceed that of ion-exchange beads.

References

GENERAL

1. F. Helfferich, *Ion Exchange*, McGraw-Hill, New York, 1962.
2. J. A. Marinsky, ed., *Ion Exchange, a Series of Advances*, Marcel Dekker, New York, Vol. I, 1966; Vol. II, 1969.
3. W. Rieman and H. F. Walton, *Ion Exchange in Analytical Chemistry*, Pergamon Press, New York, 1970.

SPECIFIC

4. T. Kawasaki and G. Bernardi, *Biopolymers*, **9**, 257 (1970).
5. H. P. Gregor, *J. Am. Chem. Soc.*, **70**, 1293 (1948).
6. D. Reiehenberg and D. J. McCauley, *J. Chem. Soc.*, **1955**, 2741.
7. G. Eisenman, *Biophys. J.*, **2**, Part 2.2, 259 (1962); G. Eisenman, in "Ion-Selective Electrodes," *Natl. Bur. Std. Special Publ.* 314 (1969).
8. D. Reichenberg, in *Ion Exchange, a Series of Advances*, J. A. Marinsky, ed., Marcel Dekker, New York, Vol. I, 1966.
9. D. H. Freeman, L. A. Currie, E. C. Keuhner, H. D. Dixon, and R. A. Paulson, *Anal. Chem.*, **42**, 203 (1970).
10. D. C. Whitney and R. M. Diamond, *J. Inorg. Nucl. Chem.* **27**, 219 (1965); *Inorg. Chem.*, **2**, 1284 (1963).
11. W. Rieman, *Record Chem. Prog.*, **15**, 85 (1954).
12. H. L. Rothbart, H. W. Weymouth, and W. Rieman, *Talanta*, **11**, 33 (1964).
13. F. H. Spedding and J. E. Powell, in *Ion Exchange Technology*, F. C. Nachod and I. Schubert, eds., Academic Press, New York, 1956.

14. K. A. Kraus and F. Nelson, *Proc. Intern. Conf. Peaceful Uses At. Energy*, Ist, Geneva, **7**, 113 (1956).

15. J. X. Khym, L. P. Zill, and W. E. Cohn, in *Ion Exchangers in Organic and Biochemistry*, C. Calmon and T. R. E. Kressman, eds., Interscience, New York, 1957.

16. Y. Arikawa and I. Makimo, *Federation Proc.*, **25**, 786 (1966)

17. R. Sargent and W. Rieman, *Anal. Chim. Acta*, **17**, 408 (1957).

18. W. Rieman, *J. Chem. Educ.*, **38**, 338 (1961).

19. O. Samuelson, *Ion Exchange, a Series of Advances*, J. A. Marinsky, ed., Marcel Dekker, New York, Vol. II, 1969.

20. M. G. Suryam and H. F. Walton, *Science*, **131**, 829 (1960).

21. P. B. Hamilton, *Anal. Chem.*, **35**, 2055 (1963).

22. S. Moore and W. H. Stein, *J. Biol. Chem.*, **192**, 663 (1951).

23. P. B. Hamilton, D. C. Bogue, and R. A. Anderson, *Anal. Chem.*, **32**, 1783 (1960).

24. D. C. Bogue, *Anal. Chem.*, **32**, 1777 (1960).

25. P. B. Hamilton, *Anal. Chem.*, **32**, 1779 (1960).

26. C. A. Burtis, *J. Chromatog.*, **51**, 183 (1970).

27. C. G. Horvath, B. A. Preiss, and S. R. Lipsky, *Anal. Chem.*, **39**, 1422 (1967).

28. C. Horvath and S. R. Lipsky, *Anal. Chem.*, **41**, 1227 (1969).

29. J. J. Kirkland, *J. Chromatog. Sci.*, **7**, 361 (1969).

30. J. J. Kirkland, *J. Chromatog. Sci.*, **8**, 72 (1970).

List of Symbols

A	counterion	R	fixed functional group
B	counterion initially		in the resin
	present in the resin	$t_{1/2}$	half time for complete
D_v	retention volume divided		conversion
	be the bed volume	v	partial molar volume
E_{Donnan}	Donnan potential	Z	electrical change of the
El	eluent ion		species
J	flux		
J_{El}	constant in Eqs. 12.23	δ	film thickness
	and 12.24	η	electrochemical potential
K_{eq}	dissociation constant	κ	constant in Eq. 12.33
K_f	complexing constant	π	osmotic pressure
K_0	equilibrium constant with	ρ	internuclear distance
	pure water in salting-out	ϕ	electrical potential
	chromatography		
L	ligand		*Superscripts*
m	number of dissociable	aq	aqueous phase
	species	r	resinous phase
M	metal	$^\circ$	thermodynamic quantity
q	resin capacity		expressed with activities

Thirteen

LIQUID-SOLID ADSORPTION CHROMATOGRAPHY

Liquid-solid (adsorption) chromatography is experimentally similar to liquid-liquid and ion-exchange chromatography, so that the equipment and techniques used in these three methods are often interchangeable (see Fig. 10.1). Liquid-solid chromatography (LSC) involves a liquid mobile phase plus a finely divided, porous solid—the *adsorbent*—as stationary phase. The adsorbent should have a relatively large specific surface area, for example, 50 to 1000 m²/g. The equilibrium that governs separation is based on the distribution of sample molecules between the *bulk liquid* phase and the *surface* of the adsorbent.

Liquid-solid chromatography shares many of the advantages and limitations of the other liquid chromatography procedures (Chapters 10, 12, and 15). Its speed and efficiency are typically somewhat less than those of gas-liquid chromatography (GLC),* but it is applicable to high-boiling and/or thermally unstable compounds. Although LSC is especially useful for the separation of less polar, water-insoluble compounds such as hydrocarbons, lipids, and steroids, it is also widely used for polar compounds. Ionic species and compounds with molecular weights over 2000, however, are usually better separated by other methods.

The separation selectivity of LSC is in many respects unique, in relation to other methods. Compound type selectivity is quite pronounced, whereas molecular weight selectivity is much reduced. This means that complex samples can be separated by LSC into *groups* of compounds of similar chemical type but varying extent of alkyl substitution; for example, monoglycerides

---

* Recent results with adsorbents of ~ 5 μ particle diameter indicate performance equivalent to GLC can be achieved (20, 21).

of varying molecular weight are separated as a group from lipid mixtures, as are other distinct chemical types (diglycerides, alcohols, hydrocarbons, etc.). For the same reason, however, LSC is not very useful for separating members of a homologous series (particularly adjacent homologs of higher molecular weight). In this respect LSC is virtually the opposite of GLC, where sample molecular weight generally plays a much more important role than chemical type in separation. Often LSC is unique in its ability to differentiate between various isomeric species. Small differences in the molecular structures of two

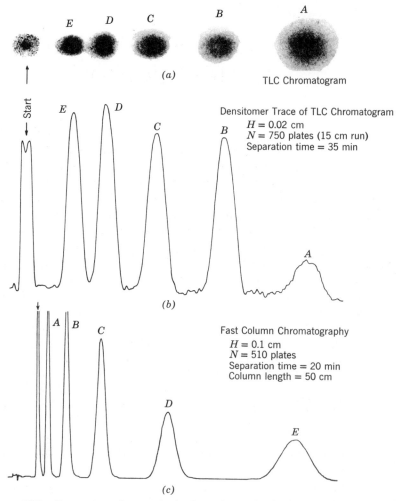

Figure 13.1. Separation of a synthetic dye mixture by both TLC (*a*) and column chromatography (*c*). Reprinted from Ref. 11, by courtesy of Friedr. Vieweg and Son, Publishers.

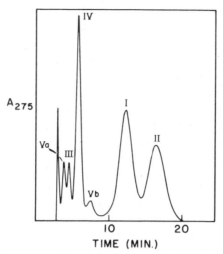

Figure 13.2. Column separation of hydrogenated quinoline mixture on silica with 35% v CH_2Cl_2–pentane solvent. Reprinted from Ref. 1.

compounds often result in large differences in relative adsorption—and in their easy separation by LSC.

As in liquid-liquid chromatography, two basic procedures exist in LSC: column chromatography and open-bed or *thin-layer chromatography* (TLC). Thin-layer chromatography is operationally similar to paper chromatography (Chapter 10). The adsorbent bed is coated as a thin layer onto an impermeable base (usually glass). Separation by TLC then proceeds as in paper chromatography with ascending development (Fig. 10.1*b*). However, TLC has an important advantage over paper chromatography in that corrosive reagents such as concentrated sulfuric acid can be used to indicate the positions of separated bands (by charring), without destroying the underlying (inorganic) adsorbent. Although LSC in columns can be carried out with quite simple equipment (e.g., Fig. 10.1*a*), fast, high-efficiency separations require more elaborate instrumentation (e.g., Fig. 10.1*c*). An example of a TLC separation is shown in Fig. 13.1, and a typical high efficiency column separation in Fig. 13.2. The TLC separation in Fig. 13.1 is shown as the actual chromatogram (*a*), a densitometer trace of the chromatogram (*b*), and a comparative separation of the same sample by column chromatography (*c*).

Liquid-solid chromatography in columns was the first practical form of chromatography. It was introduced by Tswett in 1905 and rediscovered by Kuhn and Lederer in 1931. Column LSC then underwent an explosive growth period and was eventually used to separate almost every type of sample. By the mid 1950s, the relative popularity of LSC had declined some-what because of competition from liquid-liquid, ion-exchange, and gas-liquid chromatography. This trend was sharply reversed, however, by the introduction of TLC in the late 1950s by Stahl. This technique had been invented by Kirchner in 1952, and anticipated earlier by Izmailov and Schreiber (1938). Stahl perfected and standardized TLC, so that its great value was soon widely apparent. More recently, with the introduction of "modern liquid chromatography" in columns (1), renewed interest has developed in column LSC. For a detailed discussion of LSC in general, see Refs. 1 to 3.

13.1   Theory

Any LSC theory must answer the same practical questions that arise in liquid-liquid and ion-exchange chromatography. How are the mobile phase and adsorbent selected for optimum separation in a given case? What is the effect of other experimental variables on separation? To what extent can we predict band migration rates when the structure of the migrating compound (solute) is known? As previously, the answers to these questions involve an understanding of how separation factors α, capacity factors k', and bed plate numbers N vary from one system to another. In this chapter the equilibrium aspects of LSC systems (α and k') are discussed first, followed by an examination of bed efficiency N and related nonequilibrium aspects of LSC separation. For a detailed discussion of LSC theory, see Ref. 2.

13.11   EQUILIBRIUM THEORY

General

In almost all practical LSC systems the surface of the adsorbent is com-pletely covered by a monolayer of adsorbed molecules of solute i or mobile phase M. The equilibrium for adsorption and desorption of sample mole-cules can therefore be written as follows:

$$i_{(M)} + nM_{(a)} \rightleftharpoons i_{(a)} + nM_{(M)} \tag{13.1}$$

A solute molecule initially present in the mobile phase [$i_{(M)}$] adsorbs by dis-placing some number n of mobile-phase molecules from the adsorbent surface. This then gives an adsorbed solute molecule [$i_{(a)}$] and n desorbed molecules of the mobile phase $M_{(M)}$. This process is illustrated in Fig. 13.3. The number n is determined by the relative sizes of solute and solvent molecules (i.e., their relative areas on the adsorbent surface); in the example of Fig. 13.3, $n = 2$.

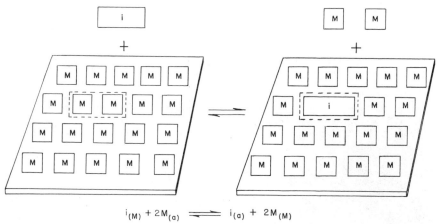

$$i_{(M)} + 2M_{(a)} \rightleftharpoons i_{(a)} + 2M_{(M)}$$

Figure 13.3. Hypothetical representation of the equilibrium between solute molecules (i) and mobile-phase molecules (M) in LSC.

A somewhat different view of the adsorption equilibrium emerges if we recognize that the adsorbed monolayer of M forms a distinct liquid phase adjacent to the adsorbent surface. In this case the adsorption equilibrium can be written as

$$i_{(M)} \rightleftharpoons i_{(a)} \qquad (13.2)$$

In other words, we have a simple partitioning of i between two distinct liquid phases: bulk M and the monolayer of adsorbed M. For infinitely dilute reference states, we have $K_x = K°$ (Section 2.11). Here K_x is the concentration distribution constant, equal to (mole fraction of i in adsorbed monolayer)/(mole fraction of i in bulk M), and $K°$ is the thermodynamic distribution constant corresponding to Eq. 13.2. Because the same liquid M forms both phases, K_x (equal to $K°$) is also equal to K, the molar concentration distribution constant.* Thus we have:

$$K = K° \qquad (13.3)$$

It is more convenient to define K in terms of the concentrations of i in the *total* adsorbent phase: $K' =$ (moles i/grams adsorbent)/(moles i/milliliters bulk M). If we define V_a as the quantity of adsorbed (monolayer) mobile phase per unit weight of adsorbent (ml/g), it is then seen that:

$$K' = K°V_a \qquad (13.3a)$$

---

* $K = [i]_a/[i]_M$ where $[i]$ refers to the concentration (mole s/ml) in adsorbed (a) or mobile (M) phases, see Section 2.11. The structure and properties of the liquid monolayer in direct contact with the adsorbent will differ from the bulk properties. However, this difference should be small, except in the case of water as mobile phase.

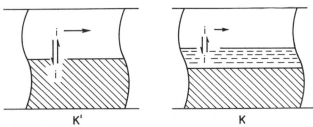

Figure 13.4. Equilibrium in LSC described in terms of bulk adsorbent (K') and adsorbed solvent monolayer (K).

This difference in the definitions of K and K' is portrayed graphically in Fig. 13.4, where i is visualized as partitioning between the mobile phase and either bulk adsorbent (K') or the adsorbed mobile-phase monolayer (K).

It is seen that K' and W (the weight of adsorbent in the bed) replace K and V_S, respectively, in Eqs. 5.6 and 5.9, giving:

$$V_R = V_M + WK' \qquad (13.4)$$

and

$$R_f = \frac{V_M}{V_M + WK'} \qquad (13.4a)$$

Similarly, $k' = WK'/V_M$.

The application of Eqs. 2.2 and 2.4 to the adsorption process (Eq. 13.1) gives $K°$ as a function of the standard chemical potentials of i and M in the mobile and adsorbed phases, respectively ($\mu_{iM}, \mu_{MM}, \mu_{ia}, \mu_{Ma}$):

$$RT \ln K° = \mu_{ia} + n\mu_{MM} - \mu_{iM} - n\mu_{Ma} \qquad (13.5)$$

An important simplification results at this point if we recognize that the mobile-phase terms μ_{iM} and $n\mu_{MM}$ cancel approximately in many LSC systems (particularly those in which the mobile phase is a nonpolar or moderately polar liquid).* Then Eq. 13.5 reduces to:

$$\log K° = \frac{\mu_{ia}}{2.3RT} - \frac{n\mu_{Ma}}{2.3RT} \qquad (13.5a)$$

Now we define the experimental parameters $\alpha'S° = \mu_{ia}/2.3RT$ and $\alpha'\varepsilon° = \mu_{Ma}/2.3RT\,A_M$, where A_M refers to the molecular area of a mobile-phase molecule M (i.e., the surface area required by adsorbed M). Substi-

---

* That is, the net adsorption terms of Eq. 13.5 ($\mu_{ia}, n\mu_{Ma}$) are generally much more important than the solution terms ($\mu_{iM}, n\mu_{MM}$). Briefly, this is so for two reasons: generally *weaker* solution interactions, relative to adsorption, and extensive *cancellation* of solution interactions (which are largely dispersion interactions in less polar liquids M).

tuting these terms into Eq. 13.5a, and recognizing that n is equal to the ratio of molecular areas for i and M, respectively $(n = A_i/A_M)$, gives

$$\log K^\circ = \alpha'(S^\circ - A_i\varepsilon^\circ)$$

Combining this relationship with Eq. 13.3a then yields an important, general relationship:

$$\log K' = \log V_a + \alpha'(S^\circ - A_i\varepsilon^\circ) \tag{13.6}$$

Equation 13.6 describes the variation of solute retention in LSC as a function of certain fundamental, measurable parameters: V_a, α', S°, A_i, and ε°.

The terms V_a and α'* represent characteristic properties of the adsorbent, varying with its type (e.g., alumina, silica, charcoal), surface area, and water content. These two parameters are approximately independent of the nature of the solute i or mobile phase M. Although V_a, the volume of adsorbed phase per unit weight of adsorbent, should vary somewhat with the nature of M (which determines the *thickness* of the adsorbed monolayer), in practice this is not significant. The thickness of the adsorbed monolayer of M is roughly constant for different solvents M. As a result, V_a is proportional to adsorbent surface area, as are solute K values (other factors being equal). The adsorbent activity parameter α' measures the ability of a unit of adsorbent surface to bind adsorbed molecules; the larger is α', the more strongly adsorbed are both i and M, and the larger is K (other factors being equal).

The parameter S° measures the adsorption energy of i onto a standard adsorbent surface (defined as $\alpha' = 1$) from a standard mobile phase M (for which $\varepsilon^\circ = 0$). Similarly ε° is the adsorption energy of M per unit area of standard adsorbent surface. We will refer to ε° as the solvent strength parameter; the larger is ε°, the smaller are all K values. In Eq. 13.6 A_i is the molecular area of the solute molecule i. The significance of the various parameters of Eq. 13.6 will be developed further in the following sections.

The Role of the Adsorbent

The effect of the adsorbent on separations by LSC is determined by several adsorbent properties, primarily chemical type, surface area, and water content. We now discuss each of these in turn.

"Chemical type" refers simply to the bulk composition of the adsorbent, for example, silica or alumina. Differences in chemical type imply differences both in surface functional groups and in the types of intermolecular interaction between adsorbent and adsorbed molecules. These in turn lead to variation in the relative migration of two solutes (adsorbent selectivity or α

---

* The adsorbent activity parameter α' should not be confused with the separation factor α. There is no connection between these two quantities.

value) as a function of adsorbent type. Thus $S°$ and $\varepsilon°$ in Eq. 13.6 vary to some extent with adsorbent type.

The common adsorbents can be divided into two major groups on the basis of type: polar adsorbents, such as silica, alumina, and other inorganic gels; and nonpolar adsorbents, such as charcoal and so-called hydrophobic (i.e., dehydroxylated) silica.* The polar adsorbents interact with adsorbed molecules via specific forces such as electrostatic attraction and hydrogen bonding. Dispersion interactions tend to cancel and hence are less important in determining relative adsorption. As a result, adsorption energy, chromatographic retention, and solvent strength tend to increase with adsorbate solubility parameter δ or "polarity," for example, saturated hydrocarbons < unsaturated and aromatic hydrocarbons < halogen derivatives < ethers < esters < ketones and aldehydes < alcohols < acids and bases. Thus the separation sequence of different compound types (i.e., by polarity) is similar for both normal LLC (polar stationary phase) and LSC on polar adsorbents.

Polar compounds are not preferentially adsorbed on the nonpolar adsorbents (most commonly charcoal). Dispersion forces are the dominant contribution to adsorption energy, so that higher-molecular-weight compounds and aromatic derivatives (i.e., more polarizable compounds—those with higher refractive indices) are preferentially retained on charcoal. For further details, see Refs. 2 and 4.

The polar adsorbents can be further differentiated as acidic or basic; for example, silica is weakly acidic because of its surface Si—OH groups, whereas alumina is strongly basic by virtue of its surface O^{2-} groups. Acidic adsorbents preferentially retain bases, in order of increasing basicity. Likewise basic adsorbents preferentially retain acidic compounds, including such weak acids as phenols and pyrroles. Significant differences in adsorbent selectivity can also be achieved by impregnating or coating an adsorbent with a complexing agent that preferentially binds a particular functional group. Silver nitrate is widely used for the selective adsorption of olefins. Olefin K values are increased relative to other compound types, and the α values of different olefin pairs are also raised. As a result the separation of olefinic derivatives is generally improved on silver nitrate-impregnated adsorbents. Inorganic borates, tungstates, molybdates, and arsenites are used similarly for the selective adsorption of *vic*-diols (e.g., glycols or sugars). Other impregnating agents that have been used in LSC include picric acid, 2,4,7-trinitrofluorenone, and related compounds for the selective separation of aromatic hydrocarbons, bisulfites for aldehydes and ketones, and mercuric ion for mercaptans and alkyl sulfides.

The surface area of an adsorbent affects V_a directly but has little effect on

---

* In terms of the solubility parameter classification of Section 10.11 (Table 10.1), polar adsorbents have nonzero values of δ_o, δ_a, and δ_b; for nonpolar adsorbents, $\delta_o = \delta_a = \delta_b = 0$.

other parameters in Eq. 13.6. Increase in adsorbent surface area therefore results in a general increase in K for all solutes (with a given mobile phase), but affects selectivity or α values only slightly. The addition of water to polar adsorbents leads to a selective covering or blocking of the most active parts of the adsorbent surface.* This causes a reduction in V_a,† because less surface is exposed for adsorption, and there is a decrease also in α', because the remaining unblocked surface interacts less strongly with adsorbing molecules. Addition of water to an initially dry adsorbent leads, therefore, to a general decrease in K for all sample components, accompanied by a modest decrease in separation selectivity. Conversely, the thermal activation of a polar adsorbent (to remove adsorbed water) results in an adsorbent with larger K values and enhanced selectivity. For a tabulation of values of V_a and α' for different polar adsorbents as a function of their water content, see Ref. 2.

The variation of adsorbent activity (i.e., V_a and α') with water content and the surface area of the starting adsorbent leads to a general problem in adsorbent reproducibility. Generally it is possible to reproduce the chromatographic properties of an adsorbent sample previously used by starting with adsorbent of the same type and similar surface area, and then controlling the water content to give K values of the right magnitude. Adsorbent of a given water content (i.e., a standardized adsorbent) rapidly equilibrates with water from the atmosphere, so that exposure of standardized adsorbents to the air must be minimized.

In selecting an adsorbent, attention should be paid to several aspects of separation: sample k' and α values, the adsorbent linear capacity (the maximum sample size that can be used without changing solute k' values; see the end of this section), the possibility of adsorbent-induced reactions of the sample, separation efficiency (see Section 13.12), and the commercial availability of the adsorbent in a useful form. Sample k' values are generally controlled by varying the mobile phase (see below) rather than by changing the adsorbent. Adsorbent water content affects both selectivity and linear capacity: selectivity increases and linear capacity decreases with decreasing water (for less than monolayer coverage of the adsorbent surface by water). Generally it is preferable to work with adsorbents that have been deactivated by at least half a monolayer of adsorbed water (i.e., 0.02 g water/100 m² of adsorbent surface; typical values are 0.03 to 0.15 g water/g adsorbent). Water-deactivated adsorbents are superior in yielding more efficient beds

---

* In most adsorption systems of chromatographic interest, both the adsorbent surface covered by water and the water itself are chromatographically inert. In other words, significant adsorption of sample onto this surface does not occur, and there is negligible *partitioning* of sample into the adsorbed water phase.

† The value of V_a can be calculated (ml/gm) follows: $V_a = 0.00035$ (adsorbent surface area) − (adsorbent water content in ml/gm).

(larger N values), in reducing the danger of adsorbent-catalyzed sample reactions, and in eliminating band tailing.

Alteration or loss of sample occurs more frequently during LSC separation than with other liquid chromatographic methods. Such reactions have been observed less often on silica than on other adsorbents; alumina is particularly prone to sample reaction in the case of base-sensitive samples. Sample oxidation seems to be promoted by adsorbents in general, so that rigorous exclusion of oxygen is required in LSC when dealing with oxidizable samples. With the advent of very fast LSC separations ("modern liquid chromatography" in columns), the danger of sample alteration has been greatly reduced, because the time available for reaction is small.

Silica and alumina are the two most widely used adsorbents, and each is commercially available in a variety of useful forms. Water-deactivated silica is a good first choice for the separation of an unknown sample, because it meets all the general requirements of an LSC adsorbent and is superior to other adsorbents in most of these properties. Alumina, Florisil, magnesia, and charcoal also find application, chiefly because of the differences in separation selectivity that they provide. Thus for quite complex samples (see Section 19.3) no single adsorbent may be capable of resolving all compounds of interest. In this situation the successive application of two or more adsorbents can be a powerful technique. Particular separation problems sometimes suggest the use of a special adsorbent (e.g., olefin separations on silver nitrate-impregnated silica).

The Role of the Mobile Phase

The mobile phase in LSC is usually selected to give solute k' values within the optimum range of 1 to 20. So-called strong solvents provide small k' values, whereas weak solvents give large k' values. Solvent strength is defined quantitatively by the parameter $\varepsilon°$ of Eq. 13.6, and a series of solvents arranged in order of increasing strength or $\varepsilon°$ values is referred to as an *eluotropic series*. Table 13.1 shows an eluotropic series for LSC separation on alumina, along with associated $\varepsilon°$ values. Relative solvent strength (i.e., the order of solvents in Table 13.1) is roughly the same for all polar adsorbents. A solvent of suitable strength for a given separation can be selected conveniently from Table 13.1 (or similar tables) by trial and error, just as in the selection of solvents for LLC from Table 10.1. In this respect, TLC is particularly useful since the relative migration rates of sample components with a given solvent can be determined quickly and easily. This information in turn indicates whether solvent strength must be increased or decreased for optimum resolution. Corresponding column separations are then carried out with the same solvent or one of similar strength.

Solvent strength is determined by how strongly solvent molecules adsorb on

Table 13.1. Solvent Strength for Alumina as Adsorbent

Solvent	ε°	Viscosity $(cP, 20^\circ)$
Fluoroalkanes	-0.25	
n-Pentane	0.00	0.23
Isooctane	0.01	
Petroleum ether, Skellysolve B, etc.	0.01	0.3
Cyclohexane	0.04	1.00
Cyclopentane	0.05	0.47
Carbon disulfide	0.15	0.37
Carbon tetrachloride	0.18	0.97
Isopropyl ether	0.28	0.37
Isopropyl chloride	0.29	0.33
Toluene	0.29	0.59
n-Propyl chloride	0.30	0.35
Benzene	0.32	0.65
Ethyl bromide	0.35	
Ethyl ether	0.38	0.23
Ethyl sulfide	0.38	0.45
Chloroform	0.40	0.57
Methylene chloride	0.42	0.44
Ethylene dichloride	0.44	0.79
Acetone	0.56	0.32
Dioxane	0.56	1.54
Tetrahydrofuran	0.57	
Ethyl acetate	0.58	0.45
Methyl acetate	0.60	0.37
Diethylamine	0.63	0.38
Nitromethane	0.64	0.67
Acetonitrile	0.65	0.37
Pyridine	0.71	0.94
Dimethyl sulfoxide	0.75	2.24
Iso-propanol, n-propanol	0.82	2.3
Ethanol	0.88	1.20
Methanol	0.95	0.60

the adsorbent surface. Consequently, for polar adsorbents ε° increases with solvent polarity. On nonpolar adsorbents such as charcoal, solvent strength is virtually the reverse of that for polar adsorbents. Solvent strength increases with the molecular size of the solvent and is greater for aromatic solvents. Thus one eluotropic series for LSC on charcoal is as follows: water (weak) < methanol < ethanol < acetone < propanol < ethyl ether < butanol < ethyl acetate < n-hexane < benzene (strong).

The strength of a solvent mixture A–B changes continuously from that of A to that of B as the proportion of B in the mixture goes from 0 to 100%. Consequently any $\varepsilon°$ value between that of pure A and B is readily obtained. Fine adjustments in solvent strength are possible with solvent mixtures, which are convenient in formulating eluotropic series; a few pure solvents can cover a wide range of $\varepsilon°$ values. Solvent mixtures also permit close control over other solvent properties, while solvent strength is held constant (see below). Table 13.2 illustrates the variation of solvent strength with composition for several binary mixtures. As a strong solvent is added to a weak one, $\varepsilon°$ for the mixture increases rapidly at first and then levels off.

Although the primary role of the solvent is the control of sample k' values, the solvent can also serve to vary separation selectivity (α values). As solvent strength is changed, Eq. 13.6 predicts that the K values of large sample molecules (i.e., those possessing large A_i values) will decrease more rapidly than those of small molecules. Consequently a change in solvent strength leads to changes in the α values of band pairs of dissimilar size. This is illustrated by the data of Table 13.3.

Equation 13.6 predicts that the α values for molecules of similar size will be unaffected by changes in solvent strength or composition. This is approximately true for weak solvent systems, but it becomes less true as solvent

Table 13.2. Solvent Strength of Binary Solutions for Alumina as Adsorbent; Solutions of Solvent B in *n*-Pentane
Values of volume per cent B shown for given $\varepsilon°$ value

				Solvent B				
$\varepsilon°$	CS_2	*i*-PrCl[a]	Benzene	Ethyl Ether	$CHCl_2$	CH_2Cl_2	Acetone	Methyl Acetate
0.00	0	0	0	0	0	0	0	0
0.05	18	8	3.5	4	2	1.5		
0.10	48	19	8	9	5	4	1.5	
0.15	100	34	16	15	9	8	3.5	2
0.20		52	28	25	15	13	6	3.5
0.25		77	49	38	25	22	9	5
0.30			83	55	40	34	13	8
0.35				81	65	54	19	13
0.40					100	84	28	19
0.45							42	29
0.50							61	44
0.55							92	65
0.60								100

[a] isopropyl chloride.

Table 13.3. Change in Solvent Selectivity with Solvent Strength for Two Solutes of Differing Molecular Sizes

Solute	K (ml/gm)	
	Pentane Solvent $(\varepsilon^\circ = 0.00)$	Benzene solvent $(\varepsilon^\circ = 0.32)$
	365	0.6
	46	2.0
	$\alpha = 7.9$	$\alpha = 0.3$

strength increases.* Consequently a change in solvent composition (particularly when $\varepsilon^\circ \gg 0$) often provides a useful change in separation selectivity or α. While changing solvent composition it is important to maintain solvent strength approximately constant, so that solute k' values remain in the optimum range of 1 to 20. For example, if two sample components are unseparated on alumina with 29% methyl acetate–pentane as solvent, a different solvent of similar ε° value (0.45; see Table 13.2) might result in a change in α, with consequent separation of the two compounds; for example, ethylene dichloride (Table 13.1) or 42% acetone–pentane (Table 13.2) might be effective. Solvents of equal strength that can be interchanged in this fashion have been referred to as *equieluotropic* by Neher (5).

Changes in separation selectivity or α among equieluotropic solvents are likely to be greatest for one of two situations. First, for solvent *blends*, maximum differences in selectivity are associated with cases in which the strong component of the solvent (e.g., chloroform in solutions of pentane–chloroform) is present in greatly different concentrations. This is illustrated in the representative data of Table 13.4, where α for the two solutes in question increases regularly as the concentration of the strong solvent component

---

* This minor (but often important) failure of Eq. 13.6 can be traced to several factors. The neglected solution terms μ_{iM} and $n\mu_{MM}$ of eq. 13.5 often do not cancel in polar solvents, leading to useful variations in separation selectivity. Similarly the adsorption energy of a very polar solvent molecule is often different on different parts of the adsorbent surface (see the discussion in Ref. 2, pp. 216–229 and Ref. 7).

**Table 13.4.   Solvent Selectivity in the Separation of
1,5-Dinitronaphthalene (DDN) from 1-Acetonaphthalene
(AN) on Alumina (7)**

| Solvent | k' | | α |
	DNN	AN	
50% Benzene–pentane	2.5	5.1	$0.5 \equiv 2.0$
23% CH_2Cl_2–pentane	5.8	5.5	0.95
4% Ethyl acetate–pentane	5.4	2.9	1.8
5% Pyridine–pentane	5.4	2.3	2.4
0.05% Dimethyl sulfoxide–30% CCl_4/pentane	3.5	1.0	3.5

decreases. A maximum change in selectivity occurs for the two extreme solvents, 50% benzene–pentane and 0.05% dimethyl sulfoxide–CCl_4–pentane. For further discussion of this effect, see Ref. 7.

The second situation in which selectivity is likely to be changed significantly by a change in solvent involves hydrogen bonding by solute and solvent. The use of more basic solvents or of strong proton-donor solvents (see Table 10.1, δ_a and δ_b values) will frequently provide a marked change in separation selectivity. For further details, see Refs. 2 and 7.

The choice of a solvent in LSC, as in LLC, is affected by other considerations than control of k' and α values. Less viscous solvents favor high separation efficiency or N values (Section 13.12). Low-boiling solvents are desirable for the recovery of sample components (preparative separation). In preparative separations the solubility of the sample in a given solvent may be an important consideration. Finally, the solvent must not interfere with the detection of separated sample bands.

Band Migration and Molecular Structure

It has already been noted that band migration rates for a given mobile phase on polar adsorbents decrease with increasing polarity of the migrating compound. Saturated hydrocarbons are the first compounds to be eluted in LSC, and very polar compounds such as acids and bases elute last. Relative sample separation order is determined mainly by the parameter $S°$ of Eq. 13.6, the adsorption energy of i in a standard adsorption system. However the size of a molecule i (its A_i value) also affects band migration rates, larger molecules (of equal $S°$) migrating more rapidly in strong solvent systems.

According to the Martin relationship (Eq. 2.47) $S°$ should be the sum of contributions $Q_i°$ from different groups j in the molecule S:

$$S° = \sum^{j} Q_j° \tag{13.7}$$

The Q_j^o values (group adsorption energies) for some common molecular groups are summarized in Table 13.5 for alumina as adsorbent. The value of Q_j^o is seen to vary with the type of groups—alkyl or aromatic—that are attached to the substituent j in the molecule i. Examination of the values of Table 13.5 reveals a number of general relationships, some of which have been referred to earlier. Alkyl groups are weakly adsorbed (small Q_j^o values and minor contribution to S^o) and have only a slight effect on sample migration rates. For example, on alumina we can calculate the following S^o values:

$S^o = 5.3$ (CO$_2$CH$_3$ structure) $S^o = 5.5$ (CO$_2$C$_{11}$H$_{23}$ structure)

Thus, there is little difference in the relative migration rates of these two compounds on alumina, despite the large difference in alkyl substitution. Because of the effect of A_i on sample migration rate (Eq. 13.6), higher homologs are generally retained *less* strongly on polar adsorbents.

The Q_j^o value of a given group is smaller when substituted onto an aromatic nucleus than onto an alkyl group, as may be seen from Table 13.5. However, the difference in Q_j^o values for an alkyl derivative of j, $R - j$, and a corresponding benzene derivative, $C_6H_5\text{-}j$, is generally about equal to the adsorption energy of a benzene ring. Consequently, the alkyl and aromatic derivatives of a given compound type tend to have similar S^o values and the same relative migration rates. Two examples of this are the aldehydes and chlorides:

C_6H_{13}—Cl   (Cl on benzene) C_6H_{13}—CHO   (CHO on benzene)

$Q_{Cl}^o = 1.82$ $Q_{Cl}^o = 0.20$ $Q_{CHO}^o = 4.73$ $Q_{CHO}^o = 3.35$
$S^o = 1.94$ $S^o = 2.06$ $S^o = 4.85$ $S^o = 5.21$

Acidic adsorbents such as silica retain basic solutes more strongly, while basic adsorbents such as alumina and magnesia preferentially adsorb acidic solutes, even those as weak as the pyrroles. According to Eq. 13.7, polyfunctional solutes (e.g., diesters, hydroxyketones) are retained more strongly than corresponding monofunctional solutes (esters, alcohols, ketones).

Equation 13.7 in conjunction with Table 13.5 provides a useful qualitative description of band migration rates in LSC as a function of sample molecular structure. However, quantitative adherence to Eq. 13.7 is the exception in LSC, mainly because of *localized adsorption* of strongly held solutes. When a sample molecule i adsorbs onto a polar adsorbent surface, polar functional

Table 13.5.   Solute Group Adsorption Energies Q_j° for Alumina as Adsorbent (2)

| Group | Q_j° | | |
	X, Y = Ar[a]	X = Al,[a] Y = Ar[a]	X, Y = Al[a]
X—CH$_3$ (methyl)	0.06	——	−0.03
X—CH$_2$—Y (methylene)	0.12	0.07	0.02
—C= (aromatic or olefinic carbon)[b]	0.31	(0.31)	(0.31)
X—Cl (chloro)	0.20	——	1.82
X—Br (bromo)	0.33	——	2.00
X—SH (mercapto)	8.70	——	2.80
X—S—Y (sulfide)	0.76	1.32	2.65
X—O—Y (ether)	1.04	1.77	3.50
X$_2$—N—Y (tertiary amine)	?	2.48	4.40
X—CHO (aldehyde)	3.35	——	4.73
X—CO$_2$—Y (ester)	2.67	3.40	5.00
X—CO—Y (keto)	3.3	3.74	5.00
X—OH (hydroxyl)	7.40	——	6.50
X—NH$_2$ (amino)	4.41	——	6.24
X—CONH$_2$ (amide)	6.2	——	8.9

[a] Ar refers to an aromatic substituent and Al to an aliphatic substituent on the solute group j.

[b] By convention it is assumed that Q_j° for an aromatic carbon atom is independent of the type of group attached to the —C= group.

groups on i tend to adsorb at specific points on the surface—so-called strong adsorption sites. This is illustrated in Fig. 13.5 for substituted benzene solutes; the groups X and Y are assumed to be polar, strongly adsorbing substituents, and the points A in the figure represent strong adsorbent sites. In the case of disubstituted compounds, the strong interaction of one substituent with a strong site (localized adsorption) generally precludes the localized adsorption of other substituents in the solute molecule, because of the rigid structure of the adsorbent surface. As a result, the adsorption energies and Q_j° values of these nonlocalized solute groups are reduced, in comparison to the case in which they are the only polar substituent in a solute molecule. This results in the effective failure of Eq. 13.7 and the Martin relationship (Eq. 2.48 or 10.4); the S° values of polyfunctional solute molecules are usually less than the values calculated from Eq. 13.7 and Table 13.5. The complication of localized adsorption can be taken into account (2), and accurate S° values can then be calculated, but this is beyond the scope of the present treatment.

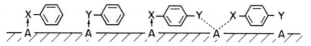

Figure 13.5. Localized adsorption of benzene derivatives; X and Y are polar (strongly adsorbing) substituents.

Apart from the complication of localized adsorption, Eq. 13.7 can fail for other reasons, some of which were discussed in Section 2.22. These effects are of interest here mainly because they lead to differences in the relative adsorption of isomeric compounds. As already pointed out, LSC is often unique in its ability to separate isomers.

Consider first the general class of disubstituted benzenes: ortho-, meta-, or para-X—C_6H_4—Y. In the case of the meta and para isomers, the two groups X and Y can interact electronically (intramolecularly), with resulting changes in the net charges on these groups. For example, if X is methyl and Y is amino, the methyl group will donate electrons to the amino group. The resulting increase in electron availability at the amino group (relative to the amino group in aniline) changes the energy of interaction of this group with a polar adsorbent surface—in general, increased electron availability means stronger adsorption. Thus m- or p-methylaniline will be adsorbed somewhat more strongly than is predicted by Eq. 13.7, since this equation does not recognize the possibility of intramolecular interaction between solute groups. More important, since electron donation by a methyl group is greater at the para than the meta positions, m-methylaniline will be adsorbed less strongly than p-methylaniline.

A similar situation arises in the case of the halogen-substituted aniline derivatives, since the adsorption energy of an aromatic halogen substituent is small (see Table 13.5), and electron donation is greater for para than for meta halogen substituents. Hence p-fluoro-, chloro-, bromo, and iodoaniline are more strongly adsorbing than their corresponding meta isomers.

In the case of o-methylaniline, intramolecular steric effects come into play along with electronic effects. The methyl group in the ortho position interferes with the interaction between the amino group and the adsorbent surface, thereby reducing the adsorption energy of the amino group. Therefore, although electronic effects in o- and p-methylaniline are probably comparable in affecting the adsorption energy of each compound, the ortho isomer is less strongly adsorbed. Similar intramolecular blocking of polar solute groups by adjacent alkyl or phenyl groups leads to reduced adsorption energies for most polar groups and to more rapid migration of the isomers in which steric hindrance is present.

Another type of intramolecular interaction that has an important effect on relative solute adsorption is hydrogen bonding. An example was given in

Section 2.22, that of *o*-nitrophenol. Here the interaction of the nitro and hydroxy groups with each other decreases the likelihood of similar interactions of each group with the adsorbent surface. Consequently *o*-nitrophenol is adsorbed less strongly than either the meta or the para isomer. It is a general rule that compounds in which strong intramolecular hydrogen bonding occurs will be adsorbed less strongly than corresponding isomers in which hydrogen bonding is absent.

The relative adsorption of isomeric compounds is also affected by *site chelation*, molecular *planarity*, and *localization* effects. When two polar substituent groups in a molecule are adjacent, but cannot hydrogen-bond with each other, the simultaneous interaction (chelation) of both groups with a single adsorbent site can lead to stronger adsorption. Thus *o*-dimethoxybenzene is more strongly adsorbed than its meta or para isomer. Because the adsorbent surface is roughly planar, solute adsorption energy decreases with increasing lack of planarity in the solute molecule. The phenylanthracene derivatives exhibit decreasing planarity (i.e., increased crowding of phenyl and anthracyl groups) in the sequence 2-, 1-, and 9-phenylanthracene, and this is also the order of decreasing relative adsorption.

Localization effects can lead to isomer separation, as is evident from the example of Fig. 13.5. Thus, if simultaneous adsorption onto strong sites is possible for *both* groups X and Y in one of the isomers of X—C_6H_4—Y, that isomer will be preferentially adsorbed. This reflects a matching of the positions of strong sites (*A* in Fig. 13.5) on the adsorbent surface to the positions of polar groups (X, Y) in the solute molecule. Such an effect is noted for the adsorption onto alumina of certain aromatic isomers (e.g., polycyclic hydrocarbons and their halogen-substituted isomers). Here strong adsorbent sites appear to be distributed linearly along the surface, leading to preferential adsorption of more nearly linear solute molecules; for example, anthracene is more strongly adsorbed than phenanthrene, and 4-chlorobiphenyl is more strongly adsorbed than 3-chlorobiphenyl.

Other Aspects of Adsorption Equilibria

In chromatographic separation it is important that band migration rates remain constant as sample size or concentration is changed. In LSC band migration rates are independent of the amount of sample for small samples, but as sample size is increased the migration rates of different bands eventually begin to increase. We can define the linear capacity θ^* of an adsorbent as the sample size (g sample/g adsorbent) at which band migration rates first begin to change significantly (e.g., by 10%) with increase in sample size. The linear capacity of some adsorbents can be impractically low (e.g.,

---

* That is, the capacity of the adsorbent for *linear* isotherm separation.

$\theta = 10^{-6}$), and some attention must usually be given to maximizing θ in LSC separation. The most important variables in this respect are adsorbent surface area and water content. Other factors being equal, adsorbent linear capacity is proportional to adsorbent surface area. The linear capacities of water-free adsorbents are usually 10 to 100 times smaller than those of corresponding adsorbents with half a monolayer of adsorbed water. Typical water-deactivated adsorbents (silica, alumina) have surface areas of 200 to 800 m²/g, with θ equal to 10^{-4} to 10^{-3}. The value of θ is greater for sample mixtures than for single components. Apart from the effect of linear capacity on band migration rates, separation usually becomes markedly poorer as sample size exceeds the linear capacity of the adsorbent (see Section 5.22).

The effect of temperature on band migration rates in LSC is similar to that in liquid-liquid or ion-exchange chromatography. Band migration is accelerated by increase in temperature, because the enthalpy of adsorption of a solute molecule is almost always negative. Increasing temperature often leads to a decrease in separation selectivity.

When binary solvent mixtures are used for the mobile phase in LSC, demixing of the solvent can occur in the initial stages of separation. The strongest solvent of the mixture is preferentially adsorbed, leaving the mobile phase depleted of this component. Solvent demixing is generally undesirable, leading to unpredictable band migration rates (because of the changing solvent concentrations within the column) and, in some cases, to poor separation. Solvent demixing is important mainly in TLC. In column LSC, solvent demixing can be avoided by initial equilibration of the column with solvent; that is, ehough solvent is passed through the column to saturate the adsorbent with the stronger adsorbing solvent component; then the sample is charged and separation is begun.

In the case of water-deactivated adsorbents, a dry (water-free) solvent will remove water from the adsorbent surface. This results in a number of adverse effects, such as lowered adsorbent linear capacity, decreased separation efficiency and hence poorer separation, and increased tendency toward adsorbent-catalyzed sample reactions. Therefore, for repeated use of the same LSC column, the water content of the solvent must be adjusted to be in equilibrium with the water on the starting adsorbent.

13.12 SEPARATION EFFICIENCY AND MOBILE-PHASE FLOW

The dynamic (as opposed to equilibrium) aspects of LSC are virtually the same as those of LLC. Consequently most of what was said in Sections 10.12 and 10.13 applies directly to separations by LSC, if we recognize the essential similarity of paper chromatography and TLC.

Equation 10.5a describes the variation of bed plate height H:

$$H = \left(\frac{1}{A} + \frac{1}{C_M v} \right)^{-1} + C_S v \tag{10.5a}$$

in both LSC and LLC; in fact, the only significant difference in the two methods is in the $C_S v$ term. In LLC the diffusion of sample into and out of the stationary phase (a thin layer of liquid) contributes to C_S, whereas in LSC this process is replaced by an adsorption-desorption process at the adsorbent surface. Often neither of these two processes makes an important contribution to H, and then Eq. 10.5a becomes essentially equivalent for both methods.

The variation of bed efficiency N with column length L, column pressure drop P, and separation time t was described in Chapter 10 in general terms. Figure 13.6 provides numerical values of this relationship in the case of a particular LSC system: a column of 20-μ silica. This figure illustrates how different experimental variables can be changed for maximum N_{eff} (Section 5) or minimum t.

With regard to other separation variables, N varies in the same manner for LSC and LLC. Fast, efficient separations are favored by nonviscous solvents, narrow columns, small particle sizes, and pellicular (coated-bead) adsorbents. Water-deactivated adsorbents generally give larger values of N than dry adsorbents. The rate of mobile-phase flow through an LSC column as a function of experimental conditions is essentially the same as in corresponding columns for LLC (Fig. 10.7).

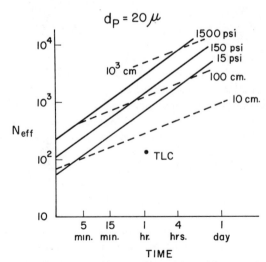

Figure 13.6. Effective plates (N_{eff}) for a column of 20-μ silica as a function of column length (------) and pressure (———), and separation time; $k' = 3$, pentane solvent. Reprinted from Ref. 18, by courtesy of the publisher of *J. Chromatog. Science*.

Bed efficiencies in TLC are normally about 1000 plates (= 150 effective plates) for separation times of $\frac{1}{2}$ to 1 hr. Comparison of this value with the data of Fig. 13.6 makes it clear that column LSC separations can be much faster and/or more efficient than separations by TLC. Similarly, we have already noted the superiority of LLC in columns versus paper chromatography in this respect.

Pellicular (surface-coated glass bead) adsorbents are commercially available; these are comparable to the pellicular supports for LLC (Section 10.12). These materials (e.g., Corasil, Vydac, Zipax) provide higher column efficiencies, at the price of lower column capacity. As yet, the relative advantage of these pellicular adsorbents with regard to column efficiency is not as pronounced as in the case of pellicular supports for LLC (see Ref. 5).

13.2 Experimental Aspects

13.21 COLUMN CHROMATOGRAPHY

The experimental aspects of LSC are closely similar to those of LLC. This is particularly true of the equipment and technique of column chromatography, where no important differences exist; see Sections 10.21 and 10.23 for a general description of the equipment and technique for LSC in columns. Similarly, for a comparison of column LSC with TLC, see the equivalent discussion in Section 10.2 for liquid-liquid column chromatography versus paper chromatography.

Columns can be packed in the same way for LSC as for LLC (Section 10.22). The selection of a given adsorbent (usually a commercial product) is governed by the factors discussed in Sections 10.12 and 13.12. Adsorbents of large particle size give poorer separations but require lower column pressures and shorter separation times (Fig. 10.7) and are more easily packed. Generally a commercial polar adsorbent is first heated at 150 to 200° to dry the adsorbent, and then a known amount of water is readded. This provides a standardized adsorbent, that is, an adsorbent that will approximately duplicate the performance of a similar material prepared previously. Sample size in LSC ranges from 1 to 50×10^{-4} g/g for analytical separations. Considerably larger sample sizes can be handled in preparative separations, particularly for easily resolvable mixtures.

Gradient elution and stepwise elution (Section 5.5) are often used in column LSC. Many sample mixtures contain components of widely varying migration rates, and successful separations can be achieved only by a change of solvent composition during separation. Ideally, for an unknown sample, the solvent gradient should be such that solvent strength $\varepsilon°$ increases linearly with eluate volume. Under these conditions (linear $\varepsilon°$) it can be shown (6) that maximum resolution is maintained throughout separation, as illustrated

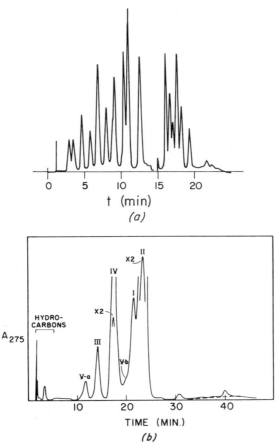

Figure 13.7. Gradient elution separations by LSC (linear change in ε° with time): (*a*) separation of a synthetic mixture of substituted aromatics on silica. Reprinted from Ref. 6, by courtesy of the publisher of the *J. Chromatog. Science.* (*b*) separation of a hydrogenated quinoline mixture on alumina. Reprinted from Ref. 1.

in Fig. 13.7 for two such gradient elution LSC separations: (*a*) a synthetic mixture of substituted aromatic hydrocarbons, and (*b*) a hydrogenated quinoline mixture. Coupled-column operation (Section 5.5) has also been applied to the separation of broad-range mixtures by LSC; for examples and comparisons with gradient elution, see Refs. 8 and 9).

13.22   THIN-LAYER CHROMATOGRAPHY

Much of the popularity of TLC can be attributed to its marked simplicity in equipment and technique. Quite elegant separations can be carried out at little expense by relatively inexperienced workers. At the same time the versatility of TLC can be significantly expanded by added equipment and a variety of special techniques. Here we emphasize only the basic experimental aspects of TLC; for a more complete treatment, see Refs. 3 and 10 and other books on this technique.

Equipment

The equipment used in TLC is related to the four main steps during separation: preparation of the bed or plate, application of the sample, development of the plate, and visualization of the final chromatogram. Adsorbent-coated rigid plates, which are used most often in TLC, can be prepared by spreading an aqueous slurry of the adsorbent onto rectangular glass plates. Although spreading can be done without the benefit of special equipment, to obtain smooth adsorbent layers of uniform thickness a mechanical spreader is required. Ready-made plates of high quality can also be purchased in a wide variety of sizes and adsorbent coatings, so that many laboratories have dispensed with the preparation of their own plates.

Sample application or "spotting" can be achieved with medicine droppers (Fig. 10.1b), capillary tubes, or micropipets; the last of these permits semiquantitative control of sample size. Special, semiautomatic applicators are commercially available that apply the sample as a uniform, narrow streak across the bottom of the plate. These are used mainly for preparative separations or in conjunction with quantitation by densitometry (see below).

The development or separation step in TLC is carried out in a vaportight chamber or tank, just as in paper chromatography (Fig. 10.1b). Ascending development is used exclusively in TLC, so that the developing chamber can be quite simple (e.g., a covered glass jar). So-called S-chambers are also used, consisting of a glass cover plate (of the same size as the TLC plate) which is clamped over the TLC plate. Spacers prevent contact of the cover plate with the adsorbent layer, leaving a small air space over the adsorbent surface. Because this air space is quite small, S-chambers do not require pre-equilibration with the mobile phase. A variety of other special developing chambers have been described (see e.g., Refs. 3 and 10) which permit TLC separations with changing solvent (gradient elution), allow operation at different temperatures, or offer other advantages.

When separation is completed in TLC, the separated (usually colorless) bands must be made visible in some manner and (in some cases) quantitated. Several different types of equipment are useful in this connection, as discussed below. For a list of commercial TLC devices, see Refs. 3 and 10 and recent buyers' guides.

Preparation of the Bed

The adsorbents used in TLC and column LSC are similar. Often TLC adsorbents contain a gypsum binder for firmer coatings, but adsorbents without binder are also used. Small amounts of a fluorescing substance can be added to the adsorbent, for the detection of ultraviolet-absorbing compounds (see below). The thickness of the adsorbent layer is usually between 0.2 and 2.0 mm: 0.2 to 0.3 mm for analytical separations, and thicker layers for preparative separations. The most commonly used plate sizes are 5 × 20 and 20 × 20 cm. Smaller slides (e.g., coated microscope slides) are useful for rapid exploratory separations to establish adequate separation conditions. After the plate is coated with adsorbent slurry, it is dried for 30 min in air and then heated to 110 to 120° to give active (water-free) absorbent.

Technique

Thin-layer chromatographic separations are carried out in a variety of different ways: one-dimensional, two-dimensional, multiple and continuous development, gradient elution and gradient layer, radial TLC, and other less important procedures. Ascending, one-dimensional development (as in Fig. 10.1*b*) with a fixed set of conditions (no solvent or adsorbent gradients) and a single solvent pass constitutes the commonest TLC technique. Two-dimensional TLC can be carried out with the same equipment. The sample is spotted at one corner of a 20 × 20 cm plate, which is developed in the normal manner. The plate is then rotated 90° so that the separated bands from the first development are positioned at the bottom of the plate, and a second development is carried out. Separation conditions are usually altered in the second development to allow separated bands to spread across the entire plate (see Section 5.53). Most commonly the solvent is changed in the second development. Alternatively, the unused part of the plate (after the first development) can be impregnated with silver nitrate for selective separation of olefins in the second development, or coated with hydrocarbon for a reverse-phase TLC separation in the second development (Section 10.4).

Multiple- and continuous-development TLC are carried out in the same way as for paper chromatography (Section 10.33). These techniques provide better resolution of samples that are difficult to separate. Gradient elution and gradient layer TLC require special equipment (see Ref. 3) and are used less often. These techniques permit the separation of complex samples containing components of widely differing migration rates (Section 5.53). Radial TLC is often used for the rapid selection of a solvent of the right strength (one that gives k' values of the right magnitude). Several portions of the sample are spotted at various positions on a TLC plate, and solvents of different strengths are added dropwise to each spot. The plate is then dried, and the resulting bands are made visible as below. A solvent of approximately the right strength will provide sample bands (rings) that are intermediate between the solvent front and the point of sample application.

Separation in normal developing chambers (e.g., Fig. 10.1b) can be carried out with or without prior equilibration of the chamber with solvent vapor. Solvent equilibration improves the reproducibility of separation but has an adverse effect on sample resolution. Plates are seldom equilibrated with solvent vapor before separation—the reverse of the usual practice in paper chromatography. Equilibration of the developing chamber with air of constant humidity is gaining favor (11), since this provides close control over adsorbent water content and gives reproducible band migration rates.

In analytical separations 25 to 100 μg of sample is spotted as above. Preparative separations can be carried out on 10 to 100 mg of sample, using 1- to 2-mm adsorbent layers and applying the sample across the entire bottom of a 20 × 20 cm or larger plate. The initial sample spot or streak should be positioned 1 to 2 cm above the solvent level in the bottom of the developing chamber.

Normally, development of the plate proceeds until the solvent front has migrated 10 to 15 cm beyond the initial sample spot. At the conclusion of development the plate is removed from the chamber and allowed to dry in the air. The plate is then ready for visualization of the separated sample bands, if these are not colored compounds that are already visible. A convenient technique for the visualization of a wide variety of organic bands is exposure of the plate to iodine vapor. A few crystals of iodine are placed in a sealed container similar to a developing chamber, and the plate is inserted into the chamber. Most organic compounds absorb iodine to give dark spots on a light adsorbent background. Since the absorption of iodine is reversible for most compounds, the positions of the separated bands are thus indicated for eventual recovery from the plate. The major limitation of iodine visualization is low sensitivity for most compounds (detection limits of 50 to 200 μg), so that sample sizes larger than normal must be used. Another popular visualization technique is the use of corrosive spray reagents (e.g., concentrated sulfuric acid), followed by heating of the plate. Almost all organic compounds char under these conditions to yield dark, easily visible bands, with detection limits down to less than 1 μg. Several hundred other spray reagents have been described (see e.g., Refs. 3 and 10) for a variety of sample types. Many of these spray reagents are specific for certain compound types (steroids, lipids, phenols, sugars, etc.), and often they give characteristic colors for different compounds. The combination of a characteristic color and a given R_f value for a band is frequently enough for its firm identification, without the need for supplementary analytical methods.

Fluorescent sample bands can be seen by exposure of the plate to ultraviolet radiation in the dark, the bands appearing as light spots on a dark background. All aromatic compounds can be visualized by fluorescent quenching. An immobile fluorescent compound is added to the plate before separation, and the developed plate is exposed to ultraviolet radiation in the dark. Aromatic bands absorb in the ultraviolet (preventing fluorescence) and appear as dark spots on a light background. Commercial TLC and adsorbents are available with or without a fluorescent additive.

Quantitative determination of the concentration of a band can be obtained directly by densitometry, for compounds that absorb in the visible or ultraviolet, or for any band after charring as above. A variety of commercial densitometers are available for this specific application.

13.3   Some Related Methods

Frontal development proceeds in LSC as described in Section 5.11. A solution is fed continuously to an adsorbent bed or column, and in the process the more strongly adsorbed components are preferentially retained on the bed. Frontal development is not well suited to the resolution of complex mixtures, however, since no sample component can be completely recovered in pure form, and only the weakest-adsorbing component can be isolated without contamination. Analytical applications of frontal development (frontal analysis) received brief attention in the 1940s, but these were soon superseded by other techniques. Frontal development is useful, however, in the purification of liquids or soluble solids, particularly when strongly adsorbing impurities are present in low concentration. Similarly, strongly adsorbing trace components are easily concentrated by frontal development, after which they can be stripped from the adsorbent bed by a strong solvent. Initial reaction products in organic syntheses are often decolorized by percolation through charcoal; this treatment selectively removes high-molecular-weight tars, since these are preferentially adsorbed on charcoal (Section 13.11). On a commercial scale, frontal development is used widely. Common examples include the decolorization of sugar and corn syrup on charcoal, the clay contacting of waxes and used oils for removal of oxidation products, and the charcoal treatment of waste water (tertiary treating). In every case the feed stream is percolated through an adsorbent bed until the bed becomes saturated with the material to be removed.

Displacement chromatography in LSC involves addition of the sample to the bed, followed by development with a very strong solvent (Section 5.11). The sample is pushed through the bed as a plug in front of the solvent, and if separation is successful the various sample components are resolved into adjacent pure bands. In most cases displacement chromatography provides poor separation of complex mixtures, and for this reason it is little used in analysis today. Larger samples can be charged in displacement chroma-

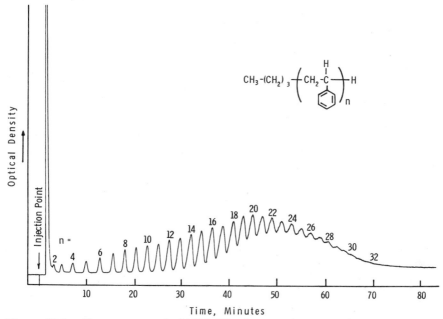

Figure 13.8. Chromatogram of a 2100 MW polystyrene by means of supercritical fluid chromatography on alumina. Reprinted from Ref. 13a, by courtesy of the publisher of the *J. Chromatog. Science.*

tography than in elution, favoring the use of displacement in preparative separations (when useful separation can be achieved by this technique).

Separations by LSC can be carried out on glass fiber sheets that have been impregnated with an adsorbent such as silica or alumina (glass fiber LSC). This technique is closely similar to TLC and offers many of the same advantages. The main differences between TLC and glass fiber LSC are greater separation speed and lower sample capacity for the latter technique. Sample resolution also tends to be poorer. Glass fiber LSC is used interchangeably with TLC in many applications, however, and is particularly useful for rapid, simple separations, such as monitoring column eluate fractions for the presence of different bands. For a review of glass fiber LSC see Ref. 12.

Supercritical fluid chromatography (SFC), described in Section 10.4 for a liquid stationary phase, can also be used with an adsorbent stationary phase (13, 13a). The resulting technique is intermediate between LSC and GSC (Section 14.1) in performance and possible applications. It would appear that SFC with an adsorbent should permit separation of higher-boiling samples than GSC, with greater speed and efficiency than are possible with LSC. This technique has been introduced only recently, however, and its full potential remains to be exploited. An illustrative separation is shown in Fig. 13.8.

13.4   Affinity Chromatography

The separation of biological macromolecules plays a very important role in the life sciences. Unfortunately the efficiency of the conventional chromatographic methods for the separation of macromolecules such as proteins, nucleic acids, and polysaccharides is rather poor, because the physicochemical properties of these substances are very similar and the specificity of the usual chromatographic systems toward these solutes is poor. A recently developed technique, however, utilizes highly specific biosorbents as the stationary phase and greatly facilitates the isolation and purification of large-molecular-weight biological substances. This technique is called *affinity chromatography* because it is based on the unique characteristic of biological macromolecules that they can bind reversibly and specifically other molecules, that is, it exploits the specific functional properties of such macromolecules. The binding of the substrate to the active site of an enzyme, the complex formation between antigen and antibody, and the hybridization of a messenger RNA with the complementary DNA are the most striking examples of such specific interactions.

The stationary phase in affinity chromatography is prepared by covalently binding an affinant (which must have high affinity for the macromolecule to be separated) to an insoluble support. The latter can be a polymer gel, such as cross-linked dextran, polyacrylamide, or agarose, or a porous material, such as porous glass. The affinant itself can be, as seen in Table 13.6, any given molecule that can bind specifically and reversibly the biological substance to be separated. It can be a competitive inhibitor like a substrate analog, an allosteric effector, or a cofactor for separating a particular enzyme. Alternatively, when the enzyme is bound to the solid matrix, it can serve as an affinant to separate these substances. Thus affinity chromatography is not restricted to the separation of macromolecules, although it is used almost exclusively in this area. The affinant can be highly specific for one substance, as in most antigen-antibody interactions, for instance, or it can have broader specificity to bind a particular group of substances. Immobilized organomercurial affinants, for example, can be used for the separation of all proteins having sulfhydryl side chains from other components of the mixture.

When the affinity of the stationary phase is specific for only one component of the mixture to be separated, affinity chromatography is carried out in the displacement mode. First the pH and the ionic strength of the solution containing that particular component are adjusted to obtain a strong binding to the biosorbent. Then the solution is percolated through the column packed with the immobilized affinant. The column retards the desired component, and other species are washed off. In the next step, the retarded

Table 13.6.   Some Applications of Affinity Chromatography

Solute to be Separated	Immobilized Affinant
Enzyme	Inhibitor, effector, or cofactor
Inhibitor, effector, or cofactor	Enzyme
Antigen	Antibody
Antibody	Antigen
—SH protein	Organomercurial compound
Enzyme subunit	Inhibitor
Polynucleotide	Complementary polynucleotide
Steroid binding protein	Steroid

solute is eluted from the column with a solution having suitable pH and ionic strength, or containing hydrogen-bond-breaking agents such as urea. In most cases the affinity constant between the solute and affinant varies sufficiently with the ionic make-up and temperature so that pH, ionic strength, and temperature can be optimized for both the binding and the elution steps. Sometimes, however, the removal of the bound solute requires a displacement agent such as a soluble affinant.

The stationary phase can retard specific groups of solutes in the adsorption step; these can be removed later by stepwise or gradient elution from the column, thus being separated into individual components. The separation of sulfhydryl proteins with a broadly specific organomercurial affinant can be carried out conveniently in this fashion, because proteins containing different numbers of SH groups have different affinities for the column material and can be eluted separately by changing the eluent buffer. Affinity chromatography can also be employed to concentrate dilute solutions of a biological substance or to remove denatured forms of macromolecules.

After the chromatographic run the column can be regenerated and reused. The great advantage of the technique is its high specificity and speed, because conventional purification methods involve many steps and concomitant loss of active material. Although affinity chromatography is preferentially carried out in the column, the use of specific biosorbents is not restricted to column operation. They can be employed to remove a specific component from a solution by single contact in a batch operation, with the advantage that the insoluble macromolecule affinant complex can be quickly isolated by filtration or centrifugation.

It should be noted that affinity chromatography is most efficient when adsorption equilibrium is reached between the solute and affinant. The rate of equilibration is rather slow, however, because a specific orientation of both the affinant and the solute is required in the binding step. This effect is also manifested in the activation energy of such specific adsorption, which is much higher than that of the unspecific adsorption involved in other chromatographic techniques. Therefore affinity chromatography usually requires relatively low flow velocities or long contact times.

13.41   PREPARATION OF COLUMN MATERIAL

Affinity chromatography is distinguished from other chromatographic techniques in that a specific column material, which has to be prepared individually, is required for any given separation problem. Therefore the preparation of the column material plays a central role in affinity chromatography. First an affinant has to be selected that has appropriate affinity to the solute to be separated. The affinant must have a functional group that permits its covalent attachment to the support, and it should retain its specific binding properties under the conditions of attachment and throughout the whole cycle of operation. Then a suitable support has to be selected. Agarose beads, as well as cross-linked dextran and polyacrylamide beads, have become the most commonly used supports, replacing the cellulose and polystyrene previously employed. An ideal support is inert so as to minimize nonspecific adsorption; it can be easily converted to react under mild conditions with the affinant, so that high immobilized affinant concentration is obtained. The support should be chemically and mechanically stable under the operating conditions, should permit high solute diffusivity inside the sorbent particles, and should allow access to all the immobilized affinant molecules. Agarose beads have a very open structure that allows the diffusion through the matrix of molecules having molecular weights in the millions. Agarose is stable and easily activated, and has a moderately high capacity. Polyacrylamide possesses a high concentration of groups modifiable for attachment of the affinant; thus it is suitable to make sorbents with very high capacity. However, the pore size of the swollen column material thus prepared is not always sufficiently large to accommodate large solute molecules. Since high adsorption capacity is required to efficiently separate solutes that have relatively low affinity to the affinant, a highly porous polyacrylamide would be a preferred support material. It should be kept in mind that unspecific adsorption of the solute on the matrix due to hydrogen-bonding, hydropholic, and ionic interactions may also play a role in the affinity chromatography and reduce the efficiency of the technique by lowering the selectivity of the stationary phase. Therefore the matrix should have a very low unspecific adsorption capacity toward the solutes that do not interact with the affinant.

In order to obtain affinity chromatographic columns with good flow characteristics, the packing should be rigid. Since the highly porous gels are very soft, mucilaginous materials, further improvement could be obtained

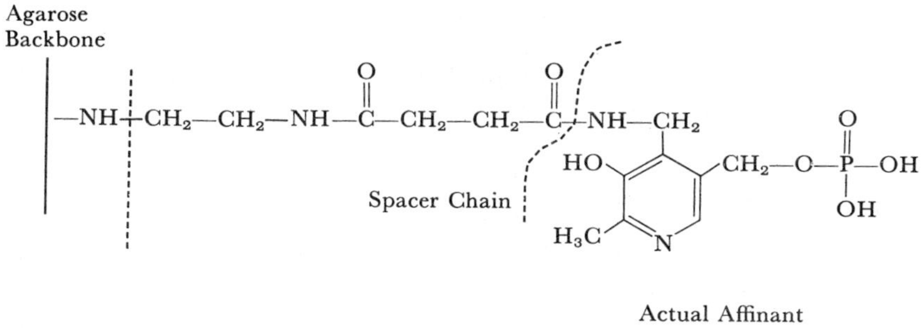

Agarose
Backbone

Actual Affinant

Figure 13.9.   Biosorbent specific for tyrosine aminotransferase of hepatome tissue culture cell. The actual affinant, pyridoxamine phosphate, is attached to the agarose matrix via a long chain that serves as spacer and facilitates binding of the large enzyme molecule. When the affinant is attached to the matrix directly, no binding occurs.

by making them in pellicular form, that is, by forming the gel layer on the surface of glass beads that would impart mechanical strength and favorable mass transfer properties to the particles. On the other hand, the use of porous glass also appears to be promising.

A variety of chemical reactions have been employed for coupling the affinant to the support. Polysaccharide type supports such as agarose and cross-linked dextran can be activated by cyanogen bromide, and the product coupled with the affinant via its amino groups under mild conditions. When the solute molecule is large or the binding between the affinant and solute is relatively weak, it is necessary to interpose a spacer between the actual affinant and the matrix, as shown in Fig. 13.9.

Polyacrylamide can be activated by converting the carboxamide side groups to hydrazide groups, which give highly reactive acylazide derivatives upon reacting with nitrous acid, as shown by the following reaction scheme:

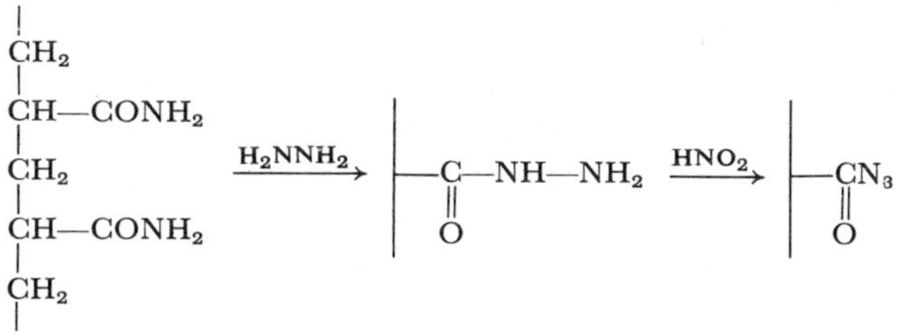

Compounds having aliphatic or aromatic primary amino groups can be coupled in a rapid reaction to the polyacrylamide matrix via the acrylazide groups. When a spacer molecule is used, it must have on one end of the chain an amino group, which is used to attach it to the matrix, and on the other end of the chain a group that can be activated to bring about the coupling of the actual affinant.

13.42   OPERATING CONDITIONS

Affinity chromatography can be used in both analytical and preparative work. As mentioned earlier, it need not be restricted to column procedures. For example, the biosorbent can be sealed into a bag or container made of a fine-mesh screen. By placing the container into the stirred solution of a particular solute or solutes, solute can be removed by adsorption and eluted separately. The advantage of this technique is that the solution can contain particulate matter which otherwise would plug the column. In addition, it permits the removal of several solutes simultaneously when containers filled with different sorbents are used at the same time. The use of the word chromatography for such single-contact batch operation is inappropriate and could well be replaced by the term affinity separation or affinity purification. In some instances the sorbtion process is carried out in a batch procedure; then the loaded sorbent is transferred into a column for washing out nonadsorbed substances and eluting the adsorbed solute or solutes in a chromatographic fashion.

Under certain circumstances the binding of a biological macromolecule to the immobilized affinant is so strong that the former cannot be eluted without causing denaturation. Then the removal of the macromolecule–affinant complex may be of value; this can be achieved under mild conditions when an appropriate link is used to couple the affinant to the matrix.

13.5   Applications

The applications of LSC are determined largely by sample type and separation goal, as indicated in the introductory pages of this chapter. Numerous examples of LSC separation are provided in Refs. 1, 3, and 14.

Separations of hydrocarbon samples by type are done almost exclusively by LSC. Petroleum samples can be separated into groups of compounds of different chemical class: saturated hydrocarbons, olefins, alkyl benzenes, alkyl naphthalenes, and higher polyaromatic hydrocarbon types; see, for example, Ref. 15. Individual polycyclic aromatic compounds (e.g., atmospheric contaminants) can be resolved on alumina, using either columns or TLC.

Liquid-solid chromatography on silica has been widely used for the separation of lipid mixtures into different chemical classes: hydrocarbons, mono-, di-, and triglycerides, alcohols, fatty acids, and so on. Magnesia as adsorbent is capable of further separation of lipid esters into steroid esters and fatty acid esters (16). Thin-layer chromatography on silica (frequently with silver nitrate impregnation for better resolution of olefinic lipids) has

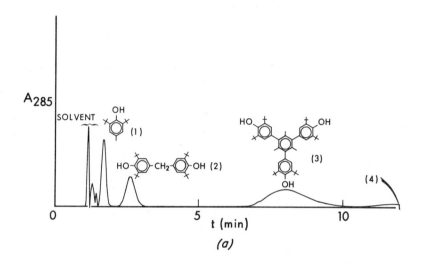

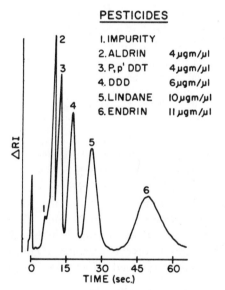

Figure 13.10. Separations by column LSC on silica: (*a*) antioxidants (1), (*b*) pesticides. Reprinted from Ref. 19, by courtesy of Waters Associates.

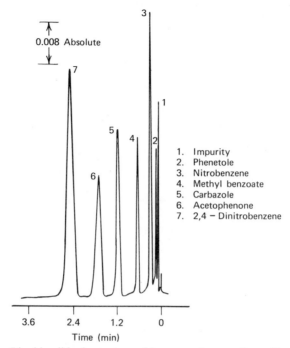

Figure 13.11. Liquid–solid chromatographic separation on 5 to 10-μm silica gel. Column, 15 cm × 2.1 mm i.d.; sample, 4 μl of a mixture of approximately 1 mg/ml of each component; flow rate, 400 ml/hr; mobile phase, *n*-hexane; detector, UV: temperature, ambient. Reprinted from Ref. 20, by courtesy of the American Chemical Society.

been an important technique for the isolation of individual lipid compounds. More often, however, the ultimate separation of a complex lipid mixture into its individual constituents requires a combination of methods (beginning with LSC), as described in Chapter 19.

Liquid-solid chromatography is widely used for the separation and analysis of other water-insoluble natural products, such as steroids, terpenes and carotenoids, and fat-soluble vitamins. Related synthetic products (e.g., pharmaceuticals, antioxidants and preservatives, and pesticides) are also successfully separated by LSC, as illustrated by the column LSC separations of Fig. 13.10.

More polar compounds, such as alkaloids, polar lipids, phenols, carbo-hydrates, amino acids, and nucleosides, can be separated by LSC, but these applications compete with similar separations by LLC (usually on paper or powdered cellulose). Basic compounds (e.g., alkaloids) are generally sepa-rated on alumina, whereas acids (e.g., phenols) are best handled with silica. Polar lipids are separated mainly by TLC on silica (see, e.g., Ref. 11), usually with mixtures of chloroform, methanol, and water as solvents, plus ammonia or acetic acid in some cases.

In the past, polysaccharides have been separated according to size on charcoal, and individual sugars have been separated on silica. Although the free amino acids can be separated on silica (TLC), this technique is more useful for the separation of less polar amino acid derivatives, such as dinitrophenyl amino acids.

Inorganic ions within a particular group (e.g., alkali cations, halogen anions,) can be resolved in some cases by TLC. Such separations are often marginal, however, and the competition of other methods has relegated LSC to an unimportant role in the analysis of inorganic ions.

Finally, Fig. 13.11 presents the high-speed separation of a group of seven polar substances on 5 to 10 μ silica gel particles (this material is identical to that used in TLC). A highly efficient, and rapid analysis is observed. Packing was achieved using a balanced slurry method. As with LLC, many new developments in this promising area can be expected during the next few years.

References

1. J. J. Kirkland, ed., *Modern Practice of Liquid Chromatography*, Wiley-Interscience, New York, 1971.

2. L. R. Snyder, *Principles of Adsorption Chromatography*, Marcel Dekker, New York, 1968.

3. E. Stahl, *Thin-Layer Chromatography: A Laboratory Handbook*, 2nd ed., Springer-Verlag, Heidelberg, 1969.

4. L. R. Snyder, *J. Chromatog.*, **36**, 455 (1968).

5. L. R. Snyder, in *Gas Chromatography, 1970*, R. Stock and S. G. Perry, eds., Institute of Petroleum, London, 1971, p. 83.

6. L. R. Snyder and D. L. Saunders, *J. Chromatog. Sci.*, **7**, 195 (1969).

7. L. R. Snyder, *J. Chromatog.*, **63**, 15 (1971).

8. L. R. Snyder, *J. Chromatog. Sci.*, **8**, 692 (1970).

9. L. R. Snyder, *Modern Practice of Liquid Chromatography*, Wiley-Interscience, New York, 1971, Chaps. 4 and 6.

10. J. G. Kirchner, *Thin-Layer Chromatography*, Wiley-Interscience, New York, 1976.

11. F. Geiss, *Die Parameter der Dünnschichtchromatographic*, Vieweg, Braunschweig, 1972, Chap. 4.

12. F. C. Haer, *An Introduction to Chromatography on Impregnated Glass Fiber*, Ann Arbor Science Pub., Ann Arbor, Mich., 1968.

13. S. T. Sie and G. W. A. Rijnders, *Separation Sci.*, **2**, 775 (1967).

13a. R. E. Jentoft and T. H. Gouw, *J. Chromatog. Sci.*, **8**, 138 (1970).

14. E. Heftmann, ed., *Chromatography*, 3rd ed., Reinhold, New York, 1974.

15. L. R. Snyder, *Anal. Chem.*, **37**, 713 (1965).

16. N. Nicolaides, *J. Chromatog. Sci.*, **8**, 717 (1970).

17. L. R. Snyder, *J. Chromatog. Sci.*, **7**, 352 (1969).

18. L. R. Snyder, *J. Chromatog. Sci.* **7**, 595 (1969).

19. Waters Associates, Framingham, Massachusetts.

20. R. E. Majors, *Anal. Chem.*, **44**, 1722 (1972).

21. J. J. Kirleland, *J. Chromatog. Sci.*, **10**, 593 (1972).

Selected Bibliography for Affinity Chromatography

F. Friedberg, *Chromatog. Rev.*, **14**, 121 (1972).

P. Cuatrecasas, in *Biochemical Aspects of Reactions on Solid Supports*, G. R. Stark, ed., Academic Press, New York, 1971, pp. 79–110.

G. Feinstein, *Naturwissenschaften*, **58**, 389 (1971).

P. Cuatrecasas and C. B. Anfinsen, in *Methods in Enzymology*, Vol. 21, S. P. Collowick and N. O. Kaplan, eds., Academic Press, New York, 1972.

R. H. Reiner and A. Walch, *Chromatographia*, **4**, 578 (1971).

List of Symbols

A surface area required by an adsorbed molecule

i solute molecule

K molar concentration distribution constant

K' distribution constant defined by Eq. 13.3a

$K°$ thermodynamic distribution constant

K_x concentration distribution constant

M solvent molecule

n number of solvent molecules displaced from the surface by a solute molecule

$Q_j°$ group adsorption energy

$S°$ experimental parameter in Eq. 13.6

V_a quantity of adsorbed mobile phase per unit weight of adsorbent

W weight of adsorbent in the bed

α' adsorbent activity parameter in Eq. 13.6

$\varepsilon°$ solvent strength parameter in Eq. 13.6

Subscripts

(a) adsorbed state

(M) in solution

Fourteen

OTHER INTERFACIAL PROCESSES

In Chapter 13 we examined equilibrium separations involving bulk liquid and the liquid-solid interface as phases. Now we will consider two related procedures that involve an interface as one of the two phases: gas-solid adsorption and gas-liquid adsorption. Although these two procedures have found their most important applications in large-scale separations, in this chapter we emphasize their analytical applications.

14.1 Gas-Solid Adsorption

Gas-solid adsorption is used in large-scale commercial separation processes to remove gases from gas mixtures (e.g., via charcoal filters) and to dry gases (e.g., via molecular sieve). Today this process is playing an increasingly important role in the control of air pollution. The adsorbent acts as a trap to collect impurities before the discharge of gases into the atmosphere (e.g., removal of SO_2 from stack gases). Alternatively, the adsorbent collects and concentrates pollutants from the atmosphere. These pollutants are then removed for analysis in order to monitor the quality of the air.

On the analytical scale, gas adsorption has found limited application, mainly in the field of gas-solid chromatography (GSC). Separations have included permanent gases, isotopic mixtures of low-molecular-weight gases, gaseous hydrocarbons, and, less frequently, high-boiling substances.

It is important to note that gas-solid adsorption shows many similarities to liquid-solid adsorption, discussed in Chapter 13. In fact, many of the adsorbent types used in liquid-solid adsorption are also employed in gas-solid

411

adsorption. In the following discussion, we point out similarities and differences between the two distribution phenomena.

As described in Chapter 13, there are fundamentally three types of adsorbents: nonpolar, acidic (electron-accepting surface groups), and basic (electron-donating surface groups), the second and third being classified as polar adsorbents. These three types of adsorbents permit the separation of a wide range of substances, both polar and nonpolar. On nonpolar adsorbents, the major force of interaction is dispersion, and, to a first approximation, adsorption coefficients are in the order of molecular weight. On polar adsorbents, dipole, hydrogen-bond, and other acid-base interactions are potential adsorption forces. It is also possible to employ chemisorption for separation in gas-solid adsorption, for example, as cited in Chapter 13, the charge transfer complexation of olefins with silver nitrate-impregnated columns.

Typical nonpolar adsorbents are graphitized thermal carbon black and polymeric saturated hydrocarbons such as polyethylene. Activated charcoal is a large-surface-area support that is quite heterogeneous in terms of the distribution of adsorption sites on its surface. However, by heating the charcoal to 3000°C a structural change occurs that converts the adsorbent into a relatively homogeneous low-surface-area support that is nonpolar. This support, graphitized thermal carbon black, has found significant application in GSC.

Adsorbents with electron-accepting surface groups (e.g., silica and porous glass) usually contain acidic hydroxyl groups (see Chapter 13). A second example is the series of crystalline aluminosilicates—the zeolites. Molecular sieves (Chapter 15) are calcium aluminum silicates in which the interatomic distances between layers of silicate are fairly constant; for instance, in the case of molecular sieve 4A the average distance is 3.8 Å. As we have already noted, these materials are important drying agents. An electron-donating adsorbent is obtained by coating a nonspecific adsorbent with an electron-rich solid, such as benzophenone. Alternatively, acidic hydroxyl groups of silica can be reacted with substances containing electron-donating functional groups (e.g., —CN). For more details concerning adsorbents, see Refs. 1, 2, and 3.

As in liquid-solid adsorption, we can describe the distribution process by an adsorption coefficient $\tilde{K}_A*$:

$$\tilde{K}_A = \frac{a}{P_i} \tag{14.1}$$

---

* As discussed in Chapter 2, the symbol $\tilde{K}$ represents a distribution coefficient that is or can be a function of sample concentration. If it is independent of concentration, over a given concentration range, the adsorption constant K_A is used.

where a = amount of solute vapor adsorbed per gram of adsorbent, and P_i = partial pressure of solute. A fundamental difference between the two adsorption processes is that in liquid-solid adsorption there is a displacement of one or more solvent molecules from the solid surface in the adsorption of the solute (Fig. 13.3), whereas in gas-solid adsorption no such competition for adsorption sites exists. As a result, the thermodynamics of the two processes differ. Thus gas-solid adsorption always involves a loss in entropy, so that the adsorption process must be exothermic and $\tilde{K}_A$ must decrease with increasing temperature. On the other hand, liquid-solid adsorption is accompanied by little if any entropy change (4, 5) because the extent of disorder remains approximately the same in both the adsorbed state and the solution. Consequently, although $\tilde{K}_A$ generally decreases with temperature, the influence of temperature is less important for liquid-solid adsorption. The principles of gas-solid adsorption have been studied intensively by a number of workers. It is not possible, within the limits of this chapter, to treat the subject in detail, however, the interested reader is referred to the selected bibliography at the end of the chapter.

As discussed in Chapter 13, separations in liquid-solid adsorption are usually achieved on the basis of class (i.e., alcohols from ketones), especially on polar adsorbents. This arises from the large differences in specific interactions for the various functional groups. Gas-solid adsorption also shows this characteristic of class separation; however, because the process includes a vaporization step, the method exhibits considerable molecular weight selectivity (see Chapter 6).

Industrial-scale application of gas-solid adsorption typically involves the removal of one or more gases from a mixture, often by irreversible adsorption. In this case the problem of reproducibility of adsorbents and nonlinearity of adsorption isotherms (see Chapter 13) is less important. However, for analytical-scale operations (frequently requiring reversible adsorption) we are usually concerned with the control of adsorbent activity. In the low surface concentration region (i.e., less than monolayer coverage) the isotherm is frequently linear. Extension of the linear range of the isotherm involves procedures similar to those described for liquid-solid adsorption. We discuss several of these procedures in the next section.

14.11 GAS-SOLID CHROMATOGRAPHY—INTRODUCTION

Gas-solid chromatography was developed before gas-liquid chromatography (6). In general, however, little work proceeded on GSC during the years of highest activity with GLC, and it is only since the early 1960s that renewed interest has been shown in this method. Reasons for the lapse of interest in GSC include (1) inconveniently large distribution coefficients and, hence, retention volumes; (2) nonlinear distribution isotherms for all but the lowest-boiling materials, leading to asymmetrical band shapes and to sample-size-dependent retention volumes; (3) poor reproducibility of adsorbent properties; and (4) catalytic activity of the adsorbent and chemisorption, that is, irreversible adsorption.

Because of the large distribution coefficients in GSC, column temperatures are often over 100°C higher than in GLC. These higher temperatures are undesirable because they increase the possibility of thermal and catalytic degradation of the sample. However, the higher operating temperatures make GSC attractive for the analysis of low-boiling mixtures; for such mixtures separation can be conveniently made at room temperature, rather than at subambient temperatures.

The long retention and nonlinearity of the adsorption isotherm are related to surface heterogeneity. Removal of the most active sites simultaneously reduces retention and improves the linearity of the isotherm, especially for polar substances.* This removal can be accomplished either chemically or physically. For example, surface silanol groups (—Si—OH) of silica type adsorbents can be chemically removed by a silanization reaction (e.g., trimethylchlorosilane). Such a procedure, however, although improving the linearity of the isotherm, also changes the chemical properties of the surface.

In general, physical adsorption, because it is simpler and reversible, is used to eliminate the most active sites. This can be done either by employing an adsorbing carrier gas or by coating the adsorbent with a small amount of polar liquid. These methods produce a new type of retention process designated as *gas adsorption-layer chromatography* (7). For example, a modifier (tail reducer) such as water can be added to the carrier gas. Since water is more polar than most solute molecules, it will be preferentially attracted to the strongest adsorption sites and will deactivate these sites. Alternatively a polar liquid, such as β,β'-oxydipropionitrile, can be coated on a polar adsorbent to produce the same effect. When deactivation occurs, there is an order-of-magnitude or greater drop in retention for polar compounds, along with improved efficiency and column speed characteristics. A particularly interesting application of gas adsorption-layer chromatography has recently been shown with spherical silica (8). We should recognize that, although gas adsorption-layer chromatography is very useful from a practical point of view, the mechanism of retention is quite complex, including adsorption at the gas-solid and gas-liquid interfaces, as well as partition in the thin layer of liquid phase.

Often, for a given polar adsorbent, the surface heterogeneity is related to

---

* By improvement of the linearity of the isotherm, we mean an increase in the range of solute partial pressures over which the adsorption coefficient is constant.

the surface area of the adsorbent. High-surface-area adsorbents (~ 500 to 1000 m²/g) contain micropores in which the surface activity is significantly greater than for other regions of the surface (1). Hence, elimination of these micropores by using smaller-surface-area adsorbents (ca. 25 m²/g) will result in improved isotherm linearity (at the expense of sample capacity for the column as a whole). The nonlinearity of the adsorption isotherm can also be overcome, of course, by using adsorbents that are homogeneous or display only a low degree of surface heterogeneity (see the discussion of adsorbents in Section 14.12).

When operating in the linear region of the isotherm, GSC possesses several advantages over GLC: (1) the already mentioned potential for low-boiling samples; (2) stable columns, that is, no bleeding of stationary phase; (3) good selectivity; and (4) rapid mass transfer for good efficiency at high velocity. These advantages are pointed out in the following sections.

14.12 THEORY OF GAS-SOLID CHROMATOGRAPHY

Retention

For the following discussion we assume linear conditions such that retention is independent of sample concentration. The retention parameter in GSC is V'_{Rg}, the net retention volume per gram of adsorbent. Since K_A, as defined in Eq. 14.1 (see footnote with this equation), is expressed in terms of solute pressure, we have

$$V'_{Rg} = K_A RT \tag{14.2}$$

Thus, from the calculation of K_A, the retention of solutes in gas-solid chromatography can be estimated. This calculation is very long and involved, with success to date being achieved only with a very well-defined surface-graphitized carbon black—and nonpolar solutes (e.g., aromatic hydrocarbons) (2, 9). We are forced, therefore, to assume simplified models and to use semiempirical equations, if we wish to have a practical means for understanding retention in GSC.

Snyder has approached this problem by assuming a model similar to that used in LSC (5, 10). The net retention volume per gram of packing is related to the adsorption energy ΔE by

$$\log V'_{Rg} = \log V_a + \Delta E \tag{14.3}$$

where V_a is the adsorbent surface volume (see Section 13.2). If we assume an inert carrier gas and ideal gas behavior, ΔE is equal to the interaction energy of the sample with the adsorbent. This energy can be divided into two quantities based on specific and nonspecific interactions, $\alpha' S^0$ and $\alpha'_d A_i$, where

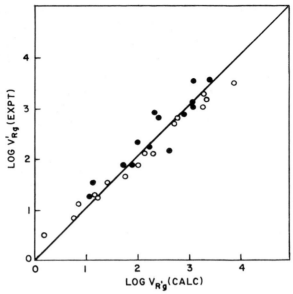

Figure 14.1. Comparison of GSC retention volumes on a column of 2.7% H_2O–Al_2O_3 (at 83°C) with values calculated from Eq. (14.4) and adsorption parameters from LSC systems. Reprinted from Ref. 10, by courtesy of Elsevier Pub. Co.

α' and α'_d can be considered as the relative adsorbent activity due to specific and nonspecific interactions, respectively; A_i is the adsorption area covered by the solute; and S^0 is the relative sample adsorption energy. In addition, an entropy term C must be included for the adsorption process. If there is strong specific adsorption, a gas molecule will lose up to 3 degrees of translational freedom. Equation 14.3 then becomes:

$$\log V'_{Rg} = \log V_a + C + \alpha'\left[S^0 - \left(-\frac{\alpha'_d}{\alpha'}\right)A_i\right] \qquad (14.4)$$

The general correctness of Eq. 14.4 is illustrated in Fig. 14.1, in which experimental $\log V'_{Rg}$ values are plotted versus those calculated assuming a value of $\alpha'_d = 0.58$, and using experimental S^0 values from LSC studies.* The column used in this case was 2.7% H_2O–Al_2O_3; however, similar agreement between experimental and calculated V'_{Rg} values was found in other gas-solid systems. This means that the extensive compilation of S^0 values in LSC can be used to help provide a first-order approximation of V'_{Rg}.

If the constant C is neglected, LSC and GSC would appear to have similar retention equations, if $-\alpha'_d/\alpha'$ is set equal to $\varepsilon°$, the solvent strength of the carrier phase (gas or liquid). However, whereas α' changes markedly with

* The constant C was assumed to be -0.65 for this comparison.

activation, the dispersion contribution α'_d remains essentially constant. This means that the term $-\alpha'_d/\alpha'$ is a strong function of surface activation, resulting in elution order changes with activation. In general, the elution order in LSC is independent of adsorbent activity. Thus a direct comparison of the GSC and LSC retention equations must be made with caution.

Column Efficiency

Column efficiency in CSC was discussed in Chapter 5. As in GLC, we can approximate efficiency by use of the classical form of the van Deemter equation:

$$H = A + \frac{B}{\bar{v}} + C\bar{v} \qquad (14.5)$$

where $\bar{v}$ is the average linear velocity of the carrier gas in the column, and the constants A, B, and C have their usual meanings (see Chapter 5). The A and $B/\bar{v}$ terms need no further comment. In GSC the resistance to mass transfer or $C\bar{v}$ term consists of three contributions: diffusion in the moving gas phase to a pore, diffusion within the stagnant gas of the pore, and kinetics of adsorption-desorption. Typically, the kinetics of the adsorption process is sufficiently fast so that gas-phase diffusion controls the mass transfer process. Thus high temperatures (fast diffusion) and smaller particle diameter ($C\bar{v} \sim d_p^2$) generally improve efficiency and permit faster analyses. Since the former results in increased potential for sample reaction and the latter in increased pressure drop for a given velocity and column length, a compromise must often be made between speed and efficiency, and pressure and sample compatibility.

Given comparable conditions, it was widely believed that GSC columns would give lower C values than GLC columns and hence better efficiency at high velocity. This belief arose from the fact that the HETP equation (Eq. 14.5) is similar for both methods except for the term accounting for resistance to mass transfer from the stationary phase. As we have already noted, the kinetics of the adsorption-desorption process can be rapid; however, diffusion in a stationary liquid phase (in GLC) can result in a much slower mass transfer. But the gain in efficiency in GSC has often not been realized in practice, because, in most cases, of surface heterogeneity of the adsorbent.

When a surface is homogeneous, impressive efficiencies and speed characteristics are observed in GSC, as Fig. 14.2 illustrates for the separation of the C_1 to C_4 hydrocarbons in 2 min on a column of alumina partially deactivated by water (to make the surface more homogeneous). Note that the average carrier gas velocity is unusually high for packed columns at 16.3 cm/sec. Other methods, previously described, for improving surface homogeneity also lead to better efficiency and faster analysis. When the surface is sufficiently homogeneous, it is possible to partially overcome the problem of high

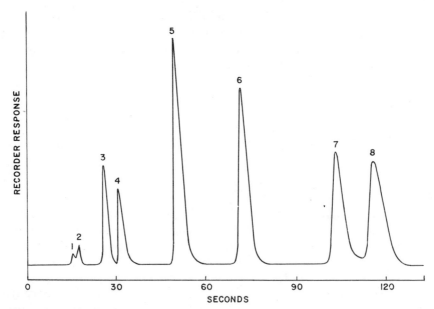

Figure 14.2. Separation of C_1 to C_4 hydrocarbons on partially deactivated alumina: $T = 80°C$, $L = 2$ m, $\bar{v} = 16.3$ cm/sec: 1 = air, 2 = methane, 3 = ethane, 4 = ethylene, 5 = propane, 6 = propylene, 7 = isobutane, 8 = butane. Reprinted from Ref. 31, by courtesy of the American Chemical Society.

column temperatures by using columns with high permeabilities. Hence packed capillaries (11), capillaries coated with adsorbent or roughened by etching (12), and aerogel columns (13) have all been used to separate components at temperatures well below those employed in packed-column GSC.

Experimental

The gas chromatographic apparatus is the same as that used for GLC (see Section 8.2); therefore many operational options in GLC (e.g., temperature programming, pressure programming) are available for GSC as well. The only added experimental factor that needs to be considered in GSC is the control of the water content in the carrier gas if deactivated polar adsorbents are used. This presents no real experimental problem, however, when suitable gas saturators and proper temperature control are employed.

Adsorbents

We have already discussed the general types of adsorbents that have been used in gas-solid adsorption and liquid-solid adsorption (see also Chapter 13). In this section we describe some individual adsorbents used in GSC.

Graphitized thermal carbon black is a relatively low-surface-area nonpolar support. To illustrate its nonpolarity, we present in Table 14.1 the retention

Table 14.1. Retention Volumes and Heats of Adsorption on Graphitized Carbon Black at $T = 100°$ (2)

Solute	δ^a	$\delta_d{}^a$	$V'_{Rg}/\bar{V}{}^b$	$-\Delta H_A{}^c$
Methanol	12.9	6.2	1.5×10^{-3}	5.3
Ethanol	11.2	6.8	2.4×10^{-3}	6.9
Acetone	9.4	6.8	3.8×10^{-3}	8.3
Acetic acid	12.4	7.0	8.1×10^{-3}	8.3
Chloroform	9.1	8.1	7.0×10^{-3}	8.9
Dioxane	9.8	7.8	7.6×10^{-3}	9.8
Carbon tetrachloride	8.6	8.6	8.5×10^{-3}	9.6
Cyclohexane	8.2	8.2	8.0×10^{-3}	8.7
Toluene	8.9	8.9	9.5×10^{-2}	11.6
Benzene	9.2	9.2	2.0×10^{-2}	9.8
n-Hexane	7.3	7.3	2.4×10^{-2}	10.4
Chlorobenzene	9.6	9.2	9.0×10^{-2}	10.8

^{a} in (cal/mole-ml)$^{1/2}$

^{b} $\bar{V}$ = molar volume; units of $V'_{Rg}/\bar{V}$ = mole/g adsorbent

^{c} in kcal/mole

volumes and heats of adsorption of a variety of compounds, along with their total solubility parameters δ and the contributions of the solubility parameter due to dispersion δ_d. To place the comparison on a common basis, we have divided the retention volume by the molar volume of each solute. It is seen that $V'_{Rg}/\bar{V}$ and ΔH_A follow roughly in the order of δ_d; there is no relationship between these two adsorption parameters and δ, an approximate measure of the polarity of the compounds (see Chapter 10).

We notice also that the relationship of either $V'_{Rg}/\bar{V}$ or ΔH_A with δ_d is not exact. Thus the δ_d value of n-hexane is significantly lower than that of benzene, yet $V'_{Rg}/\bar{V}$ and ΔH_A are slightly higher for hexane. The reason is that more polarizable groups are in close contact with carbon atoms on the graphite surface in the adsorption of hexane than of benzene (2). Consequently, hexane can be more polarized, resulting in greater dispersion forces between this solute and the graphite surface. Other examples of this behavior can be seen in Table 14.1 (e.g., benzene vs. toluene).

This stereospecific adsorption nature results in selectivity of graphitized thermal carbon black in the separation of molecules of different shapes, such as geometrical isomers (14). In addition, this support is used to separate high-molecular-weight compounds, since the minimal catalytic activity of the adsorbent permits operation at high temperatures. In recent developments graphitized carbon black has been coated on the inner walls of a glass capillary, allowing separation of components at much lower temperatures than are normally required for packed columns (15).

**Table 14.2.   Values for *trans*-3-Heptene: *cis*-3-Heptene
at 100°C on Modified Alumina** (17)

Modifier	α	Modifier	α
KCl	1.26	LiCl, 400°	1.57
CdI$_2$	1.32	LiBr, 400°	1.63
NaCl	1.34	CdBr$_2$	1.78
NaBr	1.35		

In an effort to expand the usefulness of this adsorbent, workers have coated solid substances on the surface of the support. One particularly interesting application is the use of metal phthalocyanines (e.g., copper, nickel, zinc, metal free) (16). When different metals are employed, the affinity of the adsorbent for classes of organic compounds changes markedly. Thus a variety of adsorbents are easily made by simply coating different metalo-organic complexes on the carbon black surface.

The surface properties of other adsorbents have also been modified by coating or chemically bonding substances to the surface. Phillips and Scott (17) coated adsorbents with a variety of inorganic salts, such as LiCl, NaCl, and Na$_2$SO$_4$, and observed marked changes in relative retention. One example is shown in Table 14.2 for the change in α of a cis-trans isomer separation as a function of salt coated on alumina. The selectivities of these salt-modified adsorbents have been studied in detail by Sawyer and his co-workers (18, 19). Metal salts of fatty acids and urea have also been used to modify adsorbent surfaces (20).

Recently, chemical bonding to adsorbents has also been employed to provide more homogeneous surfaces and selective separations. This work is basically an extension of the silanization reactions discussed in Chapter 8 on GLC, for the deactivation of solid supports. In one example, the acidic protons on silica surfaces (—Si—OH groups) are esterified with primary alcohols of varying carbon chain length to produce an ester bond (Si—O—C—) (21). A variety of functional groups can be placed at the end of the alcohol chain to provide different types of adsorbent surfaces (e.g., acidic, basic). Unfortunately, however, the ester bonds are thermally unstable above 200°C and are hydrolysis prone. Recently, workers have been able to synthesize chemically bonded adsorbents in which carbon is directly bonded to the silicon surface (22), and considerably more stable adsorbents are thereby produced.

Chemically bonded phases provide a fairly homogeneous surface, resulting in symmetrical peaks for samples at relatively large concentrations. The chain lengths are typically from 4 to 5 carbon bonds up to 18 to 20. Clearly, at the upper end of the scale, the properties of the support more closely resemble

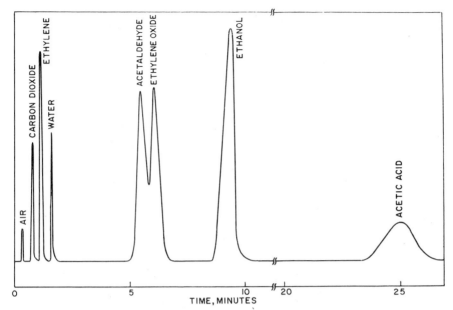

Figure 14.3. Separation of polar substances on a Porapak Q column: $T = 105°C$, $L = 2$ m, $F = 28.6$ ml/min. Reprinted from Ref. 23, by courtesy of the Institute of Petroleum.

those of a stationary liquid. Chemically bonded phases are also used in high-speed LLC (see Chapter 10), as no bleeding of liquid phase occurs.

Porous polyaromatic polymer beads have been successfully used as a support in gas chromatography (23, 24). These beads are analogous to the styrene-divinylbenzene porous gels used in exclusion chromatography (Chapter 15). Materials with high surface area (> 300 m²/g) and small pore diameter (< 100 Å) have been employed for the separation of permanent gases and simple hydrocarbons, reactive gases, isotopic mixtures, and low-molecular-weight polar materials, such as alcohols and acids. Several of these supports (e.g., Porapak Q) are characterized by their nonpolar nature such that separations are made on the basis of differences in the extent of dispersion forces between solute and porous polymer. Figure 14.3 illustrates this in the separation of low-molecular-weight polar compounds. Note the complete symmetry of the polar components and elution in the order of molecular weight, not polarity. Other porous polymers contain polar groups on the surface, for the selective retention of polar molecules. In general, only compounds up to molecular weight 150 have been chromatographed because of the low thermal stability of the beads. The mechanism of retention on this support is not fully understood at the present time. Gas-solid adsorption is clearly important; however, in certain cases solution of solute in the polymeric phase may also take place.

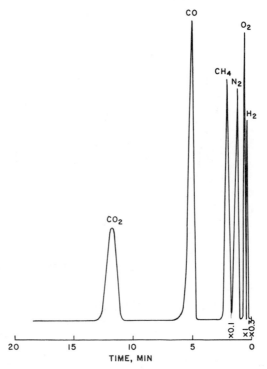

Figure 14.4. Chromatogram of a mine gas (W. G. Pye & Co., Ltd.). Column $= 60$ cm packed with molecular sieve 5A; $T =$ room temperature for 3 min, then temperature programmed at $10°/$min. Reprinted from Ref. 25, by courtesy of Pergamon Press.

Applications

The most important application of GSC is in the separation of permanent gases and low-molecular-weight hydrocarbons; GSC is almost uniquely suited for this purpose (25, 26, 27).

Isotopic separations of gases have been successfully accomplished by GSC. A particularly good example is shown in Fig. 5.3 in the separation of the nuclear spin isomers of H_2 and D_2 on an 80-meter capillary column consisting of etched glass (i.e., SiO_2 as the adsorbent) (28).

A variety of adsorbents have been used to separate the permanent gases, including alumina, activated charcoal, and fine-pore silica gel. However, zeolites and porous polymers are most frequently employed in these analyses. Figure 14.4 shows a typical separation of a mine gas on a 60-cm-long column of molecular sieve 5A.

Low-molecular-weight hydrocarbons are also easily separated by the adsorbents mentioned above. With the zeolites, Linde 5A molecular sieve has been found to be useful in the separation of straight-chain from branched hydrocarbons (29). This results from the well-defined porous structure, which is small enough to prevent the branched hydrocarbons from penetrating the porous matrix. Alumina can also be utilized for low-molecular-weight hydrocarbons; however, reproducibility is often a problem, and water should be used to obtain symmetrical peaks.

Gas-solid chromatography has been found to be a particularly valuable physicochemical tool for the study of gas-solid adsorption. For example, with highly sensitive detectors (e.g., flame ionization) heats and entropies of adsorption at zero surface coverage (analogous to infinite dilution) can readily be obtained. At such low surface concentrations, the direct influence of the adsorbent on the adsorbed solute can be determined. In addition, the adsorption isotherm can be measured by several methods (e.g., by frontal analysis, by analysis of the shape of asymmetrical bands (30)). Kiselev and his coworkers have been quite active in applying GSC to the study of adsorption. A good summary of this work can be found in Refs. 1 and 2.

14.2 Adsorptive Bubble Separation Processes

14.21 INTRODUCTION

In preceding sections we dealt with the application of several types of interfacial equilibria to separation processes—among these, liquid-solid and gas-solid adsorption. Gas-liquid adsorption equilibrium, another type that has proved useful, is the subject of this section. A series of methods have been developed, based on selective adsorption on gas bubbles in bulk liquid media (usually H_2O), hence the generic name *adsorptive bubble separation processes.*

The methods differ in the types of species being separated and the means by which the enriched interfaces are collected, the most prevalent being by production of a foam.* Ore flotation or mineral dressing is the best known of this group of processes. In this method mineral particles are made surface active (and thus floatable) by the adsorption of surface-active molecules on the mineral surface. We shall say little about this process; however, other processes have been found applicable in the analytical separation field, and these are now discussed.

---

* Gas bubbles and a foam are distinguished in this discussion. Gas bubbles occur in the aqueous medium where gas is introduced into the liquid. Although gas bubbles also are a component of a foam, a foam in actuality can be considered as a separate phase, consisting of an extended gas-liquid interface.

14.22 SURVEY OF PROCESSES

Table 14.3 shows diagrammatically the various processes that have been developed up to the present time. Adsorptive bubble separation methods can be divided into two broad classes: those in which the enriched surfaces are collected by foaming (*foam separation*), and those in which a foam is not required (*nonfoaming adsorptive bubble separation*). In foam separation, a natural subdivision is presented in terms of whether the species being adsorbed is a soluble species (*foam fractionation*) or is particulate in nature (*flotation or froth flotation*). As an example of foam fractionation, proteins or other surface-active species (detergents, dyes, etc.) can be separated by bubbling gas through the solution and collecting the foam (32).

It is useful to subdivide the flotation processes. *Macroflotation* represents the flotation of macroscopic particles, a major subdivision of this process being ore flotation. *Microflotation* deals with the flotation of microscopic particles, especially the removal of microorganisms and colloids (*colloid flotation*) (33). Colloidal species have a net charge on their surfaces when suspended in aqueous media, and addition of an oppositely charged surfactant in the required concentration permits flotation to be achieved. A subdivision, *adsorbing colloid flotation*, involves the flotation of colloidal particulate upon which dissolved material is adsorbed. In this case the major objective is the removal of the dissolved material rather than the colloidal particles. *Precipitate flotation*, which involves the removal of precipitates, offers an alternative to the settling procedures described in Chapter 11 for the removal of precipitates from solution, finding special value for precipitates that are difficult to handle because of their gelatinous nature (34). Finally, *ion flotation* involves the flotation of a nonsurface-active ion out of aqueous media via the formation of an insoluble surface-active complex with an oppositely charged surfactant. Ion flotation has found application in the removal of metallic species at the trace level (35).

Solvent sublation and bubble fractionation comprise the nonfoaming adsorptive bubble separation processes. In *solvent sublation* an immiscible organic layer is placed on top of the aqueous phase to collect the enriched material on the gas bubble surfaces (36); see Fig. 14.5. This method differs from solvent extraction in that liquid-liquid equilibrium is not achieved, and as a result the amount extracted is not the same for both processes. Solvent sublation has proved useful in the collection of low concentrations of species for subsequent analysis. Finally, in *bubble fractionation*, gas bubbles travel through an elongated column of bulk aqueous phase, transporting surface-active material to the top of the column (37). The enriched section is then removed, after achievement of steady state, by collection of the top section of the liquid. No foam is produced. This process works best for systems in

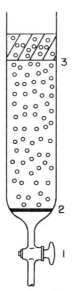

Figure 14.5. Simple extraction column for solvent sublation: 1 = two-way stopcock, 2 = porous glass frit, 3 = extraction column.

which the surface activity and/or concentration of species does not permit formation of a foam.

On an engineering scale, many of the processes listed in Table 14.3 are useful in the economic removal of low concentrations of materials from large volumes. For example, foam separation has been used for removal of pollutants from water wastes and nuclear wastes (38). Moreover, the methods —because of their effectiveness at low concentrations—find value on an analytical scale in the separation and purification of biological systems and trace analysis. For example, laurylsulfonic acid can be purified from small amounts of lauryl alcohol to a level of better than 99.9% by foam fractionation (39).

14.23 FUNDAMENTALS

Consider the foam fractionation of a simple surface-active agent. As in other phase equilibria, the distribution coefficient $\tilde{K}$ can be written as

$$\tilde{K} = \frac{\text{concentration in surface phase}}{\text{concentration in bulk phase}} = \frac{\Gamma}{C} \qquad (14.6)$$

where Γ is the surface excess of adsorbed solute (moles of solute per unit area at the defined interface in excess of moles per unit cross-sectional area in the

**Table 14.3.   Classification Scheme of Adsorptive Bubble
Separation Methods**[a]

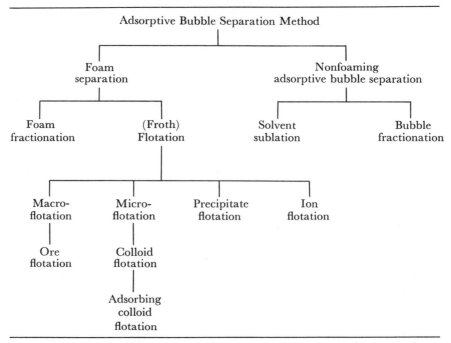

[a] B. L. Karger, R. B. Grieves, R. Lemlich, A. J. Rubin, and F. Sebba, *Separation Sci.*, **2**, 401 (1967). Reprinted by courtesy of Marcel Dekker, Inc.

bulk solution). The units of $\tilde{K}$ are centimeters, since the solute is adsorbed from a three-dimensional phase to a two-dimensional surface. The Gibbs equation can be used to relate the distribution factor to the fundamental adsorption parameter, surface tension γ:

$$\tilde{K} = \frac{\Gamma}{C} = -\frac{1}{RT}\frac{d\gamma}{dC} \tag{14.7}$$

Equation 14.7 is derived by defining the surface as a plane such that the solvent concentration is the same in both the surface and bulk phase (i.e., Γ solvent $= 0$).

Equation 14.7 permits an examination of $\tilde{K}$ with respect to surfactant concentration, from $d\gamma/dC$. A hypothetical surface tension versus concentration curve is plotted in Fig. 14.6 for a surface-active species. There are three distinct regions in this figure: one at very low concentration, a, in which the slope is nearly zero; a second region at intermediate concentrations, a to b, in

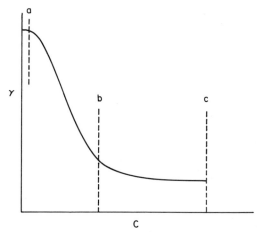

Figure 14.6. Hypothetical surface tension γ versus concentration C: plot for a surface-active agent in water. The sections a, b, and c denote specific subdivisions of the curve.

which the slope is negative; and one at higher concentrations, b to c, in which the slope again becomes close to zero. At low concentrations, ca. $10^{-6}\ M$ to $10^{-7}\ M$, little or no adsorption occurs (i.e., no separation), and the surface tension is that of the pure solvent, with $\tilde{K}$ being zero. In the intermediate region a negative slope occurs, and selective adsorption is possible, with $\tilde{K}$ greater than zero. Often a linear relationship is found over a portion of this region, with $\tilde{K}$ constant, independent of bulk surfactant concentration. Point b in the figure will usually be in the range of $10^{-4}\ M$ to $10^{-3}\ M$. In the region above b, the slope again becomes quite small, and it is expected that $\tilde{K}$ will fall off to zero from Eq. 14.7 with no resultant separation. The change in surface tension is due to the formation of micelles which are aggregates of surfactant molecules produced in the solution when the surface is completely saturated. Surprisingly, it is found in practice that adsorptive bubble separation processes work well in the micelle region (albeit not as well as the intermediate a to b region), and clearly, then, the Gibbs equation does not describe the adsorption process in this concentration region. For the most part, it is advisable to work at concentrations at which micelles do not form.

A close examination of the structure of a surface-active species reveals two distinct groups, one hydrophobic and one hydrophilic. For example, consider hexadecyltrimethylammonium bromide (HDT):

$$\left[C_{16}H_{33}-\underset{\underset{CH_3}{|}}{\overset{\overset{CH_3}{|}}{N}}-CH_3 \right]^{+} \quad Br^{-}$$

The long-chain hydrocarbon (tail) is hydrophobic and will be insoluble in

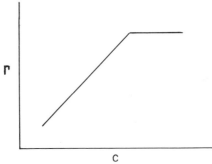

Figure 14.7. Hypothetical plot of surface excess Γ versus concentration C in a foam separation process. The example is for the case of a non-surface-active species complexing with a surfactant and being subsequently removed in a foam.

the aqueous phase. On the other hand, the charged portion (head) is quite soluble in water. With the introduction of a gas-liquid interface the surfactant migrates to the interface and aligns itself as follows:

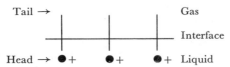

the tail portion sticking out of the water phase. The chain length is obviously important in determining the surface activity of the species. Too short a length results in complete solubility in the water, whereas an excessive length produces an insoluble species. In general, chain lengths of 12 to 18 carbon atoms are found to be best.

As noted in Section 14.2, adsorptive bubble processes are also applied to nonsurface-active species. Consider, for example, metallic ions, such as Ti^{4+} and Ni^{2+}. Addition of an anionic surfactant results in the formation of a surface-active ion-pair complex. When gas bubbles are introduced, this complex can adsorb at the gas-liquid interface and be removed by the appropriate procedure. Figure 14.7 is a plot of Γ versus C for the non-surface-active species in a foam separation removal process. The adsorption isotherm is linear until the surface is saturated, at which point Γ becomes constant, independent of the concentration of the nonsurface-active species.

The extent of removal of the nonsurface-active species can be controlled by secondary chemical equilibria, as in ion exchange. Thus the charge on a given species can exert an enormous influence on its removal and separation. This charge can be controlled chemically (by means of pH, complex formation, etc.). For example, when an oppositely charged surfactant is used, addition of a masking agent to tie up one metal species results in the selective removal of other metal species that do not react with the masking agent.

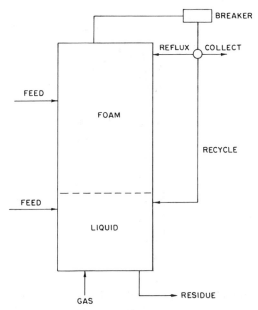

Figure 14.8. Schematic representation of several of the various modes of operation of a foam column. The several modes are discussed in the text.

14.24 EXPERIMENTAL

In Chapter 4 a number of different ways in which a separation column might operate were described. Some of these have been adapted to the foam separation system, as shown in the composite diagram of Fig. 14.8. The simplest operation is that of the batch mode, in which a bulk aqueous-phase solution (containing surfactant) is foamed by passing gas through a sparger, such as a fritted disk or a spinneret. The foam is usually collected in a collapsed liquid state. Foam breakage can be accomplished either thermally or, better, mechanically, using a spinning wire-mesh basket.

To enrich the top product, all or a portion of the collected collapsed foam liquid can be returned to the column. For example, total reflux can be employed by having the foam travel into a hot zone immediately above the column, where it is broken. The liquid then flows over foam bubbles below, with subsequent enrichment of the surface-active species. The hot zone need only be removed for final collection. This procedure can result in over a 1000-fold increase in concentration of the surface-active species in the final product. An apparatus designed for total reflux operation is shown in Fig. 14.9.

It is interesting to note that, while reflux provides for enrichment in the column, the actual mechanism of enrichment is somewhat different from

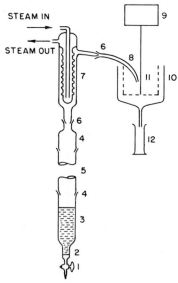

Figure 14.9. Apparatus for foam fractionation with total reflux: *1* = two-way stopcock; *2* = porous glass frit, coarse; *3* = funnel foam column section, 20 cm × 36 mm; *4* = 40/50 standard-taper joints; *5* = foam column proper, 23 cm × 34 cm; *6* = 24/40 standard-taper joints; *7* = Friedrichs condenser; *8* = tube adapter, 105°; *9* = cone drive stirrer; *10* = glass collector funnel, 20 × 13 cm, 130°, 8-mm stem; *11* = Wire basket, 7 × 9 cm; *12* = graduated cylinder, 10 ml.

that in fractional distillation. Selective adsorption results on the bubble surfaces in the solution; however, as the gas bubbles break through the solution into the foam layer, they drag along entrapped bulk liquid in the Plateau borders (interstitial regions between the gas bubbles). Since this bulk liquid dilutes the enrichment, the wetter a foam, the poorer is the separation. Return of enriched liquid in the collapsed foam to the top of the foam column (reflux) results in enrichment of the entrapped liquid, *not* the gas-liquid interface. (The entrapped liquid in the Plateau borders changes in concentration for the surface-active species from that in the bulk to a high concentration.) In addition, gas flow rates and foam column lengths should be selected to produce dry (i.e., well-drained) foams, commensurate with foam stability and separation time (37). The interfaces are quite often saturated with the surface-active material before reaching the top of the column.

Figure 14.8 shows two further possible modes of operation in which the system is operated continuously. Feeding the sample continuously into the bulk solution permits the column to act as an enricher, whereas feeding into the foam phase itself makes the column act as a stripper. Finally, as shown in

Fig. 14.8, it is possible to recycle the collapsed foam liquid into the bulk solution for the achievement of steady state in a batchwise operation. This mode is useful when distribution factors are to be measured.

14.25 APPLICATION OF FOAM SEPARATION

Table 14.4 lists some typical biopolymers that have been purified by foam separation. Many proteins are surface active and will produce a foam when gas is bubbled through the solution. Studies have revealed that the optimum pH for removal of the proteins is at the isoelectric point, where the surface tension versus concentration curve shows its largest negative slope (cf. Fig. 14.6 and Eq. 14.7). Recoveries close to 100% have often been achieved with enrichments over 20-fold. One of the key characteristics of this process is the speed with which the enrichment occurs because of the large gas-liquid interface generated in the bubbling process (40). Foam separation is a relatively mild process, a characteristic that makes it attractive for heat-sensitive species, such as biopolymers. Some enzymes may be denatured in the adsorption step on the gas bubbles; however, in many cases, enzyme activity is retained in the collected fraction. Indeed early workers (ca. 1940) were able to collect enzymes, such as cholinesterase, in crystalline form (41).

Other surface-active species have been separated by foam separation. Both anionic and cationic surfactants have been removed. In an interesting application surfactant ions have been determined quantitatively by ion flotation, in which a dye is complexed to the surfactant in a known stoichiometric amount as the surfactant is floated out of the solution (42). Colorimetric analysis of the collapsed foam liquid can be used to determine the amount of surfactant initially in the bulk aqueous phase. On the engineering scale, alkylbenzene sulfonate detergents have been removed from municipal sewage by foam fractionation (43).

Ion flotation has been applied for the removal of a variety of metal ions

**Table 14.4.   Selected Biopolymers
Separated by Foam**[a]

Bovine serum albumin	Hop resins
Albumin	Pepsin
Catalase	Lactic dehydrogenase
Urease	

[a] For references to these and other biopolymers see E. Rubin and E. L. Gaden, Jr., *New Chemical Engineering Techniques*, H. M. Schoen, ed., Interscience, New York, 1962.

from aqueous media. For example, Cs^+, Sr^{2+}, and Ce^{2+} have been removed, using a variety of surface-active agents (44). There is interest in removing these ions from aqueous media, since they occur as radioactive trace impurities in nuclear wastes. Many other metal cations have also been successfully floated, including Fe^{3+}, Fe^{2+}, Al^{3+}, and Cu^{2+}. In general best removals occur with an oppositely charged surfactant; however, in one case Fe^{3+} was floated out of aqueous media with a cationic surfactant, using an anionic chelating agent as a "bridge" (45).

Selective separations of one metal ion from another are also possible by ion flotation. In this application secondary chemical equilibria are employed to control the charge on the ions (see Chapter 2). For example, metals can be separated in hydrochloride medium, as in ion exchange, using a cationic surfactant (46). Only the anionic metal chloro complex species are removed. Thus Ni^{2+} and Fe^{3+} can be separated at $8\ N$ HCl with the flotation of $FeCl_4^-$; Ni^{2+} does not form a chloro complex. In addition, uranium has been separated from thorium by the flotation of the anionic uranyl chloro complex. Other secondary chemical equilibria have also been used; for example, aluminum has been separated from beryllium by the flotation of an oxalato-aluminate complex with long-chain fatty amines (47).

Precipitate flotation has been studied for the separation of metal ions. The formation of a precipitate and its subsequent flotation have been found to be more rapid and to require less surfactant than ion flotation (48, 49). This method can be useful in quantitative analysis, since the flotation works equally well with crystalline and gelatinous precipitates. Trace metal ions (especially radioactive isotopes) have been concentrated out of aqueous media via suitable precipitate carriers (e.g., iron hydroxide), the precipitate being removed by an appropriate surfactant collector. The latter method can be considered as being related to adsorbing colloid flotation (50), which has found recent application in the concentration of trace metals in sea water (51).

14.26   SOLVENT SUBLATION

Solvent sublation was originally introduced as an auxiliary technique to ion flotation for use in cases in which a persistent foam existed (36). Surface-active material is present in the bulk aqueous phase, on top of which is placed an immiscible liquid. Gas bubbles are generated in the aqueous phase and are buoyed upward into the organic layer, carrying the material selectively adsorbed on the bubbles. The nonaqueous layer then acts both as a collection medium and as a foam breaker. The name *solvent sublation* arises from the fact that the complex formed between a surface-active agent and an oppositely charged species is called the *sublate,* and the process of lifting the sublate by gas bubbles *sublation.*

Solvent sublation would appear to be similar to liquid-liquid extraction; however, there are several important differences. First, it has been shown that the amount of complex removed from aqueous media is independent of the volume of the organic phase in solvent sublation. For this reason, as well as others, it is clear that liquid-liquid equilibrium does not exist between the two immiscible phases (52). The consequence is that solvent sublation can be a useful process of concentration, since only a thin layer of organic phase need be placed on top of the aqueous phase. Second, in solvent extraction liquid-liquid equilibrium is rapidly achieved, even if shaking is mild. Thus it is difficult to use rates of extraction in separation. This is not the case, however, in solvent sublation; indeed, rate of gas flow and bubble size can be useful parameters to control the rate of removal. Third, emulsion formation with surface-active agents is considerably less of a problem in solvent sublation than in liquid extraction.

Solvent sublation has been used to separate the same types of species that are separated by ion flotation (53). For example, hexacyanoferrate(II) ion has been removed by dodecylpyridinium chloride into 2-octanol. As another example, using the rate of removal as a means of separation, methyl orange (MO) and rhodamine B (RB) were separated with a cationic surfactant in basic media. The MO removal was found to be rapid when an oppositely charged surfactant was used. On the other hand, at the operating pH, RB is zwitterionic, so that a surfactant–dye complex is unlikely. Yet RB is surface active and is readily sublated into the organic phase. Addition of surfactant to a mixture of the dyes suppresses the rate of removal of RB, since the surfactant effectively competes with the dye for adsorption on the bubble surface. Thus, at short time periods of gas flow, the separation of MO and RB will be large, but after longer periods separation becomes poorer. Thus control of the rate of removal can sometimes be useful in the separation of ions by solvent sublation.

Selected Bibliography

PRINCIPLES OF GAS-SOLID ADSORPTION

D. M. Young and A. D. Crowell, *Physical Adsorption of Gases,* Butterworths, London, 1962.

S. Ross and J. P. Olivier, *On Physical Adsorption,* Interscience, New York, 1964.

J. J. Bikerman, *Physical Surfaces,* Academic Press, New York, 1970, Chap. VII.

GAS-SOLID CHROMATOGRAPHY

A. V. Kiselev and Y. I. Yashin, *Gas-Adsorption Chromatography*, J. E. S. Bradley, transl., Plenum, New York, 1969.

A. V. Kiselev, in *Advances in Chromatography*, Vol. IV, J. C. Giddings and R. A. Keller, eds., Marcel Dekker, New York, 1967.

C. S. G. Phillips and C. G. Scott, in *Progress in Gas Chromatography*, J. H. Purnell, ed., Interscience, New York, 1968.

L. R. Snyder, *The Principles of Adsorption Chromatography*, Marcel Dekker, New York, 1968.

P. G. Jeffery and P. J. Kipping, *Gas Analysis by Gas Chromatography*, Pergamon Press, New York, 1964.

ADSORPTIVE BUBBLE SEPARATION

H. G. Cassidy, in *Technique of Organic Chemistry*, Vol. X, A. Weissburger, ed., Interscience, New York, 1957.

I. A. Eldib, in *Advances in Petroleum Chemical Refining*, Vol. 7, K. A. Kobe and J. F. McKetta, Jr., eds., Interscience, New York, 1963.

E. Rubin and E. L. Gaden, Jr., in *New Chemical Engineering Techniques*, H. M. Schoen, ed., Interscience, New York, 1962.

R. B. Grieves, *Brit. Chem. Eng.*, **13**, 77 (1968).

R. Lemlich, *Ind. Eng. Chem.*, **60**, 16 (1968).

B. L. Karger and D. G. DeVivo, *Separation Sci.*, **3**, 393 (1968).

R. L. Lemlich, ed., *Adsorptive Bubble Separation Techniques*, Academic Press, New York, 1971.

References

1. A. V. Kiselev, in *Advances in Chromatography*, Vol. IV, J. C. Giddings and R. A. Keller, eds., Marcel Dekker, New York, 1967.

2. A. V. Kiselev and Y. I. Yashin, *Gas-Adsorption Chromatography*, Plenum, New York, 1969.

3. J. J. Bikerman, *Physical Surfaces*, Academic Press, New York, 1970.

4. J. H. de Boer, *The Dynamical Character of Adsorption*, Clarendon Press, Oxford, 1953, p. 90.

5. L. R. Snyder, *Principles of Adsorption Chromatography*, Marcel Dekker, New York, 1968, Chap. 9.

6. S. Claesson, *Arkiv Kemi*, **A23**, No. 1 (1946); C. S. G. Phillips, *Discussions Faraday Soc.*, **7**, 241 (1949).

7. E. Cremer, *Arch. Biochem. Biophys.*, **83**, 345 (1959).

8. L. Guillemin, M. Deleuil, S. Cirendini, and J. Vermont, *Anal. Chem.*, **43**, 2015 (1971).

9. C. Vidal-Madjar and G. Guiochon, *Bull. Chim. Soc. France*, **1971**, 3110.

10. L. R. Snyder and E. R. Fett, *J. Chromatog.*, **18**, 461 (1965).

11. I. Halasz and E. Heine, in *Advances in Chromatography*, Vol. IV, J. C. Giddings and R. A. Keller, eds., Marcel Dekker, New York, 1967.
12. A. Liberti, in *Gas Chromatography, 1966*, A. B. Littlewood, ed., Institute of Petroleum, London, 1967.
13. I. Halasz and H. O. Gerlach, *Anal. Chem.*, **38**, 281 (1966).
14. G. M. Petov and K. D. Shcherbakova, in *Gas Chromatography, 1966*, A. B. Littlewood, ed., Institute of Petroleum, London, 1967.
15. C. Vidal-Madjar, J. Ganansia, and G. Guiochon, in *Gas Chromatography, 1970*, R. Stock, and S. G. Perry, eds., Institute of Petroleum, London, 1971.
16. C. Vidal-Madjar and G. Guiochon, *J. Chromatog. Sci.*, **9**, 644 (1971).
17. C. S. G. Phillips and C. G. Scott, in *Progress in Gas Chromatography*, J. H. Purnell, ed., Interscience, New York, 1968.
18. R. L. McCreery and D. T. Sawyer, *J. Chromatog. Sci.*, **8**, 122 (1970).
19. A. F. Isbell, Jr., and D. T. Sawyer, *Anal. Chem.*, **41**, 2110 (1969).
20. A. O. S. Maczek and C. S. G. Phillips, in *Gas Chromatography, 1960*, R. P. W. Scott, ed., Butterworths, London, 1960.
21. I. Halasz and I. Sebastian, *Angew. Chem., Intern. Ed.*, **8**, 453 (1969).
22. W. A. Aue and C. R. Hastings, *J. Chromatog.*, **42**, 319 (1969).
23. O. L. Hollis and W. V. Hayes, in *Gas Chromatography, 1966*, A. B. Littlewood, ed., Institute of Petroleum, London, 1967.
24. S. B. Dave, *J. Chromatog. Sci.*, **7**, 389 (1969).
25. P. G. Jeffery and P. J. Kipping, *Gas Analysis by Gas Chromatography*, Pergamon Press, New York, 1964.
26. R. J. Leibrand, *J. Gas Chromatog.*, **5**, 518 (1967).
27. G. Guiochon and C. Pommier, *La Chromatographie en Phase Gazeuse en Chimie Inorganique*, Gauthier-Villars, Paris, 1971.
28. M. Mohnke and W. Saffert, *Gas Chromatography, 1962*, Butterworths, London, 1962.
29. R. D. Schwartz and D. J. Brasseau, *Anal. Chem.*, **29**, 1022 (1957).
30. E. Cremer and J. K. F. Huber, *Angew. Chem.*, **73**, 461 (1961); J. F. K. Huber and R. G. Gerritse, *J. Chromatog.*, **58**, 137 (1971).
31. I. Halasz and E. Heine, *Anal. Chem.*, **37**, 495 (1965).
32. R. W. Schnepf and E. L. Gaden, Jr. *J. Biochem. Microbiol. Technol. Eng.*, **1**, 1 (1959).
33. R. B. Grieves and S. L. Wang, *Appl. Microbiol.*, **15**. 76 (1967).
34. R. E. Baarson and C. L. Ray, *Hydrometallurgy*, Gordon and Breach, New York, 1964, p. 656.
35. F. Sebba, *Nature*, **184**, 1062 (1659).
36. F. Sebba, *Ion Flotation*, Elsevier, Amsterdam, 1962.
37. R. Lemlich, *A.I.Ch.E. J.*, **12**, 802 (1966).
38. E. Rubin, R. Everett, J. J. Weinstock, and H. M. Schoen, *U.S. Public Health Ser. Publ.* AWTR-5 Cincinnati, 1963.
39. J. Brady, *J. Phys. Chem.*, **53**, 56 (1949).
40. S. E. Charm, J. Morningstar, C. C. Matteo, and B. Paltiel, *Anal. Biochem.*, **15**, 498 (1966).
41. R. Bader and F. Schutz, *Trans. Faraday Soc.*, **42**, 571 (1946).
42. V. M. Lovell and F. Sebba, *Anal. Chem.*, **38**, 1926, (1966).
43. E. Rubin and R. Everett, Jr. *Ind. Eng. Chem.*, **55**, 44 (1963).
44. J. J. Weinstock, S. Mook, E. Schonfeld, E. Rubin, and R. Sanford, *U.S. At. Energy Comm.* NYO 10038 (1963).
45. N. Aoki and T. Sasaki, *Bull. Chem. Soc. Japan*, **39**, 939 (1966).
46. B. L. Karger and M. W. Miller, *Anal. Chim. Acta*, **48**, 273 (1969).

47. J. A. Lusher and F. Sebba, *J. Appl. Chem.*, **15**, 577 (1965).
48. A. J. Rubin and J. D. Johnson, *Anal. Chem.*, **39** 298 (1967).
49. E. J. Mahne and T. A. Pinfold, *J. Appl. Chem.*, **18**, 52 (1968).
50. V. V. Pushkarev, Y. V. Bagretsov, and V. D. Puzako, *Radiokhimiya*, **6**, 120 (1964).
51. Y. S. Kim and H. Zeitlin, *Anal. Chem.*, **43**, 1390 (1971).
52. B. L. Karger, T. A. Pinfold, and S. E. Palmer, *Separation Sci.*, **5**, 603 (1970).
53. B. L. Karger, A. B. Caragay, and S. B. Lee, *Separation Sci.*, **2**, 39 (1967).

List of Symbols

a	amount of solute vapor adsorbed	α'	relative adsorbent activity due to specific interactions
A_i	surface area covered by adsorbed solute	α'_d	relative adsorbent activity due to nonspecific interactions
C	constant in Eq. 14.4		
ΔE	interaction energy	ε°	solvent strength of the mobile phase
ΔH_A	heat of adsorption		
$\tilde{K}_A$	adsorption coefficient	Γ	excess adsorbed solute on the surface
K_A	adsorption constant		
S^0	relative sample adsorption energy		
V_a	adsorbent surface volume		
V'_{Rg}	net retention volume per gram of adsorbent		

Fifteen

EXCLUSION PROCESSES

J. Y. CHUANG and J. F. JOHNSON

15.1 Adsorption on Molecular Sieves (Zeolites)

15.11 GENERAL

It has long been known that porous solids can separate mixtures on the basis of molecular size, shape, and polarity. Although a number of natural and synthetic materials possess this property to varying degrees (e.g., silica gel, activated charcoal), the sharpest separations according to size are obtained on natural and synthetic zeolites. The zeolites are aluminosilicates whose crystal structures are characterized by a three-dimensional array of AlO_4 and SiO_4 tetrahedra, with oxygen atoms shared by two tetrahedra. The result is a cagelike structure of precise geometry, and such adsorbents have pores of uniform shape that permeate the entire crystal. The uniform shape of the pores is the basis for separation by molecular size. Typically the adsorbent is placed in contact with a mixture of two or more components, one of which is preferentially retained. After the enriched mixture is removed, the adsorbed component is recovered from the adsorbent by heating, application of reduced pressure, or displacement with another material.

Naturally occurring zeolites were discovered and identified more than 200 years ago. Early investigations included studies on reversible dehydration by Damour in 1840, synthesis by St. Deville in 1862, and studies of preferential adsorption by McBain in 1926. The name *molecular sieve* is due to McBain.

In the late 1930s, R. M. Barrer began a systematic study of the properties of zeolites (1) and later wrote an excellent review of this subject (3). Barrer's work established many of the widespread uses of these materials in separation and purification, leading to extensive interest and research in the structure, synthesis, and applications of the zeolites in other areas such as catalysis.

15.12   DESCRIPTION OF INDIVIDUAL ZEOLITES

Over thirty zeolites occur naturally, in addition to a large number of synthetically prepared materials. Chemical modifications of natural zeolites to improve their separating ability have also been described. It is possible, however, to characterize all of these materials by a general formula:

$$Me_{x/n}[(AlO_2)_x(SiO_2)_y] \cdot mH_2O$$

where x, y, and n are integers, Me is a metal cation (e.g., Na^+, K^+, Ca^{2+}, Sr^{2+}), and m is the number of H_2O molecules. The ratio of y/x, that is, Si/Al, can be varied from 1 to 5. Varying the Si/Al ratio can change the arrangement of the SiO_4 and AlO_4 tetrahedra, thus altering the dimension of the pore structure. Readily exchangeable cations are characteristic of the zeolites. The zeolites contain intracrystalline water, which may be removed by an increase in temperature. The cations in the zeolite structure maintain electrical neutrality. Water and exchangeable cations are located in the cavities, not in the crystal framework, so that reversible ion exchange and gain or loss of water can occur.

The structures of the zeolites have been presented as various three-dimensional arrangements of SiO_4 and AlO_4 polyhedra, commonly cubes with

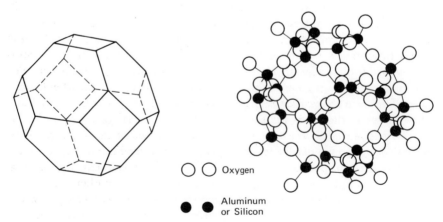

○ ○ Oxygen

● ● Aluminum or Silicon

Figure 15.1.   Schematic representation of the cubo-octahedron structure. Reprinted from Ref. 3, by courtesy of Endeavor, ICI, Inc.

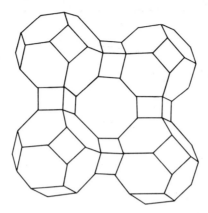

Figure 15.2. Structure of zeolite A. Reprinted from Ref. 4, by courtesy of the Indian Science News Association.

8, hexagonal prisms with 12, and cubo-octahedra with 24 tetrahedra, respectively (2). Thus, for example, the cage structure shown in Fig. 15.1 is a cubo-octahedron cage. This in turn can alternate with cubes stacked in sixfold coordination to produce the structure commonly referred to as zeolite A (Fig. 15.2). Zeolite A has an internal cavity 11.8 Å in diameter that, as the figure shows, is roughly spherical. Each of these cavities is in turn connected to six others by six "ring windows" formed by eight oxygen

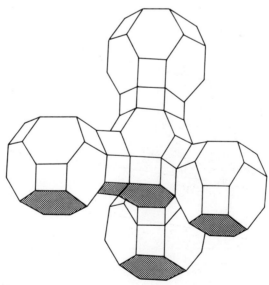

Figure 15.3. Structure of faujasite. Reprinted from Ref. 4, by courtesy of the Indian Science News Association.

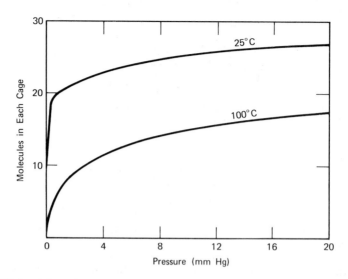

Figure 15.4. Adsorption isotherms for water on zeolite A. Reprinted from Ref. 5, by courtesy of the American Chemical Society.

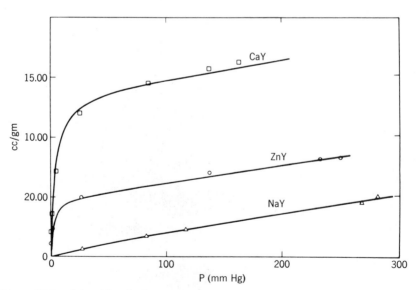

Figure 15.5. Adsorption isotherms for carbon monoxide. Reprinted from Ref. 6, by courtesy of the American Institute of Chemical Engineers.

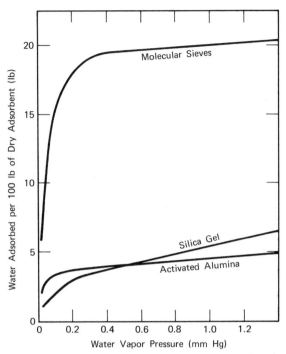

Figure 15.6. Equilibrium absorption for common absorbents. Reprinted from Ref. 7, by courtesy of Engineering, Chemical and Marine Press, Ltd.

atoms. The diameter of the ring window is 4.2 Å. Therefore a set of channels extends throughout the crystal. Figure 15.3 shows the structure of another zeolite, faujasite, which has fourfold coordination of cubo-octahedra with hexagonal structures. Here the central cavity is again 11.8 Å in diameter, but there are four 12-membered ring windows 8 to 9 Å in diameter. Since a wide variety of structures is possible, various cage and window dimensions are obtained. Table 15.1 gives representative examples of some of the more commonly used molecular sieves.

Space precludes an extensive discussion of the adsorption isotherms encountered with zeolites. Three typical sets of isotherms are given in Figs. 15.4 to 15.6. Figure 15.4 shows the number of water molecules adsorbed in the central cage of zeolite A as a function of temperature and pressure. The isotherms show a rapid initial increase with pressure, followed by a slower, almost constant increase at higher pressures. The amount of water adsorbed decreases with increasing temperature. Figure 15.5 shows similar behavior for the adsorption of carbon monoxide on three different Y zeolites. This figure demonstrates the considerable differences in adsorptive power with zeolite type. Figure 15.6 compares the amount of water adsorbed per unit

Table 15.1. Dimensions of Some Common Molecular Sieves

Name and Formula	Building Unit	Cavity Diameter (Å)	Window Diameter (Å)	Types of Adsorbate
Faujasite $(Na_2,Ca,Mg)_{30}[(AlO_2)_{60}(SiO_2)_{132}]26OH_2O$	Cubo-octahedra and hexagonal prism	11.8	8–9	H_2O, CO_2, NH_3, H_2S, SO_2, Ar, N_2, O_2, n-paraffins, isoolefins, isoparaffins, cyclic compounds
Molecular sieve X $Na_{86}[(AlO_2)_{86}(SiO_2)_{106})]264H_2O$	Cubo-octahedra and hexagonal prism	11.8	8–9	H_2O, CO_2, NH_3, SO_2, Ar, O_2, N_2, isohydrocarbons, cyclic compounds
Molecular sieve A $Na_{12}[(AlO_2)_{12}(SiO_2)_{12}]27H_2O^a$	Cubo-octahedra and cube	11.8	4.2	3A: H_2O, CO_2, NH_3, 4A: H_2O, CO_2, NH_3, CH_4, C_2H_6, C_2H_4, C_3H_6 5A: Same as 4A and the higher n-paraffins, n-olefins, and RCH_2Y (Y = OH, SH, CN, NH_2 or halogen)
Chabazite $Ca_4[(AlO_2)_8(SiO_2)_{16}]27H_2O$	Hexagonal prism	11.2 (major axis) 7.5 (minor axis)	3.7–4.2	Same as 5A

a The sodium ion in this formula (4A) can be replaced by K^+ (3A) or Ca^{2+} (5A). The effective size of windows can be various diameters (3Å, 4Å, or 5Å), depending on whether the cation is K^+, Na^+, or Ca^{2+}, respectively.

weight of adsorbent for molecular sieve, silica gel, and activated alumina. The superiority of molecular sieve is obvious. The strong affinity of the molecular sieves for water is due to the polar nature of the water molecule and the charges on the surface of the molecular sieves.

15.13 APPLICATIONS

The applications of zeolites are extensive. They can be divided into categories based on the property of the zeolite that is being utilized:

1. Separation according to molecular size and shape.
2. Water adsorption.
3. Ion exchange.
4. Catalysis.

As catalysis is not a property associated with separation, it will not be discussed further except to note that zeolites have been used as catalysts for a wide range of reactions, including oxidation, hydrogenation, isomerization, polymerization, and cracking. The possibility that the zeolite used in a separation process may also function as a catalyst must always be borne in mind.

The reversible dehydration of zeolites has been studied extensively. Heating under reduced pressure reversibly removes water present in the channels. Only a few of the many zeolites have crystal structures that remain completely unchanged during dehydration. However, even though some structural changes may occur, frequently sufficient pore size remains for the zeolite to function efficiently as a separating agent. For example, molecular sieve 4A can be used for over 1000 adsorption-desorption cycles.

The use of zeolites as water adsorbents is widespread. In addition to the high capacity shown in Fig. 15.6, the zeolites can readily reduce water concentrations to 1 ppm or less. Therefore they are commonly employed in producing very dry gas streams for argon, helium, nitrogen, carbon dioxide, air, and other gases. Zeolites can produce equivalent low levels of water in liquid hydrocarbon streams and are extensively employed for this purpose in petroleum refining and petrochemical manufacturing. An important aspect of zeolites as drying agents is that, by proper choice of critical pore diameter, simultaneous coadsorption of other materials with water can be prevented. Conversely, selection of the critical diameter may be such that drying and purification by removal of other materials may be done simultaneously if desired. An example is the removal of water, hydrogen sulfide, and mercaptans from liquified petroleum gas in a single step.

Separations according to molecular size and shape are accomplished by selection of the dimensions of the zeolite so that one type of molecule can enter the cage whereas another cannot. For example, Linde type 5A molecular sieve will adsorb *n*-butane but not isobutane. The critical dimension of a molecule is defined as the circumscribed circle of its cross section of minimum area; for example, the critical dimension is 4.89 Å for butane and 5.58 Å for isobutane. Molecular sieves can also be used to separate higher-molecular-weight *n*-alkanes from branched-chain and cyclic hydrocarbons. This has led to the utilization of zeolites for the following purposes:

1. Removal of straight-chain hydrocarbons from gasoline to increase the octane number.
2. Isolation of *n*-paraffins for feedstocks for petrochemical processes.
3. Separation of straight- and branched-chain olefins to produce biodegradable detergents.

One disadvantage of the very small zeolite pores is slow mass transfer. This is quite marked for large molecules and for liquid-phase separations, severely limiting these applications. In addition to the geometrical factor, separation is influenced also by polarity, unsaturation, temperature, and pressure. For example, ethane and ethylene have approximately the same cross section. However, on a type A zeolite 80% of the molecules adsorbed from a mixture of ethane and ethylene will be the more polar ethylene. Selection of zeolites of the proper geometry at the correct operating temperature and pressure has resulted in the following separations:

1. 4-Methylpyridine from 3-methylpyridine.
2. Double-branched from single-branched aliphatic hydrocarbons.
3. Nitrogen from oxygen.
4. Oxygen from argon and nitrogen.

15.2   Gel Chromatography

15.21   GENERAL

Section 15.1 described the use of zeolites for separation purposes. Good separations can be obtained, but only for low-molecular-weight materials. The search for materials having pore structures that would permit separation of high-molecular-weight substances by molecular size has been an extensive one and has resulted in the widely used technique of gel chromatography.

This method of separation has suffered from a diverse nomenclature. It has, for example, been referred to as *gel filtration, molecular sieve filtration, exclusion chromatography,* and *gel permeation chromatography,* as well as by the name we have chosen: *gel chromatography.* All these names refer to essentially the same process.

Basically the separation method uses a column packed with the porous separating medium. Such columns have a variety of dimensions but are typically 1 to 3 cm in diameter and 1 to 10 meters long. A number of porous materials have been used: cross-linked polymers, porous gels, ceramic materials, and others. A solvent is passed continuously through the column. A small amount of material to be separated is introduced at the head of the column, usually as a solution in the same solvent that is being pumped through the column.

The retention volume can be expressed by Eq. 5.6:

$$V_R = V_M + KV_S$$

Here the mobile-phase volume V_M and the stationary-phase volume V_S are the void volume and the total pore volume of the packing materials, respectively.

The distribution constant K depends on the molecular weight of the sample and the pore size of the gel. Small molecules that can enter into the pores freely have $K = 1$, whereas very large molecules that are completely excluded from all pores have $K = 0$. Intermediate-size molecules have access to various portions of the available pore volume and have K values between 1 and 0. Thus large molecules move through the column more rapidly than small molecules. Molecular size is associated with and identified by the particular elution volume at which a given molecule emerges. Figure 15.7 is a schematic representation of the separation process.

Historically gel chromatography dates back at least into the early 1940s. It did not receive initial widespread application because of the lack of a good chromatographic medium with a variety of pore sizes to permit separation over a wide molecular weight range. In 1959, however, Porath and Flodin introduced a gel obtained by reacting dextran and epichlorohydrin (8); it is now available commercially under the trade name Sephadex. Gel chromatography with water or buffered solutions as a solvent using such cross-linked dextran gels became one of the most important separation methods in biochemistry. An excellent review with many references to separations by this type of gel is given in *Gel Chromatography* by Determann (9).

Although success in separating hydrophilic materials was rapidly achieved, the development of successful separating media for hydrophobic polymers proceeded more slowly. Early workers, for example, Vaughan (10) and Brewer (11) in 1960, demonstrated the applications of this technique, using organic solvents. Again widespread application was not forthcoming because of lack of a suitable wide range of separation gel. In 1962 Moore made two major improvements for separations involving hydrophobic polymers (12). One improvement involved the use of macromolecular polymers, which have a moderately rigid gel structure and a wide range of permeabilities. As a

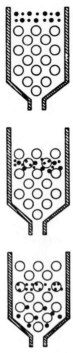

Figure 15.7. Schematic representation of gel chromatography. Reprinted from Ref. 9, by courtesy of Springer-Verlag.

result components having molecular weights from a few thousand to several million could be separated by the proper choice of gel porosity. The other advance consisted of connecting the column to a detector, which monitored the solute concentration of the effluent continuously. Moore used a differential refractometer for this purpose. A review of applications of separations using organic solvents is given in Ref. 13.

The normal aim of separation is the more or less complete resolution of distinct species. A special case exists, however, for high-molecular-weight polymers. Here there is frequently a continuous distribution of molecular weights. Although complete separation by oligomers is usually impossible, sufficient separation can be achieved to determine *molecular weight distributions* by gel chromatography. Previous separation methods for this purpose required many man-days of effort per sample. Gel chromatography make these determinations automatically within a few hours. This application of gel chromatography is widely used, and the special requirements for it are described in subsequent sections.

It should be emphasized, however, that gel chromatography can also be used in the conventional manner for analysis and collection of pure components, both in the low- and the high-molecular-weight range.

The scale of gel chromatography varies widely. For analytical purposes, sample sizes are small, usually of the order of a few milligrams. For preparatory separations, larger-size columns are used and sample sizes range up to 1 g or more.

15.22 MECHANISM AND SEPARATION

One of the unique characteristics of gel chromatography is the close relation between molecular size and elution volume. The distribution constant in Eq. 5.6 depends on the molecular size, not on the chemical nature of the sample. For a homopolymer, molecular weight determines the molecular size uniquely. It has been found experimentally that a plot of elution volume versus log (molecular weight) is a straight line (within a suitable molecular weight range). The relation can be expressed by the following equation:

$$V_R = A + B \log (\text{mol. wt.}) \qquad (15.1)$$

where A and B are constants. The curve obtained by this method (e.g., Fig. 15.8) is used for calibration.

Although the detailed separation mechanism is still not clear, various

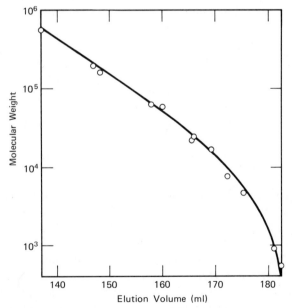

Figure 15.8. Typical calibration curve. Reprinted from Ref. 43.

theories have been suggested to explain how molecular size determines the distribution constant. Among these theories the most important are steric exclusion and restricted diffusion.

Steric Exclusion

Flodin (15) explained the separation mechanism as follows: the pore size of some region of the packing material is so small that the large molecules cannot diffuse into this region. By considering the pore size distribution of the column bed, assigning to each pore size a specific size of polymer, and assuming that molecules can enter pores larger than their size but not smaller, one can derive a relation between elution volume and log (molecular weight) (16). For flexible-chain molecules, Cassasa and Tagami suggest that it is the reduction of conformation entropy of molecules in the pores that influences separation.

Restricted Diffusion

In the gel phase, the diffusion of the polymer is slowed down by steric hindering and by friction. In the chromatographic process, small molecules, because of their higher diffusion constants, can diffuse into the pores of the packing materials readily and spend more time in the stationary phase than large molecules; hence they have larger elution volumes (44). Under equilibrium conditions, most experiments show that the elution volume is insensitive to the flow rate change. This appears to contradict the diffusion mechanism, which predicts a dependence of elution volume on diffusion rate. In the case of nonequilibrium conditions, Yau and Malone have shown that differences between the diffusion rates of different-sized molecules should contribute to the separation (45).

Most experimental evidence now available suggests that steric exclusion is the most important separation mechanism. The diffusion effect must also play some role in the separation, however, particularly under nonequilibrium conditions.

There are two differences between molecular sieves and gel chromatography:

1. The determining factor for permeation of solute is different: minimum cross section for molecular sieves, and hydrodynamic radius for gel chromatography. For example, in the separation of isomers of different branching, molecular sieves prefer the less branched isomer, whereas the branched isomer has a *larger* elution volume than the nonbranched isomer in gel chromatography (Section 15.13).

2. The application range of molecular weights is different: small molecules for molecular sieves, wide range of molecular weights for gel chromatography —from several hundreds to several millions (Sections 15.13 and 15.21).

To convert elution volume to molecular weight requires a calibration curve. This is obtained by preparing fractions of narrow molecular weight distribution and determining their molecular weights by an independent method. Typically the absolute methods of light scattering, ultracentrifugation, or osmometry can be used, or the secondary method of solution viscometry if a known relationship between limiting solution viscosity and molecular weight is available. The fractions are then injected into the chromatograph under operating conditions identical to those employed for the sample, and a calibration curve is constructed by measuring the elution volume to the center of the peak. For most columns a plot of log (molecular weight) versus elution volume has a linear portion with curvature at both the low- and high-molecular-weight portions. A typical curve is shown in Fig. 15.8.

The disadvantage of this technique is that it requires determination of a calibration curve for each type of polymer study. The advantages of a "universal" calibration curve (i.e., a single curve applicable to all samples) are obvious, and a number of approaches to achieve this have been tried. Benoit et al. (30) and numerous others have shown that the elution volume is a function of the hydrodynamic volume. As the hydrodynamic volume is proportional to the product of molecular weight and limiting solution viscosity, narrow-molecular-weight fractions should fall on the same curve regardless of polymer type. Figure 15.9 shows such a plot for a wide variety of polymers and copolymers of widely different shapes. If a viscometer is used to measure the limiting solution viscosity, a single calibration curve of this type can be used for all types of polymers. However, even in the absence of an automatic viscometer, the hydrodynamic volume concept greatly facilitates calibration for polymers for which characterized fractions are not available. Two general calibration methods have been described.

In the first it is assumed that the intrinsic viscosity $[\eta]$-molecular weight relationship is known and can be described by a Mark-Houwink equation:

$$[\eta] = KM^a$$

where the constants K_s, a_s and K_x, a_x for the standard polymer (usually polystyrene) and the polymer under study, respectively, are known. Then use of the relationship

$$\log M_x = \left(\frac{1}{1 + a_x}\right) \log \left(\frac{K_s}{K_x}\right) + \left(\frac{1 + a_s}{1 + a_x}\right) \log M_s$$

establishes the calibration curve for polymer x (31).

If the Mark-Houwink parameters are not known, but a sample is available for which any two of the three parameters $\bar{M}_n$, $\bar{M}_w$ (see Sect. 15.26), and $[\eta]$ are known, the method of Pickett et al. (29) can be used to compute a

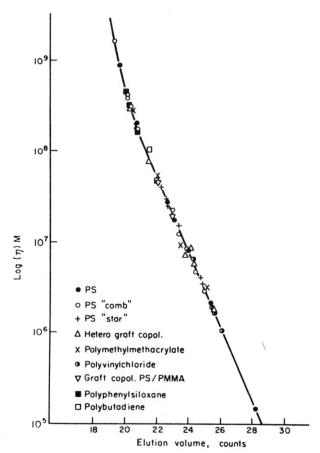

Figure 15.9.   Universal calibration plot. Reprinted from Ref. 42.

calibration curve. Provder et al. (32) have applied this technique extensively and concluded that, when two or three characterized samples that may have broad molecular weight distributions are used, the results are comparable to calibration curves generated by the more tedious methods on fractions.

15.23   COLUMN MATERIALS AND SOLVENTS

A variety of materials have been used to effect separation by gel chromatography, including organic and inorganic substances. Ideally a packing should be chemically inert and stable both thermally and mechanically, give good resolution, have a low resistance to flow, and be inexpensive. Table 15.2 lists a number of the more widely used materials and their characteristics. It must

Table 15.2. Column Packings for GPC

Material	Solvents	Separation Range of Molecular Weights	Trademark and Manufacturer[a]
Styrene–divinylbenzene gel	Organic	$5 \times 10^3\text{–}10^6$	1
Styrene gels	Organic	$5 \times 10^2\text{–}1.5 \times 10^3$	2
Polyvinylacetate gel	Polar organic	$5 \times 10^2\text{–}1 \times 10^5$	3
Porous silica beads	Organic–aqueous	$5 \times 10^2\text{–}2 \times 10^6$	4
Porous glass	Organic–aqueous	$5 \times 10^2\text{–}1.5 \times 10^6$	5
Porous glass	Organic–aqueous	$2 \times 10^3\text{–}9 \times 10^6$	6
Dextran gels	Aqueous	$2 \times 10^2\text{–}2 \times 10^7$	7
Acrylamide–methylene bisacrylamide	Aqueous	$2 \times 10^2\text{–}4 \times 10^5$	8
Agarose	Aqueous	$10^4\text{–}1 \times 10^8$	9

[a] 1: trademark "Styragel," Waters Associates, Inc., Framingham, Mass.; 2: trademark "Bio-Beads S," Bio-Rad Laboratories, Richmond, Calif.; 3: trademark "EM-Gel-OR," Waters Associates; 4: trademark "Porasil," Waters Associates; 5: Corning Porous Glasses, Lab Products, Corning, N.Y.; 6: trademark "Bio-Glass," Bio-Rad Laboratories; 7: trademark "Sephadex," Pharmacia Fine Chemicals, Piscataway, N.J.; 8: trademark "Bio-Gel P," Bio-Rad Laboratories; 9: trademark "Bio-Gel A," Bio-Rad Laboratories.

be emphasized that this list is not intended to cover all packing materials and that the separation ranges are only approximate. A list of trade names and suppliers is included.

Probably the most widely used column packing are the Styragels. These are cross-linked gels made from styrene, ethylvinylbenzene, divinylbenzene, vinylethylbenzene, and diethylbenzene; the amount of cross linking is controlled by the composition and the amount of diluent present. They give good resolution and separate over a wide range of molecular weights. Styragel columns can be used with most organic solvents and, with suitable operating techniques, with some alcohols and acids.

The Bio-Beads S are softer polystyrene gels that give excellent resolution in a lower-molecular-weight range. Their lack of mechanical rigidity limits the operating pressures that can be used. The cross-linked polyvinyl acetates, EM-Gel-OR, are especially useful with polar organic solvents.

The porous silica and glass beads have the advantages of excellent mechanical stability and high thermal stability. Polar sites are present and must be deactivated to prevent tailing.

The cross-linked dextrans, Sephadex, and the agarose gels, Sepharose, have been extensively employed for materials soluble in water or buffered solutions. Such materials include proteins and nucleic acids. An excellent review of work on proteins by Ackers has been published (14).

The choice of solvent is usually based on solubility, detector response, and compatibility with packing materials. Solvents that are effective at ambient or near-ambient temperatures are frequently chosen for ease of operation. For differential refractometer detectors, solvents with a large difference in refractive index from the solute are desirable to increase the detector response. Similarly, nonabsorbing solvents are needed with spectroscopic detectors.

Low-viscosity solvents permit the use of longer columns for a given pressure drop and increase column efficiency, as discussed in Section 10.23.

Mixed solvents have been employed both to increase solubility and to decrease adsorption on the interface. The advantages of using theta solvents, particularly for mechanism studies, have been demonstrated by Moore and Arrington (23).

15.24   INSTRUMENTATION

The essential parts of a gel permeation chromatograph are the solvent-handling system (i.e., solvent reservoir, degasser, and pump), sample-injection system, fractionation columns, and detector systems to measure and record the elution volume and concentration of sample in the eluate. Figure 15.10 shows a schematic diagram of a commercially available gel permeation

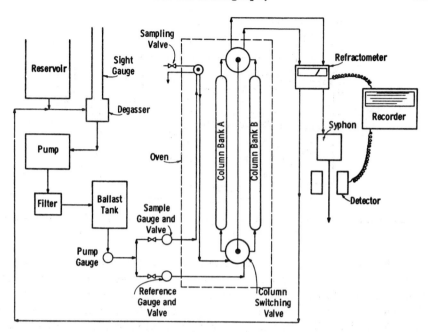

Figure 15.10. Schematic diagram of a gel permeation chromatograph. Reprinted from Ref. 46, by courtesy of the publisher of the *J. of Paint Technology*.

chromatograph. The instrumental requirements are essentially the same as described in Section 10.2.

The most frequently used technique for measuring and recording elution volume is by use of a siphon. Each time the siphon discharges, a photoelectric device detects the discharge and, through appropriate circuitry, a signal is produced and recorded. Usually this signal is superimposed briefly on the signal from the recorder, producing a rapid pen deflection. Two factors can change the volume discharged (17): evaporation of solvent during filling and discharge, and flow from the column during the discharge of the solvent. Methods to minimize evaporation have been described (17), but for precise results the siphon must be calibrated at the flow rate at which the column is being operated.

The resolution of a gel permeation column can be improved by increasing the column length. Bombaugh et al. (24) used a column 160 ft in length having 180,000 plates and achieved separation on materials that differ in molecular weight by less than 10%.

Increased column length is accompanied by larger elution volumes and hence longer analysis time. Additionally, to maintain a constant flow rate the pressure drop increases approximately linearly with column length (Section

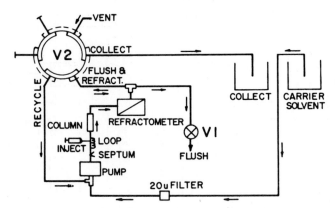

Figure 15.11. Schematic diagram of recycle operation. Reprinted from Ref. 25, by courtesy of Elsevier Publishing Company.

10.23). The pressure drop effect can be reduced by recycling the solute through the column. Figure 15.11 shows a schematic diagram of recycle operation (25). Since the available volume in a closed system is limited, and the peak width increases with the number of cycles, care must be taken to prevent overlap. This is accomplished by removing separated species while continuing to recycle those not separated. Figure 15.12 shows the improvement in resolution due to the recycling operation.

15.25   SEPARATION VARIABLES

The controllable experimental variables that affect separation are (*a*) nature of the packing material, (*b*) column length, (*c*) flow rate, and (*d*) sample size. Column packing materials were discussed in Section 15.23. Column packing techniques, in as much as they control column efficiency, must meet the usual

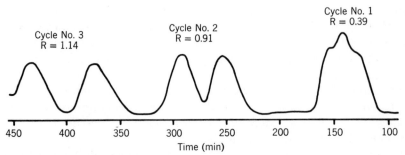

Figure 15.12. Effect of recycle on resolution. Column = styrgel 2.5×10^4 Å (4 ft); flow = 14.4 ml/min; solvent = toluene. Reprinted from Ref. 27, by courtesy of Marcel Dekker, Inc.

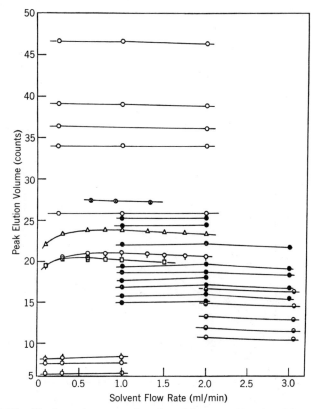

Figure 15.13. Elution volume as a function of flow rate. Reprinted from Ref. 26.

requirements of any chromatographic column (i.e., uniform packing density, avoidance of void space, etc.) and will not be described in detail.

The most important variable available to control separation, once the nature of the column packing and the column length have been established, is the flow rate. The effect of flow rate on elution volume has been the subject of many studies, the results of which are not totally in agreement. However, from the preponderance of evidence it appears that elution volume is affected very little, if any, by flow rate; see, for example, Ref. 21 and Fig. 15.13 (26).

The effect of flow rate on separation efficiency was discussed extensively in Sections 5.3 and 5.4. Generally speaking, the efficiency decreases with increasing flow rate. However, the quantitative relationship has not been definitely established and may be a function of the nature and type of column packing. Figure 15.14 shows data from a number of sources, collected by Duerksen and Hamielec (26). In liquid chromatography the effect of

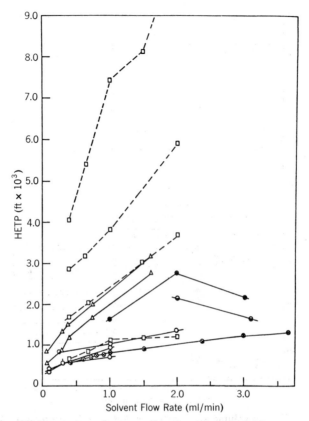

Figure 15.14. Efficiency versus flow rate. Reprinted from Ref. 26.

flow rate on efficiency has been found to fit the empirical expression (Section 10.12)

$$H = Dv^n \tag{15.2}$$

where H is plate height, v is solvent velocity, and D and n are constants for some packings. Typical values for n are 0.3 for Styragel and 0.6 for Porasil at velocities below 2 cm/sec (25). Similarly, Table 15.3 gives results on flow rate versus resolution. As the flow rate increases by a factor of 8, the resolution goes down by one half. The results to date suggest that minimum analysis time for a given resolution will be obtained by use of longer columns operating at relatively high flow velocities.

Sample size can also affect both elution volume and efficiency in exclusion chromatography. In general, larger sample sizes result in an increase in elution volume, although the opposite effect has been observed under conditions of larger sample overloads at fast flow. The dependence of elution

Table 15.3.   Flow Rate versus Resolution

Flow Rate (ml/min)	Resolution
14.4	1.06
29.6	0.95
59.1	0.81
92.0	0.69
121.0	0.57

Reprinted from Ref. 27, by courtesy of Marcel Dekker, Inc.

volume on sample size is a function of molecular weight; the higher the molecular weight, the greater is the dependence.

The length of time during which the sample is injected must be considered in calculating the elution volume for highest precision. One method is to measure the elution volume from the midpoint of the injection. Another procedure is to add the injection volume to the volume of the connecting tubing, detector volume, and so forth, and subtract from the observed volume. Table 15.4 illustrates this procedure. Both methods have small deviations at long injection times. With the sensitive detectors now in use sample sizes can usually be sufficiently small so that concentration effects are negligible.

The effects of temperature on the performance of gel permeation chromatographs have not been extensively investigated. Frequently temperature is considered only as it affects the solubility of the high-molecular-weight material. Thus the polyolefin polymers are chromatographed at 135 to 150°C only because this is a temperature range in which they will go into solution. As the operation of most, if not all, of the detectors becomes more difficult at high temperatures, they are usually avoided if possible. However, Little et al.

Table 15.4.   Injection Time versus Elution Volume

Injection Time (sec)	Connection Volume (ml)	Observed Volume (ml)	Corrected Volume (ml)
10	1.85	38.60	36.75
20	1.95	38.60	36.65
50	2.25	39.05	36.80
90	2.25	40.05	37.50

Reprinted from Ref. 28, by courtesy of Kobunshi Kagaku.

Table 15.5. Effect of Temperature

Polystyrene Molecular Weight	23° C			80° C		
	Elution Volume	Pressure Drop	Resolution	Elution Volume	Pressure Drop	Resolution
411,000	76.77	2200	0.63	75.1	1350	0.84
19,850	92.95		1.30	92.0		1.83
Acetonitrile	115.32			110.2		

Reprinted from Ref. 21, by courtesy of Marcel Dekker, Inc.

(21) found a marked increase in resolution with rise in temperature, as shown in Table 15.5. In addition to the improved resolution, the lowered viscosity permits operation at the same flow rate with a lower pressure drop at the higher temperature. The increase in efficiency can be attributed to better mass transfer.

With increasing temperature, elution volume decreases for a given polymer–solvent system, as illustrated in Fig. 15.15. Presumably this is due to the expansion of the polymer chains at higher temperature, with perhaps a small contribution from the expansion of the gel.

15.26 DETERMINATION OF MOLECULAR WEIGHTS AND MOLECULAR WEIGHT DISTRIBUTION

For polymers, there is usually a distribution of molecular weights that covers various molecular weight ranges. If the number of polymer molecules which have the molecular weight M_i is N_i, then the various averages of molecular weight are defined as follows.

Number average of molecular weight, $\bar{M}_n$:

$$\bar{M}_n = \frac{\sum_i N_i M_i}{\sum_i N_i} \qquad (15.3)$$

Weight average of molecular weight, $\bar{M}_w$:

$$\bar{M}_w = \frac{\sum N_i M_i^2}{\sum N_i M_i} \qquad (15.4)$$

Viscosity average of molecular weight, $\bar{M}_v$:

$$\bar{M}_v = \left(\frac{\sum N_i M_i^{a+1}}{\sum N_i M_i} \right)^{1/a} \qquad (15.5)$$

where a is the Mark-Hauwink constant.

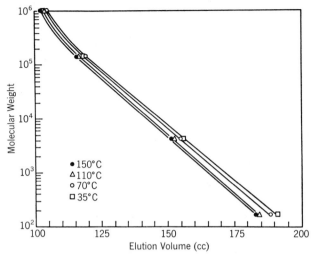

Figure 15.15. Dependence of elution volume on column temperature. Reprinted from Ref. 22.

Z-average of molecular weight, $\overline{M}_z$:

$$\overline{M}_z = \frac{\sum N_i M_i^3}{\sum N_i M_i^2} \tag{15.6}$$

The relation between these averages is as follows:

$$\overline{M}_n < \overline{M}_v \leqslant \overline{M}_w \leqslant \overline{M}_z$$

If $\overline{M}_w/\overline{M}_n \simeq 1$, the polymer has a narrow molecular weight distribution; as this ratio increases, there is an increase in the breadth of the molecular weight distribution.

Before the introduction of gel chromatography, the determination of these molecular weight averages and the molecular weight distribution required many days of work. It was necessary to fractionate the sample into a suitable number of fractions in which the molecular weight distributions were narrow; then, from the amount of polymer and the molecular weight of each partition, the molecular weight averages and molecular weight distribution could be calculated.

It is relatively easy to determine the molecular weight averages and distribution of a polymer by gel chromatography. Only the elution curve and the calibration curve of the sample are required. The construction of calibration curves was discussed in Section 15.22.

From the calibration curve one determines the molecular weight M_i of polymer eluted at elution volume V_i. The peak height h_i at V_i gives the

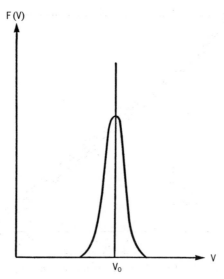

Figure 15.16. Chromatogram of a monodisperse sample. Reprinted from Ref. 33, by courtesy of Marcel Dekker, Inc.

relative concentration C_i of molecular weight M_i, that is, the relative number n_i of M_i:

$$n_i = \frac{C_i/M_i}{\sum C_i/M_i} = \frac{N_i}{\sum N_i}$$

Substituting these data into Eqs. 15.3 to 15.6 gives

$$\overline{M}_n = \frac{\sum h_i}{\sum h_i/M_i} \tag{15.7}$$

$$\overline{M}_w = \frac{\sum h_i M_i}{\sum h_i} \tag{15.8}$$

$$\overline{M}_v = \left(\frac{\sum h_i M_i{}^a}{\sum h_i}\right)^{1/a} \tag{15.9}$$

$$\overline{M}_z = \frac{\sum h_i M_i{}^2}{\sum h_i M_i} \tag{15.10}$$

A problem in determining molecular weight and molecular weight distribution from gel permeation chromatograms is due to the presence of band spreading. A monomeric compound will appear as a curve of finite width; see Fig. 15.16. The width of the curve depends on a variety of band-spreading mechanisms both internal and external to the column (Section

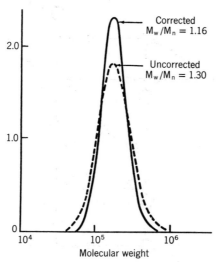

Figure 15.17. Effect of band spreading. Reprinted from Ref. 33, by courtesy of Marcel Dekker, Inc.

5.3). As a result the height of the curve does not represent the number of compounds present at the molecular weight corresponding to any given elution volume, because this height is influenced by the number of neighboring compounds. The computed molecular weight distribution will be significantly larger, therefore, than the actual distribution (see Fig. 15.17).

A variety of computational methods have been employed to correct for the effect of band broadening; these were reviewed recently by Tung (33). A commonly encountered difficulty is oscillation, which has been discussed in detail by Duerksen (34). Choice of an optimum specific method is still uncertain.

15.3 Clathration

15.31 GENERAL

Clathrates are inclusion compounds characterized by a "cage" in which one molecule or a framework made up of one type of molecule encloses a second. The second molecule cannot escape, therefore, even though it is not chemically bonded. Figure 15.18 shows the cage structure of ammonia, nickel cyanide enclosing benzene. A wide variety of clathrating agents of both organic and inorganic structure are known. The size of the cage is determined by the clathrating agent. The size, shape, and location of the entrapped

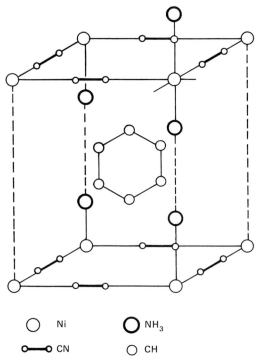

O   Ni   ⬤   NH₃

O—O   CN   O   CH

Figure 15.18. Cage lattice structure of ammonia–nickel cyanide–benzene [Ni(CN)₂·NH₃·C₆H₆]. Reprinted from Ref. 35.

molecule at the time of crystallization determine whether or not clathrates will be formed. Typical nomenclature describes the two components of the single-phase solid as *cage* or *guest* and *entrapped atom* or *host*.

The utility of clathrates as a separation medium is due to the fact that clathration depends primarily on the size and shape of the guest molecule. Thus selective clathration can afford a route to separation. The method is adaptable to continuous processing. A brief description of one of the early proposed process separations will be given as an example.

Jones and Fay (36) patented a process for recovering high-purity benzene from refinery naphtha stocks that are relatively low in benzene. The naphtha stock was continuously blended with nickel cyanide–ammonia and homogenized to form a solid clathrate compound with benzene. The clathrate was separated from the other compounds by centrifugation and then was slurried in benzene and treated with steam and ammonia in a countercurrent process, so that it decomposed into benzene and nickel cyanide–ammonia. Benzene, steam, and ammonia were recovered at the top of the stripping column, and nickel cyanide–ammonia at the bottom for recycling.

**Table 15.6. Inorganic Complex Compounds That Form
Clathrates with Organic Molecules**[a]

$Min(4\text{-MePy})_4(SCN)_2$	$Ni(4\text{-Ac-Py})_4(SCN)_2$
$Mn(4\text{-EtPy})_4(SCN)_2$	$Ni(3\text{-NH}_2Py)_4(SCN)_2$
$Mn(4\text{-EtPy})_4(CNO)_2$	$Ni(4\text{-NH}_2Py)_4(SCN)_2$
$Mn(4\text{-EtPy})_4Cl_2$	$Ni(3\text{-BrPy})_4(SCN)_2$
$Mn(Isoq)_4(SCN)_2$	$Ni(3\text{-CyPy})_4(SCN)_2$
$Ni(Py)_4(SCN)_2$	$Ni(4\text{-CyPy})_4(SCN)_2$
$Ni(Py)_4Cl_2$	$Ni[4(Hyme)\text{-Py}]_4(SCN)_2$
$Ni(3\text{-MePy})_4(SCN)_2$	$Ni(Nicotinamide)_4(SCN)_2$
$Ni(4\text{-MePy})_2(SCN)_2$	$Ni(Me\ isonicotinate)_4(SCN)_2$
$Ni(4\text{-MePy})_4(SCN)_2$	$Ni(Et\ isonicotinate)_4(SCN)_2$
$Ni(4\text{-MePy})_4(N_3)_2$	$Ni(i\text{-nicotinic acid thioamide})_4(SCN)_2$
$Ni(4\text{-MePy})_4Cl_2$	$Ni(isonicotimanide)_4(SCN)_2$
$Ni(4\text{-MePy})_4(formate)_2$	$Ni(Isoq)_4(SCN)_2$
$Ni(4\text{-MePy})_4(No_2)_2$	$Co(Py)_4(SCN)_2$
$Ni(4\text{-EtPy})_4(SCN)_2$	$Co(4\text{-MePy})_4(SCN)_2$
$Ni(4\text{-EtPy})_4(formate)_2$	$Co(4\text{-EtPy})_4(SCN)_2$
$Ni(4\text{-ViPy})_4(SCN)_2$	$Fe(4\text{-MePy})_4(SCN)_2$
$Ni(4\text{-}n\text{-PrPy})_4(SCN)_2$	$Fe(4\text{-EtPy})_4(SCN)_2$
$Ni(3\text{-Et-4-MePy})_4(SCN)_2$	$Cu(Py)_4(SCN)_2$
$Ni(4\text{-}n\text{-BuPy})_4(SCN)_2$	$Cu(4\text{-MePy})_4(SCN)_2$
$Ni(4\text{-BzPy})_4(SCN)_2$	$Zn(4\text{-MePy})_2Cl_2$

[a] (The following abbreviations are used: Py, pyridine; Isoq, isoquinoline; Me,
methyl; Et, ethyl; Pr, propyl; Bu, butyl; Bz, benzyl; Ac, acetyl; NH_2, amino;
Cy, cyano; Hyme, hydroxymethyl; Vi, vinyl; i, iso; n, normal.)
Reprinted from Ref. 37, by courtesy of the American Chemical Society.

15.32 SPECIFIC CLATHRATING AGENTS

As several thousand specific clathrates have been reported in the literature,
only selected representative compounds that are most frequently encountered
will be mentioned here. Table 15.6 lists a number of complex inorganic
compounds known to form clathrates with organic molecules. In addition, a
considerable number of clathrates are formed using Hofmann-type clathrating agents: $Cu(NH_3)_2M'(CN)_4 \cdot 2G$, where $M' = Ni$, Pd, or Pt, and
$G = C_6H_6$, C_4H_4S, or $PhNH_2$.

Casellato and Casu (38) have reported clathrate formation with benzene,
toluene, m- and p-xylene, styrene, bisphenyl, and so on, using Werner
complexes of the type ML_4SCN, where $M = Co$ or Ni, and $L = \gamma$-picoline.

Among the organics one of the most widely studied groups is the hydroquinone clathrates. Hydroquinone forms clathrates with various molecules,

including sulfur dioxide, methanol, methyl cyanide, formic acid, carbon dioxide, hydrogen chloride, acetylene, argon, krypton, and xenon.

Phenol forms clathrates with a wide variety of gases, including halogenated low-molecular-weight compounds.

Many organic molecules clathrate with tri-*o*-thymolide. Among these are methyl, ethyl, and propyl iodide and bromide and dibromopropane.

For a more extensive discussion of the various clathrating agents, see Refs. 39 and 40.

15.33   APPLICATIONS

A variety of organic compounds can be purified or separated by the use of clathrates. Commercially the most interest has been shown in the separation of benzene and the various C_8 aromatic hydrocarbons. The separation of these isomers constituted one of the earliest applications of clathrates and has been used extensively ever since (37).

Mixtures of inert gases can be separated by adjusting the pressure at which clathration takes place. Thus, for example, argon can be separated from neon and krypton from xenon.

A number of racemic mixtures have been separated by clathration. At least 100 such mixtures have been partially or completely resolved (see, e.g., Ref. 41).

The ability of clathrates to store inert gases has been utilized in a number of ways. Textiles, dyes, paints, and other products can be labeled by including a clathrate containing a labeled gas. This can then be studied to determine the age and average temperature of storage.

Aromatic compounds as guest substances in clathrates have been useful in mechanism studies. Irradiation of such enclosed compounds permits studies of short-lived radicals by electron spin resonance at relatively high temperatures.

Acknowledgment

This work was supported in part by the National Science Foundation Grant GP 28618.

References

1. R. M. Barrer, *Proc. Roy. Soc. (London)*, **A167**, 392 (1938).
2. R. H. Duffett and G. J. Minkoff, *Discovery*, **25**, 32 (1964).

3. R. M. Barrer, *Endeavor*, **23**, 122 (1964).

4. B. M. Deb, *Sci. Culture*, **32**, 279 (1966).

5. D. W. Breck, *J. Chem. Educ.*, **41**, 678 (1964).

6. J. A. Rabo, C. L. Aupell, P. H. Kasai, and V. Schoemaker, *Chem. Eng. Progr. Symp. Ser.*, **63**, 31 (1967).

7. R. W. Burnett and W. T. Turnbull, *Brit. Chem. Eng.*, **11**, 261 (1966).

8. J. Porath and P. Flodin, *Nature*, **183**, 1657 (1959).

9. H. Determann, *Gel Chromatography*, Springer-Verlag, New York, 1968.

10. M. F. Vaughan, *Nature*, **188**, 55 (1960).

11. P. I. Brewer, *Nature*, **188**, 934 (1960).

12. Anon., *Chem. Eng. News*, Dec. 17, 1962.

13. J. F. Johnson and R. S. Porter, in *Progress in Polymer Science*, Vol. II, A. D. Jenkins, ed., Pergamon Press, New York, 1970, pp. 203–256.

14. G. K. Ackers, *Advan. Protein Chem.*, **24**, 343 (1970).

15. P. Flodin, Dissertation, Uppsala, 1962.

16. M. J. R. Cantow and J. F. Johnson, *J. Polymer. Sci.*, Part A-1, **5**, 2835 (1967).

17. W. W. Yau, H. L. Suchan, and C. P. Malone, *J. Polymer Sci.*, Part A-2, **6**, 1349 (1968).

18. G. Meyerhoff, *Makromol. Chem.*, **118**, 265 (1968).

19. G. Meyerhoff, *Separation Sci.*, **62**, 239 (1971).

20. D. Goedhart and A. Opschoor, *J. Polymer Sci.*, Part A-2, **8**, 1227 (1970).

21. J. N. Little, J. L. Waters, K. J. Bombaugh, and W. J. Pauplis, *Separation Sci.*, **5**, 765 (1970).

22. M. J. R. Cantow, R. S. Porter, and J. F. Johnson, *J. Polymer Sci.*, Part A-1, **5**, 987 (1967).

23. J. C. Moore and M. C. Arrington, *Preprints 3rd Intern. Seminar on GPC*, Geneva, May 1966.

24. K. J. Bombaugh, W. A. Dark, and R. F. Levangie, *J. Chromatog. Sci.*, **7**, 42 (1969).

25. K. J. Bombaugh, *J. Chromatog.*, **53**, 27 (1970).

26. J. H. Duerksen and A. E. Hamielec, *J. Appl. Polymer Sci.*, **12**, 2225 (1968).

27. K. J. Bombaugh and R. F. Levangie, *Separation Sci.*, **5**, 751 (1970).

28. Y. Ishida, S. Veda, K. Kawai, and I. Kimura, *Kobunshi Kagaku*, **27**, 33 (1970).

29. H. E. Pickett, M. J. R. Cantow, and J. F. Johnson, *J. Appl. Polymer Sci.*, **10**, 917 (1966).

30. H. Benoit, Z. Grubsic, P. Rempp, D. Decker, and J. G. Zilliox, *J. Chem. Phys.*, **63**, 1507 (1966).

31. H. Coll and L. R. Prusinowski, *J. Polymer Sci.*, Part B, **5**, 1153 (1967).

32. T. Provder, J. C. Woodbrey, and J. H. Clark, in *Gel Permeation Chromatography*, K. H. Altgelt and L. Segal, eds., Marcel Dekker, New York, 1971, pp. 493–528.

33. L. H. Tung, in *Gel Permeation Chromatography*, K. H. Altgelt and L. Segal, Marcel Dekker, New York, 1971, pp. 73–80.

34. J. H. Duerksen, in *Gel Permeation Chromatography*, K. H. Altgelt and L. Segal, eds., Marcel Dekker, New York, pp. 81–102.

35. V. M. Bhatnagar, *J. Chem. Educ.*, **40**, 646 (1963).

36. A. L. Jones and P. S. Fay, *Chem. Abstr.*, **50**, 10392f (1956).

37. W. D. Schaeffer, W. S. Dorsey, D. S. Skinner, and C. G. Christian, *J. Am. Chem. Soc.*, **79**, 5870 (1957).

38. F. Casellato and B. Casu, *Ind. Chim. Belg.*, **32** (Spec. No), Part 2, 707 (1967).

39. M. Hagan, *Clathrate Inclusion Compounds*, Reinhold, New York, 1962.

40. A. Sopkava and J. Jaluvka, *Chem. Listy*, **63**, 237 (1969).

41. D. Lawton and H. M. Powell, *J. Chem. Soc.*, 2339 (1958).

42. Z. Grubisic, P. Rempp, and H. Benoit, *J. Polymer Sci.*, Part B, 753 (1967).

43. T. C. Kendrick, *J. Polymer Sci.*, Part A-2, **7**, 297 (1969).

44. G. K. Ackers, *Biochemistry*, **3**, 723 (1964).

45. W. W. Yau and C. P. Malone, *J. Polymer. Sci.*, Part B, **5**, 663 (1969).

46. R. L. Bartosiewicz, *J. Paint Technol.*, **39**, 28 (1967).

List of Symbols

a	constant in the Mark-Houwink equation	$\bar{M}_w$	weight average of molecular weight
A	constant in Eq. 15.1	$\bar{M}_z$	Z average of molecular weight
B	constant in Eq. 15.1		
D	constant in Eq. 15.2	n	constant in Eq. 15.2
K	constant in the Mark-Houwink equation	$[\eta]$	intrinsic viscosity
M	molecular weight		*Subscripts*
$\bar{M}_n$	number average of molecular weight	i	polymer being eluted from the column
$\bar{M}_v$	viscosity average of molecular weight	s	standard
		x	unknown

Part Three

Other Separation Methods

Sixteen

BARRIER SEPARATION PROCESSES

R. A. CROSS and H. STRATHMANN

16.1 Introduction

In barrier separation processes, mixtures of various chemical species are transported under the driving force of gradients in their chemical or electrochemical potentials through a restrictive interphase. A separation of components is caused by their different transport rates in the interphase. Most barrier separation processes of practical importance utilize semipermeable membranes as the restrictive interphase. During recent years processes referred to as reverse osmosis, ultrafiltration, dialysis, and electrodialysis have gained increasing attention as convenient methods for concentration, purification, and separation of molecular solutions. Synthetic membranes have been developed that can successfully be used, for example, to produce potable water from the sea, to recover helium from natural gas, to remove sulfur dioxide from stack gases, to separate azeotropic mixtures such as alcohol–water, or to remove urea from blood. Semipermeable membranes may also be applied as separators in batteries and fuel cells and as ion-selective analytical tools. Synthetic membranes and membrane separation processes have been extensively studied (1–7). Certain membrane and transport models have been developed, and processes such as reverse osmosis and dialysis have been described quantitatively with phenomenological equations (8–10).

16.2   Structure and Transport Properties of
Semipermeable Membranes

The key part of a practical membrane separation process is the actual membrane itself. To serve its purpose successfully a membrane should be semipermeable, that is, it should allow the passage of certain chemical species completely while preventing or strongly retarding the permeation of others. At the same time the transport rate for the permeating species must be high enough so that reasonably fast separation can be achieved. The membrane should have good mechanical and chemical stability. Three basic membrane structures are used for most of today's separation processes. For the understanding of the various membrane separation processes, a brief discussion of the structure of semipermeable membranes and the transport mechanism by which they are believed to function is necessary.

16.21   MICROPOROUS MEMBRANE

A microporous membrane is in its structure and function very similar to a traditional filter. It has a rigid, highly voided structure with randomly distributed interconnected pores, the size of which is extremely small, that is, 10 to 1000 Å in comparison to conventional filters, which have a pore size greater than 10,000 Å. Discrimination is achieved by pore size. All particles or molecules larger than the largest pores are completely rejected by the membrane; all particles smaller than the smallest pores pass the membrane completely. All particles of intermediate size are partially rejected in correspondence to the pore size distribution of the membrane. With a microporous membrane only particles that differ considerably in size can be separated.

16.22   HOMOGENEOUS MEMBRANE

The homogeneous membrane is essentially a homogeneous film or interphase through which a mixture of chemical species is transported via molecular diffusion. The separation of various components in a mixture is directly related to their transport rates within the interphase, which again are determined by the diffusivities and the concentrations of the individual components in the film. With homogeneous diffusive type membranes, particles of exactly the same size may also be separated when their concentrations (i.e., their solubilities in the film) differ significantly.

16.23   ELECTRICALLY CHARGED MEMBRANE

In general, electrically charged membranes are porous in structure. The pore walls carry fixed ions, which may be positively or negatively charged.

In the case of positive fixed ions the membrane is referred to as an *anion-exchange membrane*; with negative fixed ions it is called a *cation-exchange membrane*. Separation in an electrically charged membrane is achieved not only by pore size, but also by exclusion of coions (i.e., ions of the same charge as the fixed ions) from the membrane (see Chapter 12). Membranes that consist of a mixture of negatively and positively charged macroscopic segments are referred to as *mosaic membranes*.

16.24 THE ASYMMETRIC "SKIN" TYPE MEMBRANE

Transport rates through membranes are inversely proportional to membrane thickness. To achieve high transport rates it is desirable, therefore, to build the membranes as thin as possible. Present-day polymer film fabrication technology does not permit manufacture of defect-free films thinner than about 20 μ. However, for some membrane separation processes, such as reverse osmosis or ultrafiltration, significantly thinner membranes are desirable. This has led to a novel membrane fabrication procedure resulting in asymmetric membrane structures (11, 12). Such a structure consists of an extremely thin (0.1 to 1 μ) so-called *skin*, which represents the actual discriminating membrane, supported on a 0.1- to 1-mm-thick microporous substructure. Figure 16.1 shows schematically a cross section of an asymmetric membrane. The "skin" in an asymmetric membrane may be a porous or homogeneous membrane and may be neutral or charged. The microporous support allows rapid transport of all species, relative to the "skin."

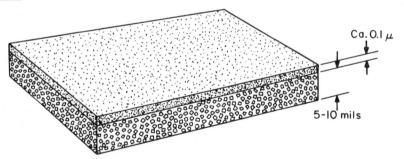

Ca. 0.1 μ

5-10 mils

Figure 16.1. Pictorial representation of asymmetric diffusive membrane.

16.3 Fluxes and Driving Forces in Membrane Separation Processes

Separation in membrane processes is the result of different transport rates of various chemical species, under one or more driving forces, through an interphase referred to as a membrane. The transport rates are determined by the driving forces acting on the individual component and by its mobility within the membrane. Driving forces in membrane separation processes result from the differences in (a) hydrostatic pressure, (b) chemical or electrochemical potential, or (c) temperature between two subsystems separated by the membrane. All transport processes, regardless of whether energy, electricity, or matter is transported, take place in a nonequilibrium state. They cannot be described adequately, therefore, by thermodynamics in the classical sense, which is concerned almost exclusively with equilibrium states. Instead, transport processes are generally described by various phenomenological laws that relate the various flows to the corresponding driving forces in the form of proportionalities. For example, Fick's law describes the relation between flow of matter and a concentration gradient. Ohm's law describes the relation between electrical current and a potential gradient. Fourier's law describes the relation between heat transport and a temperature gradient.

In membrane separation the driving force of a concentration gradient not only may result in a flow of matter but also can cause the buildup of a hydrostatic pressure gradient, a phenomenon referred to as *osmosis*. Analogously a hydrostatic pressure gradient not only may result in a volume flow but also can lead to the buildup of a concentration gradient, referred to as *reverse osmosis*. A gradient in the electrical potential across a membrane may lead not only to an electrical current but also to a transport of matter, as in electrodialysis, or to the buildup of a hydrostatic pressure gradient referred to as *electro-osmosis*. A temperature gradient across the membrane not only creates a flow of heat but also can result in a flow of matter. This process is called *thermodiffusion*; in contradistinction a mass flow due to a concentration gradient may cause a temperature gradient. This phenomenum is referred to as the *Dufour effect*. Furthermore, the transport of certain components can also be the result of a kinetic interaction with the flow of other components. These so-called *cross* or *coupling effects* are in general mathematically described by the phenomenological laws mentioned above. In membrane separation processes only driving forces that can lead to significant mass transport, such as gradients in the hydrostatic pressure, in the concentration, and in the electrical potential, are of practical importance.

16.4   Theory and Practice of Selected Membrane Separation Processes

Although, as mentioned above, there are many membrane separation processes and combinations of various such processes, for the purposes of this book only five of the most useful processes will be covered. These are ultrafiltration, reverse osmosis, gas diffusion, dialysis, and electrodialysis.

16.41   ULTRAFILTRATION

Theory

Separation of a macromolecular solution by filtration under an applied hydrostatic pressure is referred to as *ultrafiltration*. The basic process is rather simple and differs from ordinary filtration only in the size of the particles that are separated. Differentiation between the two processes is in large measure arbitrary. The term ultrafiltration is generally used to describe separation involving solutes of molecular dimensions greater than 20 Å in diameter. Separation of particles whose molecular dimensions are of the same order of magnitude as those of the solvent is referred to as *reverse osmosis*. Separation of particles larger than 1 μ is called *filtration*. In addition to the size of the particles to be separated from a solution, the three processes also differ significantly in the hydrostatic pressure applied as driving force for the separation process. In ordinary filtration the applied hydrostatic pressure ranges from a fraction of an atmosphere to an atmosphere. In ultrafiltration 1 to 10 atm and in reverse osmosis 10 to 100 atm of hydrostatic pressure are applied as the driving force. Ultrafiltration thus encompasses all membrane-moderated, pressure-activated separations involving solutions of medium-molecular-weight (ca. 1000 and up) solutes, macromolecules, and colloids at hydrostatic pressures between 1 and 10 atm.

Ultrafiltration membranes are microporous in their structure. Separation is achieved by the pore size distribution. As stated in Section 16.21, all particles larger than the largest pores in the membrane matrix are completely retained, and particles smaller than the smallest pores completely pass the membrane. All particles smaller than the largest pores but larger than the smaller pores are rejected in correspondence with the pore size distribution. The so-called cutoff level of the membrane (i.e., the minimum size or molecular weight of a particle to be rejected) is thus determined by its mean pore size. The sharpness of the cutoff of a membrane depends on its pore size distribution (cf. "Gel Chromatography," Section 15.2). In Fig. 16.2 the rejection characteristics of an ultrafiltration membrane with a sharp and a diffusive cutoff are shown schematically.

The rejection of an ultrafiltration membrane for a certain solute is generally expressed as follows:

$$R = \left(1 - \frac{C_p}{C_f}\right) \times 100 \qquad (16.1)$$

where R is the rejection in per cent, and C_p and C_f are the concentrations of the solute in the product and in the feed solution, respectively.

The product solution concentration can be expressed by

$$C_p = \frac{J_s}{J_v} \qquad (16.2)$$

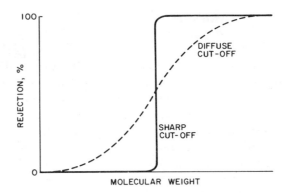

Figure 16.2. Rejection characteristics of ultrafiltration membrane—sharp versus diffuse cut-off.

where J_s is the solute flux, and J_v is the solvent flux. The solvent flux J_v is given by

$$J_v = \frac{\varepsilon r^2}{8\eta} \frac{\Delta P}{\tau\lambda} \tag{16.3}$$

where ε is the membrane porosity, r is the pore radius, η is the solution viscosity, ΔP is the pressure difference across the membrane, τ is the tortuosity factor (the actual pore length divided by the membrane thickness), and λ is the thickness of the membrane.

Since the solute flux is affected by a pressure *and* a concentration driving force, it is given by

$$J_s = J_v C_f (1 - \sigma) - D_s{}^m \frac{dC_s{}^m}{dx} \tag{16.4}$$

where $C_s{}^m$ is the concentration of solute in the membrane, $D_s{}^m$ is the diffusivity, $dC_s{}^m/dx$ is the concentration gradient of solute across the membrane, and σ is the fraction of pure solvent flux which passes through pores smaller than those retaining solute molecules.

For most ultrafiltration systems of practical importance, the convective flux of solute, the first term on the right-hand side of Eq. 16.4, is much larger than the diffusive flux, the second term of the right-hand side. Considering this factor and substituting Eqs. 16.2, 16.3, and 16.4 into Eq. 16.1, one obtains

$$R = \sigma \times 100 \tag{16.5}$$

Equation 16.5 shows that in ultrafiltration the rejection of the membrane for a certain solute is, to a first approximation, independent of the hydrostatic pressure, whereas the solvent flux (Eq. 16.3) is directly proportional to the hydrostatic pressure.

For economical reasons the ultrafiltration rate should be as high as possible. Equation 16.3 shows the factors that influence this rate. Since the major factor is the thickness λ, nearly all membranes utilized in ultrafiltration are asymmetric membranes, where the effective value of λ is the thickness of the "skin."

Applications

Ultrafiltration can be used for concentration, purification, or separation. It is a particularly attractive concentration procedure for such compounds as proteins, nucleic acids, polypeptides, enzymes, antibodies and antigens, viruses, and other substances whose chemical structures or biological activities are likely to be altered by common concentrative procedures such as precipitation, evaporation, or freeze drying. Ultrafiltration can be conducted rapidly at low temperatures. This process can be used to quantitatively recover specific compounds, or to concentrate specific compounds existing at low concentrations and thus bring them into detectable range for analytical purposes.

Ultrafiltration can be utilized for purification when the membrane either retains or passes the compound to be purified. When a high-molecular-weight compound or colloidal species exists in a mixture with low-molecular-weight organic compounds or salts, the high-molecular-weight species can be purified by continuously "washing" the species with solvent (usually water). The mixture containing the species to be purified is placed in a membrane cell, and solvent is forced into the cell from a pressurized reservoir. The purification is governed by the relationship

$$\frac{C_0}{C_f} = e^{V_f/V_0} \qquad (16.6)$$

where C_0 is the initial impurity concentration, C_f is the final impurity concentration, V_0 is the volume of the initial mixture in the cell, and V_f is the volume of solvent or wash solution used. When the species to be purified is the low-molecular-weight component of a mixture, purification is accomplished simply by ultrafiltration—the impurity remains in the cell.

Separation of two or more species differing in molecular weight can be effected by ultrafiltration if the molecular weight difference is fairly large (perhaps a factor of 10), and if a membrane exists with the proper molecular weight cutoff. As a rule the ultrafiltration membranes in use today have fairly diffuse cutoffs, and only slight separations can be obtained if the molecules of interest are close in molecular weight.

Equipment

For small-volume separations, laboratory ultrafiltration cells such as those shown in Fig. 16.3 are used. The membrane is supported at the bottom of the

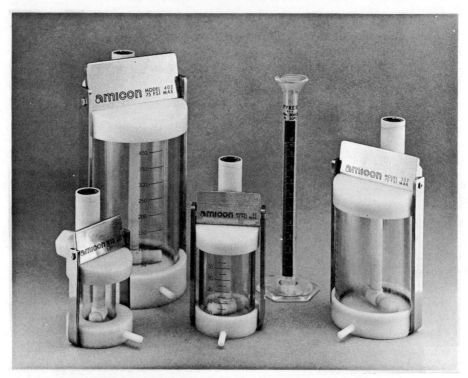

Figure 16.3.   Laboratory ultrafiltration cells. Reprinted by courtesy of Amicon Corporation.

cell, and the solution to be ultrafiltered is placed in the cell and pressurized. A magnetic stirrer prevents excessive buildup of a layer of rejected material on the membrane surface.

Other types of equipment are available for larger-scale applications. Figure 16.4 is a membrane stack employing spacers to provide thin, spiral channels for continuous ultrafiltration. This device has over 2 ft² of membrane and is suitable for large-scale laboratory operations. For still larger-scale industrial applications hollow-fiber membrane systems and various tubular membrane modules are available from Romicon, Abcor, and Dorr-Oliver.

Operational Problems

The major factor limiting the speed and effectiveness of an ultrafiltration process is the buildup on the upstream surface of the membrane of a layer of solute molecules which are rejected or retained by the membrane in the course of ultrafiltration. This is called *concentration polarization*. Since the

Figure 16.4. Thin-channel ultrafiltration system containing 15 membranes; total area-2.2 ft². Reprinted by courtesy of Amicon Corporation.

layer serves to reduce the rate of ultrafiltration, control of the thickness of the layer is of major importance if ultrafiltration is to be a practical separation process (13). If a high shear is not maintained at the membrane surface, this layer will increase in thickness until the flux rate drops to a very low level. High shear can be obtained by using either high flow rates across the membrane surface or rapid stirring. A more satisfactory method for many applications, especially when delicate solutes such as enzymes, some proteins, or hormones are involved, is to confine the flowing liquid in a very thin channel (0.2 to 0.6 mm). Even with laminar flow this method can produce the required shear rate, since this rate is proportional to the ratio of the fluid velocity to the channel depth.

Membranes

A wide variety of membranes for ultrafiltration are now available; those supplied by Amicon Corporation are shown in Table 16.1. The nominal molecular weight cutoffs available range from 500 to 300,000. Typical of ultrafiltration membranes is the high water flux obtainable at modest pressures.

Table 16.1.   Ultrafiltration Membranes[a]

Membrane Designation	Nominal Molecular Weight Cutoff	Water Flux at 55 psi (ml/min cm²)
UM05	500	0.025
UM2	2,000	0.06
UM10	10,000	0.17
PM10	10,000	0.40
PM30	30,000	1.4
XM50	50,000	0.70
XM100A	100,000	1.9
XM300	300,000	3.6

[a] Amicon Corporation, Lexington, Mass.

16.42   REVERSE OSMOSIS

Theory

The basic difference between reverse osmosis and ultrafiltration lies in the size of the solute that is separated from a solvent under the driving force of a hydrostatic pressure. As pointed out before, the molecular size of the solute is of the same order of magnitude as that of the solvent, that is, reverse osmosis is used for separation with solutes up to a molecular weight of about 500. This low molecular weight of the solute has some significant consequences for the performance of the process. To separate particles that are of about the same order of magnitude a microporous membrane is not suitable. Therefore most reverse osmosis processes utilize a homogeneous membrane or an asymmetric membrane with a homogeneous skin (14,15). In this case molecular diffusion is the transport process in the membrane. A solution of particles that have a molecular weight of 500 or less may have a significant osmotic pressure, perhaps as high as 100 atm, depending on the concentration of the solution. To yield any separation the osmotic pressure difference between the feed solution and the filtrate has to be overcome by an applied hydrostatic pressure; hence significantly higher hydrostatic pressures are required in reverse-osmosis processes than in ultrafiltration.

When a solution is confined under pressure on one side of a reverse osmosis membrane, the solvent and solute transport rates across the membrane can be approximated by the relationships

$$J_v \simeq \frac{K_1}{\lambda} (\Delta P - \Delta \pi) \qquad (16.7)$$

$$J_s \simeq \frac{K_2}{\lambda} (C_b - C_p) \qquad (16.8)$$

where J_v and J_s are the solvent and solute fluxes across the membrane, ΔP is the hydraulic pressure difference between the upstream solution and the product, $\Delta \pi$ is the osmotic pressure difference between the two solutions, C_b is the upstream solute concentration, and C_p is the solute concentration in the product. Also, K_1 and K_2 are the transport coefficients of the membrane to solvent and solute, respectively, and λ is the effective membrane thickness.

Mass conservation requires that

$$C_p = \frac{J_s}{J_v} \qquad (16.9)$$

and simultaneous solution of Eqs. 16.7, 16.8, and 16.9 yields

$$R = \left(1 - \frac{C_p}{C_b}\right) \times 100 = \frac{K_1/K_2(\Delta P - \Delta \pi)}{1 + K_1/K_2(\Delta P - \Delta \pi)} \times 100 \qquad (16.10)$$

where R is the rejection of the membrane. It can be noted from Eq. 16.10 that here, unlike the ultrafiltration case, the rejection increases with increasing effective pressure difference across the membrane. This is true because the solvent flux is pressure dependent, whereas the solute flux is nearly independent of pressure.

Applications

The most common application for reverse osmosis is the large-scale purification of saline water sources. Although reverse osmosis is generally not an economical process for sea water desalination, it can be practical in certain areas for brackish water purification. Reverse osmosis is currently being investigated for several pollution control areas, for example, tertiary treatment of sanitary sewage, purification of acid mine waters, and concentration of whey streams from the milk and cheese industry.

In the laboratory, water purification by reverse osmosis would probably be of only limited feasibility. Although reverse osmosis may be more economical than distillation, the wide availability and acceptance of distillation make it generally the process of choice. However, if low-molecular weight solutes such as salts, sugars, and other low-molecular weight organic compounds are to be concentrated, reverse osmosis can be an effective tool.

Although the flux rates are generally lower than those for ultrafiltration, reverse osmosis has the same advantages as ultrafiltration in situations where an alternative concentration process might damage the solute of interest.

Equipment

Small stirred high-pressure cells suitable for use with membrane discs in the laboratory are available from Amicon Corporation, and a small flow cell can be obtained from Abcor, Inc. If it is desired to feed a solution into the cell, a high-pressure pump or a bladder type accumulator operating from a gas cylinder may be used. Accumulators are available from Greer Corporation. Larger membrane cells and complete systems can be purchased from such companies as Calgon-Havens, Gulf General Atomics, Permutit, Philco-Ford, and Du Pont. In all of these systems pressures ranging from 200 to 800 psi are used.

Membranes

Sheet cellulose acetate membranes for use with the small cells mentioned above are available from Eastman Chemical Products, Inc. These membranes are of the asymmetric type and have the characteristics given in Table 16.2.

Membranes with different rejection characteristics are also available in many of the larger systems mentioned above.

Table 16.2. Cellulose Acetate Reverse-Osmosis Membranes[a]

Membrane Designation	NaCl Rejection (%)	Flux at 600 psi (gal/ft² day)
KP 98	98	9–14
KP 96	96	14–22
KP 90	90	22–30

[a] Eastman Chemical Products, Inc.

Operational Problems

As is true also of ultrafiltration, concentration polarization in a reverse-osmosis process serves to reduce overall efficiency. This is due to the convective transport of solutes to the membrane, where they are rejected and increase in concentration until an equilibrium is set up by back diffusion into the solution. In this case a reduction in flux is due primarily to the higher osmotic pressure difference across the membrane, which serves to reduce the effective driving force for solvent transport. The effective salt rejection is also decreased, since the high salt concentration at the membrane surface increases the salt flux through the membrane. As in the ultrafiltration process, the concentration polarization can be controlled by employing turbulent flow or laminar flow in thin channels (16).

With reverse osmosis and also with ultrafiltration, an important point to bear in mind when the concentration of a particular solute is of interest is that, even with a reasonably high rejection membrane, solute losses can be significant when a severalfold concentration is carried out. The relationship governing this loss is

$$\delta = 1 - (1 - \Delta)^{1-R} \qquad (16.11)$$

where δ is the fractional loss of solute, Δ is the recovery rate or the volume of solvent removed through the membrane divided by the initial volume, and R is the intrinsic rejection of the membrane. In Fig. 16.5 the fractional loss is plotted versus recovery with the membrane rejection as a parameter.

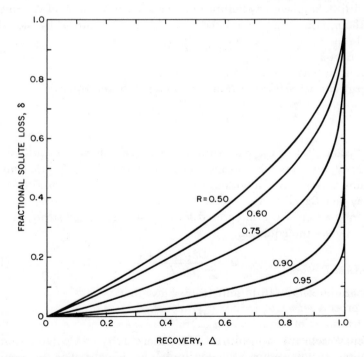

Figure 16.5. Fractional solute loss as a function of recovery with rejection as a parameter for batch or one-pass reverse osmosis or ultrafiltration.

16.43   GAS DIFFUSION

Theory

The differences in permeability exhibited by a polymer membrane to different gases can serve as the basis for separating the components of a gaseous mixture. The permeation of a polymer membrane by a gas is considered to take place in three steps: the dissolution of the gas at one interface, the diffusion of the gas through the membrane, and the evolution of the gas at the other interface.

The diffusion constant is defined in terms of a concentration driving force by Fick's first law (see Section 3.21):

$$J = -D\frac{\partial C}{\partial x} \tag{16.12}$$

where J is the flux rate per unit area, D is the diffusion constant, and C is the concentration of gas sorbed in the polymer at distance x from a reference plane.

If in the permeation process the dissolution and evolution of gas are assumed to take place at equilibrium, the concentration C of the dissolved gas at the interface may be related to the partial pressure P by the relation

$$C = kP \tag{16.13}$$

where k is the Henry's law solubility constant.

A permeability coefficient $\bar{P}$ may be defined by the equation

$$J = \frac{\bar{P}(P_1 - P_2)}{L} \tag{16.14}$$

where P_1 and P_2 are the partial pressures at the inflow and outflow faces, respectively, and L is the thickness of the membrane. For permanent gases $\bar{P}$ remains constant over a wide range of pressures and at constant temperature may be referred to as a constant.

A combination of the integrated form of Eq. 16.12 for steady flow and Eq. 16.13 gives the equation

$$J = \frac{Dk(P_1 - P_2)}{L} \tag{16.15}$$

A comparison of Eqs. 16.14 and 16.15 shows that

$$\bar{P} = Dk \tag{16.16}$$

Therefore $\bar{P}$, which is a measure of the penetrability of a polymer film to a gas, consists of the product of k, which may be thought of as the amount of dissolved gas present in the membrane, and D, which is a measure of the rate at which an individual molecule will move through the membrane.

The above relationships are valid only if the diffusion coefficient D is independent of concentration, a condition that holds for permanent gases well above their critical temperature. Other cases have been covered in the literature (17).

For permanent gases in a mixture the flux of each component is generally governed by Eq. 16.14, that is, it is not dependent on the presence of the other components. In order to discuss the use of membranes to separate the components of a mixture it is useful to define another transport parameter, called the apparent selectivity $S_{A/B}$, as

$$S_{A/B} = \frac{(J_A/J_B)}{(X_A/X_B)} \tag{16.17}$$

which has the same significance in a transport rate-moderated membrane separation process as the separation factor does in an equilibrium separation process. In this equation the J's are fluxes of individual components in a two-component system through the membrane, and the X's are the mole fractions of the two components in the retentate.

If the total pressure downstream of the membrane is very low relative to the upstream pressure, or in the limit zero, then the apparent selectivity reduces to another parameter, the membrane permselectivity $S_{A/B}^0$, which is given by

$$S_{A/B}^0 = \frac{\bar{P}_A}{\bar{P}_B} \tag{16.18}$$

Here $\bar{P}_A$ and $\bar{P}_B$ are the permeabilities of component A and component B, respectively.

For any real separation, however, the downstream pressure will be finite, in which case the apparent selectivity is given implicitly by

$$S_{A/B}^0 = S_{A/B} \frac{X_B + S_{A/B}X_A - \pi_2/\pi_1}{X_B + S_{A/B}X_A - (\pi_2/\pi_1)S_{A/B}} \tag{16.19}$$

where π_1 and π_2 are the upstream and downstream total pressures, respectively.

Table 16.3 gives some representative values of membrane permeabilities and permselectivities for two pairs of gases for several commercially available homogeneous polymer films.

For the purposes of this chapter two types of membrane separation stages as shown in Fig. 16.6 will be considered: a well-mixed stage and an unmixed crossflow stage. In the well-mixed stage the permeate and retentate are of constant composition. In the unmixed crossflow stage the feed moves in plug flow (no longitudinal mixing) and the permeate is removed at the point in which it passes through the membrane.

Table 16.3. Permeabilities and Permselectivities of Commercially Available Polymer Films

Membrane	Permeability[a] at 25°C (SCF × mil)/(1000 ft² × day × atm)					Permselectivity	
	$\bar{P}_{CO2}$	$\bar{P}_{H2}$	$\bar{P}_{O2}$	$\bar{P}_{N2}$		$S^0_{CO2/H2}$	$S^0_{O2/N2}$
Silicone rubber	23,000	5,520	5,010	2,300		4.2	2.2
Natural rubber	1,130	425	196	71.4		2.7	2.7
Ethyl cellulose (Ethocel)	365	27.2	226	72.4		13.5	3.1
Polycarbonate (Lexan)	67.9	102	11.9	2.54		0.67	4.7
Polystyrene	47.0	126	10.6	2.01		0.37	5.3
Acetal (polyformaldehyde) (Delrin)	—	—	3.22	0.187		—	17
Polyvinyl fluoride (Tedlar)	0.764	2.97	0.170	0.057		0.26	3.0

[a] Most permeability data were obtained from *Engineering Design for Plastics*, Eric Baer, ed., Reinhold, New York, 1964. Other sources may give slightly different values for the polymers listed.

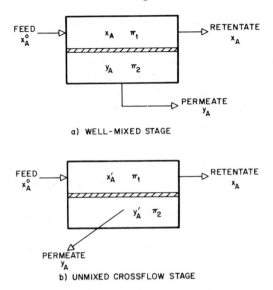

a) WELL-MIXED STAGE

b) UNMIXED CROSSFLOW STAGE

Figure 16.6. Membrane permeation stages for gas separation.

In the well-mixed stage the apparent selectivity required to perform a given separation can be expressed by

$$S_{A/B} = \left(\frac{1 - X_A}{X_A}\right)\left(\frac{X_A^0}{1 - X_A^0}\right)\frac{R_B}{1 - R_B} \tag{16.20}$$

where R_B is the recovery based on component B in the retentate, X_A^0 refers to the feed composition of the more permeable constituent, and X_A designates the retentate composition.

The unmixed crossflow stage is a much more efficient design (as is an unmixed countercurrent stage). Unfortunately, a theoretical treatment is beyond the scope of this chapter. The subject has been partially covered in the literature (18, 19).

Gas separation can also be obtained under hydrostatic gradients in microporous membranes with uniform pores at pressures low enough so that the mean free path of a molecule is large compared to the pore diameter. Again, a theoretical treatment is beyond the scope of this book but is given elsewhere (20).

Application, Equipment, and Membranes

In spite of an extensive amount of laboratory effort on membrane separation of gaseous mixtures very few commercial processes have resulted from the work. This comparative lack of success is due primarily to two major problems:

1. The transmission fluxes attainable through most polymeric films of reasonable thickness are quite low, necessitating the use of extremely large membrane areas to achieve realistic capacity.
2. Most polymeric materials have only modest selectivity ratios for gases of commercial interest. This means that many separations of interest are not attractive with the polymeric materials commercially available today.

The only two present or past commercially available items of which the writer is aware are a helium purifier employing glass membranes, sold by Electron Technology, Inc., and several types of hydrogen purifiers employing palladium membranes; for example, one is sold by Matheson Company. Large plants are operated by the Atomic Energy Commission to separate uranium isotopes by diffusion through microporous membranes. The separation factor, however, is very small, and numerous stages are necessary.

The researcher who wishes to investigate membrane separation of gases is in general faced with searching for gas permeation data for polymer films in the literature, making permeability measurements if the data are not available, and constructing a permeation cell to carry out the required separation. References 21, 22, and 23 provide a reasonable quantity of data on gas permeability coefficients. Dow Chemical Company sells a small hollow-fiber cell for small-scale air enrichment.

16.44 DIALYSIS

Theory

A dialyzer is a device in which one or more solutes are transferred from one fluid to another through a membrane under a concentration driving force. For most efficient operation the fluids are moving, and for the purposes of this discussion the solution to be depleted of solute will be called the feed and the fluid receiving the solute will be termed the dialysate. The overall efficiency of a dialyzer is governed by two independent factors: the ratio of the flow rates of the two fluids, and the rate constant for solute transport between the fluids (which is determined by the properties of the membrane, the membrane area, the fluid channel geometry, and the local fluid velocities). The nomenclature of artificial kidney workers is used in this section, and the theory is developed as in a paper by Michaels (24).

With reference to Fig. 16.7 a material balance can be expressed as

$$N = Q_B(C_{B_i} - C_{B_o}) = Q_D(C_{D_o} - C_{D_i}) \qquad (16.21)$$

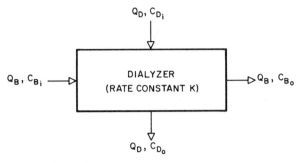

Figure 16.7. Dialyzer flow parameters.

where Q represents the volumetric flow rate; C is the solute concentration; subscripts B and D refer to feed and dialysate, respectively; subscripts i and o designate inlet and outlet conditions, respectively; and N is the overall solute transport rate.

The overall rate of solute transport N is also expressed by

$$N = k^+ A \, (\overline{\Delta C_M}) \tag{16.22}$$

where $\overline{\Delta C_M}$ is the average solute concentration difference between the fluids, A is the membrane area, and k^+ is the overall rate constant. The proper average solute concentration difference to use is the logarithmic mean of the inlet and outlet differences. The three most common cases to be considered are parallel flow of both fluids, countercurrent flow, and flow with the dialysate completely mixed.

The efficiency of a dialyzer is expressed in terms of its "dialysance" D_B, which is defined as

$$D_B = \frac{N}{C_{B_i} - C_{D_i}} \tag{16.23}$$

The dimensionless ratio D_B/Q_B can be regarded as a convenient efficiency parameter, since it represents the fraction of maximum attainable solute depletion in the feed that is actually achieved in the device.

By simultaneous solution of Eqs. 16.21, 16.22, and 16.23 and the use of the logarithmic mean concentration driving force, the dialysance ratio D_B/Q_{B_i} can be expressed as follows.

Parallel flow:

$$\frac{D_B}{Q_B} = \frac{1}{1 + Z} \left[1 - e^{-R(1 + Z)} \right] \tag{16.24}$$

Mixed-dialysate flow:

$$\frac{D_B}{Q_B} = \frac{1 - e^{-R}}{1 + Z(1 - e^{-R})} \tag{16.25}$$

Countercurrent flow:

$$\frac{D_B}{Q_B} = \frac{1 - e^{R(1-Z)}}{Z - e^{R(1-Z)}} \tag{16.26}$$

If $Z = 1$,

$$\frac{D_B}{Q_B} = \frac{R}{R + 1}$$

where $Z = Q_B/Q_D$ and $R = k^+A/Q_B$.

The above relationships can be used to analyze dialyzer performance and to calculate expected performance for different flow rates for a given dialyzer.

The overall rate constant k^+ is influenced by the properties of the membrane and of the fluid boundary layers on each side of the membrane. These factors were discussed by Michaels (24) and will not be covered here.

Applications

Dialysis is a comparatively "ancient" process, used for years to remove salts and low-molecular-weight solutes from solutions. One early method of conducting a dialysis process was to fill cellophane tubing or sausage casing with the solution to be purified, seal the tubing, and then immerse it in a large volume of water. Although this process was a rather inefficient one, the low-molecular weight solutes slowly diffused from the tubing, and the solution was eventually purified.

As is true of all membrane separation methods, other factors than the membrane contribute to the rates and efficiency of dialysis. One pertinent factor is the resistance due to the boundary layer between the solution to be purified and the membrane. The low-molecular-weight solutes diffuse through the membrane and are not replaced as fast as they are removed from the bulk of the solution. The other factor occurs on the "downstream" side of the membrane, where a concentration gradient exists between the membrane and the bulk of the solution "washing" the membrane. These factors are two additional resistances in series with the membrane and must be carefully considered when designing a dialysis process.

Craig was one of the first investigators to recognize the inefficiencies inherent in the conventional dialysis methods (25). He designed and used a thin-film dialysis cell, shown in Fig. 16.8. The thin films on either side of the membrane reduce the lengths of the diffusion paths and result in much lower boundary layer resistances.

Craig was also instrumental in developing two other techniques of particular interest to the analytical chemist. In one of these, which he called *stepwise countercurrent dialysis* (26), he took advantage of the different dialysis

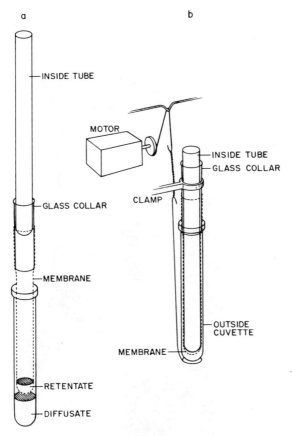

Figure 16.8. Schematic drawing of an analytical thin-film dialysis cell. Reprinted from Ref. 25, by courtesy of Marcel Dekker, Inc.

rates of different-size molecules for separating complex mixtures into families of restricted sizes. In the other he "calibrated" cellophane membranes by measuring the dialysis rates of a series of carefully purified compounds of known configurations. He could then compare the diffusional size of a compound of unknown configuration with the values for standard compounds (26).

The principles discussed above have been used to design highly efficient artificial kidneys based on dialysis. In this process the blood from a patient suffering from acute or chronic kidney failure is passed into a dialyzer from a connection to one of his arteries. Low-molecular-weight toxins in the blood, such as urea, creatinine, and uric acid, pass across the membrane into a "dialysate" solution of such composition that the osmotic pressure is the

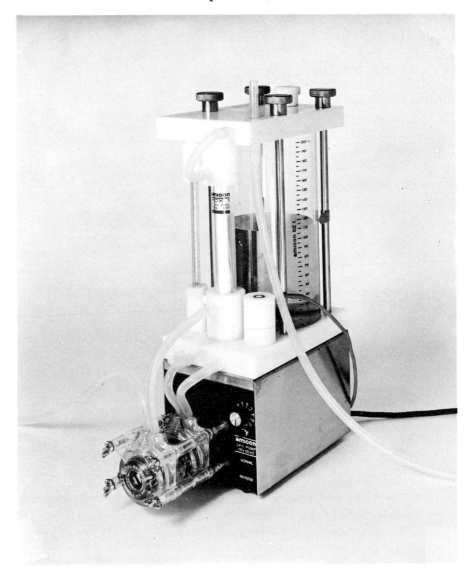

Figure 16.9. Hollow-fiber laboratory dialyzer. Reprinted by courtesy of Amicon Corporation.

same as that of the blood; the rate of transport of certain salts is thus controlled. The blood then returns to the patient through a connection to one of his veins. Patients without functioning kidneys have been kept alive in reasonably good health for over 10 years by means of this process. The interested reader is referred to Refs. 24 and 27 for more details.

Efficient dialyzers are now available from Amicon Corporation and from Dow Chemical Company for use in the laboratory. Figure 16.9 is a photograph of one such device, which can remove 99% of the salt from a liter of solution in less than 4 hr.

Membranes

The cheapest and most satisfactory membrane for general use is uncoated cellophane, as manufactured by Union Carbide, Du Pont, or Bemberg (Germany).

16.45   ELECTRODIALYSIS

Theory

Electrodialysis is a process in which electrically charged membranes are utilized to separate components of an ionic solution under the driving force of an electric current. The principle of the process is illustrated in Fig. 16.10, which shows a schematic diagram of a typical electrodialysis cell. This cell consists of a series of anion- and cation-exchange membranes, arranged in an alternating pattern between an anode and a cathode. An ionic solution (e.g., an aqueous salt solution) is pumped through the membrane stack. When a direct-current potential is applied between an anode and a cathode, the positively charged ions in the solution migrate toward the cathode. Whereas the ions pass easily through the cation-exchange membrane, they are retained by the anion-exchange membrane. Likewise, the negatively charged ions in the solution migrate toward the anode, passing through the anion-exchange membrane, and are retained by the cation-exchange membrane. The overall result is an ion concentration increase in alternate compartments; simultaneously the other compartments become depleted of ions. To minimize the cell resistance the distance between the membranes is low (ca. 0.1 cm). External pumps are used to obtain high flow velocities in the compartments bound by the membranes, so as to minimize problems associated with concentration polarization.

For a membrane to be useful in this process, the principles of electrodialysis require a high selectivity between ions of opposite charge and a good electric conductivity, that is, high current efficiency.

Membrane selectivity. The selectivity of an anion- or a cation-exchange membrane can be related to the ion transport numbers (28) by

$$\psi_a = \frac{\overline{T}_- - T_-}{T_+} \tag{16.27}$$

$$\psi_c = \frac{\overline{T}_+ - T_+}{T_-} \tag{16.28}$$

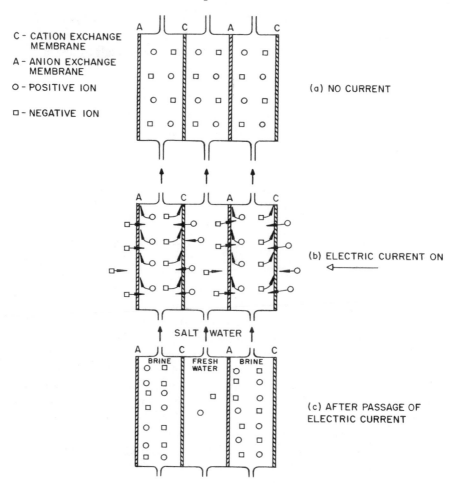

Figure 16.10. Principle of electrodialysis.

where ψ_a and ψ_c are the permselectivities of an anion- and a cation-exchange membrane, respectively; $\overline{T}_-$ and $\overline{T}_+$ are the transport numbers of anion and cation within the membrane; and T_- and T_+ represent anion and cation transport numbers in the solution. The transport numbers are dimensionless variables that can be related to the ion transference number by

$$T_+ = zt_+ \tag{16.29}$$

and

$$T_- = zt_- \tag{16.30}$$

where z is the valence in equivalents per mole, and t_+ and t_- are the transference numbers, defined as the numbers of moles transferred by 1 faraday of electricity through a stationary cross section. The permselectivity is not a membrane constant but depends on the concentration of the feed solution.

Current efficiency. The efficiency with which current transports salt through any set of membranes is defined as (29)

$$\eta_s = \frac{n_c T_- \psi_c + n_a T_+ \psi_a}{n_c T_- + n_a T_+} \qquad (16.31)$$

where η_s is the current efficiency, and n_c and n_a are the numbers of cation- and anion-exchange membranes in a stack.

The current efficiency can also be affected by current leakage through the stack manifolds and by electroendosmosis. The efficiency loss due to current leakage through the stack can easily be controlled by proper design. The efficiency loss due to water transfer, however, can reach a significant value when solutions with high salt concentrations are processed. For a given desalting process the effect of water transport on the current efficiency can be expressed by:

$$\eta_w = \left(1 - n\bar{t}_w \times 0.018 \frac{m_i}{\eta_s}\right) \qquad (16.32)$$

where η_w is the current efficiency as related to water transfer, n is the number of membranes in a stack, $\bar{t}_w$ is the transference number of water in the membrane, and m_i is the molality of the feed water.

Current utilization. The amount of salt removed from the process stream by a given current depends on the number of selective membranes in the stack and on the differences between the transport numbers of the ions in the membranes and in the solutions. The net removal of salt from a process stream is given by

$$\Delta(N_p F_p) = \frac{I \xi_{\mathscr{F}}}{\mathscr{F}} \qquad (16.33)$$

where $\Delta(N_p F_p)$ is the decrease in normality and the flow rate of the process stream passing through the stack, $\mathscr{F}$ is the Faraday constant, I is the total current, and $\xi_{\mathscr{F}}$ is the so-called Faraday utilization factor and represents the amount of salt removed per unit of current.

Power requirements. The minimum theoretical energy required to separate salt from a saline is given by (29, 30)

$$u = 5.21 \times \Delta N \frac{\ln \beta}{\beta - 1} - \frac{\ln \alpha}{\alpha - 1} \qquad (16.34)$$

where u is the energy in kilowatt-hours per 1000 gallons of product, ΔN is the concentration difference between feed and product solution, N is the concentration in equivalents per liter, $\beta = N_f/N_c$, $\alpha = N_f/N_p$, and the subscripts f, p, and c identify feed, product, and concentrate, respectively.

In practice the energy consumed is typically 10 to 20 times the minimum theoretical requirement. This increased consumption is caused by irreversible processes in the system, such as concentration polarization. In any case the power requirements strongly increase with an increase in the concentration of the feed solution and the amount of salts removed from a given amount of feed solution.

Electrodialysis is presently utilized on an industrial scale for desalination of brackish waters, and at low feed water concentration (< 3000 ppm) is superior in terms of cost to comparable processes such as distillation and reverse osmosis.

Applications, Equipment, and Membranes

Electrodialysis is being used or has been evaluated for desalination of water, preparing boiler feed makeup and chemical process water, recovery of brine from sea water, deashing of sugar solutions, deacidification of citrus juices, and metathesis.

The authors do not know of a small electrodialysis cell for laboratory work that is commercially available. Larger equipment and membranes are available from such companies as American Machine and Foundry, Ionics, and Permutit.

References

1. S. B. Tuwiner, *Diffusion and Membrane Technology* (ACS Monograph 156), Reinhold, New York, 1962.
2. U. Merten, *Desalination by Reverse Osmosis*, MIT Press, Cambridge, Mass., 1966.
3. A. S. Michaels, in *Advances in Separation and Purification*, E. S. Perry, ed., John Wiley, New York, 1968.
4. N. Lakshminarayanaiah, *Transport Phenomena in Membranes*, Academic Press, New York, 1969.
5. R. E. Kesting, *Synthetic Polymeric Membranes*, McGraw-Hill, New York, 1971.
6. S. Sourirajan, *Reverse Osmosis*, Academic Press, New York, 1970.
7. N. N. Li, R. B. Long, and E. J. Henley, *Ind. Eng. Chem.*, **57** (3), 29 (1965).
8. A. S. Michaels, *Fundamentals of Membrane Permeation*, Proceedings of the Symposium Membrane Processes for Industry, Southwest Research Institute, 1966.
9. R. Schlögl, *Stofftransport durch Membranen*, Steinkopff Verlag, Darmstadt, 1964.
10. R. N. Rickles, *Ind. Eng. Chem.*, **58** (6), 19 (1966).
11. S. Loeb and S. Sourirajan, *Advan. Chem. Ser.*, **38**, 117 (1962).

12. A. S. Michaels, *Ind. Eng. Chem.*, **57**, 32 (1965).

13. W. F. Blatt, A. Dravid, A. S. Michaels, and L. Nelsen, in *Membrane Science and Technology*, Plenum, New York, 1970.

14. W. Banks and A. Sharples, *J. Appl. Chem.*, **16**, 153 (1966).

15. P. Meares, *Eur. Polymers J.*, **2**, 241 (1966).

16. H. Strathmann, "Control of Concentration Polarization in Reverse Osmosis Desalination of Water," Report 336, Office of Saline Water, 1968.

17. J. Crank, *The Mathematics of Diffusion*, Oxford University Press, London, 1964.

18. S. Weller and W. Steiner, *J. Appl. Phys.*, **21**, 279 (1950).

19. S. Weller and W. Steiner, *Chem. Eng. Progr.*, **46**, 585 (1950).

20. M. Benedict and T. Pigford, *Nuclear Chemical Engineering*, McGraw-Hill, New York, 1957.

21. E. Baer, ed., *Engineering Design for Plastics*, Reinhold, New York, 1964.

22. J. Brandrup and E. Immergut, eds., *Polymer Handbook*, Interscience, New York, 1966.

23. A. Lebovits, *Mod. Plastics*, March, 139 (1966).

24. A. S. Michaels, *Trans. Am. Soc. Artificial Internal Organs*, **12**, 387 (1966).

25. L. C. Craig et al., *J. Macromol. Sci—Chem.*, **A3**(1), 133 (1969).

26. L. C. Craig, in *Analytical Methods of Protein Chemistry*, Vol. 1, Pergamon Press, New York, 1960.

27. C. K. Colton, "A Review of the Development and Performance of Hemodialyzers," U.S. Public Health Service, 1967 (Federal Clearinghouse Accession PB182–281).

28. A. Winger, G. Bodamer, and R. Kunin, *J. Electrochem. Soc.*, **100**, 178 (1953).

29. L. Shaffer and M. Mintz, in *Principles of Desalination*, Academic Press, New York, 1966.

30. K. S. Spiegler, in *Ion Exchange Technology*, Academic Press, New York, 1956.

Selected Bibliography

W. K. Chen, "Electrodialysis," in *Kirk-Othmer Encyclopedia of Chemical Technology*, 2nd ed., Vol. 7, Interscience, New York, 1965.

R. E. Kesting, *Synthetic Polymeric Membranes*, McGraw-Hill, New York, 1971.

U. Merten, *Desalination by Reverse Osmosis*, MIT Press, Cambridge, Mass., 1966.

E. S. Perry, ed., *Progress in Separation and Purification*, Vol. 1, Interscience, New York, 1968.

R. N. Rickles, *Membranes—Technology and Economics*, Noyes Development Corp., Park Ridge, N.J., 1966.

List of Symbols

A	membrane area	C_p	product concentration
C_b	upstream solute concentration	$C_s^{\,m}$	solute concentration in the membrane
C_f	feed concentration	$\Delta \bar{C}_M$	average solute concentration difference
C_f	final concentration		
C_0	initial concentration	D_B	dialysance

$D_s{}^m$ solute diffusivity in the membrane

I total electric current

J flux

J_s solute flux

J_v solvent flux

k Henry's law solubility constant

k^+ overall rate constant

K_1 transport coefficient of the membrane to solvent

K_2 transport coefficient of the membrane to solute

L thickness of membrane

m_i molality of feed water

n number of membranes in a stack

N overall solute transport rate

P partial pressure

ΔP pressure difference across the membrane

P permeability coefficient

Q volumetric flow rate

r pore radius

R rejection of a solute by the membrane

R_B recovery of component B in the retentate

S apparent selectivity

S^0 membrane permselectivity

t ion transference number

$\bar{t}_w$ transference number of water in the membrane

T transport number in solution

$\bar{T}$ transport number in the membrane

u power requirement in electrodialysis

V_f volume of solvent used in ultrafiltration

V_0 initial volume of solution in ultrafiltration

X mole fraction in the retentate

X° mole fraction in the feed

z valence in equivalents per mole

α N_f/N_c Eq. 16.34

β N_f/N_p Eq. 16.34

δ fractional loss of solute

Δ recovery rate

ε membrane porosity

η_s current efficiency

η_w current efficiency of water transfer

λ membrane thickness

$\xi_{\mathscr{F}}$ Faraday utilization constant

π osmotic pressure

π_1 total upstream pressure

π_2 total downstream pressure

τ tortuosity factor

Subscripts

a anion-exchange membrane

B feed

c cation-exchange membrane

c concentrate

D dialysate

i inlet

o outlet

Seventeen

ELECTROPHORESIS

M. BIER

17.1 Introduction

Electrophoresis is the most useful separation technique for the analysis and characterization of complex biological mixtures of proteins. Although the phenomenon of electrophoretic transport in an electrical field has been known for over 150 years, its singular usefulness for the analysis of protein mixtures was first established by Tiselius in 1937. His separation of serum proteins into four major components—serum albumin and α-, β-, and γ-globulins—is one of the cornerstones of the modern development of protein chemistry. The elegance of the Tiselius equipment was matched by equally sophisticated theories concerning the phenomenology of the separation process, such as boundary anomalies, boundary spreading, and interpretation of patterns of rapidly reacting systems.

The Tiselius apparatus was, however, expensive, exacting as to experimental procedures, and limited in resolving power. Subsequently, parallel to the development of paper chromatography, paper electrophoresis and other forms of zone electrophoresis were introduced. These led to a better understanding of the principles involved, and to a liberation from the constraints of orthodoxy of the moving-boundary technique. A proliferation of techniques and applications followed, resulting in (a) extreme simplification of equipment and procedures, (b) an unexpected increase in resolving power by such procedures as high-density gel electrophoresis and immunoelectrophoresis, (*c*) ease of assay of biological activity directly on the supporting medium, and (*d*) extrapolation of the technique for proteins to substances of medium or low molecular weight, either organic or inorganic.

17.11   DESCRIPTION AND CLASSIFICATION

Electrophoresis is defined as the transport of electrically charged particles in a direct-current electric field. The particles may be simple ions, complex macromolecules and colloids, or particulate matter—either living cells, such as bacteria or erythrocytes, or inert material, such as oil emulsion droplets and clay (1). Migration can occur also in a nonuniform alternating current, provided the particles are polarizable, and this transport phenomenon is called *dielectrophoresis* (2). In the present discussion we are interested only in electrophoresis in liquid media, preferably aqueous solutions or suspensions. In gaseous systems, electrophoretic transport is used in many so-called electrostatic processes, on a broad industrial scale. Applications may range from smoke abatement and accelerated smoke-curing of fish to ore refinement and concentration. Xerography is a most recent application of such electrostatic principles (2).

Electrophoretic separation is based on differential rates of migration in the bulk of the liquid phase and is not concerned with reactions occurring at the electrodes. Such processes as electrodeposition of paints and other coatings and electrolytic refinement of metals all require electrophoretic transport of active components to the electrodes, but it is the electrode reactions that are of critical importance.

We are restricted, therefore, to a discussion of separation processes achievable in the bulk liquid phase and based on electrical transport. Highest resolution is obtained if an element of discontinuity is introduced into the liquid phase, such as a pH gradient, or the sieving effect of high-density gels. Membrane barriers may also be introduced into the pathway of migrating particles. Within this realm, a number of different techniques have evolved, which can be classified according to such functional parameters as the use of anticonvective means, the shape of the vessel, and the superimposition of pH gradients. Probably the most useful criterion for classification is whether electrophoresis is carried out in free solution or on supporting media. An enumeration of techniques in free solution is presented in Table 17.1, and the distinction between the techniques is relatively clear cut. Use of anticonvective supporting media permits a much greater latitude of techniques and instruments, as is shown in Table 17.2. Some of the above techniques result in complete separation of the components into distinct zones; this is referred

Table 17.1. Free Electrophoresis Techniques

1. Microscope electrophoresis
2. Tiselius moving-boundary electrophoresis
3. Column electrophoresis—isoelectric focusing
   (density and pH gradient electrophoresis)*[a]
4. Free-flowing curtain electrophoresis*
5. Electrodecantation and electrophoresis-convection*
6. Forced-flow electrophoresis*

[a] Asterisk denotes techniques used mainly for preparative purposes.

to as *zone electrophoresis*. In other techniques, complete separation is not obtained, and measurement is based on visual observation of individual particles or the analysis of boundary movement of partially overlapping zones.

There is also no clear-cut distinction between preparative and analytical techniques, as many of the analytical techniques may be usable for the preparation of small quantities of the individual components. Nevertheless, the techniques used mainly for preparative purposes are identified with an asterisk in Tables 17.1 and 17.2.

Table 17.2. Electrophoresis on Supporting Media-Zone Electrophoresis

Process	Support	Apparatus	Additional Features
Continuous-flow*[a]	Paper	Moist chamber	Immunoelectrophoresis
Batch operation	Membranes	Trough*	pH gradient
	Powders*	Column*	High-density gels
	Gels	Curtain*	High voltage
	Sponge*		

[a] Asterisk denotes techniques used mainly for preparative purposes.

17.12 USE IN BIOCHEMISTRY AND TECHNOLOGY

Most applications of electrophoresis are related to the analysis of proteins in serum or other body fluids. For a simple analysis, suitable for detection of major components, zone electrophoresis on cellulose acetate membranes is sufficient. A variety of staining procedures is available, and the biological activity of enzymes can be directly observed, using specific substrates. Automatic electronic analyzers have greatly simplified the quantitation of the electrophoretic patterns. For more discriminating analysis, that is, higher resolution, high-density gel electrophoresis or immunoelectrophoresis is necessary. Immunoelectrophoresis became easily accessible with the commercial availability of many broad-spectra or single-protein antisera.

The above techniques have proved to be invaluable to much of biochemical research in proteins, as they serve not only to analyze complex mixtures and prepare purified fractions, but also to define and characterize the individual protein species and to determine their isoelectric points. They constitute the best proof of the purity of an isolated fraction and have led to the discovery that many proteins are present in several genetic variants. Genetic mutants have been found for many of the normal components of human or animal sera, and their study has given a strong new impetus to the whole field of human genetics.

In clinical medicine, electrophoresis is used for the diagnosis of certain specific diseases, characterized by the presence of abnormal proteins, either genetically conditioned or induced by the disease. A well-known example of a genetic molecular disease is sickle-cell anemia, due to the abnormal hemoglobin S in the red blood cells of the affected individuals. In other diseases, a normal protein may be completely absent, as in agammaglobulinemia, or present in too low or too high concentrations. These striking applications of electrophoresis notwithstanding, the general diagnostic usefulness of electrophoresis in clinical laboratories has fallen short of the more enthusiastic expectations.

Tiselius moving-boundary electrophoresis remains the method of choice for the determination of absolute electrophoretic mobilities. It is also useful for the study of rapidly interacting protein systems and for other similar physicochemical research. For the determination of the electrophoretic mobilities of particulate matter, such as erythrocytes and bacteria, the only method available is microscope electrophoresis. This is the oldest of the analytical techniques and is based on direct observation of migrating particles under a microscope. For preparative fractionation of these cellular elements, various forms of curtain electrophoresis without supporting media are available. High-voltage electrophoresis is used for the analysis of amino acids, peptides, and other nonprotein mixtures. The wide scope of applications of this technique has been reviewed in detail (3).

It is most unfortunate for protein chemistry that the outstanding achievements of electrophoresis for the analysis of protein mixtures are not matched by equally satisfactory preparative methods. On a laboratory scale, for micropreparative purposes, several techniques are available, for instance, various forms of column electrophoresis with density and/or pH gradients, and gels or powder support. Other techniques applicable are those on starch blocks and other supporting media, where separation is achieved by simple sectioning of the supporting matrix. Filter-paper-supported curtain electrophoresis has been largely replaced by instruments whereby stabilization is achieved by the viscosity of a thin film defined by two parallel glass plates.

It is difficult to scale upward any of these techniques, and for the separation of larger volumes of proteins various forms of chromatography on Sephadex, modified celluloses, and other materials are often preferable. On a commercial scale, most large-volume protein fractionations are still carried out by fractional precipitation with salts, alcohol, and so on, rather than by either electrophoretic or chromatographic techniques.

Electrodecantation and forced-flow electrophoresis are the only two techniques adaptable for large-scale industrial applications. Neither procedure possesses the resolving power of the analytical techniques, and only one purified protein fraction can be obtained at a time. On the other hand, these methods are applicable to a wide range of purely technological problems not usually associated with electrophoresis. Electrodecantation, the older of the two techniques, has been applied to such problems as the creaming of rubber latex and the concentration of Teflon latex. Forced-flow electrophoresis is a more versatile and newer technique, adaptable to a variety of electrokinetic processes, such as electrophoretic separation, electrofiltration, and electroosmotic concentration (4, 5).

17.2   Theory

The object of the theory of electrophoresis is to relate the experimentally measurable electrophoretic velocity to more fundamental parameters of ion interaction, charge, and structure in the neighborhood of charged particles or surfaces. The theory of the electric double layer deals with these parameters and is necessitated by the fact that colloidal particles create their own ionic environment, being much larger and having a higher charge than small ions.

17.21   ORIGIN OF CHARGE

In proteins and many other macromolecules, charge originates from the ionization of functional groups that are an integral part of their molecular structure. In proteins, these groups can give rise to ions of both polarities; at acid pH, proteins are usually positively charged, whereas in basic media they are negatively charged. The pH at which the net charge due to ionization is zero is called the *isoionic point*. This may not coincide with the pH of zero electrophoretic mobility, called the *isoelectric point*, because of possible adsorption of free ions such as Na^+ or Cl^-.

Ion adsorption modifies the net charge of proteins and other macromolecules and is also responsible for the electrical charge of many hydrophobic surfaces such as hydrocarbon droplets or air bubbles. Most adsorption effects are due to anions, which convey a negative character to these surfaces. Cations, being more hydrated than anions, are less likely to be adsorbed; ionic detergents, because of their sizable hydrophobic portion, are particularly strongly adsorbed.

Colloidal particles can also acquire charge by virtue of unequal dissolution of the oppositely charged ions of which they are composed. For example, colloidal silver iodide particles may acquire a negative charge in the presence of an excess of the negative iodide ion, and a positive charge in the presence of an excess of the silver ion. Hydrogen and hydroxyl ions play a similar role for metal oxide and hydroxide colloids. Such ions are referred to as the *potential-determining ions*.

17.22   ZETA POTENTIAL AND THE ELECTRIC DOUBLE LAYER   (6)

The theory of the electric double layer deals with the structure of the boundary between two phases, considering it as a layer of finite dimensions rather than as a mathematical plane. If ions of one sign are an integral part of, or are adsorbed by one of the phases, ions of the opposite sign will be attracted by the resulting electric field and will accumulate near the phase boundary, though still remaining subject to Brownian movement. The fixed charges of one phase plus the relatively mobile counterions of the other phase constitute the *electrical double layer*. The total charge of the double layer is zero, but the spatial distribution of charges is not random and gives rise to an electric potential. The potential is defined as the work necessary to bring a unit charge of the same sign from infinity to a given point in space.

The double layer for a spherical particle is schematically illustrated in Fig. 17.1. Its total negative charge Q is imagined to be symmetrically distributed over its surface but could also be considered to be at its center. The shaded area surrounding the sphere represents the solvent molecules, which are bound so firmly to the surface that they move with the particle during electrophoresis, defining a surface of shear distinct from the actual particle surface. The potential ψ_0 at the surface of the particle is given by $\psi_0 = Q/\varepsilon a$, where ε is the dielectric constant of the medium, and a the radius of the sphere. This potential will be different, however, from the potential at the surface of shear, the distance at this point being $x = a + \Delta a$, where Δa is the statistical average of the thickness of the hydration shell. It is the potential at this shear boundary that governs the electrophoretic mobility of the particle and is therefore given special recognition, being called the *electrokinetic* or *zeta potential* ζ.

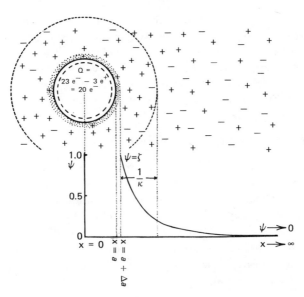

Figure 17.1. Schematic diagram of a large spherical particle with its ionic environment. The dotted area indicates the hydration shell. The potential $\psi = \zeta$ at the shear boundary and falls to $\psi = 0$ at infinite distance. Values of ψ are in relative units and presumed to be negative. Outer dashed circle indicates the thickness of the electric double layer (7). Reprinted from Ref. 7, by courtesy of Hofner Publishing Co.

Surrounding the negatively charged spherical particle there is a statistical accumulation of its positively charged counterions, the distribution of co- and counterions falling to random levels only at infinite distance. According to the Gouy-Chapman treatment of the double layer, the potential decreases with distance x according to

$$\psi_x = \psi_0 \exp\left(-\kappa x\right) \tag{17.1}$$

where κ is the familiar Debye-Huckel constant:

$$\kappa = \left(\frac{8\pi e^2 n_0 z^2}{\varepsilon k T}\right)^{1/2} \tag{17.2}$$

Here e is the electronic charge, n_0 the bulk concentration of each ionic species, z the valence of the (symmetrical) electrolyte, and k the Boltzmann constant. It is customary to refer to $1/\kappa$, which has the dimension of length, as the "thickness of the double layer." This is indicated in Fig. 17.1 by the outer circular dotted line and is the distance over which the potential decreased by one exponential factor. The zeta potential, as well as the thickness of the double layer, $1/\kappa$, decreases rapidly with an increase in concentration or valence of the electrolytes in the medium.

Quantitative treatment of the double layer is extremely complex and in some respects still unsolved. The Gouy-Chapman treatment outlined above is the simplest, as it assumes that the space charges of the ionic cloud are carried by dimensionless ions. Stern, Graham, Overbeek (6), and others have refined this treatment, by considering the finite size and hydration of the electrolyte ions, as well as the geometry of the charged surface.

17.23   MIGRATION IN AN ELECTRIC FIELD (6)

The zeta potential cannot be measured directly but can be calculated only on the basis of the theories of the electric double layer. There is, therefore, no direct way to check the accuracy of the calculations, and their reliability depends on the adequacy of the physical model used in the treatment of the double layer.

A charged particle in an electric field is exposed to four forces, the resultant of which is a steady state, characterized by a constant electrophoretic migration velocity. The main two forces are the *electrophoretic attraction* K_1:

$$K_1 = QE \qquad (17.3)$$

where Q is the charge of the electrokinetic unit and E the potential gradient applied, and the *Stokes friction* K_2:

$$K_2 = -f_c U \qquad (17.4)$$

where U is the electrophoretic velocity and f_c the friction coefficient of the colloid particle. For a rigid sphere, large in comparison to the molecules of solvent,

$$f_c = 6\pi\eta a \qquad (17.5)$$

where η is the viscosity of the solvent.

One needs also to estimate two additional forces, arising from the presence of small electrolytes in the medium. The electric field exerts an attraction on these ions, which is transferred to the solvent, causing an electro-osmotic flow of the liquid with respect to the particles. This gives rise to the so-called *electrophoretic retardation* K_3. The last force, K_4, known as the *relaxation effect*, is brought about by the distortion of the ionic atmosphere, due to differences in movement of the counterions and the colloid particles, as a consequence of which each particle is no longer in the center of its atmosphere.

In the steady state, achieved nearly immediately upon application of the electric field, the sum of these four forces is equal to zero, and the electrophoretic velocity U becomes

$$U = \frac{QE + K_3 + K_4}{f_c} \qquad (17.6)$$

There are several solutions to Eq. 17.6. Using the Debye-Hückel theory of strong electrolytes, Hückel calculated the electrophoretic retardation force

$$K_3 = (\varepsilon \zeta a - Q)E \qquad (17.7)$$

and, by neglecting K_4, one obtains the electrophoretic mobility μ as the ratio of velocity over field strength:

$$\mu = \frac{U}{E} = \frac{\varepsilon \zeta}{6\pi\eta} \qquad (17.8)$$

This equation was derived on the assumption that the lines of forces were not deformed by the presence of the charged particles. It is applicable, therefore, only for spheres of small diameter, and at low ionic strength, where the product κa is small compared to 1. The expression κa is the ratio of particle diameter to double layer thickness.

In the Smoluchowski treatment, the assumption is made that the entire double layer is parallel to the direction of the dc field and is applicable only to large particles, where κa is also large, preferably above 100. The resulting Helmholtz-Smoluchowski equation is

$$\mu = \frac{U}{E} = \frac{\varepsilon \zeta}{4\pi\eta} \qquad (17.9)$$

Equations 17.8 and 17.9 were reconciled by the calculations of Henry, where

$$\mu = \frac{U}{E} = \left(\frac{\varepsilon \zeta}{6\pi\eta} \right) f(\kappa a) \qquad (17.10)$$

and where $f(\kappa a)$ varies from 1 to 1.5, depending on the dimensions of κa. It is this equation that ought to be used for calculations of the zeta potential when κa falls in the range between $\frac{1}{10}$ and 100, that is, intermediary to the domains of the Hückel and Helmholtz-Smoluchowski equations.

In all of the above equations, the relaxation effect K_4 is neglected. A more detailed treatment of the relation between electrophoretic mobility and zeta potential is available in an excellent presentation by Overbeek and Wiersema (6).

Much of protein electrophoresis is carried out under conditions where Henry's equation is applicable. Experimentally, however, electrophoretic data usually yield somewhat lower values for protein charge than direct titration data, and there is, at present, no fully satisfactory theory.

17.3 Techniques

The many techniques presently available are the result of the varying needs of individual experimenters. In analytical electrophoresis, the primary classification of protein components is due to their relative positions in a developed electrophoretic pattern. The notion of α-, β-, and γ-globulins, although no longer identifying specific protein entities, has remained useful as indicative of the general mobility range of a given protein. Components are sometime identified by their electrophoretic mobilities μ, expressed as migration velocity per unit field strength, the dimensions being (cm/sec)/ (V/cm), or, simplified, square centimeter per volt per second. In microscope electrophoresis, however, mobilities for particulate matter such as microorganisms or blood cells are expressed in microns per centimeter, the dimensions being $(\mu/\text{sec})/(\text{V/cm})$. When mobilities are used for calculations of zeta potential, the centimeter-gram-second electrostatic unit should be used, which is (cm/sec)(abV/cm) = 300 (cm/sec)/(V/cm).

It should be emphasized, however, that mobilities as well as overall electrophoretic patterns depend on the buffer employed, and in moving-boundary electrophoresis, for most protein analyses, they usually relate to the customary 0.05 N sodium diethyl barbiturate–diethyl barbituric acid (veronal) buffer, pH 8.6. Any change in composition or pH of the buffer will alter the mobilities. In electrophoresis on supporting media, the medium itself has an effect on the electrophoretic patterns and can result in changes in the relative positions of various components, because of electro-osmosis and molecular sieving effects in high-density gels. The electric field E (V/cm) is usually not measured directly but is calculated using Ohm's law:

$$E = \frac{IR}{A}$$

where I is the current in amperes, R the specific resistivity of the buffer employed, and A the cross area of the cell at the point of measurement. The specific resistivity of the buffer R has to be determined with a platinum-electrode conductivity cell, by means of an alternating current, using a Wheatstone resistance bridge. Volt-ohm-meters, used in usual electric work, are not applicable.

17.31 MICROSCOPE METHOD OF ELECTROPHORESIS (7, 8)

This method is the only technique available for the determination of electrophoretic mobilities of large particles, such as blood cells, microorganisms, and clay particles. It is based on direct microscopic observation of the migration of the particles, suspended in an appropriate buffer and exposed to a dc current. This is the earliest form of analytical electrophoresis, and before the introduction of the moving-boundary technique it was applied for the study of colloidal materials of submicroscopic size (by adsorbing them onto carriers such as collodion). Electrophoresis, being largely a property of surface charge, that is, zeta potential, is relatively insensitive to particle size of the carrier employed.

The apparatus for this technique is relatively simple. It requires a cell into which a microscope can be focused, an electrode system and power supply, and provisons for emptying and filling the cell. Several cells, flat or cylindrical, have been developed, and they can be oriented horizontally, laterally, (for flat cells), or vertically. The most common cell is that of Abramson (7), illustrated in Fig. 17.2. No matter which cell is used, it is essential to avoid any convective flow of liquid not due to electro-osmosis. Thermal convection is avoided by using low current density and heat filters in the microscope illuminator. If reversible electrodes, such as copper-copper sulfate or mercury-mercuric nitrate are employed, gas formation at the electrodes is avoided. None of these electrolytes, of course, should be allowed into the measuring cell itself.

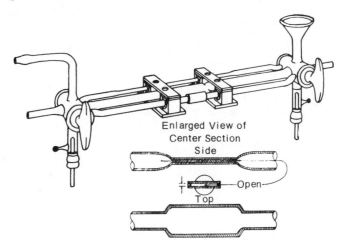

Enlarged View of
Center Section
Side
Open
Top

Figure 17.2. Flat horizontal cell for microscope electrophoresis according to Abramson (7). Migration of particles within the center section is measured by direct microscopic observation. Reprinted from Ref. 7, by courtesy of Hofner Publishing Co.

The microscopic observation of the migrating particles is complicated by the presence of a strong electro-osmotic flow of buffer within the cell, caused by the electrical charge of the glass walls. Hydrodynamic considerations of flow within flat cells of the Abramson type indicate that there are only two stationary levels within the cell (at 0.21 and 0.79 of the cell thickness), where the electro-osmotic flow is reduced to zero. The microscopic measurements of migration velocity have to be carried out at these two levels: several measurements must be taken, and the results averaged.

In comparison to other methods of electrophoresis, the microscope technique is not currently in widespread usage. From a historic point of view much of the theory of electrokinetic phenomena has been developed on the basis of the microscope technique. Its primary present use is the determination of the mobilities and isoelectric points of various micro-organisms and cellular components of blood. It is most interesting, for instance, that the electrophoretic mobility of red blood cells is very constant for individuals of the same animal species, but varies greatly from species to species. Mobility measurement can also be used to obtain indirect information on the structure of the charged surface. Most important is the dependence of the mobility on pH and salt concentration. Other factors affecting the mobility may be specific adsorption of biologically active materials, chemical treatment such as lipid extraction, or biological effects such as virus infection. In bacterial populations, strain differences often result in mobility differences.

17.32 MOVING-BOUNDARY ELECTROPHORESIS (9)

The moving-boundary technique, the first method to demonstrate the great resolving power of electrophoresis, remains the method of choice for the determination of absolute mobilities of soluble proteins. It is an elegant technique in which several logistics problems were solved simultaneously in a most ingenious (if complicated) way by achieving (*a*) initial formation of sharp boundaries, (*b*) automatic stabilization of boundaries by density gradients, (*c*) good heat dissipation, and (*d*) optical visualization of components during migration. It is embodied in the Tiselius cell, with its U-shaped channel of rectangular cross section and optically flat fused-glass windows. The cell consists of three sections, which can slide independently of each other on lubricated glass surfaces, as illustrated schematically in Fig. 17.3. The bottom section and one limb of the cell are first filled with protein;

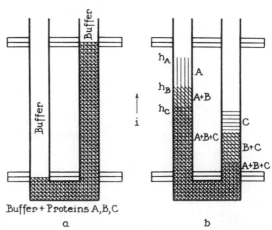

Figure 17.3. Schematic presentation of the center section of a Tiselius electrophoresis cell (19). (*a*) Initial boundaries position. (*b*) Upon electrophoresis, separation of boundaries has taken place in the two limbs of the U-shaped cell.

the remainder of the cell is then filled with buffer. When an electric field of proper polarity is established, migration of proteins will cause the initially sharp boundaries to separate into several subboundaries. The greater density of the protein solution stabilizes its boundaries and prevents convective mixing with the lighter buffer but also prevents complete separation of protein components. The components remain overlapping, and the analysis is based on the migration velocity of the density boundaries in the ascending and descending limbs. For various theoretical reasons, these are rarely symmetrical.

In early work using simplified U-tube arrangements the migration of colloids was followed either by visual color and turbidity or by ultraviolet light absorption. These techniques have been completely displaced by schlieren and interference optics, which give much better resolution and quantitation of the boundaries. Two schlieren systems are available, based on the scanning and the cylindrical lens cameras. Both depend on the deflection of light as it passes through the gradient of index of refraction existing across each boundary, and give the easily recognizable patterns illustrated in Fig. 17.4. The plots actually represent dn/dh, the change of index of refraction n with the height h of the electrophoretic channel. For quantitation, one has to recur to planimetry, that is, compare the areas under each

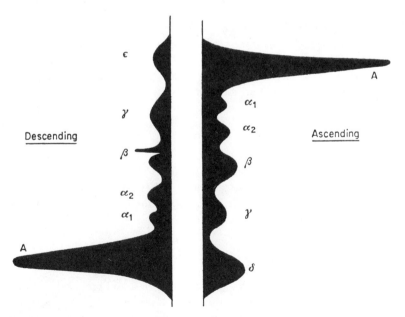

Figure 17.4. Descending and ascending patters of a serum sample photographed with a schlieren scanning camera. Reprinted from Ref. 20, by courtesy of Academic Press, Inc., London.

peak. The specific refractive index increment of most proteins is about 0.00188, defined as $(n_{\text{solution}} - n_{\text{solvent}})/p$, where the concentration p is expressed in grams dry weight of protein per 100 ml solution.

Interference methods depend on the retardation of the wave front as it passes through two solutions of different optical density. They are considerably more sensitive to low protein concentrations than the schlieren methods. A typical comparison of schlieren and interference recordings for a dilute protein sample is presented in Fig. 17.5. Quantitation of the interference patterns is obtained by simple counting of fringes, or fractional fringes, within a boundary.

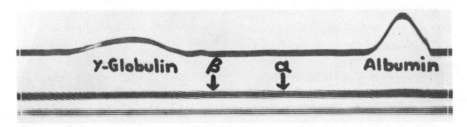

Figure 17.5. Comparison of interferometric and schlieren patterns of a dilute albumin–globulin mixture. Traces of α and β-globulins are visible only by interferometry.

Before electrophoresis, the protein sample is equilibrated against the buffer by prolonged dialysis. Nevertheless, the pure buffer will usually have slightly higher conductivity than the protein solution, causing the appearance of boundary anomalies. In the ascending limb, the leading, fastest protein is moving into a region of higher conductivity, which tends to slow down the migration velocity. This results in a sharpening of the ascending boundary. The opposite effect is noticed in the descending boundary, where the trailing boundary moves into a region of lower conductivity, resulting in a diffusion of its pattern. In addition, the original discontinuity of electrolyte concentration across the boundary causes salt boundaries to appear in the cell, which are not related to the presence of a protein. These two boundaries, denoted as δ in the ascending and ε in the descending limb, remain relatively stationary during electrophoresis.

17.33   ZONE ELECTROPHORESIS (1)

The term zone electrophoresis refers to a group of loosely related techniques, characterized by complete separation of components into separate zones. This is in distinction to moving-boundary electrophoresis, where complete separation is not achieved, and the analysis is based on movement of zone boundaries only. Zone techniques have become the most commonly used ones for both analytical and preparative purposes and are applicable not only to proteins but also to all other charged soluble substances. Their advantages are many: excellent resolution, extreme simplicity of techniques and equipment, speed, capability for handling many samples simultaneously, possibility of identifying specific components by staining characteristics and/ or biological activity, and so on.

The first technique of this type was electrophoresis on filter paper, which developed parallel to paper chromatography. Both of these techniques had an enormous impact on biochemistry. In particular, the resolution of serum proteins on paper electrophoresis was in all respects as good as the results with moving-boundary electrophoresis. Consequently, it became evident that the sophistication of the Tiselius cell was not indispensable, and that the main problems involved stabilization of boundaries, adequate heat dissipa- tion, and visualization of separated components.

Filter paper—and, by extension, other porous media—proved to provide an ideal solution to these problems. Boundary stabilization is automatic, as within its capillary spaces gross hydraulic flow is prevented by viscous forces, while molecular processes, such as diffusion and electrophoretic migration, proceed unhindered. The small cross section of the paper provides for easy cooling and makes high-voltage separations possible. Finally, visualization is achievable by simple staining techniques, rather than complex optical systems. Stains specific for proteins, lipids, carbohydrates, and other com- pounds permit further characterization of the separated components or detection of enzymic activity.

It is difficult to estimate absolute mobilities from zone electrophoretic migration velocities. The electrical charge of the supporting media is usually not negligible and may cause considerable electro-osmotic flow of buffer, usually in the direction opposite to the migration of the negatively charged solutes. This can be ascertained by using markers of electrically neutral solutes, such as glucose or dextran, not affected by the electrical current. Evaporation of buffer from exposed supporting media may also cause con- vective flows, as well as contribute to local changes of conductivity, and molecular sieving too can affect the migration rate. Moreover, there are some simple structural parameters to be considered, such as the ratio of the volume of solid support to total volume, and the effective path length of migration if the nature of the supporting medium imposes a tortuous path- way on the migrating particles. These parameters have been considered at length by Kunkel and Trautman (10), Waldmann-Meyer (11), and others.

Electrophoresis on Filter Paper or Cellulose Polyacetate Membranes (12)

These techniques are used mainly for routine analyses of protein solutions, where high resolution is not necessary. The most common apparatus for filter paper is the Durrum cell, illustrated in Fig. 17.6. A series of paper strips

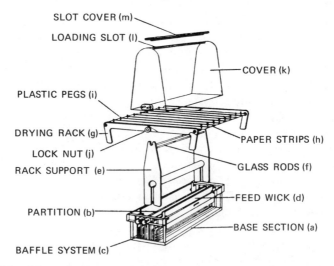

SLOT COVER (m)
LOADING SLOT (l)
COVER (k)
PLASTIC PEGS (i)
DRYING RACK (g)
PAPER STRIPS (h)
LOCK NUT (j)
RACK SUPPORT (e)
GLASS RODS (f)
FEED WICK (d)
PARTITION (b)
BASE SECTION (a)
BAFFLE SYSTEM (c)

Figure 17.6. Schematic diagram of the Durrum paper electrophoresis cell. In use, the paper strips on the drying rack are folded over the rack support and come into contact with the feed wicks. Reprinted from Ref. 12, by courtesy of Academic Press, Inc.

is draped over central glass rods, the ends being in contact with wicks bathing in the electrode buffer. The most common buffer is 0.05 N veronal, pH 8.6, similar to that used in Tiselius cell.

Cellulose acetate membranes have largely displaced filter paper as a support medium. Their main advantages are (a) even simpler equipment is sufficient; (b) more uniform porosity results in sharper boundaries and permits better resolution and shorter migration times; and (c) they can be

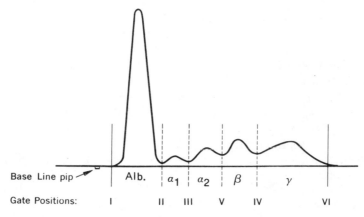

Base Line pip   Alb. α_1 α_2 β γ

Gate Positions: I   II   III   V   IV   VI

Figure 17.7. Densitometer tracing of a serum pattern on cellulose acetate membrane.

rendered completely transparent after staining, thus permitting more accurate quantitation. These membranes are somewhat delicate to handle, being brittle when dry, and limp when wet, and the recent tendency is to use them in a combined form, the membrane being laminated on a rigid sheet of transparent plastic. A variety of densitometers are commercially available for the quantitation of the stained and cleared membranes. Most convenient are instruments that give automatic integration of areas under each peak. A typical pattern of normal human serum is shown in Fig. 17.7.

High-Voltage Electrophoresis (3, 13)

Filter paper has retained its unique usefulness as a supporting medium for high-voltage electrophoresis. High voltage is particularly useful for separations of low-molecular-weight substances, such as amino acids, peptides, and many other organic and inorganic substances having charged groups. The application of high voltage requires maximum cooling of the paper, which can be accomplished in two ways. In flat-plate electrophoresis the paper is sandwiched between two flat plates, at least one of which is mechanically cooled. Typically, the bottom plate may be an insulated metal block with cooling coils; the top, a glass plate. In tank electrophoresis, the filter paper is immersed during electrophoresis in a cooled liquid medium, usually a hydrocarbon. This type of apparatus is illustrated schematically in Fig. 17.8. The voltage applied may be as high as several thousand volts, and the paper sheets as large as 18 × 46 in.

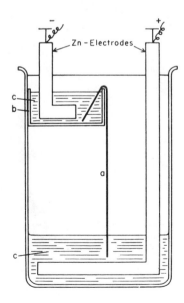

Figure 17.8. High-voltage electrophoresis apparatus of the tank type. The filter paper (*a*) is suspended in carbon tetrachloride, and its ends are dipping into the electrode compartments (*c*). Reprinted from Ref. 3, by courtesy of Academic Press, Inc.

The list of substances amenable to high-voltage analysis is too long to be reviewed here (13). One particular area of usefulness of high-voltage electrophoresis is in the "fingerprinting" of proteins. In this procedure, a protein hydrolyzate, consisting mainly of amino acids or low polypeptides, is subjected to two-dimensional separation, usually by electrophoresis in one direction of the filter paper, and chromatography in the other. A uniquely characteristic distribution of amino acids or peptides is obtained on the paper, as illustrated in Fig. 17.9. This procedure is particularly useful to pinpoint differences between two similar proteins—hence the term fingerprinting.

A recent improvement in this field is thin-layer electrophoresis (TLE), in which separations are carried out on cellulose or silica-gel-coated glass plates.

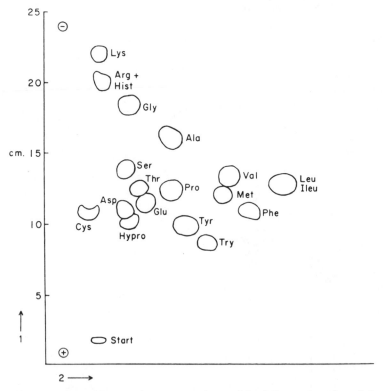

Figure 17.9. Amino acid separation on paper by combined electrophoresis and chromatography: 1 = electrophoresis at pH 1.9, 60 V/cm, 90 min; 2 = ascending chromatography in *sec*-butanol–formic acid-water (75:15:10). Reprinted from Ref. 3, by courtesy of Academic Press, Inc.

The glass plates rest on a cooled surface, and a voltage of about 900 V is impressed across. The main advantages of TLE are reduction in size of plates (8 × 8 in.), increase in speed of resolution, and higher sensitivity.

Electrophoresis on Granular Support Media (10)

This form of electrophoresis is most commonly used for preparative purposes. A slab or block of packed granules is prepared by pouring a slurry of granules onto a flat plate to form a uniformly thick layer imbibed with buffer. The granular material may be powdered cellulose, granular starch, synthetic resins such as polyvinyl chloride, or small glass beads. After electrophoresis, the positions of the various bands are located by staining a narrow strip of the block, and then the rest of the block is physically cut and separated. The components can be effectively recovered from the segmented sections by simple filtration.

Immunoelectrophoresis (14, 15)

Immunoelectrophoresis is usually carried out in agar gel. Although the resolving power of electrophoresis is no better on agar than on cellulose acetate, electrophoresis becomes a tool of most exquisite sensitivity when combined with immunodiffusion. The principle of immunoelectrophoresis is schematically illustrated in Fig. 17.10. A layer of agar, 1 to 2 mm in thickness, is poured onto a glass plate, frequently a microscope slide, and allowed to solidify. The protein sample is introduced into a small well cut in the agar, and upon electrophoresis it forms a rather ill-defined pattern, illustrated in Fig. 17.10b. A trough is then cut parallel to the path of the migrated proteins and is filled with a suitable antiserum, and the slide incubated for 24 to 48 hr in humid atmosphere. Immunoprecipitin arcs are formed, each arc corresponding to a pair of antigen-antibody proteins present in the system. These arcs are visible directly but can also be stained for greater emphasis or differentiation. In electrophoresis of human serum, the antiserum can contain antibodies against all components of the serum, resulting in a bewildering array of arcs, illustrated schematically in Fig. 17.11. Very often, however, there is advantage in using a monovalent antiserum, reacting with one specific protein only.

The above description illustrates only one of the many ways in which electrophoresis and immunodiffusion can be used in concert. Agarose gels, prepared from a purified agar fraction, are preferred in some instances. Cellulose acetate membranes can also be used for immunoelectrophoresis, but as a rule offer poorer resolution.

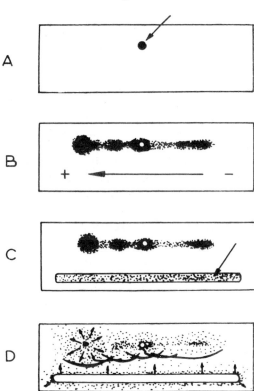

Figure 17.10. The principle of immunoelectrophoresis. (*a*) The antigen mixture (serum) is introduced into a well punched out in a layer of buffered agar gel. (*b*) Application of the current (4 V/cm) effects electrophoretic separation of the components of the mixture. (*c*) Antiserum is introduced into a longitudinal trough alongside the path of electrophoretic migration. (*d*) Antigens and antibodies meet by diffusion through the gel and form precipitin patterns of the Ouchterlony type. Reprinted from Ref. 22, by courtesy of Elsevir Publishing Co.

High-Density Gel Electrophoresis (16)

It is possible to prepare high-density gels with effective pore sizes approaching the dimensions of the protein molecule one wishes to separate. In such gels, small molecules migrate unhindered; the bigger the molecule, or the more asymmetric, the greater is the retardation due to molecular sieving. This results in great enhancement of the resolving power, protein molecules being characterized by widely differing parameters of size and shape. This principle of separation is also used in gel filtration chromatography (see Chapter 15).

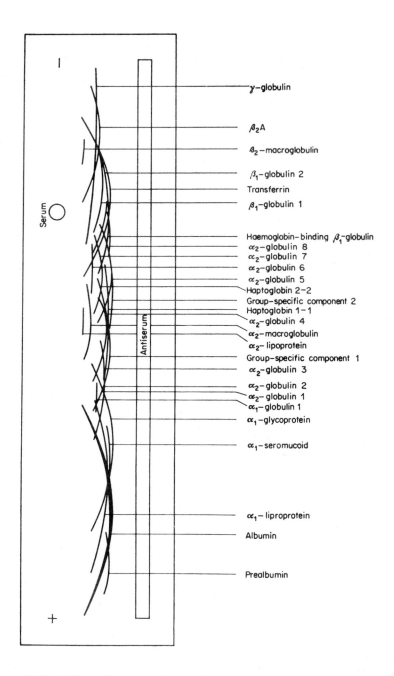

Figure 17.11. Schematic presentation of the precipitin arcs obtained on immuno-electrophoresis.

517

The first high-density gels were prepared from specially hydrolyzed starch, up to 12 to 15% in content. Because of their opaque nature, quantitation was difficult. Their resolving power was comparable to that obtainable with immuno-electrophoresis, as shown in Fig. 17.12. Presently in more widespread use are gels of polyacrylamide, a synthetic polymerization product of complete transparency and clarity. Because of limited stability, performed gels are not commercially available, and each user has to prepare them individually. Concentration of acrylamide can vary from 3 to 30%, and varying degree of cross linking can be obtained by adding a divalent monomer, N,N'-methylene-bisacrylamide. Polymerization can be induced by photochemical means or by the addition of peroxides. The gels can be prepared in vertical troughs, similar to those used for starch gel electrophoresis. More often, however, the gels are formed in small glass tubes, each tube accommodating only one sample. The protein components appear in the gel as a series of stacked discs, and the method is aptly referred to as disc electrophoresis. The amazing resolving power of this technique is illustrated in Fig. 17.13. Polyacrylamide electrophoresis is also adaptable for preparative purposes, yielding fractions of excellent purity, but in small quantities only.

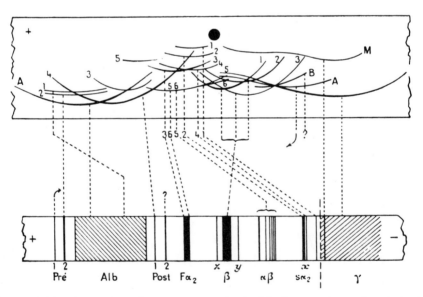

Figure 17.12. Comparison of serum patterns obtained by immunoelectrophoresis (top) and starch gel electrophoresis. Because of molecule sieving, the relative positions of some α and β-components are changed in starch gels. Reprinted from Ref. 16, by courtesy of Academic Press, Inc.

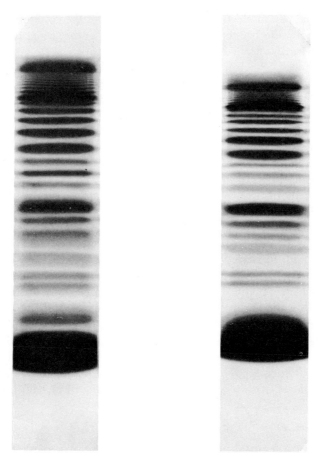

Figure 17.13. Disc electrophoresis patterns in polyacrylamide gels of two serum samples (Canalco, Rockville, Md.).

17.34 COLUMN ELECTROPHORESIS—ISOELECTRIC FOCUSING (17)

Stabilization of protein boundaries in a liquid column can be achieved by generating within the column a density gradient by means of an electrically neutral solute, usually sucrose. The sucrose gradient, to be effective, has to be steeper than the gradient caused by the protein-buffer boundaries. This usually requires rather high concentrations of sucrose, up to 50%. The columns may be glass-jacketed, and a refrigerant circulated through the annular space. Upon electrophoretic fractionation, the column can be emptied into a suitable fraction collector, and the sucrose separated from the purified protein fractions by dialysis or ultrafiltration.

This version of electrophoresis has acquired greatest usefulness in isoelectric focusing, where a pH gradient is superimposed on the density gradient. Each protein component will migrate electrophoretically till it encounters the pH of its isoelectric point. At that level within the column, it will cease to migrate, as, by definition, it is not affected by the electrical field at the isoelectric condition. The pH gradient cannot be formed artificially, by simply mixing suitable electrolytes, as such a procedure will result in continuous migration of the electrolytes as well, and no stable pH gradient will be formed. Instead, it is necessary to generate a natural pH gradient; such a gradient is obtained by prolonged electrolysis of a suitable mixture of amphoteric substances, ampholytes, having a wide range of isoelectric points. As a result of the electrolysis reactions at the electrodes, the anode will become most acid, the cathode most alkaline, with a continuous pH gradient in between. Each species of ampholyte present will accumulate at its respective isoelectric point. The final shape of the pH gradient in the column will depend on the nature of the ampholytes present, and the LKB Instrument Corporation has developed a special mixture called Ampholine, containing a large number of aminocarboxylic acids, in the molecular range of 300 to 600, covering the most useful pH range (pH 3 to 10). Used in combination with sucrose, it makes possible the formation of stable pH and density gradients, within which isoelectric focusing of protein mixtures can be carried out with ease.

17.35 CONTINUOUS-FLOW CURTAIN ELECTROPHORESIS (18)

This technique of preparative electrophoresis was a logical extension of paper electrophoresis. It is schematically illustrated in Fig. 17.14. A continuous downward flow of electrolyte is established on a hanging curtain of filter paper, and the sample to be separated continuously added as a small spot in the flowing buffer. The electrical field is perpendicular to the flow of buffer, and the components are separated, forming an angular pattern. The angle of deflection α is obtained from the following relation:

$$\text{Tangent } \alpha = \frac{\text{rate of electrophoretic migration}}{\text{rate of buffer flow}}$$

The separated components are collected at the drip points from the scalloped bottom edge of the paper. The main drawback of the method is the relatively

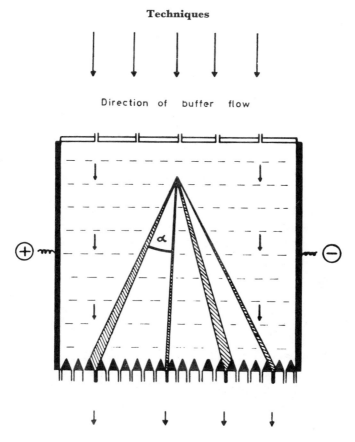

Direction of buffer flow

Figure 17.14. Principle of continuous flow-curtain electrophoresis. Protein is continuously applied on top center of the paper, and components are separated tangentially. Reprinted from Ref. 18, by courtesy of Academic Press, Inc.

limited carrying capacity for the buffer of most papers. Nevertheless, for a while this method enjoyed popularity, and a number of instruments were commercially available.

It has been largely displaced by a more sophisticated arrangement whereby separation is obtained within a free film of buffer, without any supporting medium. This thin film is contained between two parallel glass plates and is stabilized by viscosity only. The schematics of such an apparatus are illustrated in Fig. 17.15. With such free-flowing films, greater throughput can be achieved than on paper. Moreover, the method lends itself also to separation of particulate matter, for example, mitochondria and microsomes, or even bacteria, and red and white cells.

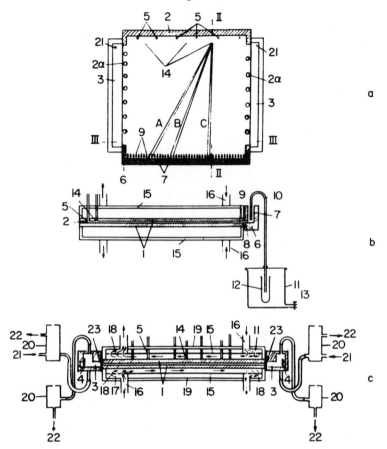

Figure 17.15. Cross section of the separating chamber of the free-flowing curtain type electrophoresis apparatus. *1* = Glass plates (500 × 500 mm); *2* = gaskets; *3* = electrode compartments; *4* = Pt electrodes; *5* = buffer ports; *6* = exit ports; *7, 8* = receiving capillaries; *9, 10* = suction tubes; *11* = storage cabinets; *12* = collection tubes; *13* = magnetic valve; *14* = sample entrance ports; *15* = cooling jacket; *16* = cold air ducts; *17* = cold air distribution vents; *18* = opening for *17*; *19* = insulation; *20* = overflow tubes; *21, 22* = entrance and exit apertures. Reprinted from Ref. 18, by courtesy of Academic Press, Inc.

17.36   ELECTRODECANTATION AND FORCED-FLOW ELECTROPHORESIS (4, 5)

The phenomenon of electrodecantation may occur whenever a semipermeable membrane is interposed into the pathway of electrophoretically migrating colloids. Semipermeable membranes, such as cellophane or parchment paper, allow free passage of electrolyte ions but retain colloids by virtue of their greater particle size. The migrating colloids accumulate, therefore, in front of the membrane, where they form a layer of increased concentration. This colloid-enriched layer will either decant to the bottom or float to the top, depending on whether the colloid is heavier or lighter than the solvent. In a static cell, eventually all the charged colloid will be stratified at the bottom or top, in a concentrated form. Isoelectric components, if present in the original colloid mixture, will not decant, as by their very definition they are not affected by the electrical field. The technique can be used, therefore, either for the purification of isoelectric components in a mixture or for the concentration of electrically charged components. The efficiency of the process depends on the number of membranes in a given volume, and these are frequently stacked at close proximity to each other. A commercial version of such a multimembrane electrodecantation apparatus, designed by Polson, is shown in Fig. 17.16. Its membranes are divided into three sections and are immersed in a cooled buffer for better heat dissipation.

Electrodecantation is capable of very large throughputs and has been studied most extensively as a possible substitute for centrifugation in the creaming of natural rubber latex. The installation in a rubber plantation in

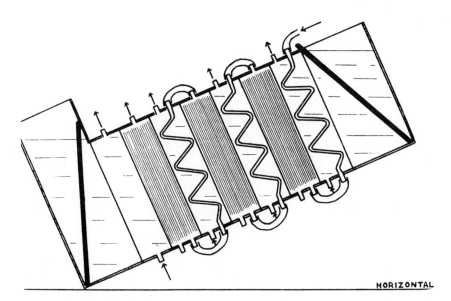

Figure 17.16. Multimembrane electrodecantation apparatus of Polson. The supernatant from the first separation section is passed into the bottom of the second, and the supernatant of this section into the bottom of the third. The supernatant from this section passes into a receiver. Reprinted from Ref. 4, by courtesy of Academic Press, Inc.

Malaya was probably the most ambitious attempt to industrialize the electrophoretic process (4). Electrodecantation is also useful for the purification of
certain protein fractions, particularly γ-globulin from human serum, where
the desired purified end product can be rendered isoelectric. The applications
to protein purification have been studied most extensively by Kirkwood and
his associates, using a slightly modified technique known as electrophoresis-
convection (4).

Forced-flow electrophoresis is a newer and more versatile technique,
capable of fractionations similar to those obtained with electrodecantation,
but applicable also to other electrokinetic membrane processes, such as
electrodialysis, electro-osmotic concentration, or electro-filtration (5) (see
Chap. 16). In its simplest configuration, it is illustrated in Fig. 17.17. Like
electrodecantation, it utilizes a system of parallel semipermeable membranes,
except that a filter element is introduced between each two membranes,
separating the intramembrane spaces into input and output compartments.
The colloidal solution to be processed is continuously circulating through the
input compartments, and part of the liquid is forced to flow through the filter
into the output compartments. The liquid flowing through the filter is exposed
to the electric field, forcing the colloids to migrate in countercurrent direction.
In Fig. 17.17 this is shown by the broken and solid arrows, respectively. If the

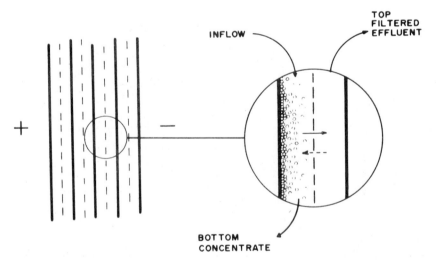

Figure 17.17. Schematic side view of a forced-flow electrophoresis cell assembly.
Solid lines = semipermeable membranes; broken lines = microporous filters; solid
arrows = direction of liquid flow; broken arrow = direction of electrophoretic migration.
Particles are presumed to be negatively charged. Reprinted from Ref. 5, by courtesy of
Plenum Publishing Corporation.

electrophoretic vector is equal to or larger than the linear rate of liquid flow in the opposite direction, no charged colloid will go through the filter. The filter acts also as a frictional boundary between the downward-flowing layer of increased colloid density and the upward-flowing colloid-impoverished layer. This renders the whole process independent of the unpredictable element of decantation, and complete control of all operational parameters is possible through appropriate regulation of flow rates.

Forced-flow electrophoresis has been shown to be applicable for large-scale water and waste purification through direct removal of colloid impurities. Continuous filtration of algae, clay, and other gelatinous materials is possible, without any formation of filter cake, its deposition being prevented by the electric field. The technique has been also applied to a number of biological processes, such as blood electrodialysis or *in vivo* fractionation of blood proteins (5, 18, 19).

References

1. M. Bier, ed., *Electrophoresis*, Academic Press, New York, Vol. I, 1959; Vol. II, 1967.

2. H. A. Pohl and W. F. Pickard, eds., *Dielectrophoretic and Electrophoretic Deposition*, Electrochemical Society, New York, 1969.

3. T. Wieland, "Applications of Zone Electrophoresis," in *Electrophoresis*, M. Bier, ed., Academic Press, New York, 1959, Vol. I, p. 493.

4. M. Bier, "Preparative Electrophoresis," in *Electrophoresis*, M. Bier, ed., Academic Press, New York, 1959, Vol. I, p. 263.

5. M. Bier, "Electrokinetic Membrane Processes," in *Membrane Processes in Industry and Biomedicine*, M. Bier, ed., Plenum, New York, 1971, p. 233.

6. J. Th. G. Overbeek and P. H. Wiersema, "The Interpretation of Electrophoretic Mobilities," in *Electrophoresis*, M. Bier, ed., Academic Press, New York, 1967, Vol. II, p. 1.

7. H. A. Abramson, *Electrophoresis of Proteins*, Hafner, New York, 1964.

8. C. C. Brinton and M. A. Lauffer, "Microscope Method of Electrophoresis," in *Electrophoresis*, M. Bier, ed., Academic Press, New York, 1959, Vol. I, p. 427.

9. L. G. Longsworth, "Moving Boundary Electrophoresis," in *Electrophoresis*, M. Bier, ed., Academic Press, New York, Vol. I, pp. 91, 137.

10. H. G. Kunkel and R. Trautman, "Zone Electrophoresis," in *Electrophoresis*, M. Bier, ed., Academic Press, New York, Vol. I, p. 225.

11. H. Waldmann-Meyer, *Chromatogr. Rev.*, 5, 1 (1963).

12. C. Wunderly, "Paper Electrophoresis," in *Electrophoresis*, M. Bier, ed., Academic Press, New York, 1959, Vol. I, p. 179.

13. R. Clotten and A. Clotten, *Hochspannungselektrophorese*, Georg Thieme Verlag, Stuttgart, Germany, 1962.

14. L. P. Cawley, *Electrophoresis and Immunoelectrophoresis*, Little, Brown, Boston, 1969.

15. P. Burtin and P. Grabar, "Nomenclature and Identification of the Normal Human Serum Proteins," in *Electrophoresis*, M. Bier, ed., Academic Press, New York, 1967, Vol. II, p. 109.

16. H. Bloemendal, "High Resolution Techniques," in *Electrophoresis*, M. Bier, ed., Academic Press, New York, 1967, p. 379.

17. H. Haglund, *Sci. Tools*, **14**, 17 (1967).

18. K. Hannig, "Preparative Electrophoresis," in *Electrophoresis*, M. Bier, ed., Academic Press, New York, 1967, Vol. II, p. 423.

19. M. Bier et al., *Trans. Am. Soc. Artificial Internal Organs*, **16**, 325 (1970).

20. D. J. Shaw, *Electrophoresis*, Academic Press, New York, 1969.

21. H. E. Schultze and J. F. Heremans, *Molecular Biology of Human Proteins*, Elsevier, 1966.

List of Symbols

a	radius of sphere	R	specific conductivity of buffer
A	cross-sectional area	U	electrophoretic velocity
e	electronic charge	Z	valence
E	potential gradient		
f_c	friction coefficient	α	angle of deflection
h	height of electrophoretic channel	δ	solute boundary in the ascending limb
I	electric current	ε	solute boundary in the descending limb
K_1	electrophoretic attraction		
K_2	Stokes's friction	ε	dielectric constant
K_3	electrophoretic retardation	ζ	zeta potential
K_4	relaxation effect	κ	Debye-Hückel constant
n	refractive index	μ	electrophoretic mobility
n_0	bulk concentration of ionic species	ψ_0	electric potential at the surface
p	protein concentration	ψ_x	electric potential at distance x from the surface
Q	electric charge		

Eighteen

MISCELLANEOUS SEPARATION PROCESSES

In addition to the separation methods of preceding chapters, there exist a large number of separation processes that cannot be discussed in detail within the framework of this book. Some of these separation techniques have only very special applications or are merely laboratory curiosities; others may have great technological significance, for example, sedimentation in waste water treatment. In this chapter a few of those remaining separation methods are discussed that, in the judgment of the authors, deserve mention but do not qualify as subjects of individual chapters. Consequently the separation processes grouped together here lack a common denominator with respect to the underlying physicochemical principles.

18.1 Ultracentrifugation

The ultracentrifuge has become increasingly important as a separation tool in the field of polymer science, since Svedberg introduced it nearly half a century ago for the determination of the particle size of gold sols. This technique has played a central role in the study of proteins and nucleic acids as a tool to obtain physicochemical data. Recent developments indicate, however, that ultracentrifugation is gaining as a separation method on the preparative scale for the isolation and purification of biopolymers, viruses, and cell fragments. According to the instrumentation employed, ultracentrifugation is generally divided into two classes. Analytical centrifuges are

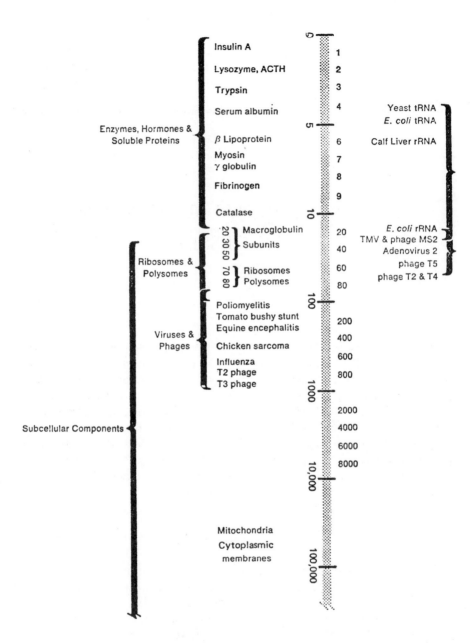

Figure 18.1. Sedimentation coefficients of certain biological particles. The scale is in svedberg units, and the sedimentation coefficients of the individual particles are given by the S_{20}, w values, that is, corrected for water and 20°C. Reprinted by courtesy of Beckman Instruments.

528

used to determine the physical properties of solutes or particles; preparative centrifuges, to separate them on the basis of those physical properties.

The theory of ultracentrifugation is now well developed, and the mathematical treatment can be found in textbooks on physical biochemistry, as well as in the books and reviews listed at the end of the chapter. Therefore only a few basic concepts are reviewed here.

The centrifugal field strength is given by $x\omega^2$, where x is the radial position, and ω is the angular velocity in radians per second. One revolution equals 2π radians. In practice, the centrifugal field generated in a centrifuge is usually expressed in relation to the earth's gravitational field: $x\omega^2/g$.

The centrifugal force is obtained by multiplying the effective mass of the sedimenting particle by the accelerating field. The effective mass is the actual mass minus a buoyancy correction to allow for the solvent mass displaced. Thus the centrifugal force Φ is given by

$$\Phi = (m - m\bar{v}\rho)\omega^2 x = m\omega^2 x(1 - \bar{v}\rho) \tag{18.1}$$

where m is the mass of the particle, ρ is the density of the solution, and $\bar{v} = 1/\rho_p$, the partial specific volume of the particle as given by the reciprocal particle density ρ_p.

It is convenient to define a *sedimentation coefficient* s as the sedimentation velocity per unit of centrifugal field strength:

$$s = \frac{dx/dt}{\omega^2 x} \tag{18.1a}$$

which has the dimension of time. Since s is usually of the order of 10^{-13} sec, it is measured by convention in svedbergs (S); 1 S equals 10^{-13} sec. The sedimentation coefficients, calculated for 20°C and water, s_{20w}, are shown for a few biologically important substances and particles in Fig. 18.1.

The molecular weight M and the diffusivity D of a solute are related to the sedimentation coefficient by the Svedberg equation:

$$M = \frac{sRT}{D(1 - \bar{v}\rho)} \tag{18.2}$$

This equation states that one can calculate the absolute molecular weight of a solute if s, D, and $\bar{v}$ can be measured.

18.11 THE ULTRACENTRIFUGE

An ultracentrifuge provides a high and precisely controlled rotor speed, at uniform temperature. In most modern ultracentrifuges an electric motor drives a rotor situated in an evacuated and cooled chamber. Analytical ultracentrifuges can be operated at speeds up to 72,000 rpm, so that centrifugal fields of several hundred thousand times gravity can be obtained. These instruments have a small cell in the rotor and are equipped with an optical system that facilitates the measurement of the concentration or the concentration gradient of the solute or solutes at different positions in the cell. Analytical ultracentrifuges are used mainly to determine the precise molecular weights of polymers and to gain information on their molecular conformations and dispersities. In addition they can serve for quantitative analysis of mixtures. Preparative centrifuges normally have no optical system, but they have a much higher cell capacity than analytical units. In recent years preparative ultracentrifuges have become the most important tool for the separation of large-molecular-weight biopolymers and cell fragments, particularly as a result of the introduction of density gradient techniques.

Optical systems

The determination of the solute concentration at different radial positions in the cell of the ultracentrifuge requires elaborate optical systems that permit photographic measurements at time intervals or chart recorder readout during centrifugation.

Absorption systems measure light absorption, which is proportional to concentration if Beer's law is valid. Modern systems utilize monochromators so that the optimal wavelength can be selected for a particular component of a mixture. The light is passed through the cell, and a camera lens focuses an image of the cell onto a photographic plate, or the image is scanned directly by a traveling slit or photomultiplier assembly. Because of the very high specific absorptivity of nucleic acids at 260 mμ this system permits detection of concentrations as low as 0.001% of DNA.

The Raleigh interferometer measures differences in refractive index between solution and solvent. In dilute solutions the refractive index difference is proportional to the concentration expressed as weight per volume of solution. A double cell is used, with solvent in one side and solution in the other. An interference pattern is obtained on a photographic plate, allowing a very easy and accurate measurement of differences in the refractive index and hence in the concentration.

Schlieren systems are widely used, particularly for sedimentation velocity measurements, but they have the lowest sensitivity among the three optical systems. The method is based on the schlieren effect. When parallel light passes through a region where there is a refractive index gradient, it is deflected as it would be by a prism and the angle of deflection is proportional to the cell thickness and to the refractive index gradient. By a complex optical arrangement a profile of the refractive index gradient as a function of the axial position in the cell is obtained. Thus the schlieren method is suited for the measurement of concentration gradient, as opposed to the other two methods described, which permit the measurement of solute concentration.

The principle of measurement is depicted in Fig. 18.2. As discussed later, the two basic measurements by analytical ultracentrifuges concern the sedimentation velocity and sedimentation equilibrium. The results obtained by using different optical systems are illustrated by the tracings on the right-hand side of Fig. 18.2. The curves in the upper diagrams, which show the solute concentration as a function of the position, can be obtained directly from light absorption or from the interference pattern. The lower diagrams show the concentration gradient as a function of position as obtained by schlieren systems.

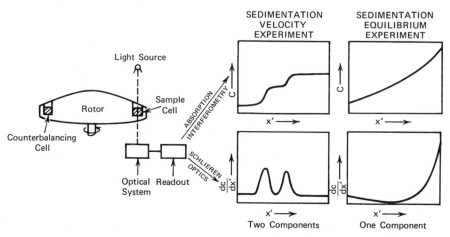

Figure 18.2. Principle of measurements with the ultracentrifuge. Records of sedimentation data as obtained by the different optical systems are illustrated at the right. The two upper graphs show the concentration; the lower graphs, the corresponding concentration gradient as a function of the position in the cell at a given time.

18.12 SEDIMENTATION VELOCITY

When a centrifugal field is applied to a homogeneous solution of large molecules, solute molecules begin to migrate, and a region near the meniscus is cleared entirely. Thus a moving boundary is formed, and by measuring the rate of motion of this boundary, v_b, from Eq. 18.1a, we can calculate the sedimentation coefficient since

$$v_b = \frac{dx_b}{dt} = x_b\omega^2 s \tag{18.3}$$

where x_b is the position of the boundary at time t. By integrating Eq. 18.3 we obtain

$$\ln \frac{x_b(t)}{x_b(t_0)} = \omega^2 s(t - t_0) \tag{18.4}$$

where $x_b(t)$ is the radial position of the boundary at time t. The sedimentation coefficient is conveniently obtained graphically when $\ln [x_b(t)/x_b(t_0)]$ is plotted against $t - t_0$. The slope of the straight line is given by $\omega^2 s$, and ω is known. The sedimentation coefficient is, however, dependent on the concentration C and it has been found empirically that the relationship can be approximated by

$$s = \frac{s^0}{1 + kC} \tag{18.5}$$

where s^0 would be the value at zero concentration, and k is an empirical constant that is always positive and greater for synthetic polymers than for globular proteins. Thus s^0 may be determined from the intercept of a $1/s$ versus C plot. From the slope of the curve the shape of the solute molecule can be inferred. The effect of solvent viscosity, density, and temperature should also be taken into account, in order to obtain comparable sedimentation coefficients. Usually s^0 is converted to $s^0_{20,w}$, that is, to the value of the sedimentation coefficient that would be obtained in a solvent having the viscosity and density of water at 20°C.

If the diffusivity of the solute were zero, the boundary would be flat as it traverses the cell. Such an ideal situation is approximated only by very large molecules at high rotor speed. The actual shape of the boundary is influenced by diffusion resulting from the concentration gradient at the boundary and by the polydispersity of the solute. Both cause the boundary to become less sharp; the separation of these effects is possible but rather difficult. On the other hand, the concentration dependence of s has the opposite effect because in most cases the solute molecules in the low-concentration region of the boundary sediment faster than those in the high-concentration region. Thus the resulting boundary sharpening may be misleading because the shape of the boundary suggests a lower diffusivity and more uniform dispersity for the solute than is actually the case. When a mixture of two or more solutes is investigated, and their sedimentation coefficients are mutually concentration dependent, each component sediments more slowly in the presence of the other than it would alone. Therefore an erroneously low s value is observed for the faster component, which is always passing through a solution of the slower component. In turn the slow component tends to pile up behind the fast boundary and thus behaves as if its sedimentation coefficient were smaller than the actual value. Since most optical techniques show only the total concentration change across each boundary, this Johnston-Ogston effect can cause large errors in the quantitative analysis of mixtures.

Nevertheless, measurements of sedimentation velocity not only can provide reliable physical data for macromolecular or supermolecular structures, but also represent a method for separation and quantitative analysis.

18.13   SEDIMENTATION EQUILIBRIUM

Sedimentation occurs until a state of equilibrium is reached, that is, the potential of each species becomes uniform in the cell. Then at equilibrium in such a system, subjected to a field of force, the total potential of the solute is the same at every point. The total potential μ^* is the sum of the chemical potential μ and the potential energy per mole. In a centrifugal field this is given by

$$\mu^* = \mu - \tfrac{1}{2}M\omega^2 x^2 \tag{18.6}$$

where M is the molecular weight. Equilibrium is reached when

$$\frac{\partial \mu^*}{\partial x} = \frac{\partial \mu}{\partial x} - M\omega^2 x = 0 \tag{18.7}$$

and under such conditions there is no net flow of solute. In practice, equilibrium measurements are carried out at relatively low rotor speeds, in order to avoid accumulation of solute at the bottom of the cell.

Sedimentation equilibrium experiments allow very precise molecular weight determination of large molecules, since diffusivity and molecular shape do not affect the results. For a single homodisperse solute the molecular weight is expressed under ideal conditions by

$$M = \frac{2RT \ln (C_1/C_2)}{\omega^2(1 - \bar{v}\rho)(x_2{}^2 - x_1{}^2)} \tag{18.8}$$

where x_1 and x_2 are different radial positions in the cell, and C_1 and C_2 are the respective solute concentrations. Hence, by measuring the concentration as a function of x, the molecular weight can be determined, as Eq. 18.8 predicts that a plot of $\ln C$ versus x^2 should give a straight line.

In the literature on ultracentrifugation cited at the end of this chapter a number of methods are described for using sedimentation equilibrium measurements to determine molecular weight in practical situations. A notable exception is the Archibald approach to equilibrium method. It is based on the fact that throughout any sedimentation experiments there is no flow at the meniscus and at the cell bottom; therefore the ratio of the concentration gradient to the product of the radial position and the concentration is a constant at these two points at all times at fixed ω. This enables one to determine the molecular weight rapidly by measuring the concentration gradients by schlieren optics and the concentration by another optical system.

18.14   ZONAL ULTRACENTRIFUGATION

It has already been mentioned that a mixture can be analyzed for its components by differential centrifugation in a sedimentation velocity experiment. Solutes that are initially uniformly distributed sediment with different velocities according to their s values. The sedimentation process is most conveniently followed by the progress of the schlieren peaks. This method, which resembles frontal chromatography (Section 5.22) is not suitable, however, for the separation of all components, because only the slowest component can be recovered in pure form and the analytical results may be in error because of the Johnston-Ogston effect.

By using density gradient techniques, however, it is possible to obtain distinct solute bands for both analytical and preparative purposes. In these methods the medium, in which the solutes sediment, has a density gradient due to a concentration gradient of an inert substance. The separation is achieved by one of the following three techniques.

Equilibrium Isodensity Method

Initially, the sample is uniformly distributed in a solution of a low-molecular-weight salt, such as CsCl. Upon centrifugation a density gradient is established, that is, the density of the solution increases from the top to the bottom of the cell because of the sedimentation equilibrium distribution of the salt. Solute molecule bands move to the positions that correspond to the density of the molecules and associated solution components, that is, the macromolecular solutes occupy their isodensity positions.

Rate Zonal Method

This differential migration technique resembles elution chromatography. The cell or tube is filled with solvent–medium so that a density gradient is preformed. A frequently used medium is sucrose solution. The gradient is formed by decreasing the sucrose concentration from the bottom to the top of the tube; then a layer of the sample solution is placed at the top, and spinning begins. The components of the mixture sediment in zones roughly in the order of their sedimentation coefficients. The density gradient serves to give sharp bands by preventing convectional mixing.

Isopycinic or Preformed Gradient Isodensity Method

A gradient that covers the range of particle densities of interest is preformed. During spinning the particles move up or down until they come to rest at the position where the particle density matches the density of the medium.

Figure 18.3 shows separations obtained by density gradient ultracentrifugation and illustrates the suitability of this technique for separating macromolecules. The resolution of unlabeled and ^{16}N-labeled DNA molecules played a particularly important role in determining the mechanism of DNA replication and verifying the Watson-Crick DNA model. Zonal ultracentrifugation constitutes one of the most important tools of molecular biology and is also used widely for the separation of viruses.

Zonal ultracentrifugation has significant advantages in analytical work and is frequently performed using analytical cells. Actually, the equilibrium isodensity method is confined to analytical cells. The main analytical advantages are as follows: (*a*) preliminary dialysis is not needed; (*b*) the individual components of multicomponent systems are physically separated, and no Johnston-Ogton effect occurs; and (*c*) less sample is needed.

The advantages of density gradient ultracentrifugation arise from the actual physical separation of the components, which cannot be achieved otherwise. By scaling up the rotor the loading capacity of the technique can be raised to a reasonable level. A variety of rotors have been used in preparative density gradient ultracentrifugation. Most commonly angle rotors and swing-out bucket rotors, which are depicted in Fig. 18.4, have been employed. A variety of devices have been described for producing density gradients; these are similar in principle to those used in gradient elution chromatography. The detection and collection of the separated zones can be achieved by a variety of means. The tube content can be withdrawn layer by layer into a fraction collector; or, when thin cellulose nitrate tubes are used, the content can be sliced upon freezing. By piercing the base of the tube with a fine syringe needle the liquid can be withdrawn through the flow cell of an analyzer, for example, a scanning spectrophotometer. Alternatively the tube content can be displaced by pumping a dense solution from the bottom, using a precision metering pump.

The scaling up of zonal ultracentrifugation was greatly advanced by the introduction of the Anderson rotor, which is shown schematically in Fig. 18.5. Fluid lines to the center and the outer edge of the rotor consisting of four sector-shaped compartments facilitate filling and emptying while spinning. The rotor is first loaded with the preformed gradient at low speed (3000 rpm), starting with the light end of the gradient being introduced to the rotor edge, followed by a "cushion" of a denser solvent to expel air. Then the direction of fluid flow is reversed, the sample is introduced at the center,

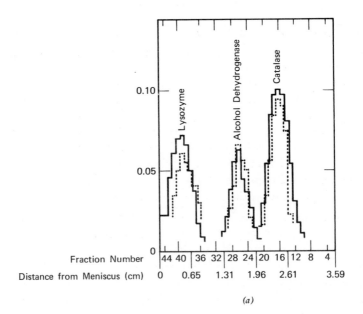

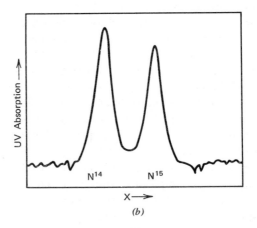

Figure 18.3. Separations by zonal ultracentrifugation. (*a*) Separation of three enzymes by rate zonal ultracentrifugation with sucrose gradient. The solid lines and dotted lines represent enzyme activities, using tris buffer and a crude bacterial extract as the solution medium, respectively. Conditions: 37,700 rpm, 12.8 hr, 3°C. Reprinted from Ref. 1, by courtesy of The American Society of Biological Chemists, Inc. (*b*) Separation of ^{15}N- and ^{14}N-substituted bacterial DNA molecules by equilibrium isodensity ultracentrifugation with CsCl gradient. Reprinted from Ref. 2, by courtesy of the National Academy of Sciences.

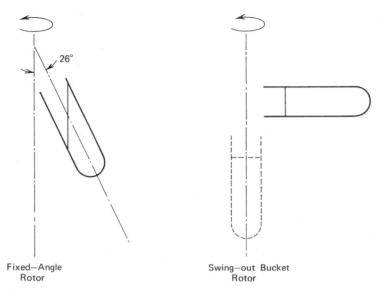

Figure 18.4. Rotor types frequently used in preparative ultracentrifugation.

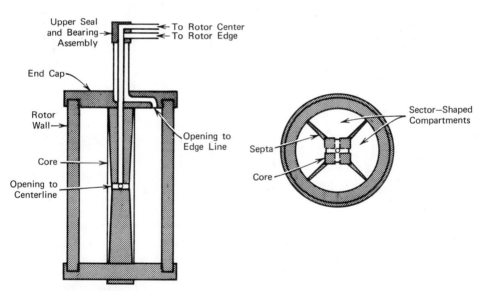

Figure 18.5. The Anderson rotor: schematics of the B IV rotor used at intermediate speeds. Total volume = 1725 ml for use with 1000 to 1200 ml of gradient. The hollow cylinder is driven from below and rotated about the core axis. The upper shaft of the rotor extends through the upper bearing, which also serves as a vacuum seal, and terminates in the special enclosed fluid line seal. Reprinted from Ref. 3, by courtesy of Beckman Instruments.

537

and an equal volume of "cushion" is displaced from the rotor edge. The sample is followed by a buffer solution termed "overlay," which is lighter than the sample and pushes the latter into the rotor. The rotor is accelerated to operating speed (up to 40,000 rpm), which is maintained until the desired separation is attained. After the run, the rotor is decelerated and its content is displaced at low speed without mixing by pumping a dense cushion solution to the outer edge. Thus the overlay followed by the sample layer is displaced through an ultraviolet detector into a fraction collector. By stream splitting, a part of the effluent can be pumped through other monitoring devices, such as an automatic analyzer. This last stage of the process is illustrated in Fig. 18.6.

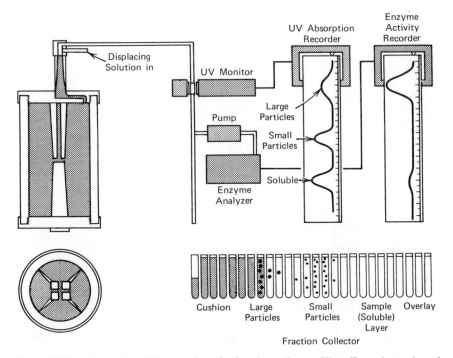

Figure 18.6.   Unloading of the rotor into the fraction collector. The effluent is monitored both by a UV detector and an enzyme analyzer. Reprinted from Ref. 3, by courtesy of Beckman Instruments.

The Anderson rotor can also be used in continuous-flow operation when the sample solution flows through the rotor and the particles of interest sediment into a gradient from which they are fractionated and recovered. Because of the difficulties of sealing the fluid lines at high speeds, however, the operational mode described above is preferred and is used for large-scale separations. The Anderson rotor can be employed for both the rate zonal and the isopycinic methods. Most commonly sucrose solutions are used, but in titanium rotors cesium chloride can also be employed.

Density gradient ultracentrifugation is widely applied for the separation of subcellar fractions, viruses, nucleic acids and polynucleotides, enzymes, and other biological materials. The technique is also useful to determine binding constants, buoyant densities, sedimentation coefficients, and other parameters.

18.2 Particle Classification by Size

The classification of particulate matter is of great technological importance. In the general sense, particle classification is any process whereby particles that have a singular distinguishing property such as size, shape, chemical composition, density, or color are separated. The most widely used classification process is the sizing of particles, that is, their separation into groups having different size ranges. On an analytical scale this technique is used to determine the particle size distribution of a given sample. In technology, particle size separation is employed to obtain particles having a given size range, or to remove particles that are smaller (undersize, fines) or larger (oversize) than a given value. Figure 18.7 shows a comprehensive chart concerning particulate matter of scientific and technological interest.

A primary problem in particle classification is how to define particle size. The answer is simple when the particles have a unique dimension (e.g., spheres). The size of irregular particles is usually characterized by the diameter of equivalent spheres which have either the same sedimentation velocity or the same volume or the same projected area, or just pass through the same square aperture of screens as the irregularly shaped particle.

The efficiency of a binary particle separation process that results in a coarse and a fine fraction is commonly expressed by the so-called Newton classification efficiency η_N, which is given by

$$\eta_N = 100\left(\frac{C_c}{C_S} - \frac{F_c}{F_S}\right) \tag{18.9}$$

Here C_c and F_c are the quantities of true coarse and true fine particles, respectively, in the coarser collected fraction, and C_S and F_S are the quantities of the true coarse and fine particles, respectively, in the starting material. Complete separation of coarse and fine particles yields an η_N value of 100, and no separation gives an η_N value of zero.

The goal of particle separation on an analytical scale is the determination of the particle size distribution, which is a significant characteristic of a particulate sample. There is a great diversity in the representation of the

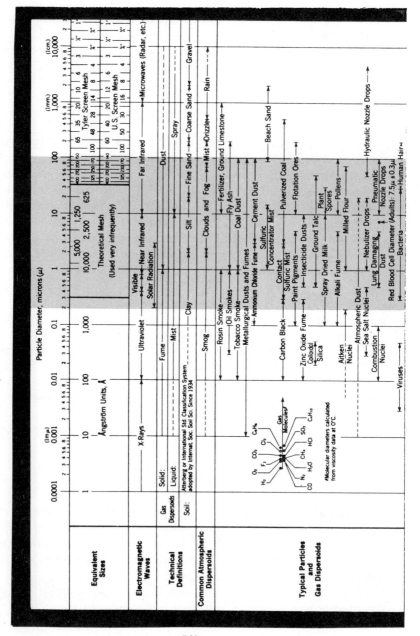

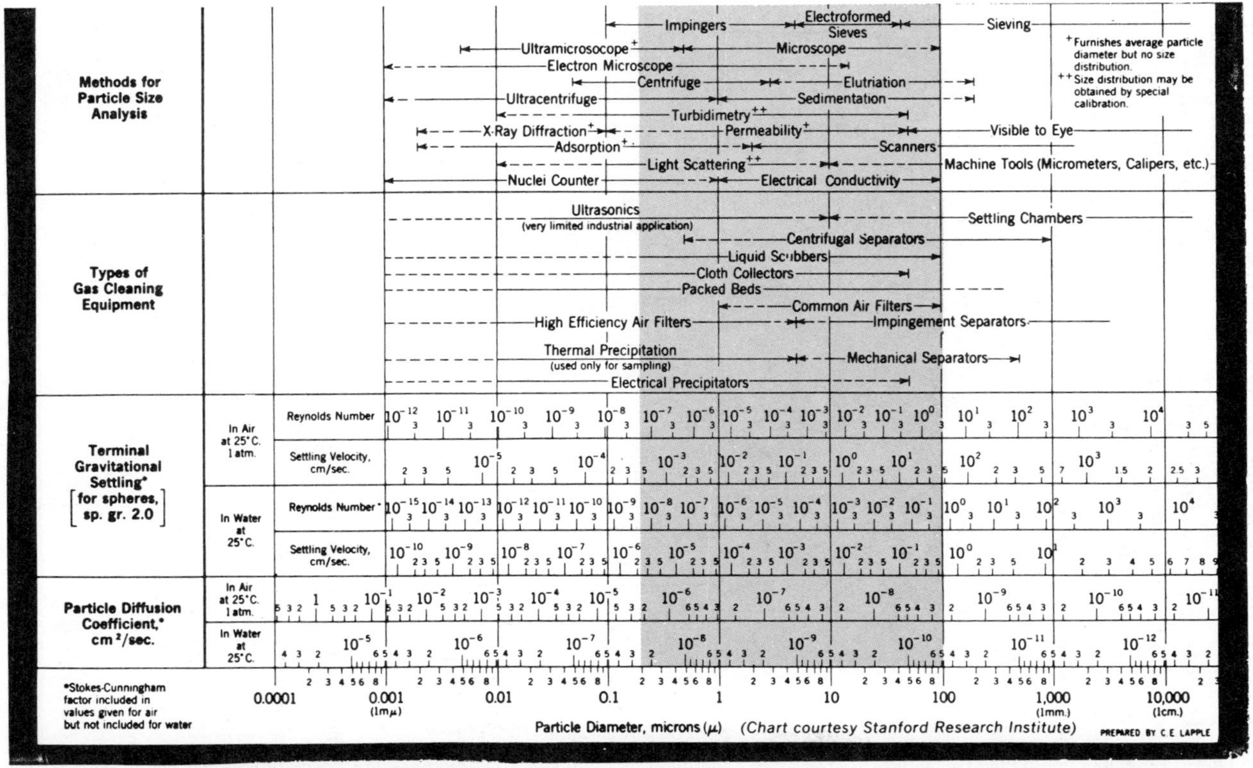

Figure 18.7. Particle characteristics. The shaded area covers the particle size range from 0.02 to 100 μ, representing fine powders.

result of size analysis. Statements such as "not more than 5% above 100 μ" convey little information. Graphical representation is used most frequently. Although plots of frequency against particle diameter would be most informative, cumulative plots are easier to prepare and are employed most commonly. In such plots the cumulative percentage of oversize (or undersize) particles is plotted against particle diameter. The percentage can refer to number, weight, or volume and has to be specified. Such a cumulative plot for glass microspheres is shown in Fig. 18.8. From the cumulative distribution curve the ogives, which are frequently employed to represent the distribution, are easily obtained. The median (50%) diameter and the lower and upper ogives (25% and 75%, respectively) are used most frequently, but sometimes decile values (10, 20, ..., 90% diameters) are also helpful. The median diameter is a measure of the central tendency, and the spread is given by $(d_{75\%} - d_{25\%})/2$.

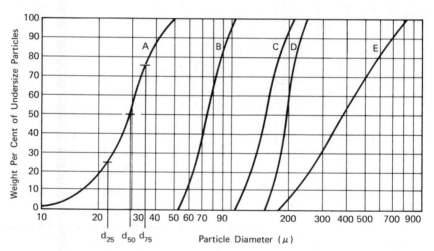

Figure 18.8. Cumulative plots representing the size distribution of fine glass beads. The cumulative undersize in weight per cent is plotted against the particle diameter. For the finest product (A) the ogives (d_{25}, d_{50}, and d_{75}) are also shown. It is seen that the ogives do not give information about the large amount of fines present in A.

A parametric representation of the size distribution is frequently achieved by fitting the actual distribution curve either to the normal (Gaussian) distribution function or the log-normal distribution functions, which are depicted on a cumulative plot of percentage oversized particles versus particle diameter in Fig. 18.9. The Gaussian distribution function is written for particle diameter as

$$N = \frac{\sum N}{\sigma\sqrt{\pi}} \exp\left[-\frac{1}{2}\left(\frac{d - \bar{d}}{\sigma}\right)^2\right] \tag{18.10}$$

$\sum N$ is the total number of particles, where N is the number of particles having diameter d, $\bar{d}$ is the arithmetic mean particle diameter, and σ is the standard deviation. The log-normal distribution function is given as

$$N = \frac{\sum N}{\log \sigma_g \sqrt{\pi}} \exp \left[-\frac{1}{2} \left(\frac{\log d - \log \bar{d}_g}{\log \sigma_g} \right) \right] \qquad (18.11)$$

where the subscript g refers to the geometric mean values. When the actual distribution can be fitted to one of these functions, the particle size distribution can conveniently be expressed by the corresponding mean values and standard deviations.

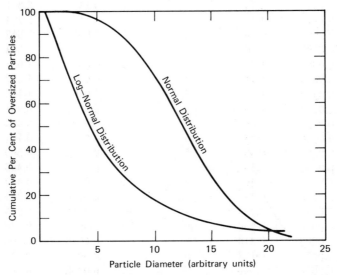

Figure 18.9. Cumulative plots representing normal (Gaussian) and log-normal distributions. Per cent oversize is plotted against the particle diameter.

18.21 SIEVING

Sieving is one of the most common methods for separating particles according to size, both in the laboratory and in industry. It is the most convenient way to grade powders above 38 μ in diameter, by using standard screens; see Table 18.1. Sieving can be extended to 10 μ or less by the use of recently introduced micromesh sieves. A single screen produces only two fractions, of course, but with a series of screens in tandem, with mesh sizes decreasing in the direction of material movement, any desired number of fractions can be obtained.

We shall discuss briefly the fractionation of powders. This is of interest

Table 18.1.   Sieve Specifications

	U.S. Standard Screens				Tyler Screens		
	Aperture		Tolerance on Average Aperture		Aperture		Double Tyler Series
No.	in.	mm	($\pm$ %)	Mesh	in.	mm	Mesh
3	0.265	6.73	3	3	0.263	6.680	
3½	0.223	5.66	3		0.221	5.613	3½
4	0.187	4.76	3	4	0.185	4.699	
5	0.157	4.00	3		0.156	3.962	5
6	0.132	3.36	3	6	0.131	3.327	
7	0.111	2.83	3		0.110	2.794	7
8	0.0937	2.38	3	8	0.093	2.362	
10	0.0787	2.00	3		0.078	1.981	9
12	0.0661	1.68	3	10	0.065	1.651	
14	0.0555	1.41	3		0.055	1.397	12
16	0.0469	1.19	3	14	0.046	1.168	
18	0.0394	1.00	3		0.039	0.991	16
20	0.0331	0.84	5	20	0.0328	0.833	
25	0.0280	0.74	5		0.0276	0.701	24
30	0.0232	0.59	5	28	0.0232	0.589	
35	0.0197	0.50	5		0.0195	0.495	32
40	0.0165	0.42	5	35	0.0164	0.417	
45	0.0138	0.35	5		0.0138	0.351	42
50	0.0117	0.297	6	48	0.0116	0.295	
60	0.0098	0.250	6		0.0097	0.246	60
70	0.0083	0.210	6	65	0.0082	0.208	
80	0.0070	0.177	6		0.0069	0.175	80
100	0.0059	0.149	6	100	0.0058	0.147	
120	0.0049	0.125	6		0.0049	0.124	115
140	0.0041	0.105	8	150	0.0041	0.104	
170	0.0035	0.088	8		0.0035	0.088	170
200	0.0029	0.074	8	200	0.0029	0.074	
230	0.0024	0.062	8		0.0024	0.061	250
270	0.0021	0.053	8	270	0.0021	0.053	
325	0.0017	0.044	8		0.0017	0.043	325
				400	0.0015	0.037	

either to determine particle size distribution (e.g., in quality control) or to obtain a narrow cut of particles (e.g., column packing materials for gas and liquid chromatography).

Standard screens are woven wire sieves having the specifications shown in Table 18.1. The screens of the older Tyler system are designated by the number of meshes per linear inch, whereas a screen of the U.S. Standard Sieve series is designated by a number that does not necessarily correspond to the mesh number of the Tyler screen scale. In practice a sieve fraction is best expressed by the apertures of the lower and upper screens, for example, sieve fraction 88 to 105 μ, instead of 150 to 170 mesh or U.S. No. 140 to 170 (which are equivalent, as seen from Table 18.1).

The screening of small particles is often more difficult than it would appear. The particles may stick to each other, and clogging of the screen can easily occur. Therefore screening with a liquid flow through the screens (wet screening) is often employed. The removal of fines usually takes place rapidly, but "near-mesh" particles are eliminated at a gradually diminishing rate. Different powders exhibit different screening properties, which also depend on the humidity; therefore the optimum loading and the screening time must be determined experimentally.

In the laboratory the following three screening devices are employed most commonly:

1. Ro-tap testing sieve shaker (Tyler Co.), which reproduces the circular tapping motion given testing sieves in hand sieving, but with a uniform, mechanical action. This device can be used for both dry and wet screening.

2. Sonic sifters (Allen-Bradley Co.), which use a vertical oscillating column of air to move the particles through a single sieve or set of sieves. Since the timed mechanical action reduces blinding of the screens and particle agglomeration, this sifter can be used for dry screening of particles smaller than 27 μ.

3. Air-jet sifter (Alpine Co.), which utilizes the principle of a rotating jet blowing air up from beneath an enclosed sieve. This keeps the powder moving on the sieve, while the undersize particles are drawn through the rest of the sieve by the air returning, via a suitable filter, to the suction side of a fan. This device can also be used for dry sieving of particles finer than 37 μ.

18.22   SEDIMENTATION

The sedimentation of particles in a static fluid offers a simple method for size fractionation and is widely exploited for the determination of particle size distribution in the laboratory.

When small particles move through a fluid in laminar motion under the influence of gravity, only their constant (so-called terminal) velocity is of interest, if the settling process is observed for any appreciable time. The force F exerted at terminal velocity upon a spherical particle having diameter d is

$$F = \frac{\pi}{6} d^3 g(\rho_p - \rho) \qquad (18.12)$$

where g is the gravitational constant, and ρ_p and ρ are the densities of the particle and fluid, respectively. The resistance to motion, R, is given by

$$R = 3\pi \, du\eta \qquad (18.13)$$

where u is the velocity, and η is the viscosity of the fluid. Thus the terminal settling velocity u_t of the particle is described by $F = R$, or

$$u_t = \frac{d^2 g(\rho_p - \rho)}{18\eta} \qquad (18.14)$$

which is known as Stokes law. This relationship is valid when the Reynolds number is less than about 0.1, the particles are uniformly dispersed, and their concentration is low enough so that they do not interfere with each other's motion.

In order to apply Stokes law for the determination of particle size distribution, any convectional motion of the fluid due to temperature gradients or disturbances of the normal flow pattern (e.g., wall effects) must be avoided. It should also be considered that Brownian motion may cause deviations from Stokes law when the particle diameter is less than 3 μ.

A variety of methods have been developed for particle size determination by sedimentation. The simplest methods employ a cylindrical vessel filled with a uniform dispersion. Alternatively a layer of dispersion is placed on the top of the pure liquid in the cylinder. After a given time period measurements are made to determine the amounts or concentrations at different levels or at the same level at different times.

For incremental methods of analysis the samples are taken from a very thin layer of suspension in the cylinder and the weight or number of particles is measured. For cumulative methods of analysis the amount settling in a particular plane in a given time, normally at the bottom of the cylinder, is measured. Pipet methods (e.g., the Andreasen pipet), sedimentation columns, and sedimentation balances are frequently used devices for the determination of particle size distribution. Several instruments are available in which the changes in particle concentration at various depths below the surface in a sedimentation cell are followed, for example, by measuring the changes in light transmission. Particle size determination by centrifugal methods is based on the same principle.

Sedimentation is often used in the laboratory to remove fines or coarse particles from a powder. It also can be employed to obtain different fractions. The powder is dispersed in liquid in a tall cylinder and allowed to settle for a time corresponding to the smallest particle size selected. After this time the supernatant is decanted without disturbing the sediment. Ideally, all particles removed by decantation would be smaller than the selected size. In order to obtain a reasonably efficient separation, however, the procedure has to be repeated several times. When this fractional decantation method is used for different limiting particle sizes, a fractionation of the powder can be obtained.

18.23 ELUTRIATION

Elutriation differs from sedimentation in that the fluid moves upward and thereby carries with it all particles whose settling velocity by gravity is less than the fluid velocity. Stokes law cannot always be used to calculate the conditions, as the flow rates are often higher than would correspond to the maximum permissible Reynolds number. In addition, velocity profile and wall effects, as well as convectional mixing, introduce complications.

In elutriation a series of chambers is connected so that the cross section (i.e., the volume) of the vessels gradually increases in the direction of flow. The shape of the chambers is an inverted cone with very steep walls, in order to obtain uniform flow. In such elutriation systems the mixture of particles is introduced into the largest chamber, where the coarsest particles settle out against the rising current of liquid or gas. The overflow containing the smaller particles is carried to the next chamber, where the same process takes place again. Thus the particles are progressively fractionated in the sequence of chambers. The accumulated material is removed from each chamber periodically. Although precise classification is never attained, elutriation is a useful tool in the laboratory to obtain narrow size ranges of subsieve particles.

Centrifugal methods of particle separation are based on the same principle as elutriation, but gravity is replaced by a centrifugal field. Although the time to achieve separation is thereby greatly reduced, elaborate machines are needed. The elutriator is air in such centrifugal separators, which are often called air classifiers. Two examples of these counterflow equilibrium classifiers using a centrifugal field are the Bahco-Spiral classifier and the Zig-Zag classifier (Alpine). Recently a centrifugal transverse-flow classifier was introduced by Alpine. This is not an elutriator in the classical sense but can yield very narrow cuts of fine particles.

18.3　Electromagnetic Separation (Mass Spectrometry)

At the beginning of this century J. J. Thomson demonstrated that heavy charged particles can be separated with respect to their mass and charge by the following principle. A particle of mass m and charge e moving in the x direction with a velocity v *in vacuo* is deflected by an electric field E_y. If the angle of deflection is small, the displacement along the y coordinate is expressed by

$$\Delta y = kE_y \frac{e}{mv^2} \qquad (18.15)$$

If a magnetic field B is applied, it causes the particle to deflect along the z coordinate, the displacement being given by

$$\Delta z = k* \frac{Be}{mv} \qquad (18.16)$$

The constants k and $k*$ depend on the dimensions and geometry of the apparatus proper. From Eqs. 18.15 and 18.16 it follows that particles of different mass-to-charge ratios will have different parabolic trajectories.

The first mass spectrograph, which is depicted schematically in Fig. 18.10, was constructed by Aston. The positive ions produced by high-voltage discharge were first collimated by two slits. The ion beam passed through the electric field maintained by two parallel plate electrodes so that they were deflected according to their kinetic energy and charge. A portion of the beam was sampled by a suitable aperture and then passed through the magnetic field and deflected in the opposite direction. By this arrangement the particles having different masses moved in different trajectories and were focused at different points on a plane. Thus they were separated and could be conveniently detected on a photographic plate.

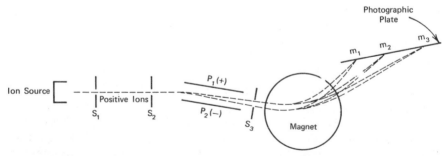

Figure 18.10. Schematic illustration of the Aston mass spectrograph. The positive ions generated in the ion source are collimated by the slits S_1 and S_2. By passing through the electric field, obtained by the plates P_1 and P_2, and slit S_3 the ions are separated in the magnetic field according to their mass.

Since then, a variety of sophisticated mass spectrometers based on this principle and called magnetic analyzers have been developed. The major distinguishing design feature of these devices is the method of focusing the individual mass beams. The focusing is necessary because ions having the same mass do not have the same energy; therefore they have different velocities and trajectories. Single focusing involves focusing either for velocity or for direction and as a result provides much lower resolution than double focusing, which compensates for both variations. The resolution of modern magnetic instruments with double focusing is about one part in 100,000. The resolution is expressed by the ratio of the mass of a particle to the mass difference of two separated particles, which is measured in several different ways in practice.

The schematics of a double-focusing mass spectrometer is shown in Fig. 18.11. It is based on the so-called Mattauch-Herzog geometry and comprises a radial electrostatic field which allows the transmission of ions having narrow energy distribution to the subsequent magnetic analyzer and yields a highly resolved mass spectrum. Another design uses Nier geometry, which makes it possible to bring ions of nearly equal masses sequentially into focus along identical trajectories. The first type of geometry permits focusing of individual ion beams either in a plane simultaneously or at a point successively. The Nier geometry allows single-point focusing only. When the ion

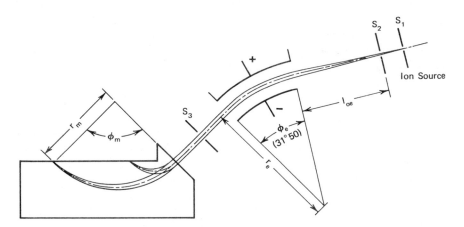

Figure 18.11. Double-focusing system with Mattauch-Herzog geometry. The ion beam is collimated by slits S_1 and S_2 and passes through the radial electrostatic field, characterized by the sector angle ϕ_{ee} and by the mean radius r_e. Thus, after passing through slit S_3, a monoenergetic ion beam enters the magnetic field with fixed ϕ_m and r_m, where focusing according to mass occurs. The resolution is adjusted by changing the slit widths, and the mass range is increased by increasing the magnetic field strength.

beams are focused in a plane, the mass spectrum of the sample can conveniently be obtained on a photographic plate; such instruments are frequently called mass spectrographs. In the case of single-point focusing the mass spectrum is generated by changing the magnetic field strength and detecting the intensity of the individual ion beams with a fixed electron multiplier, that is, by scanning the mass range of interest. The resolution is improved by narrowing the width of the slits in the instrument at the cost of sensitivity.

Mass spectroscopy was originally applied to determine the mass of individual isotopes and the composition of isotopic mixtures, and it has become an invaluable tool for elementary analysis of solids. Today mass spectrometry is widely used for the identification and structure determination of complex volatile organic molecules as well as for their quantitative analysis.

Several other types of mass spectrometers have also been developed. The time-of-flight mass spectrometer utilizes a pulsating electric field; this causes the components of an injected mixture of ions that have the same energy but different masses to move with different velocities through an ion drift region. The velocity of the light species is greater than that of the heavy ions. The particles of different masses impinge upon a high-gain electron multiplier and are detected. Ions of a given mass traverse the drift tube with a velocity v_d, given by

$$v_d = \left(\frac{2neV}{m} \right)^{1/2} \tag{18.17}$$

where V is the accelerating potential for an ion of charge ne and of mass m. If the length of the drift tube is L, the difference in migration time Δt for two ions of masses m_1 and m_2 is calculated by

$$\Delta t = \frac{L}{2neV} (m_1 - m_2) \tag{18.18}$$

It follows that the resolution increases with the tube length and decreases when the accelerating potential is increased. A time-of-flight mass spectrometer provides rapid measurements with a relatively simple instrument. In practice up to 100,000 spectra can be obtained per second; hence this instrument can be used to study reactions on the microsecond scale.

Quadrupole mass spectrometers are nonmagnetic devices. They utilize a quadrupole mass filter consisting of four rods to which a radio-frequency potential and a superimposed dc potential are applied. The potential field that is periodic in time and symmetric with respect to the axis along which the ions are injected will transmit a selected mass group and cause other ions to be deflected. The transmitted ions are collected by the ion detector, which is usually an electron multiplier. The upper limit of the mass range and the resolution of the quadrupole mass spectrometer is lower than that of the magnetic instruments. As a relatively inexpensive device, however, it finds wide application for analysis, particularly in the detection of ions that are not heavier than 500 mass units.

Since all instruments separate only ionized species, the generation of ions is a very important aspect of the technique. Generally, positive ions are formed, and the most common ionization methods are spark ionization of solid samples and electron bombardment of vaporized samples. Recently, however, several gentle ionization methods such as photo-, field, and chemi-ionization have been introduced to ionize organic molecules specifically or without much fragmentation.

The use of mass spectrometers has been greatly expanded by their tandem operation with gas chromatographs. The actual separation of the sample components takes place in the gas chromatograph, and the individual peaks are identified by their mass spectra. In order to maintain the low pressure in the mass spectrometer the carrier gas ballast has to be removed before the solute peak is admitted to the ionization source. This is achieved by a separator which concentrates the sample in the column effluent with a minimum loss of sample molecules. The gas chromatograph–mass spectrometer system has become the most powerful tool for analyzing complex mixtures of volatile and volatilizable components both quantitatively and qualitatively.

Although electromagnetic separations have attained prominence in analytical work and mass spectroscopy has become common in the laboratory, hopes for production-scale applications have not materialized. The large electromagnetic separation device, CALUTRON, built at Oak Ridge National Laboratory for the separation of isotopes, has not proved to be economical.

18.4   Thermal Diffusion

Separation by thermal diffusion is based on the concentration gradient that is set up within a fluid mixture in the presence of a thermal gradient. Maximum separation occurs in a convection-free system, when the effect of thermal diffusion is exactly counterbalanced by the remixing effect of molecular diffusion. The great attention bestowed on this process has been due to the fact that it constitutes a relatively simple method for the application of irreversible thermodynamics, rather than to its efficiency, which is generally low as mentioned in Chapter 3.

Because of its simplicity, thermal diffusion has found laboratory application for isotope separations in gaseous mixtures, using the thermal diffusion column developed by Clusius and Dichel. In its simplest form it utilizes a vertical glass tube 1.0 to 1.5 cm in diameter with a heated wire along its axis and a cooled wall. The effect of thermal diffusion is multiplied by a countercurrent flow upward at the hot surface and downward at the cold surface. Several columns can be combined into a cascade to enhance the separation. Isotopes of Ar, Ne, Kr, Xe, N_2, and O_2 have been separated by this technique. A significant improvement has been achieved by adding to the binary isotope mixture a third component that has a molecular weight between the weights of the isotopes to be separated, but can later be removed from the product by chemical means. For example, Ar isotopes could be obtained in pure form by using deuteriated HCl in the process and removing it from the products by means of NaOH.

Liquid mixtures have also been separated in a similar fashion on the basis of the Soret effect. It was found that the separation factor is determined by the difference in the molecular configurations, rather than in the molecular weights, of the components. For example, *cis* and *trans*-1,2-dimethylcyclohexane isomers could be separated by this method in batch operation. Although a variety of equipment design has been suggested for the separation of both gaseous and liquid mixtures, thermal diffusion has not become a widely used separation process, mainly because of its low thermodynamic efficiency and high energy requirement. Yet, for the separation of biopolymers such as enzymes, this technique still may represent a relatively simple and useful method.

18.5   The Use of Enzymes for Separation

In section 13.4 on affinity chromatography we discussed the isolation and purification of enzyme inhibitors by using immobilized enzymes. The employment of enzymes for separations goes back, however, to the mid-1800s, when Pasteur first separated the racemic mixture of DL-ammonium tartarate by letting yeast grow on it. Actually, this was achieved by enzymatic degradation of the L-isomer, which made possible the isolation of the remaining D-isomer.

Since then, the stereospecific nature of enzyme-catalyzed reactions has been used with some success to obtain optically pure amino acids or other substrate isomers. Since the isolation of suitable enzymes was often beset with difficulties, in many cases such reactions were accomplished through the action of micro-organisms or tissue fractions.

For the resolution of amino acid racemates, enzymatic oxidation, hydrolysis, and decarboxylation are the most commonly used techniques. For example, D-amino-oxidase of mammalian kidney and L-amino acid oxidase of snake venom have been used to obtain optically pure amino acids by the following reaction:

$$\text{DL-amino acids} + O_2 \xrightarrow{\text{oxidase}} NH_3 + H_2O_2 + \alpha\text{-ketoacid} + \text{unsusceptible isomer}$$

Of course the use of one enzyme leads to the isolation of only one of the two isomers. In order to obtain both, the racemic mixture is divided into two parts, which are then reacted with L-oxidase and D-oxidase, respectively. Naturally, the maximum yield for each isomer is 50%, and this is a serious drawback of such methods. In addition the oxidation is not complete with every amino acid. For example, L-threonine and L-proline are poor substrates for the L-oxidase of *Crotalus adamanteus*.

Only very recently have Tosa et al. (4) succeeded in developing an economical process for obtaining optically pure L-amino acids. L-Aminoacylase immobilized on DEAE–cellulose is used in a packed-bed reactor. The reaction steps are as follows.

First the racemic amino acid mixture is acetylated:

$$\text{DL-amino acids} \xrightarrow{\text{acetic anhydride}} N\text{-acetyl-DL-amino acids}$$

Subsequently, the acetyl group is removed from the L-isomer in a reaction catalyzed by the immobilized L-aminoacylase:

$$N\text{-acetyl-DL-amino acids} \xrightarrow{\text{L-aminoacylase}} N\text{-acetyl-D-amino acid}$$
$$+ \text{ L-amino acid} + \text{acetic acid}$$

After isolating the L-amino acid the remaining D-isomer is racemized:

$$N\text{-acetyl-D-amino acid} \xrightarrow{\text{racemization}} N\text{-acetyl-DL-amino acid}$$

This scheme permits the recovery of L-amino acids in high yield from racemic mixtures. The final step is the separation of the product, the L-amino acid, from the acetyl-D-amino acid. This is easily carried out by crystallization from alcohol. The advantages of this process were first demonstrated by obtaining L-methinoine in almost 100% yield in a continuous process that is schematically illustrated in Fig. 18.12. It is applicable, however, to a variety of amino acids. Such ingenious utilization of enzymic degradation can replace physical separation methods for a variety of racemic mixtures. The employment of immobilized enzymes is particularly promising, because these enzymes do not contaminate the product.

Enzymatic reactions play an increasing role also in analytical work. Difficult separations can be circumvented by removing an interfering component by a specific enzyme reaction, so that the remaining component can be determined. Or, more frequently, a component is subjected to chemical

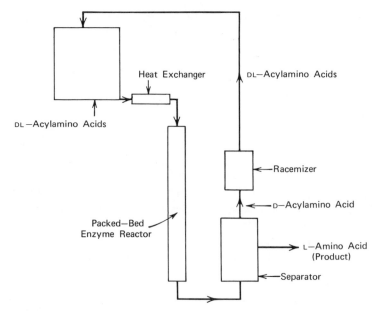

Figure 18.12. Flow sheet of the continuous process for the isolation of L-amino acids developed by Tosa et al.

change very specifically by enzyme action so that it can be determined in the presence of closely related substances without separation (e.g., by a color reaction).

References

1. R. B. Martin and B. Amers, *J. Biol. Chem.*, **236**, 1372 (1961).

2. M. Meselson, F. W. Stahl, and J. Vinograd, *Proc. Natl. Acad. Sci. U.S.*, **43**, 581 (1957).

3. N. G. Anderson, *Fractions* (Beckman Instruments), **1**, 2 (1965).

4. T. Tosa, T. Mori, N. Fuse, and I. Shibata, "Studies on Continuous Enzyme Reactions. II. Preparation of DEAE-Cellulose-Aminoacylase Column and Continuous Optical Resolution of Acetyl-D,L-Methinoine," *Enzymologia*, **31**, 225 (1966).).

Selected Bibliography

ULTRACENTRIFUGATION

T. Svedberg and K. O. Pedersen, *The Ultracentrifuge*, Oxford University Press, London, 1940.

H. K. Schachmann, *Ultracentrifugation in Biochemistry*, Academic Press, New York, 1959.

T. J. Bowen and A. J. Rowe, *An Introduction to Ultracentrifugation*, Wiley-Interscience, New York, 1970.

P. M. Browning, *Preparative Ultracentrifuge Applications: an Annotated Bibliography*, Spinco Division, Beckman Instruments, Palo Alto, Calif., 1969.

W. J. Archibald, in *Ultracentrifugal Analysis in Theory and Experiment*, J. W. Williams, ed., Academic Press, New York, 1963, p. 39.

H. Fujita, *Mathematical Theory of Sedimentation Analysis*, Academic Press, New York, 1962.

J. Vinograd, in *Methods of Enzymology*, S. P. Colowick and N. O. Kaplan, eds., Vol. 6, Academic Press, New York, 1963, pp. 854–870.

W. G. Flamm, M. L. Birnstiel, and P. M. B. Walker, in *Subcellular Components—Preparation and Fractionation*, G. D. Birnie and S. M. Fox, eds., Butterworths, London, 1969.

N. G. Anderson, in *Methods of Biochemical Analysis*, D. Glick, ed., Vol. 15, Wiley-Interscience, New York, 1967, pp. 271–310.

N. G. Anderson, ed., "The Development of Zonal Centrifuges and Ancillary Systems for Tissue Fractionation and Analysis, *Natl. Cancer Inst. Monogr.* 21 (1966).

J. H. Coates, in *Physical Principles and Techniques of Protein Chemistry*, Part B, S. J. Leach, ed., Academic Press, New York, 1970, pp. 2–95.

V. N. Schumaker, *Advan. Biol. Med. Phys.*, **11**, 245 (1967).

J. Vinograd and P. Bruner, *Biopolymers*, **4**, 131 (1966).

PARTICLE CLASSIFICATION BY SIZE

G. Herdan, *Small Particle Statistics*, Butterworths, London, 1960.

R. D. Cadle, *Particle Size: Theory and Industrial Application*, Reinhold, New York, 1966.

C. Orr, Jr., *Particulate Technology*, Macmillan, New York, 1966.

R. R. Irani and C. F. Callis, *Particle Size: Measurement, Interpretation and Application*, John Wiley, New York, 1963.

C. Orr, Jr., and J. M. Dalla Valle, *Fine Particle Measurement*, Macmillan, New York, 1959.

ELECTROMAGNETIC SEPARATIONS (MASS SPECTROMETRY)

F. A. White, *Mass Spectrometry in Science and Technology*, John Wiley, New York, 1968.

H. C. Hill, *Introduction to Mass Spectrometry*, Heyden, London, 1966.

R. I. Reed, ed., *Modern Aspects of Mass Spectrometry*, Plenum, New York, 1968.

A. E. Cameron, in *Physical Methods in Chemical Analysis*, Vol. 4, W. G. Berl, ed., Academic Press, New York, 1961, pp. 119–132.

THERMAL DIFFUSION

K. Cohen, *The Theory of Isotope Separation*, Natl. Nucl. Energy Ser., Division III, Vol. 1B, McGraw-Hill, New York, 1951.

H. London, ed. *The Separation of Isotopes*, G. Newnes, London, 1961.

P. H. Abelson, N. Rosen, and J. I. Hoover, "Liquid Thermal Diffusion," *U.S. At. Energy Comm. Rept.* TID-5229 (1958).

THE USE OF ENZYMES FOR SEPARATION

R. Bently, *Molecular Asymmetry in Biology*, Vol. 1, Academic Press, New York, 1969, pp. 69–147.

J. P. Greenstein and M. Winitz, *Chemistry of the Amino Acids*, Vol. 1, John Wiley, New York, 1961, p. 729.

List of Symbols

B	magnetic field	R	resistance to motion
C_c	quantity of coarse particles in the coarse fraction	s	sedimentation coefficient
		s^*	sedimentation coefficient extrapolated to zero concentration
C_s	quantity of coarse particles in the starting material		
d	particle diameter	s_{20w}	sedimentation coefficient calculated for water and 20°C
$\bar{d}$	arithmetic mean particle diameter		
		S	Svedberg unit
d_g	geometric mean particle diameter	u_t	terminal settling velocity
		$\bar{v}$	specific volume
e	electronic charge	$\bar{v}_b$	velocity of boundary
E	electric field	v_d	drift velocity
E_c	quantity of fine particles in the coarse fraction	V	accelerating potential
		x_b	position of boundary
F_s	quantity of fine particles in the starting material		
g	gravitational constant	η_N	Newton classification efficiency
k	constant in Eq. 18.5 or in Eq. 18.15	ρ_p	density of particle
k^*	constant in Eq. 18.16	δ_g	standard deviation with respect to the geometric mean
L	length of tube		
m	mass of particle	μ^*	total potential
M	molecular weight	Φ	centrifugal force
N	number of particles	ω	angular velocity

Nineteen

MULTISTEP SEPARATION SCHEMES FOR COMPLEX SAMPLES

Chapters 2 through 18 provide a systematic basis for approaching most analytical separation problems. Beginning with what we know about a given sample, and depending on our separation goal, a promising method is selected, and experimental conditions are optimized for the sample in question. Often this is as far as we need to go. In some cases, however, the sample proves to be sufficiently complex so that it is inconvenient—or even impossible—to achieve the desired separation by means of a single method. In such cases a *combination* of methods or separation steps is necessary. The design of such multistep separation schemes requires, first, a knowledge of the relative capability of each individual method. Second, we need to know what factors determine how many separation steps are required in a given scheme, how the individual separations are to be selected, and in what order they are to be combined. Toward this end, the present chapter discusses some of the considerations involved in designing a separation scheme and reviews some examples from the literature of successful multistep separation schemes. The characteristics of individual separation methods were summarized and contrasted in Chapter 6.

19.1 Designing a Multistep Scheme

A good separation scheme consists of some minimum number of individual steps, each of which contributes to the overall separation goal. How we approach a given separation problem is determined by what we know about the sample and by our separation goal. The sample may come to us as a totally unknown mixture that must be completely characterized. The problem is simpler when we know something about the sample (number of different compounds, functional groups present, etc.), or if we are interested in only some of the compounds present. An even easier separation job is involved when a mixture of known compounds must be resolved for their individual quantitation, or when a single, known compound must be isolated from a mixture. Often both our knowledge of the composition of the sample and our analysis goal can be greatly clarified by initial attempts at separating or otherwise analyzing the mixture in question.

Having assessed the nature of the sample and decided on our separation goal, we are ready to consider the design of a separation scheme. This involves a number of important considerations:

- What methods (and how many) will be required?
- In what sequence will these methods be used?
- What scale of separation will be used (how much sample must be separated?)

Finally, we must decide at what stage to terminate the separation. In some cases the isolation of individual pure compounds will be required for further characterization of each compound. In other cases, a selective detector can be used to analyze sample fractions that are only partially resolved.

19.11   SELECTING AN INDIVIDUAL SEPARATION METHOD

The factors that must be considered in choosing one or more methods for a separation scheme are summarized in Table 6.1: adaptability, selectivity, fraction capacity, load capacity, speed, and convenience. In this chapter these topics, as they relate to the design of a separation scheme, are examined in turn. We can dispense with adaptability by noting that the components must be compatible with the conditions of the separation method.

Separation Selectivity

This is a major consideration in the choice of individual methods for a separation scheme. The selectivity required is in turn determined by the nature of the sample mixture. If we know that the sample contains compounds of widely different molecular weights, a separation step that exhibits pronounced molecular weight selectivity will probably be required, for example, exclusion chromatography, dialysis, or distillation. If we suspect the presence of different compound types in the mixture, one or more steps that show functional group selectivity will be needed, for example, ion exchange or liquid-solid chromatography. If closely related compounds are likely to occur together in the mixture (and their separation is required), methods that provide shape or isomer selectivity are likely to be useful, such as fractional crystallization, liquid-solid chromatography, or thermal diffusion. When we are faced with a highly complex sample that includes all of these separation requirements, each type of selectivity must be represented in the final separation scheme.

In addition to the need for adequate separation of the sample at hand, another aspect of separation selectivity should be considered: predictable separation and the use of separation per se for compound identification. We saw in preceding chapters that tentative identification of a particular sample component is often possible, by comparing its separation behavior within a given separation system (e.g., retention time, R_f value, electrophoretic mobility, distillation temperature) with that of a pure reference compound of known structure. Even in the absence of suitable reference compounds, a separation scheme can provide information concerning the structure of an unknown sample component. In this case we must know how separation behavior is related to molecular structure. The Martin equation discussed in Section 2.22 provides a quantitative basis for such comparisons, but for many reasons is not broadly applicable. However, similar qualitative or semiquantitative deductions of molecular structure from separation behavior are widely used. Suppose that we have a separation scheme consisting of two successive steps (methods). Assume that the first method (A) separates strictly on the basis of sample molecular weight, while the second method (B) separates strictly according to sample functionality (type and number of functional groups). The separation behavior of a given compound in system A defines its molecular weight within narrow limits, and its behavior in system B similarly limits the nature and number of functional groups within the molecule. This information, in conjunction with additional analytical data on the separated compound, will often permit its identification. An example of this approach is given in Section 19.2 for the analysis of petroleum heterocompounds.

The use of separation behavior to infer molecular structure (as described above) is less successful when each of the individual separation methods exhibits more than one type of selectivity. If in the above example each method (A and B) exhibits simultaneous molecular weight and functional group selectivity, separation behavior in each system does not uniquely define any aspect of molecular structure. Instead various combinations of molecular weight and substituent functional groups can give equivalent separation behavior. Consequently it is desirable to select separation methods each of which exhibits primarily one type of selectivity. In some cases, however, we must compromise, because only rarely does a given method exhibit a single type of selectivity, to the *complete* exclusion of all others. As an example, we may want to take advantage of the ability of exclusion chromatography to separate on the basis of shape, despite the normal predominance of molecular weight selectivity. We can achieve our goal in this case by first narrowing the molecular weight distribution of the starting sample, using another separation method that provides predominantly molecular weight selectivity (e.g., distillation). Then distillation fractions of narrow molecular weight range can be further separated according to molecular shape by exclusion chromatography, without interference from the normally pronounced molecular weight selectivity of exclusion chromatography. An example is provided by the separation of petroleum hydrocarbons into alkanes, cycloalkanes, and polycycloalkanes. Distillation fractions of narrow molecular weight are further separated according to shape (into alkanes, monocycloalkanes, etc.) by means of gel permeation chromatography (1) or thermal diffusion (2).

Fraction Capacity

The overall fraction capacity required in a separation scheme is determined by the sample complexity, that is, the number of individual components, and the relative concentrations of the species in which we are interested. Mixtures that contain hundreds or thousands of individual compounds require one or more separation steps of high fraction capacity ϕ. If we are interested only in compounds present in relatively high concentration (i.e., major components), the minor constituents can be ignored, and the value of ϕ required is correspondingly reduced. Where there is an interest in trace components, on the other hand, the overall value of ϕ required by the separation scheme is normally increased. The *maximum* fraction capacity ϕ_{max} of a separation scheme is given as the product of ϕ values for the successive steps in the scheme. For example, if the separation scheme consists of an exclusion chromatography step ($\phi = 10$), followed by electrophoresis ($\phi = 25$) and finally by ion-exchange chromatography ($\phi = 100$), the maximum peak capacity of the overall separation scheme is given as

$$\phi_{max} = 10 \times 25 \times 100 = 25,000$$

From this example we can immediately appreciate the enormous advantage of serial separations that combine two or more individual methods. However, two qualifications should be noted before our enthusiasm prompts us to the blind pursuit of large ϕ_{max} values. First, ϕ_{max} is calculated on the assumption that there is no serious duplication of selectivity in the various methods which form the overall scheme. Second, large values of ϕ_{max} (relative to ϕ values for the individual steps) generally imply a large number of individual separation operations, which in the limit (as $\phi_{max} \to \infty$) means an impractically large separation effort.

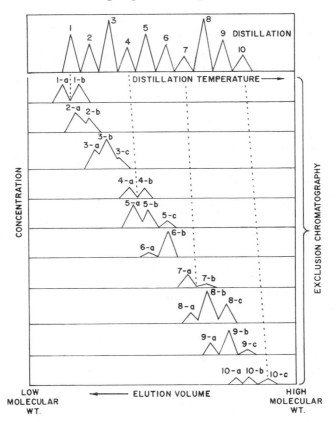

Figure 19.1. Hypothetical illustration of reduced peak capacity in a two-step scheme that involves methods of similar selectivity.

A duplication of selectivity in two or more steps of a separation scheme leads to a loss in potential fraction capacity ϕ_{max}, as is illustrated in Fig. 19.1. Here we have the hypothetical example of a preliminary distillation step ($\phi = 10$) followed by a separation by exclusion chromatography ($\phi = 10$).* Generally the separated compounds or fractions will not be distributed equally, as shown for the distillation step of Fig. 19.1, but will involve partially separated compounds and widely separated compounds (see, e.g., Fig. 5.16). If each of the two separation steps or methods of Fig. 19.1 exhibited only molecular weight selectivity, each compound or fraction from the distillation step would yield a single compound or unresolved fraction in the exclusion step (i.e., no further separation). Actually, however, the existence of some functional group selectivity in distillation, combined with

---

* In general, the application of a separation method (of fraction capacity ϕ) to a given sample or sample fraction will yield up to ϕ separate fractions, as illustrated in Fig. 19.1.

some shape selectivity in exclusion chromatography, leads to a further splitting of distillation fractions into individual compounds or less complex fractions, as seen in Fig. 19.1. It is apparent from this figure, however, that the maximum fraction capacity of the separation scheme ($\phi_{max} = 100$) greatly exceeds the actual fraction capacity ($\phi \approx 20$ to 30). This emphasizes the general importance of avoiding (if possible) duplication of separation selectivity in the methods selected for a separation scheme.

As already mentioned, very large values of ϕ_{max} require a correspondingly large number of individual separation operations, necessitating an impractical amount of time and work. Thus in the example of Fig. 19.1 a fraction capacity of $\phi = 10$ in the first separation step yields 10 fractions for further separation in the second step. One separation by distillation and 10 separations by exclusion chromatography are therefore required, for a total of 11 separation operations. Similarly, in the above example of a separation scheme based on exclusion chromatography ($\phi = 10$), electrophoresis ($\phi = 25$), and ion-exchange chromatography ($\phi = 50$), with $\phi_{max} = 25,000$, one exclusion chromatography step yields 10 fractions, each of which must be separated by electrophoresis to yield $10 \times 25 = 250$ fractions, and each of these must in turn be separated by ion exchange. Thus 1 exclusion, 10 separations by electrophoresis, and 250 separations by ion exchange are required, for a total of 261 separation operations. In designing a separation scheme for large ϕ_{max}, the correspondingly large number of individual separation operations should be kept in mind.

It is possible, however, to reduce this multiplication of separation operations in a two-step scheme, through the use of two-dimensional open-bed chromatography (Section 5.53). Unfortunately the ϕ values of open-bed methods (in one dimension) are normally small, restricting the potential value of this technique. Duplication of selectivity in the two directions of migration (as in two-dimensional thin-layer chromatography) is also difficult to avoid. Nevertheless these two-dimensional methods represent a convenient route to moderately large ϕ_{max} values ($\phi_{max} \approx 100$) via two-step operation.

Load Capacity, Speed, and Convenience

These aspects of individual separation methods are readily evaluated in the course of designing a separation scheme. After the basic requirements of selectivity and peak capacity (as well as adaptability) have been met, the final choices among usable methods will be made on the basis of maximum convenience and minimum separation time, with load capacities large enough to allow separation on a convenient scale. In preparative separations and trace analysis, load capacity will receive increased emphasis at the expense of other separation characteristics.

19.12   THE SEQUENCE OF METHODS IN A SEPARATION SCHEME

Several considerations exist that make one sequence of separation steps superior to another arrangement of the same individual methods. We will review a few of the more obvious considerations.

As a Function of Fraction Capacity

In general it is advantageous to arrange successive separation steps so as to minimize the total number of separation operations and thus reduce the total work required. A minimum number of separation operations usually corresponds to a scheme in which the individual steps are arranged in order of *increasing fraction capacity*. This follows from the rough equivalence between ϕ for a given method and the number of separate fractions that method can yield, for example, 50 fractions for a method with $\phi = 50$. As an example, consider again a scheme based on exclusion chromatography ($\phi = 10$), electrophoresis ($\phi = 25$), and ion-exchange chromatography ($\phi = 50$). When these separation steps are arranged in this order, a total of 261 separation operations was calculated above. If the exclusion chromatography and ion-exchange steps are reversed, however, we can similarly estimate that 2751 separation operations will be required ($1 + 250 + 2500$). Beginning a separation scheme with a method of large fraction capacity usually guarantees that the total separation will be a laborious, time-consuming project.

As a Function of Load Capacity

In general, the initial separation(s) should be done by methods of relatively high load capacity. The reason is that the average size of each fraction becomes smaller after each step in the separation scheme, and the need for large load capacities in later steps decreases. For example, in the above case of a scheme based on exclusion chromatography, electrophoresis, and ion exchange, the total sample is charged to the first step (exclusion chromatography), an average of $\frac{1}{10}$ of the total sample is charged to each of the electrophoresis separations in the second step, but only $\frac{1}{250}$ of the total sample is charged to the average ion-exchange separation in the third step. Separations with very low load capacities, therefore, belong at the end of the separation scheme. One cannot always obtain a sequence of steps with simultaneously increasing fraction capacity and decreasing load capacity, but in practice this is not a common problem. Methods of high load capacity tend to have lower fraction capacities, and (for a given method) increased loading generally decreases the effective fraction capacity of a method.

   The advantage of arranging separation steps in the order of decreasing load capacity is based on the assumptions that the various fractions are of approximately equal size, and that we are interested in all sample components. In trace analysis, where only a single, minute fraction of the total sample may be of interest, it is possible to begin a separation scheme with a method of low load capacity if that method is unaffected by the remainder of the sample (i.e., the compounds other than the trace components of interest). For example, consider the determination of low concentrations of a polar antioxidant in a hydrocarbon oil. Separations by thin-layer chromatography normally are limited with respect to load capacity (0.01 to 1 mg), but much larger loadings (e.g., 0.1 to 1 g) are possible with the sample in question. The hydrocarbon oil is not strongly adsorbed and does not affect the separation of the polar antioxidant; in this case load capacity is determined only by the amount of total antioxidant, not the amount of total sample.

As a Function of Speed and Convenience

As can be seen from the above discussion, steps that occur later in a separation scheme usually involve more individual operations (more applications of a given method). Therefore it is preferable to place slow, inconvenient methods early in the separation scheme, rather than at the end.

As a Function of Selectivity

If a given step (e.g., gel chromatography, dialysis) involves a method of pronounced molecular weight selectivity, it may be advantageous to begin the separation scheme with this method, other factors permitting. One reason is that most separation methods exhibit some molecular weight selectivity, and narrowing the molecular weight range of fractions for subsequent separation (in the scheme) eliminates this complication. Another reason applies for samples that contain components of both low and very high molecular weight. Many separation methods (gas chromatography, liquid-solid chromatography, etc.) are not suitable for the separation of high-molecular-weight (e.g., > 1000 to 2000) compounds, and the presence of such compounds in fractions that are to be separated by these methods leads to poor results.

It is also desirable that any separation into water-soluble and water-insoluble fractions occurs early in the separation scheme. The reason is similar to that given above for methods which exhibit molecular weight selectivity; that is, some methods do not work well with water-soluble samples, whereas others are unsuitable with water-insoluble fractions.

19.13 SEPARATION SCALE

The amount of sample separated is determined by the quantity of each final fraction that will be required, and by the concentrations of these fractions in the starting sample. In unfavorable cases (e.g., certain hormones in blood, sex attractants in insect extracts) where the concentrations of components of interest are quite small (10^{-6} to 10^{-12} g/g), and milligram amounts must be recovered for complete characterization, large quantities of starting sample will obviously be required. In other cases a sensitive detector that follows the final separation step (e.g., electron capture in gas chromatography) may be able to detect picograms or less of the separated component, or the compound of interest may be present in the sample in high relative concentration. In any case, the scale of the overall separation scheme must be planned to yield adequate amounts of the final compounds of interest.

It should be emphasized, however, that in complex separation schemes involving many steps and/or operations, to make the separation scale much larger than necessary is a mistake. Ideally the final scheme will furnish amounts of each fraction (or of the fraction of lowest concentration) that are *just* adequate. Accurate estimates of the necessary separation scale can often be obtained by preliminary separations on a much smaller scale than will eventually be required, with preliminary analysis of separated fractions. The extra work involved in such trial separations will often prove to be small in comparison with the possible ultimate savings in effort. If it happens in the final separation that one or two fractions (out of a large number) are in insufficient supply, that *portion* of the separation scheme required to prepare just these fractions can often be duplicated in a small fraction of the time required for the total separation. Thus the overall time required for separation will still be less than for a scheme whose initial scale was too large.

19.2   Some Examples of Separation Schemes

Separation schemes that illustrate the systematic application of the preceding general principles are not easy to find. Two possible explanations can be offered in defense of the many workers engaged in analytical separation. First, practitioners of analytical separation have tended to be technique oriented, that is, to regard themselves as gas chromatographers, electrophoresis experts, or ultracentrifuge specialists, rather than separation scientists. Instead of considering a given problem in terms of all possible separation methods, solution is often attempted by the exhaustive application of a single method, with only occasional use of other techniques as a necessary evil. Second, relatively few separation problems have emerged that demand the *systematic* exploitation of several specially chosen methods. And, as we will see, the *random* application of successive methods to a separation problem *can* yield satisfactory results.

19.21   ROUTINE DETERMINATION OF ALDOSTERONE IN HUMAN URINE

The following analytical scheme by Aakvaag (3) was designed to determine a known compound (aldosterone) in a highly complex but reasonably well-defined mixture (urine), at the part per billion level. The separation of a trace component from a mixture of a great many compounds requires either a multistep separation scheme or a method of extreme specificity (e.g., radioimmunoassay, based on separation by antibody binding or affinity chromatography; Section 13.4). The multistep procedure of Aakvaag is diagrammed in Fig. 19.2.

The first step of the procedure involves acid hydrolysis, which is not a separation per se, but which frees the aldosterone from its water-soluble conjugates (sulfates, glucuronides, etc.) and allows free aldosterone to be concentrated by methylene chloride extraction (step 2). Since the bulk of urine consists of water plus hydrophilic substances, this simple extraction allows an enormous enrichment of aldosterone in the extract (1000 × plus).

The methylene chloride fraction is next separated by paper chromatography (step 3). The moderately high fraction capacity of this step automatically ensures a significant further purification of the resulting aldosterone fraction, separating the mixture on the basis of differing solubilities in the two liquid phases (moving and stationary). Separation selectivity is similar for both steps 2 and 3, but step 2 provides large load capacity, whereas step 3 offers greatly increased fraction capacity.

Step 4 involves oxidation of the aldosterone fraction, with conversion of aldosterone to the γ-lactone (A.γ-lactone) and presumably some oxidation of other compounds in this fraction. Two possible purposes are served by this step. The aldosterone is derivatized to a more stable compound for eventual separation by gas chromatography (step 7), and the narrow range of compounds from step 3 is broadened by adding different functional groups to some of the components of this fraction. This then increases the selectivity of step 5 (extraction with ethyl acetate) for differentiation between the A.γ-lactone and contaminating compounds in this fraction. In the absence of oxidation, the similar selectivities of steps 2, 3, and 5 would lead to only limited further concentration of A.γ-lactone in step 5.

Step 6, separation by thin-layer chromatography, introduces a new type of selectivity. This, combined with the moderate fraction capacity of TLC, leads to significant further enrichment of the A.γ-lactone fraction. Finally, step 7 (gas chromatography) combines still another type of selectivity with large fraction capacity and direct detection of the separated A.γ-lactone.

This example illustrates the successive application of a variety of separation methods for the isolation of a single known compound. When properly chosen with regard to selectivity and applicability for the sample in question,

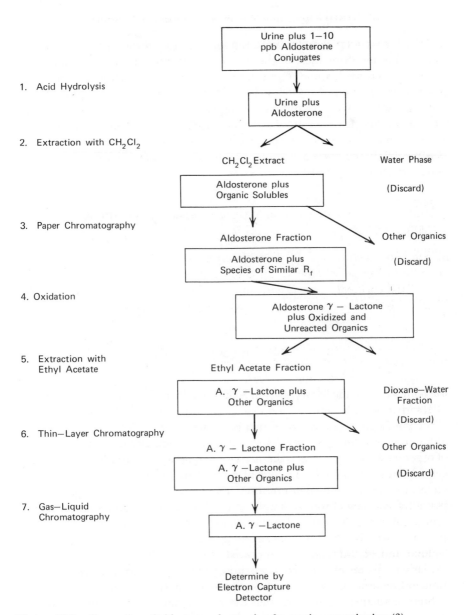

Figure 19.2. Separation of aldosterone from urine for routine quantitation (3).

each successive step can eliminate a major portion of the undesired species, while retaining all or most of the compound of interest. In a routine determination such as this one, quantitative recovery is not necessary (recovery of aldosterone here is $37 \pm 5\%$). Correction can be made (to a 100% basis) if the actual recovery is sufficiently repeatable, or through the use of an internal standard. Often chemical reaction (e.g., steps 1 and 4 above) can serve to good advantage to advance the separation objective, because the response of the compound of interest toward reaction is easily determined (for a known compound).

19.22   ISOLATION OF A BIOLOGICAL STIMULANT FROM INSECT FECES

Beroza and his coworkers (4) have described the isolation and characterization of a biologically active compound—a host-seeking stimulant—from the feces of the corn earworm. The compound in question, which proved to be 13-methylhentriacontane (**1**), attracts a parasitic predator (*Microplitis*

$$\overset{\displaystyle CH_3}{\underset{\displaystyle}{\overset{\displaystyle |}{CH_3(CH_2)_{11}CH(CH_2)_{17}CH_3}}}$$

I

croceipes). The scheme for separation is outlined in Fig. 19.3.

Most biologically active compounds that function as insect attractants have been found to be lipophilic; this suggests the initial concentration of the compound of interest by organic extraction. Step 1 of the Beroza scheme was observed to concentrate the active compound into the hexane extract (activity was followed by biological testing, using the parasite that is attracted by the compound of interest). The hexane extract was separated in step 2 by column chromatography on silica, a logical follow-up step for organic-soluble materials of intermediate molecular weight. The active compound was found to be concentrated into the first fraction, which, being weakly retained, was assumed to consist of hydrocarbons.

A small portion of the active fraction from step 2 was analyzed by gas-liquid chromatography (GLC) and found to consist of 6 to 8 compounds (bands). A further separation (step 3) on charcoal as adsorbent was therefore undertaken. The selectivity of charcoal is quite different from that of silica, permitting separation of aliphatic from aromatic compounds, as well as separation by molecular weight (as opposed to the predominant functional group selectivity of silica). The active fraction from step 3 was found to be considerably reduced in complexity (3 GLC bands versus 8 in the preceding fraction). Charcoal chromatography is normally limited by incomplete

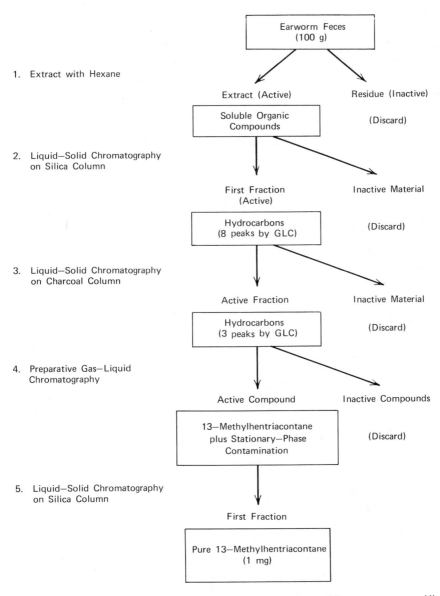

Figure 19.3. Isolation of 13-methylhentriacontame from feces of the corn earworm (4).

sample recovery and variable performance of different charcoal batches. These limitations make charcoal less suitable for routine separations in which complete or constant sample recoveries may be required. In the present case of a one-time isolation of a compound of interest, however, this consideration proved to be unimportant.

The small quantity of sample in the active fraction from step 3 was next separated by preparative GLC, step 4 (a technique of more limited load capacity than steps 1 to 3). A single, active band was recovered, which was found to be contaminated by stationary-phase bleed from the GLC column. This fraction was then reseparated by the procedure of step 2, which gave the pure, active compound in the first fraction. Normally, repetition of the same step at different points in a separation scheme would serve little purpose, because separation selectivity would be exactly the same in each case. Here, however, the introduction of a new compound (the GLC stationary phase) to the mixture in step 4 alters this logic.

The pure, active material from step 5 was eventually characterized by other analytical means and found to be 13-methylhentriacontane (I).

19.23   SEPARATION AND ANALYSIS OF INDIVIDUAL SKIN LIPIDS

Figure 19.4 outlines a considerably more ambitious separation–analysis scheme than either of the two foregoing examples: the determination by Nicolaides and his coworkers (5) of individual compounds in the lipids associated with human skin. The skin lipids represent an extremely complex mixture (thousands of individual compounds), which it is desired to analyze in as much detail as possible. The final characterization of each of these many compounds is greatly simplified by achieving compound type separations early in the scheme. In this way functional group determinations need be made for only a few preliminary fractions, rather than for each compound after its final isolation.

Step 1 of Fig. 19.4 is a simple ether extraction of the skin, followed by filtration to eliminate solid debris. The skin lipids from step 1 are separated in steps 2 and 3 into ten distinct compound type fractions: saturated hydrocarbons, olefins, monoesters, diesters, triglycerides, fatty acids, diglycerides, free sterols, monoglycerides, and a small quantity of more polar compounds. The possibility of such compound type separations by liquid-solid chromatography was noted in Section 13.1; LSC is used routinely to analyze lipid samples in terms of compound type percentages. The conventional liquid-solid separation of step 2 gives eight distinct fractions, two of which are mixtures. The triglyceride–fatty acid fraction can be separated in a variety of ways; LSC on the adsorbent Florisil (magnesium silicate) was chosen in the present scheme (step 3). The diglyceride–free sterol fraction from step 2

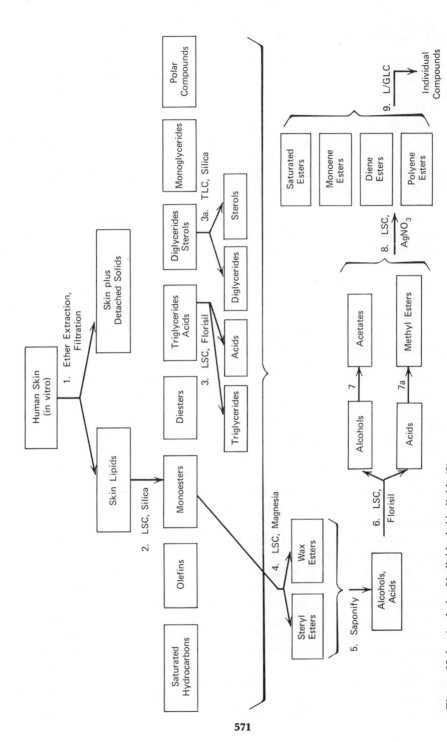

Figure 19.4. Analysis of individual skin lipids (5).

was likewise separable by thin-layer chromatography on silica (step 3a). This separation may appear puzzling, after incomplete resolution in step 2 (an essentially equivalent chromatographic system). Apparently, however, the TLC separation was more efficient (larger number of theoretical plates) than the column separation of step 2.

Each of the ten fractions resulting from steps 2, 3, and 3a was further separated (work on some of these fractions is not yet complete, however). In Fig. 19.4 we follow the further handling of the monoesters through the rest of the separation scheme, since this fraction is more or less typical of the problems involved with all of the various fractions. The total monoester fraction consists of both fatty esters (esters of fatty acids plus aliphatic alcohols) and steryl esters (esters of steryl alcohols, e.g., cholesterol). This is a difficult separation, because the two compound subtypes do not differ with respect to functional groups, and each group consists of numerous individual compounds of varying molecular weights and differing alkyl substitutions. Reasonably clear-cut separation of the monoesters into fatty esters and steryl esters was obtained, however, by LSC on the adsorbent magnesia (step 4). This adsorbent is apparently unique in providing marked *shape* selectivity, with preferential retention of the more *planar* steryl esters. It should be noted that this separation could not be obtained without prior fractionation of the original sample into a monoester fraction, since in the absence of steps 1 to 3 the fatty esters and steryl esters would have overlapped other compound types (on magnesia).

At this point in the separation scheme it was recognized that complete separation of individual esters would probably not be possible. A compromise was therefore adopted: the saponification (step 5) of all esters into their constituent acids and alcohols, followed by the isolation and characterization of these compounds. Each of the saponified fractions from step 5 was resolved into acid and alcohol fractions by LSC on Florisil (step 6).

Derivitization of the various acid and alcohol fractions was found useful at this point (steps 7 and 7a). The resulting alcohol acetates and acid methyl esters are more stable and give better separation characteristics in subsequent separations (steps 8 and 9). Each of the four ester fractions from steps 7 and 7a is now separated by silver ion complexation (secondary chemical equilibria) into fractions that differ in the number of double bonds contained within each molecule. Thin-layer chromatography on silver nitrate impregnated-silica allows separation of each ester fraction into saturated esters, esters with one double bond, esters with two double bonds, and so forth. Again, as in the case of the separation on magnesia (step 4), such a separation would not be possible (without overlap of different fractions) if the various fractions at this point contained more than a single compound type in each (fatty alcohol acetates, steryl alcohol acetates, etc.). This points up a further criterion for the relative ordering of different separation methods within a multistep scheme: Methods that simultaneously exhibit both major selectivity (functional group selectivity in the present case) and minor selectivity (shape selectivity or double-bond selectivity in steps 4 and 8) should be placed relatively late in the separation scheme, when it is desired to take advantage of the minor selectivity (a similar example is cited in Section 19.11 for petroleum hydrocarbons). This is particularly true for samples that are quite complex.

In the final step (step 9) of the scheme of Fig. 19.4, each fraction from step 8 is separated into individual compounds by gas-liquid chromatography. Individual compounds are then characterized by mass spectrometry. The separation scheme at this point (prior to step 9) has defined the functionality and the number of olefinic double bonds for each of the compounds to be separated in step 9. The mass spectrometer, in conjunction with relative retention data from the GLC separation, completes the analysis of each compound by specifying the carbon number (C_{10} to C_{30} for the acids, C_{10} to C_{35} for the alcohols) and the type of chain branching (if any).

The scheme of Fig. 19.4 provides a good example of the combination of well-known techniques of specified separation selectivity (steps 1, 2, 8, 9), supplemented by new procedures (steps 3, 4, 6) where necessary to provide the required resolution of a highly complex sample. This approach, which is typical of that generally needed for the solution of such separation–analysis problems, is further illustrated by the example of Section 19.24.

19.24 SEPARATION AND ANALYSIS OF THE OXYGEN- AND/OR NITROGEN-CONTAINING COMPOUNDS IN PETROLEUM

Mixtures of natural products (e.g., the skin lipids, body fluids such as blood and urine) offer the greatest challenge to the separation specialist, by virtue of the great diversity, complexity, and number of their components; the presence of biologically important compounds in both major and trace concentrations; and the special handling problems that result from the instability of these delicate biomolecules. Petroleum shares many of the same features of the natural products, a phenomenon attributable to its formation from ancient plant deposits. Our knowledge of the composition of petroleum has advanced only with corresponding developments in separation technology, and previous attempts at petroleum analysis (see, e.g., Refs. 6 and 7) provide excellent examples of the art and science of separation. Now we examine a recent separation scheme that has contributed to our present knowledge of the molecular structure of petroleum: the analysis of the oxygen- and/or nitrogen-containing (N–O) compounds in petroleum by Snyder and coworkers (8).

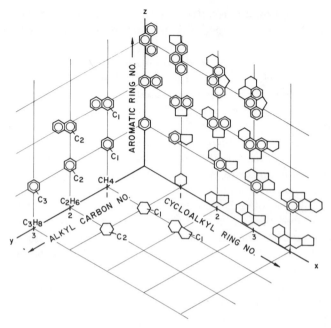

Figure 19.5. Petroleum hydrocarbons: their structural range. Reprinted from Ref. 8, by courtesy of the American Chemical Society.

Before the study in question, it was known that petroleum hydrocarbons (the main constituents of petroleum) occur as a more or less continuous distribution of compounds which vary in aromatic ring number, cycloalkane ring number, and total carbon number (or length and number of alkyl substituents). The compositional range of these petroleum hydrocarbons is schematically represented in Fig. 19.5 as a three-dimensional plot. Furthermore, it was known that other petroleum constituents (e.g., the N–O compounds) show a similar structural diversity. This can be illustrated for the basic nitrogen compounds of petroleum by substituting a nitrogen atom into the various aromatic nuclei of Fig. 19.5, for example,

The difficulty in separating and analyzing these petroleum N–O compounds is much greater than in the case of the hydrocarbons, for a number of reasons. The N–O compounds constitute only a small fraction of most petroleum distillates, and many individual compound types occur at the trace level. Also, a very large number of different compound types make up the N–O compounds. Finally, these compounds are much less stable than the hydrocarbons, so that decomposition and loss during separation can be a serious problem.

The separation scheme outlined in Fig. 19.6 was developed for the isolation of individual N–O compound types for final analysis by high-resolution mass spectrometry. Since mass spectrometry can differentiate (i.e., separate) the constituents of a sample in terms of alkyl and cycloalkyl substitution, it was not required that the preceding separation scheme itself resolve compounds differing only in these structural features. Consequently, molecular weight and shape selectivity were not necessary in the separation scheme; indeed it was desirable to choose methods that minimized these types of

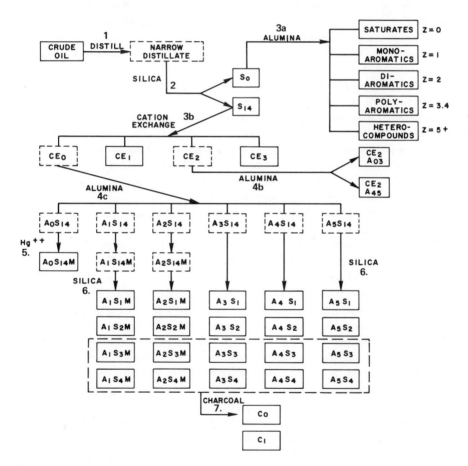

Figure 19.6. A separation scheme for petroleum nitrogen and oxygen compounds. Reprinted from Ref. 8, by courtesy of the American Chemical Society.

separation selectivity. As in the case of the skin lipids analysis, the separation scheme was designed to furnish as much information as possible on the molecular structures of compounds found in the various fractions. In this connection preliminary studies (9) established the separability of different compound types within the scheme. It was possible, therefore, to predict within narrow limits what compound types could or could not occur in a given final fraction. This in turn allowed a number of structural assignments (particularly for some of the minor N–O compounds) that would not otherwise have been possible.

In step 1 of Fig. 19.6, 1 kg of the petroleum sample for analysis is distilled into several narrow-boiling fractions. This step aids the following separation steps but does not contribute to separation per se. Thus narrowing the molecular weight spread of each of these fractions before further separation overcomes the slight molecular weight selectivity of the following steps and improves the sensitivity (per compound) in the final analysis by mass spectrometry. In step 2 of this scheme the sample is separated by liquid-solid chromatography on silica into two fractions: S_0, a nonpolar fraction that contains all of the hydrocarbons plus the nonpolar furan derivatives, and S_{14}, a hydrocarbon-free N–O compound concentrate. Silica provides unique selectivity at this step, in that polycyclic aromatic hydrocarbons do not significantly overlap the elution of the polar N–O compounds.

In step 3a the S_0 fraction is separated by LSC on alumina into fractions of differing aromatic ring numbers: saturated hydrocarbons, zero rings; monoaromatic hydrocarbons, one ring; and so forth. The furan derivatives are similarly separated by aromatic ring number. Subsequent mass spectrometer analysis of each of these fractions (step 3a, not shown) separates each furan type from the corresponding hydrocarbons and permits the determination of individual furan derivatives according to aromatic ring number, cycloalkyl ring number, and total alkyl substitution.

In step 3b the main N–O compound fraction S_{14} is separated by ion-exchange chromatography into fractions of varying basicity: CE_3, pyridine derivatives ($pK_a \approx 5$); CE_2, indoles and amides ($-2 < pK_a < 2$); CE_1, carbazole derivatives ($pK_a \approx -3$); and CE_0, neutral compounds. The CE_3 fraction, consisting of pyridines, quinolines, and benzoquinolines, was further separated (step 4a, not shown) according to aromatic ring number by LSC on charcoal. Charcoal chromatography has already been noted as having pronounced aromatic selectivity. The pyridine, quinoline, and benzoquinoline fractions from step 4a were further separated (and analyzed) according to cycloalkane and alkyl carbon substitution by mass spectrometry.

In step 4b the indoles and amides of the CE_2 fraction are separated from each other by LSC on alumina. These two fractions and the CE_1 fraction are finally analyzed for their individual constituents by mass spectrometry. The CE_0 fraction is separated (step 4c) into a number of fractions, also by LSC on alumina. This adsorbent shows selectivity according to both functional group polarity and molecular acidity.

At this point, aliphatic sulfides which contaminate the fractions from step 4c are removed and discarded in step 5: chromatography on cation-exchange resin impregnated with mercuric ion. The mercuric ion forms a strong selective complex with alkyl sulfides (secondary chemical equilibrium) permitting their removal from the N–O compounds of interest. Each of the sulfide-free fractions from steps 4c and 5 ($A_0S_{14}M$, $A_1S_{14}M$, A_3S_{14}, etc.) is further separated by LSC on silica (step 6). This adsorbent provides significantly different selectivity than does alumina, exhibiting no acidity selectivity and providing enhanced adsorption of aliphatic compounds. As a result, the remaining N–O compounds in the 20 fractions from the dual alumina-silica separation are separated into individual fractions with relatively little overlap of major compound types at this point. The aliphatic N–O compounds present in the more strongly retained silica fractions (enclosed in the dashed box in Fig. 19.6) are separated from the aromatic N–O compounds by LSC on charcoal (step 7).

The final fractions from steps 6 and 7 are now analyzed by mass spectrometry and other analytical methods. The composite mass spectral analyses for all fractions, supplemented by other analyses and data inferred from the separation scheme per se, yield a total analysis of the various N–O compounds present in the starting petroleum distillate. Repetition of this procedure for the other petroleum distillates yields a final analysis for the total crude oil.

The separation scheme of Fig. 19.6 evolved from systematic studies of various separation methods, preliminary, small-scale petroleum separations with partial characterization of resulting fractions, and (finally) successive application of the resulting scheme to each crude oil distillation fraction. The experience gained with each successive application was used to modify the separation scheme for application to the next distillation fraction. For a full discussion of the strategy involved in the design and ultimate development of this scheme, see Refs. 9 and 10.

19.3 Conclusion

The discussion and examples of this chapter are intended to provide some insight into the design and application of successful separation schemes. It should be apparent to the reader, however, that no hard and fast rules applicable to every case can be given. Traditionally, analytical separation has been more art than science, and the empirical development of suitable schemes for a given problem has been the general rule. We believe, however, that it is possible (and desirable) to approach separation problems with a little more science and a little less art. Nevertheless the experimental study of how a given sample responds to different separation methods, plus the innovative application of old methods in new ways, will continue to play a major role in the development of individual separation schemes and the advancement of separation science. Similarly it should be clear by now that ancillary techniques (e.g., derivative formation), the capabilities of supporting analytical methods, and the separation goal itself play important roles in the design of a successful separation scheme. Successful attempts at separation—at least for more difficult cases—require a well-rounded combination of skills, including knowledge of specific separation methods and principles, as well as of supporting analytical and general chemical inputs, plus a clear appreciation of overall strategy (i.e., a philosophy of analytical separation).

References

1. H. N. Oelert and J. H. Weber, *Separation Sci.*, **5**, 669 (1970).
2. F. W. Melpolder, R. A. Brown, T. A. Washall, W. Doherty, and C. R. Headington, *Anal. Chem.*, **28**, 1936 (1956).
3. A. Aakvaag, *Clin. Chim. Acta*, **34**, 197 (1971).
4. R. L. Jones, W. J. Lewis, M. C. Bowman, M. Beroza, and B. A. Bierl, *Science*, **173**, 842 (1971).
5. N. Nicolaides, in *Progress in the Chemistry of Fats and Oils*, in press.
6. F. D. Rossini, B. J. Mair, and A. J. Streiff, *Hydrocarbons from Petroleum*, A.C.S. Monograph 121, Reinhold, New York, 1953.
7. H. L. Lochte and E. R. Littman, *The Petroleum Acids and Bases*, Chemical Publishing Co., New York, 1955.
8. L. R. Snyder, *Accounts Chem. Res.*, **3**, 290 (1970).
9. L. R. Snyder and B. E. Buell, *Anal. Chem.*, **40**, 1295 (1968); *J. Chem. Eng. Data*, **11**, 545 (1966).
10. L. R. Snyder, B. E. Buell, and H. E. Howard, *Anal. Chem.*, **40**, 1303 (1968); L. R. Snyder, *Anal. Chem.*, **41**, 314, 1084 (1969).

Selected Bibliography

C. J. King, *Separation Processes*, McGraw-Hill, New York, 1971.
C. J. O. R. Morris and P. Morris, *Separation Methods in Biochemistry*, Wiley-Interscience, New York, 1964, Chap. 14.

INDEX